Electrical Railway Transportation Systems

IEEE Press
445 Hoes Lane
Piscataway, NJ 08854

Electrical Railway Transportation Systems

Morris Brenna
Federica Foiadelli
Dario Zaninelli

WILEY

Published by John Wiley & Sons, Inc., Hoboken, New Jersey.
Published simultaneously in Canada.

Library of Congress Cataloging-in-Publication Data is available.

ISBN: 978-1-119-38680-3

Printed in the United States of America.

10 9 8 7 6 5 4 3 2 1

Table of Contents

Foreword

Having read this extraordinary compendium of electric traction, inevitably, I recall the lessons that I attended 30 years ago at the "Politecnico di Milano" in the classrooms within the Department of Electrotechnics, as a student among others of Prof. Dario Zaninelli. Just by reading this text, I have now realized what it was like, then how it is today, to use an organic and complete description of the many and different components of an electric traction system, from the electrical substation to the onboard systems, a "Testo Unico" of traction electricity.

What strikes the reader is the clear terms of clarification, inevitably resulting from a profound knowledge and expertise of the authors on the subject matter. Clarity that allows you to meet at different levels of curiosity and learning as the articulation of the text lends itself to easily navigate in system descriptions, reconstructing both the historical evolution and the different technological solutions adopted in different countries around the world, to drop both vertically and very analytically into the study of the subsystems and the operation of all components. At the same time, this text will be useful to those who, like myself, have taken advantage of the work of the authors for a useful and organic review of the subject and for the designers and researchers to work to make electrical traction ever more efficient starting from state of the art.

It is striking, and it is one of the reasons for the remarkable size of the work, including the incredible number of technical solutions adopted in different countries, sometimes deliberately preventing interoperability, for example, for reasons of war, as it was the case in Europe in the first half of the twentieth century. As the UIC chairman, I declared the need for interoperability between rail networks, which UIC supports the production of standardization leaflets. In addition, I emphasized for continuous research to improve the overall performance of traction electrical systems, as it will guarantee in the future both the extension of high-speed networks and the electrification of many more diesel lines. UIC, together with its members, have been committed to increase energy efficiency by 2030 and to reduce emissions per unit of traffic by 50% compared to 1990, and to further improve them by 2050.

It is therefore my pleasure to present and support this compendium that will help the railways of all countries of the world to integrate and to grow even more, in a responsible and clean manner.

Renato Mazzoncini
President of the International Union of Railways (UIC)
CEO of Ferrovie dello Stato Italiane

Acknowledgments

First and foremost, we would like to thank our families for standing beside us throughout our careers and for having the patience with us in writing this book that decreased the amount of time we could have spend with them.

We would like to thank all the students and experts of railway companies that over the years allowed us to stimulate new researches and to deepen the industry innovations in an exciting sector as railway world.

An additional special thanks to our colleague Michela Longo for her support.

Chapter 1

Introduction to Railway Systems

Electric traction has become increasingly important for the collective transport of people and goods, since it effectively contributes to the mitigation of congestion and pollution caused by road traffic. In its long history, which began at the end of the nineteenth century, it has experienced remarkable development and, in every era, it promptly made the most of progress in electrical engineering, mechanical engineering, power electronics, and also automation, often creating an incentive for new technology research and a valuable testing ground.

Electric traction has undisputed advantages in areas where levels of performance, safety, environmental compatibility, and economy of service must be guaranteed, such as the rapid transit of urban and suburban populations, long-haul journeys, high-speed rail, and in traversing mountain passes and underwater tunnels.

It should also be noted that the huge investments in infrastructure, equipment, and rolling stock that are required make it very costly to upgrade rail systems with the rhythm that rapid technological progress entails. Moreover, it is exceedingly difficult to radically transform those that may appear "outdated" with others that are modern and efficient. Within these objectives, difficulties arise because of the existence of a multiplicity of types of systems and materials, which place technical and, especially, economic obstacles in the path of a fully interoperable rail. This variety, moreover, makes it difficult to define a culture replete with the present and diffused solutions at an international level.

In the following sections, we will introduce the main features of a rail system that are the basis of differentiation of the various lines around the world.

1.1 TRACTION ELECTRIFICATION SYSTEMS

The term "traction" is intended to indicate the set of phenomena, equipment, and systems that contribute to cause the movement of vehicles; the "electric"

Electrical Railway Transportation Systems, First Edition. Morris Brenna, Federica Foiadelli, and Dario Zaninelli.

(a) (b)

Figure 1.1 First railroad presented by von Siemens in Berlin in 1879 (a); Lichterfelde tram in 1881 (b). (Reproduced with permission of Siemens AG, Munich/Berlin.)

attribute specifies the mechanical strength for traction that is produced by one or more electric motors. The idea of using electric traction (ET), instead of the thermal characteristic of the steam engine, dates back to the end of the last century with the first direct current motors (DC) derived from the Pacinotti ring (1863). Shortly after the first ET applications, diesel traction proved possible with gasoline, diesel, and gas engines. Diesel engines (diesel cycle) allowed for the construction of powerful locomotives, whereas electric engines had not yet reached the performance necessary to drive heavy vehicles, particularly on long routes.

The first implementation of electric traction motor drive was the small electric railway built, in 1879, by Werner von Siemens for the Berlin Industrial Exhibition (Figure 1.1a): the DC locomotive, with 2.2 kW power, was powered at 150 V and was pulling three small wagons in the exhibition, each with six seats. Two years later, in 1881, Siemens & Halske put into service, at their own expense, the first electric streetcar in Lichterfelde near Berlin, on a line approximately 2.5 km long; the vehicle had an output of 7.5 kW (Figure 1.1b).

Within the span of a few years, there were electric trams in Vienna, Frankfurt, Switzerland, France, and then in the United States, where, in 1886, Van Depoele built a tramway network in Montgomery (Alabama). He accomplished an important step because he used a simple overhead power supply wire on which a metallic contact slid affixed to a wooden grip. The following year in Richmond (Virginia), Sprague perfected the system, using a tubular metallic rod current outlet fitted with a grooved wheel. Besides, he introduced the "nose suspension" for traction drive motors to reduce their mechanical stress. Traction motors were fixed brushes and transmitted the motion to the wheelsets through adapter gear units; the speed adjustment was effected by rheostat and field variation, and a drum "controller" fitted with sliding contacts.

The electric tram, whose power supply had now been increased to 500–600 V, had thus found its almost final configuration, which was used, in fact, for a half century. Around 1890, the system had gained approximately 20 other U.S. cities and was spreading all over the world; in Italy, the Firenze-Fiesole tram was created in 1890. In railways, where steam power still ruled unchallenged, increasingly systemic issues began to surface in the tunnel routes because of the locomotive fumes, which sometimes caused tragic accidents. The problem was acute where the traffic was most intense. This is why electric traction was introduced in 1890 in the urban railways of London and Liverpool and, more specifically, in railways in Baltimore in 1895, on an underground stretch of approximately 5 km between two stations. The electrification system was the same as the tram system, but in view of the high currents involved, it was preferable to replace the contact wire with a large cross-sectional steel wire, arranged in different ways; a solution that quickly resulted in the lateral "third rail" that, particularly in the case of metros, allowed for the adoption of smaller models. The achieved performance was now considerable: in Baltimore, electric locomotives had mass of 90 t and power of 1060 kW. Traction at low-voltage direct current spread in the span of a few years, in addition to the tram and metro segment, to suburban and regional railways, reaching an overall coverage of approximately 20,000 km toward the end of the century.

Therefore, already before 1900, with the development of the railways and, in particular, with the rise of tunnel segments (mountain tunnels, metros, etc.), the introduction of electric traction became necessary to replace steam power and its inherent economic (high cost of coal) and passenger safety (poisoning caused by smoke in the tunnel) issues.

At that time, there were two types of electric motors equipped with the best mechanical properties suitable for traction drive: those with direct current with serial field excitation and those in single-phase alternating current with commutator. The first solution to be adopted was direct current at low voltage (less than 500 V) that did not demonstrate high levels of difficulties. The motors in alternating current, powered at mains frequency, had switching problems at the commutator. Therefore, in order to be used, AC systems had to be powered at lower frequencies (in the United States, the 25 Hz frequency was adopted, whereas in Germany the 15 Hz, that is, a third of the power frequency that at that time was 45 Hz).

The main advantage of alternating current consists in the possibility of adopting a higher voltage supply because interrupting an alternating current is simpler compared with interrupting a direct one. With equal power consumption, this implies not only lower current and therefore lower losses but also lower costs of installation of the contact line (because of the possibility of reducing the section of the conductor, the costs for conductor materials and those for masts, which may be less robust, are also reduced).

Having an AC voltage at a reduced frequency, the reactance of the lines is less than that with mains frequency and the resistance of the circuit is similar to that in direct current, thus guaranteeing low voltage drop. By contrast, however, due to the frequency being different from that of the mains, the power supply needs generating stations exclusively dedicated to the rail or, for interconnection with the grid, converter stations. Years ago, rotating conversion groups were used because power electronics were not sufficiently developed to provide static converters able to cope with this need. However, prior to 1900, the electrical grid was weak and not meshed like the current one and, furthermore, very large loads did not exist, thus the construction of generating stations and high-voltage power lines exclusively dedicated to the railway system was warranted.

Another type of motor that was used for traction was the three-phase asynchronous motor. However, a railway frequency voltage was still required for its operation, namely, a third of the mains frequency. In fact, since high capacity gear units were not available at that time, motors had to be connected to the wheels via kinematic connection with connecting rods and cranks that originated from those of steam locomotives. It followed that the motor rotation speed had to be low and equal to that of the wheels, from which stemmed the need to reduce the power supply frequency. The advantages of this power system were the same as for single-phase electrification in alternating current but, by contrast, had the major disadvantage of needing two overhead conductors (the third phase was provided by the rails), thus introducing major complications especially in proximity of exchanges where there was a need to isolate the overhead conductors to avoid creating short circuits between the phases.

The direct current system had the advantage of having simple and versatile motors and, given the negligible power in use at the time, the fact of having low power supply voltages did not constitute a problem. It is no coincidence that the first applications of DC traction were in tramways and urban transportation, whereas for the electrification of crossing tunnels, where the power involved was greater, the AC system was preferable with much higher rated voltages of approximately 3–4 kV.

Thanks to the progress of power electronics and electromechanical technology, an undulatory current commutator motor was developed after the Second World War by which the vehicle could be powered by a single-phase alternating current line at mains frequency, and the transformation and conversion to direct current were activated onboard. It was therefore possible to break free from the railway frequency and introduce the single-phase electrification system at mains frequency (50 Hz in Europe and 60 Hz in the United States).

Currently, the motors used in modern vehicles are all in alternating current of an induction or permanent magnet synchronous type with inverter drive, for which, in essence, the need to have a suitable voltage for the motor was eliminated. The vehicle is equipped with converters that adapt the line voltage at the inverter input, which controls the traction drive motor; this results in the advantage of interoperability between the various electrification systems.

Today, the most efficient system is the mains frequency single-phase system, suitable for both the regional and suburban transport lines and for high-speed (HS) lines. For the latter, given the high power used, it is essential to have high voltages, in order to limit the absorbed current. If these are too high, they generate uptake problems between the pantograph and overhead line preventing the speed exceeding 250 km/h.

DC electrification systems are still widely used in road-bound urban transport systems, such as metros, streetcars, trolley buses, and so on, with voltages of approximately 750–1500 V and preferable to those at alternating current for their lower impact on the medium-voltage power supply networks. They are also used at higher voltages even for regional railways.

Given the wide use of DC traction, and not just in the railway area, the search for innovative systems that are able to ensure the good power supply quality of traction vehicles, as well as the reduction of interference on the AC network, is increasingly important. In addition to these objectives of purely technical nature, energy savings linked to the recovery of energy during braking play an increasingly important role. Therefore, it is necessary to provide bidirectional power supply systems, also able to receive power from the traction vehicles and to release it back into the electrical network or store it in appropriate storage systems to then allow it to be reused.

1.1.1 DC Electrification

The first major applications of DC electrification date back to 1860 when the first electric machines for tramway traction, with voltage of approximately 500 V DC, were built. After 1890, DC electric traction was also applied to metro railways, with gradually increasing voltages (750, 1500, and 3000 V), with priority to line development in tunnels, where diesel traction had serious problems due to the persisting fumes.

The power supply systems in direct current gave the possibility to derive the power supply directly from the primary lines at mains frequency without introducing unbalances, with contained power factors and distortions, and without the risk of unwanted flux on the contact lines. In addition, the limitation of voltage drop due only to the resistive components of the line impedances and the simplicity of the parallel operation of substations with the bilateral power supply of the segments, combined with the absence of induced voltages in the neighboring lines of the rail network, were advantages that, to date, still make this kind of power supply preferable in many applications.

The success of DC for the power supply of railway lines is due, among other things, to the unique tractive effort of the commutator motor with series excitation.

The first applications of systems with ground electric power supply were those of tramways in Paris and Berlin in 1881. In 1890, it was also introduced in the London metro with power supply from two additional rails compared with line rails.

In order to have a traction power greater than that sufficient solely for the urban and suburban systems, it was necessary to elevate the values of the power supply voltage up to DC 1500 V, and subsequently up to DC 3000 V. The voltage of DC 1500 V was found to be the maximum cost-effective voltage to be generated with the rotating conversion systems installed in the power supply substations of the railway networks. This also prevented an excessively complex connection of the traction drive motors to enable correct operation in the starting phase.Thanks to the experience gained in the United States since 1914 by the Chicago and Milwaukee and St. Paul railroads on electrification at DC 3000 V; in Italy, the secondary line to the standard Torino-Ceres gauge was electrified at DC 4000 V in 1920. Not everyone believed that it was the appropriate time for the continuous use of such high voltage. France, which already had some lines supplied with a low-voltage third rail, preferred not to exceed 1500 V in the electrification of the Pau–Tarbes of Paris–Orléans segment, and adapted the third rail at 1500 V from Chambéry to Modane. Similarly, in the Netherlands, Japan, Australia, and elsewhere, DC at 1500 V was adopted.

In other systems, such as in Italy, with the aim of also allowing relatively heavy traffic on electrified lines, the DC 3000 V was adopted right from the start, as soon as the overall electrification policy of the entire network was implemented. This system has thus supplanted other systems that were developed, such as the three-phase railway frequency, which, particularly following the destruction caused by the Second World War, was reconstructed in DC 3000 V. The DC 3000 V system occurred although, due to the needs of the installed onboard motors, it was necessary to keep two traction motors connected in series, and at least four connected during the starting conditions. These problems were then overcome by the advent of electronic drives that permitted the adjustment of the motor voltage regardless of the power supply. In addition to this, in the first applications, the conversion substations, which were made with rotating systems, were particularly burdensome and complex. The first example of this voltage level was created in the United States by the aforementioned Chicago–Milwaukee–Saint Paul railway line, in which a synchronous motor that was configurable as asynchronous, two 1500 V dynamos connected in series, and the exciter of the synchronous motor and that of the two dynamos were all assembled on the same axis, all at a rated power of 2 MW with ample possibility of overload.

The main power supply limit in direct current consists of the maximum applicable voltage limit. Currently, in fact, it is technically difficult to succeed in developing switches capable of withstanding continuous reestablishment voltages higher than 6 kV. Applications have been researched requiring the implementation of a circuit breaker using some SCR static switches at the AC/DC interface point. In this case, in fact, it is sufficient to stop the control pulses of the thyristors in order to break the current. Such systems, regularly used in systems for high-voltage DC electricity transmission, have not found practical application in the railway area, essentially due to the simultaneous development of single-phase AC systems equally suitable for linear density high power applications (Figure 1.2).

Figure 1.2 DC railway electrification system in Italy.

1.1.2 Single-Phase Electrification at Railway Frequency

In electric rail traction, it is very important to be able to adopt high line voltages because only then can adequate power with sufficiently low current values be transmitted to trains. During the early decades of the twentieth century, the problem was solved with various solutions that, in individual nations, were affected by the influence of special guidelines and technical and economic interests.

In German-speaking countries and initially in the United States, single-phase alternating current was chosen, which made it possible to reach voltages of 10–15 kV and power the traction motors, with conveniently reduced voltage, with a transformer installed onboard the locomotives. On the other hand, the advantage of being able to use single-wire contact lines influenced the choice of the single-phase commutator motor as a traction vehicle motor. This motor, created for DC applications, when operating in alternating current manifests transformative electromotive forces that make it difficult to switch to the commutator. Therefore, in an attempt to reduce the effects, it was necessary to adopt power supply frequencies lower than the mains frequency. As a result, 16 and 2/3 Hz systems were developed in systems with mains frequency of 50 Hz, and 20–25 Hz in systems of 60 Hz.

(a) (b)

Figure 1.3 AC railway electrification system at railway frequency in Sweden: (a) primary lines and (b) railway line.

In the United States, in 1906, the New York–New Haven line chose the 25 Hz frequency, with 11 kV. In Europe, the single-phase traction at 6300 V–25 Hz was applied to the Hamburg suburban network and the 5500 V–15 Hz to a private Bavarian railway.

In 1911, the Prussian railways electrified the Dessau–Bitterfeld segment at 10 kV–15 Hz; the tests were successful and were the basis for the agreement of 1912 between the German railway authorities for the adoption of single-phase traction at 15 kV and 16 and 2/3 Hz, which was then introduced in Switzerland, Austria, and, subsequently, in Sweden and Norway (Figure 1.3).

1.1.3 Single-Phase Electrification at Mains Frequency

The use of single-phase alternating current allows for the voltage level of the contact lines to be increased, thereby reducing the current values drawn by the pantograph, and at the same time ensuring the most appropriate values for the power supply of the motors through the simple use of transformers. Furthermore, the somewhat limited current values of the trains in such systems allow for the single-wire contact line to be maintained and for the creation of much lighter and cost-effective contact lines than those for direct current that fit more easily with the mechanical requirements for good collection of current. The adoption of the mains frequency makes the direct connection of the power supply lines to the mains network possible without having to use conversion systems that are not simple transformers. On the other hand, the AC current power supply at mains frequency has demonstrated problems with unbalances induced on the power supply rail from the rail load, which by its nature is a single-phase load. To reduce these unbalances, however, various substation connection

Figure 1.4 AC railway electrification system at mains frequency in (a) the United States and (b) Japan.

specifications have been suggested that, however, result in the loss of the bilateral power supply possibilities and require connection to high-voltage systems, therefore not suitable in urban areas.

The great advantage that the adoption of single-phase mains frequency systems have in relation to the possibility of greatly simplifying substations was advantageously used only when, thanks to the advent of electronic systems, it was possible to overcome the problems relating to the use of a commutator motor. Essentially, after Second World War, an impetus led to the systematic development of networks supplied by AC power at mains frequency. Therefore, the system, which has now also been relaunched in countries that previously adopted DC power, has now reached greater development than that of DC power at 3000 V, covering approximately a third of electrified lines in the world (Figure 1.4).

1.1.4 Three-Phase Electrification at Railway Frequency

The use of the three-phase alternating current has substantially been justified by the possibility of using the three-phase asynchronous motor for traction, with its high qualities of strength and economic feasibility and its potential to be powered directly even at voltages of a few kilovolts.

In 1895, the tramways in Lugano, Switzerland, experimented with the three-phase low frequency system. The locomotives were equipped with motors built according to the rotating field patent of Galileo Ferraris, thus asynchronous three-phase squirrel cage motors. The system was also proposed for railway traction with power plants and primary lines operating at 16.7 Hz, called railway frequency. The voltage of approximately AC 3000 V was selected, even if restricted to crossing lines, in Europe and in the United States. During the early years of the twentieth century, after some successful experiments in Hungary and successful applications in Switzerland, in 1896–1899, on the other segment of Lugano and in the Burgdorh–Thun line, the latter at 750 V–40 Hz, three-phase electric traction was also adopted in Italy. The choice of using the high line voltage, set at the very high value, for those times, of AC 3000 V, mainly fulfilled the will to disengage all the electric traction passenger and goods services, which required the use of locomotives with power up to 1000–1200 kW. The limit of the current collected from an overhead contact line, estimated at approximately 300 A, required the above-stated voltage.

The special frequency solutions were chosen to obtain running speed of 50–60 km/h and even lower, with motors having an acceptable number of poles (in practice 6 or 8), without resorting to adapter gear units, still not considered reliable for high powers. Therefore, motion transmission was via rod–crank kinematic connection, which was due to torque ripple and, thus, from power absorbed by the motors. This ripple sometimes caused interference in the three-phase power supply system and the telecommunication lines parallel to it.

Already in the early 1900s, however, the three-phase system showed its limits, particularly regarding the difficulties of locomotive speed control (being rigidly fixed to the rotation speed of the asynchronous motor running speed) and maintenance of the two-wire contact line that, moreover, made it problematic to overcome the speed of 100 km/h due to mechanical issues. In fact, the "rigidity" of the tractive efforts of the three-phase locomotives and the inability to adjust the speed to the actual needs of the service and the track were one of the main disadvantages of the system, not offset by the robustness and reliability of the asynchronous traction motors and ease by which downhill regenerative braking could be carried out, at speeds slightly higher than those for synchronism. The other weak point, namely, a bipolar contact line, did not allow the speed of 90–100 km/h to be exceeded; this constituted a heavy limit compared with the same steam locomotives that reached a speed of 120–130 km/h.

For these reasons, all countries that had initially adopted the three-phase system abandoned it within a few years, with the exception of Italy that, given the need to take advantage of the large hydroelectric resources available and at the same time reduce coal imports, continued with the application of this power supply system. Therefore, in the early decades of the 1900s, while the three-phase alternating current 3400 V electrified network was expanding in Italy, the single-phase AC with voltages ranging from 11 to 15 kV at 16.7 Hz became predominant in the other countries that had chosen alternating current.

Today, this type of power supply system has been completely abandoned and it will not be described in the remainder of the book for this reason. Only particular applications still survive supplied with three-phase systems at mains frequency (Figure 1.5).

Figure 1.5 Particular three-phase AC railway electrification system at mains frequency for the Corcovado Railway in Rio de Janeiro, Brazil.

1.2 TYPES OF ELECTRIC POWER SUPPLY IN RAILWAY LINES

The technologies available in the various eras of development of electric traction, especially regarding the transmission of power, voltage levels, and speed control of traction drive motors, have led, over time, to the adoption of different technological solutions that persist to this day.

In classic urban transport trolley buses, trams, and metros, DC traction at 600–750 V or 1500 V is always used. The 750 V system is also used in metros on tires adopted in Paris, Montreal, Mexico City, Santiago, Lyon, and Marseille.

On the other hand, for the above-stated reasons, the railway area has a wide variety of systems; the most significant are the following:

- DC at 750 V in Britain (2000 km) with third rail power supply;
- DC at 1500 V in Japan, France (6000 km), the Netherlands, and so on;
- DC at 3000 V Russia (27,600 km), Italy (10,500 km), Poland, Spain (6400 km), South Africa, Brazil, Czech Republic, Slovakia, Belgium, and so on;
- Single-phase alternating current at the special frequency of 16.7 Hz at 15 kV, in Germany, Sweden, Switzerland, Norway, Austria: in total, the single-phase European network at 16.7 Hz covers over 33,000 km. In the United States, the 25 Hz frequency with 11–12 kV was adopted for single-phase alternating current electric traction;
- Single-phase alternating current at mains frequency of 50 or 60 Hz, at 25 kV, in Russia (26,800 km), France (7000 km), Japan, India, former Yugoslavia, China, Great Britain, Hungary, Finland, and so on, and on European high-speed networks where the railway frequency does not exist.

Around the world, the development of electrified railways is greater than 200,000 km (Table 1.1), namely, 17.2% of the global railway network.

Table 1.1 Electric Traction Railway Coverage Around the World

DC mains		
Up to 1 kV	7,650 km	
1–2 kV	20,440 km	96,980 km (47.3%)
Over 2 kV	68,890 km	
AC single-phase mains		
15 kV–16 and 2/3 Hz	32,940 km	
25 kV–50/60 Hz	72,110 km	105,050 km (51.2 %)
Other systems	3,000 km	3,000 km (1.5%)

1.3 TRACK AND TRAIN WHEEL

Figure 1.6 schematically represents a track consisting of two rails supported on sleepers (treated wood, concrete, or steel) and anchored to them by means of appropriate baseplates. In railway lines, the sleepers are in turn anchored in a ballast of crushed stone or on concrete platforms equipped with elements that limit the transmission of vibrations. When crossing bridges, the tracks may be laid directly on the structure, with appropriate measures implemented to allow for controlled thermal expansion.

A typical rail cross section is shown in Figure 1.7: Recall that there is a set of standardized profiles, characterized according to linear mass, ranging from light rails with linear mass from 21 to 36 kg/m, to heavy ones from 50 to 60 kg/m used in trunk railway and metro lines.

The track is made from sections of rail of length l connected by joints that allow for some movement due to thermal expansion.

Figure 1.6 Basic representation of a track.

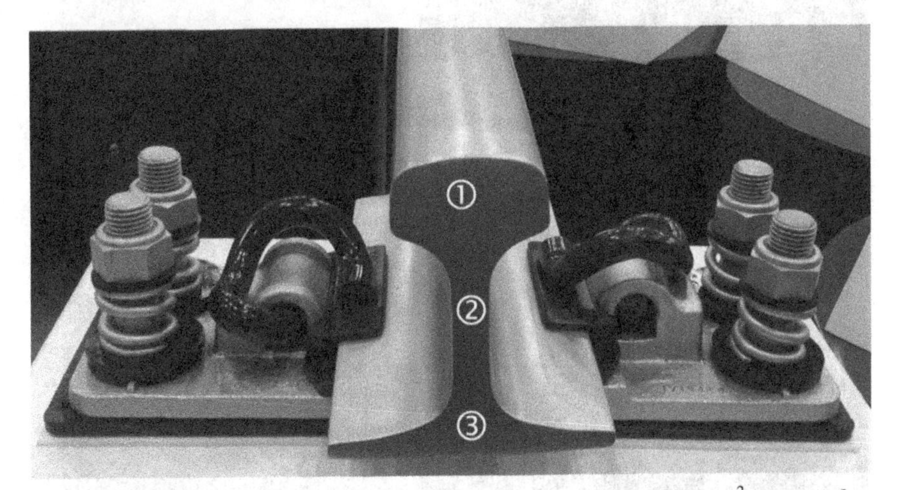

Figure 1.7 Section of a rail with linear mass of 60 kg/m (cross section = 7886 mm^2; moment of inertia $J_x = 3055\ \text{cm}^4$; moment of inertia $J_y = 513\ \text{cm}^4$) clamped on a sleeper: (1) head, (2) web, and (3) foot.

Table 1.2 Distribution of World Railway Networks According to Gauge

Gauge	km	Extension %
Narrow, up to 914 mm	21,900	1.8
Narrow, 950–1067 mm	213,300	17.9
Normal, 1432–1445 mm	703,200	59.1
1498–1524 mm (Russia)	150,200	12.6
Broad, 1580–1676 mm	102,100	8.6
	1,190,700	100

Source: From Railway Directory, London 1993.

The track foundation can be significantly improved by reducing the distance between the sleepers and increasing their mass, or by laying them in concrete; this allows for the joints to be eliminated, as the rails can be butted and welded in place, thus achieving long uninterrupted welded rail sections with significant benefits for the ride quality of the rolling stock.

In electric traction lines, the track is normally used as the negative conductor of the power supply line; since the joints give rise to a significant additional electric resistance, they are short-circuited by means of flexible copper cable ties

Figure 1.8 Broad gauge in regional Spanish railway lines.

welded to the adjacent rail pieces. There is a clear advantage, also from this point of view, in eliminating the joints or, at least, in reducing their number.

The distance between the inner faces of the rails at the head is called the gauge: The most common one used in approximately 60% of the world's railways (Table 1.2) is 1435 mm; at the extremes, there are narrow gauge and broad gauge railways: 1520 mm is used in Russia; 1600–1668–1676 mm in Spain (Figure 1.8), Portugal, Latin America, and India.

Wheels have a steel rim that rests on the upper surface of the rail head, and it has a guiding flange (Figure 1.9). The wheels are normally of rigid construction, consisting of a rim (tire) fitted to a steel body as shown in Figure 1.10, or forged in a single solid unit.

Figure 1.9 Steel wheels with rounded flange for railway and subway applications.

Figure 1.10 Wheelset for railway applications.

Figure 1.11 Elastic wheel with flat flange for tramway applications.

In trams elastic wheels may be used with rubber segments fitted between wheel and tire. Two wheels and an axle constitute a wheelset, as shown in Figure 1.11; the wheelset is automatically guided by the tire flanges; this system is very effective and enables the formation of convoys that may be very long, as they are drawn or pushed on the tracks.

Chapter 2

Basic Notions for the Study of Electric Traction Systems

2.1 THE PARK TRANSFORM

The study of electronic power converters, such as AC motor drives or converters connected to an electrical network, requires the use of mathematical methods that allow for their correct description, also under dynamic conditions.

The most commonly used tool is the Park transform, which allows us to more effectively describe three-phase instantaneous magnitudes such as voltage, current, magnetic flux, and so on.

This transformation is particularly convenient and meaningful for the analysis of electromagnetic phenomena, both during transient and sinusoidal or distorted steady states. Furthermore, the formal description of the Park transform by means of space vectors allows us to rediscover classical methods for the study of three-phase systems, such as phasors and symmetrical components.

The main advantage of the Park transform in the study of rotating machines is that it eliminates dependence on the reciprocal angular coupling displacement between stator and rotor. Space vectors are also extremely useful for developing three-phase converter theory and studying its applications. In three-phase electrical systems, the space vector is significant for its role in unifying analytical formulations, both in stationary and dynamic conditions and at component and system levels.

Electrical Railway Transportation Systems, First Edition. Morris Brenna, Federica Foiadelli, and Dario Zaninelli.

2.1.1 The Stationary Reference Frame Park Transform[1]

The stationary reference frame Park transform is a linear transformation with constant real coefficients, where the transformation matrix $\boldsymbol{T}_0$ is as follows:

$$\boldsymbol{T}_0 = \begin{bmatrix} \sqrt{\frac{2}{3}} & -\frac{1}{\sqrt{6}} & -\frac{1}{\sqrt{6}} \\ 0 & \frac{1}{\sqrt{2}} & -\frac{1}{\sqrt{2}} \\ \frac{1}{\sqrt{3}} & \frac{1}{\sqrt{3}} & \frac{1}{\sqrt{3}} \end{bmatrix} = \sqrt{\frac{2}{3}} \begin{bmatrix} 1 & -\frac{1}{2} & -\frac{1}{2} \\ 0 & \frac{\sqrt{3}}{2} & -\frac{\sqrt{3}}{2} \\ \frac{1}{\sqrt{2}} & \frac{1}{\sqrt{2}} & \frac{1}{\sqrt{2}} \end{bmatrix} \tag{2.1}$$

This matrix when applied to a triplet of instantaneous phase values $\boldsymbol{v}_f(t)$ (e.g., voltages)

$$\boldsymbol{v}_f = \begin{bmatrix} v_a(t) \\ v_b(t) \\ v_c(t) \end{bmatrix}$$

gives rise to the three Park components: $\boldsymbol{v}_p(t)$ components on the α and β axes and the zero-sequence component: $\boldsymbol{v}_p = \begin{bmatrix} v_\alpha(t) \\ v_\beta(t) \\ v_0(t) \end{bmatrix}$ So, the relationship between the Park magnitudes and their original three-phase counterparts is expressed as follows:

$$\begin{bmatrix} v_\alpha(t) \\ v_\beta(t) \\ v_0(t) \end{bmatrix} = \boldsymbol{T}_0 \begin{bmatrix} v_a(t) \\ v_b(t) \\ v_c(t) \end{bmatrix}, \quad \boldsymbol{v}_p = \boldsymbol{T}_0 \boldsymbol{v}_f \tag{2.2}$$

Matrix (2.1) is orthogonal since it has the property that its inverse is $\boldsymbol{T}_0^{-1}$ equal to its transpose. The orthogonality property implies that the modulus of the determinant is unitary and that the inner products are preserved.[2] This ensures, as will be seen, invariance of power, energy, and vector moduli.

[1] *Notes on the symbols used*
Overscore symbols: variables or complex constants.
Normal character symbols: variables or real constants and moduli of complex values.
Capitalized symbols: real or complex constants.
$\overline{a}^*$ $\overline{a}^*$ $\overline{a}^*$ complex conjugate of $\overline{a}$.
Re(·): real part, *Im* (·): imaginary part.
j: imaginary unit.

[2] The preservation of inner product consists of the following property. Let $\boldsymbol{a}_p = \boldsymbol{T}_0\boldsymbol{a}_f$, $\boldsymbol{b}_p = \boldsymbol{T}_0\boldsymbol{b}_f$, then $\boldsymbol{a}_p^t\boldsymbol{b}_p = \boldsymbol{a}_f^t\boldsymbol{b}_f$.

$$T_0^{-1} = T_0^t = \begin{bmatrix} \sqrt{\frac{2}{3}} & 0 & \frac{1}{\sqrt{3}} \\ -\frac{1}{\sqrt{6}} & \frac{1}{\sqrt{2}} & \frac{1}{\sqrt{3}} \\ -\frac{1}{\sqrt{6}} & -\frac{1}{\sqrt{2}} & \frac{1}{\sqrt{3}} \end{bmatrix} = \sqrt{\frac{2}{3}} \begin{bmatrix} 1 & 0 & \frac{1}{\sqrt{2}} \\ -\frac{1}{2} & \frac{\sqrt{3}}{2} & \frac{1}{\sqrt{2}} \\ -\frac{1}{2} & -\frac{\sqrt{3}}{2} & \frac{1}{\sqrt{2}} \end{bmatrix}$$

Thanks to the orthogonality property of the matrix, the inverse transformation is obtained by applying the transpose T_0^t to the Park variables, thus:

$$\begin{bmatrix} v_a(t) \\ v_b(t) \\ v_c(t) \end{bmatrix} = T_0^t \begin{bmatrix} v_\alpha(t) \\ v_\beta(t) \\ v_0(t) \end{bmatrix}, \quad v_f = T_0^t v_p \tag{2.3}$$

2.1.2 Representation of Space Vectors

The components v_α and v_β define the complex variable known as a space vector or Park vector

$$\overline{v}(t) = v_\alpha + jv_\beta, \quad v_\alpha = Re(\overline{v}), \quad v_\beta = Im(\overline{v}) \tag{2.4}$$

while the term v_0 represents the zero-sequence component defined as

$$v_0(t) = \frac{v_a + v_b + v_c}{\sqrt{3}}$$

Once the space vector has been defined, the transformation (2.2) can assume a more direct and usable form, and the Fortescue operator[3] $\overline{\alpha}$ is introduced:

$$\overline{v}(t) = \sqrt{\frac{2}{3}}\left[v_a + \overline{\alpha} v_b + \overline{\alpha}^2 v_c\right] \tag{2.5}$$

in which $\overline{\alpha} = e^{j\frac{2}{3}\pi} = -\frac{1}{2} + j\frac{\sqrt{3}}{2}, \quad \overline{\alpha}^2 = \overline{\alpha}^* = e^{-j\frac{2}{3}\pi} = -\frac{1}{2} - j\frac{\sqrt{3}}{2}$

which can, in turn, be expressed in matrix form as follows:

$$\overline{v}(t) = \sqrt{\frac{2}{3}}\begin{bmatrix} 1 & \overline{\alpha} & \overline{\alpha}^2 \end{bmatrix} \begin{bmatrix} v_a(t) \\ v_b(t) \\ v_c(t) \end{bmatrix}, \quad v_0(t) = \frac{1}{\sqrt{3}}\begin{bmatrix} 1 & 1 & 1 \end{bmatrix} \begin{bmatrix} v_a(t) \\ v_b(t) \\ v_c(t) \end{bmatrix}$$

[3] Let $\overline{\alpha}$ indicate the 120° rotation operator, for consistency with the symbology of symmetrical components. It should not be confused with the index (not overlined) of the first component on the stationary reference frame.

The complex variable $\overline{v}$ (space vector) and the scalar v_0 (zero-sequence component) fully identify the three-phase system. Note that the space vector, invariant with respect to an additive term that is common to the phase variables, constitutes the pure three-phase component of the system, that is, independent of the zero-sequence component (the voltage vector does not depend on the voltage phase reference). The angle, in value and sign, of the space vector depends on the (arbitrary) order and rotation sense assigned to the phase variables.

The phase variables, once the space vector $\overline{v}$ and the zero-sequence component v_0 are known, can, in turn, be expressed by the vector form (2.3):

$$
\begin{aligned}
v_a &= \sqrt{\frac{2}{3}} Re(\overline{v}) + \frac{v_0}{\sqrt{3}} \\
v_b &= \sqrt{\frac{2}{3}} Re(\overline{\alpha}^2 \overline{v}) + \frac{v_0}{\sqrt{3}} \\
v_c &= \sqrt{\frac{2}{3}} Re(\overline{\alpha} \overline{v}) + \frac{v_0}{\sqrt{3}}
\end{aligned}
\tag{2.6}
$$

The expressions (2.6) correspond to a significant geometrical interpretation for a pure three-phase system, then having a null zero-sequence component. By associating the phase variables with the three directions in the complex plane, with phase a matched to the α axis and the others at 120° increments in the direction of positive rotation, the projection of the vector onto these oriented directions provides (excluding the coefficient $\sqrt{2/3}$) the instantaneous value of the phase variables (Figure 2.1).

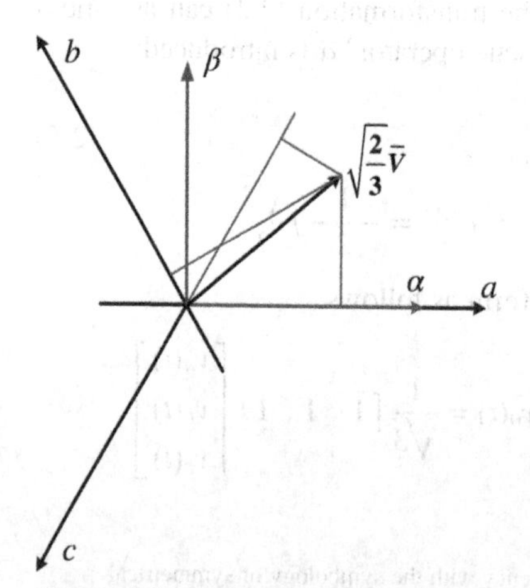

Figure 2.1 Representation of a pure three-phase system through the stationary reference frame Park transform.

Since the line-to-line voltages are always measurable for a three-phase section, it is useful to define the space voltage vector of the line-to-line voltages alone. Given that these have an overall zero-sum, there are numerous equivalent expressions for the calculation of $\overline{v}$. For example,

$$\begin{aligned} v_\alpha &= \frac{v_{ab} - v_{ca}}{\sqrt{6}} \\ v_\beta &= \frac{v_{bc}}{\sqrt{2}} \\ \overline{v} &= \frac{v_{ab} - v_{ca} + j\sqrt{3}v_{bc}}{\sqrt{6}} = \frac{v_{ab} - v_{ca} + j\sqrt{3}(v_{ab} + v_{ca})}{\sqrt{6}} \\ &= \frac{2v_{ab} + v_{bc} + j\sqrt{3}v_{bc}}{\sqrt{6}} = \frac{-2v_{ca} - v_{bc} + j\sqrt{3}v_{bc}}{\sqrt{6}} \end{aligned}$$

Hence, the phase voltages depend only on the space vector and can be expressed as follows:

$$\begin{aligned} v_{ab} &= \frac{1}{\sqrt{2}} Re\left[\left(\sqrt{3} + j\right)\overline{v}\right] \\ v_{bc} &= \sqrt{2} Re[-j\overline{v}] \\ v_{ca} &= \frac{1}{\sqrt{2}} Re\left[\left(-\sqrt{3} + j\right)\overline{v}\right] \end{aligned}$$

Figure 2.2 illustrates the relationships between the voltage space vector and the instantaneous values of the line-to-line voltages. The instantaneous values are the projections onto the directions indicated by the space vector multiplied by the coefficient $\sqrt{2}$.

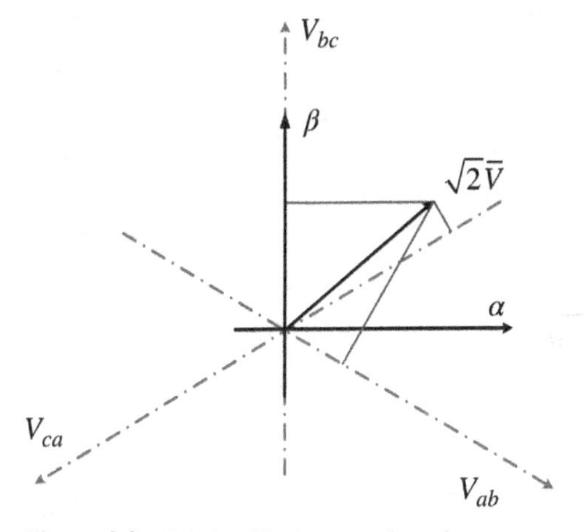

Figure 2.2 Relationships between the voltage space vector and the instantaneous values of the line-to-line voltages.

2.1.2.1 Special cases

There are several special cases in which the Park transform leads to having a space vector that is simple and practical to use.

Symmetrical Three-Phase Sinusoidal System Consider a symmetrical sinusoidal system of phase voltages with direct cyclic phase sequence, as is the case in most distribution systems (the subscript f indicates phase or physical dimensions):

$$\begin{aligned} v_a &= \sqrt{2}V_f\cos(\omega t + \varphi) \\ v_b &= \sqrt{2}V_f\cos\left(\omega t - 2\pi/3 + \varphi\right) \\ v_c &= \sqrt{2}V_f\cos\left(\omega t + 2\pi/3 + \varphi\right) \end{aligned}$$

Applying (2.2), a space vector with constant modulus is obtained that has uniform angular velocity and angular frequency ω in a positive sense (Figure 2.3):

$$v_\alpha = \sqrt{3}V_f\cos(\omega t + \varphi), \quad v_\beta = \sqrt{3}V_f\sin(\omega t + \varphi)$$

$$\overline{v} = \overline{V}_1 e^{j\omega t}, \quad v_0 = 0, \quad \text{with } \overline{V}_1 = \sqrt{3}V_f e^{j\varphi}$$

Reversed Symmetrical Three-Phase Sinusoidal System Considering a symmetrical three-phase system, but with reversed phase sequence:

$$\begin{aligned} v_a &= \sqrt{2}V_f\cos(\omega t + \varphi) \\ v_b &= \sqrt{2}V_f\cos\left(\omega t + 2\pi/3 + \varphi\right) \\ v_c &= \sqrt{2}V_f\cos\left(\omega t - 2\pi/3 + \varphi\right) \end{aligned}$$

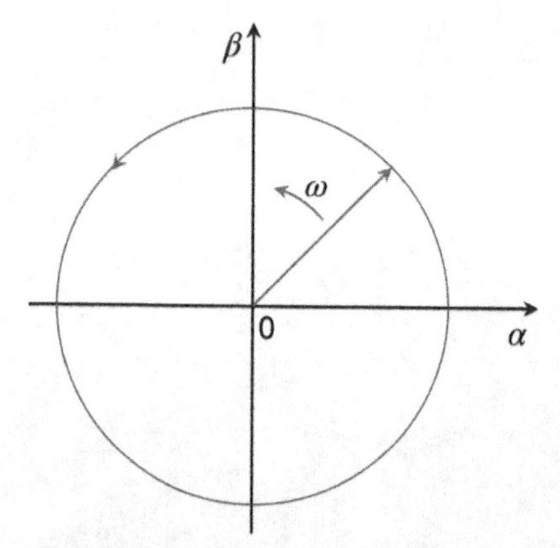

Figure 2.3 Representation of the voltage space vector corresponding to a symmetrical direct three-phase system.

the result will again be a space vector with constant modulus and uniform angular velocity at angular frequency ω, but rotating in the opposite direction:

$$\overline{v} = \overline{V}_{-1}e^{-j\omega t}, \quad v_0 = 0, \quad \text{with } \overline{V}_{-1} = \sqrt{3}V_f e^{-j\varphi}$$

Generic Sinusoidal Three-Phase System A generic sinusoidal triplet gives rise to a space vector resulting from the sum of two terms: one has constant modulus with positive rotation, while the other has constant modulus but rotation in the negative direction (this may also be derived by applying the principle of superposition):

$$\overline{v} = \overline{V}_1 e^{j\omega t} + \overline{V}_{-1}e^{-j\omega t}$$

Furthermore, the zero-sequence term is also present if the sum of the instantaneous phase values is not zero. In this case, the Park plane vector describes an elliptical trajectory (Figure 2.4). The case of the generic sinusoidal triplet will be addressed again later.

2.1.2.2 Constituent relationships for generic bipoles by means of the Park transform

As stated, the Park transform is applicable to triplets of variables such as voltages, currents, linked fluxes, electric charges, and so on. We now address how these are interrelated (impedances or admittances in a generalized sense).

For example, consider a set of three linked fluxes expressed as a function of the currents. In matrix form, indicating a 3×3 inductance matrix with $\boldsymbol{L}_f$,

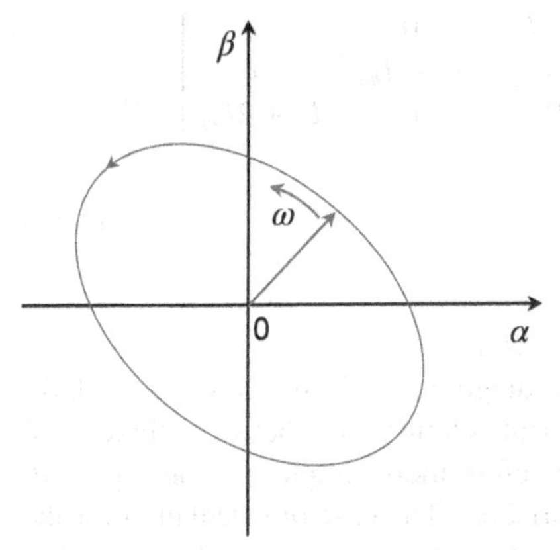

Figure 2.4 Representation of the voltage space vector corresponding to a generic sinusoidal three-phase system.

we have

$$\boldsymbol{\Psi}_f = \boldsymbol{L}_f \boldsymbol{i}_f \tag{2.7}$$

Applying the Park transform (2.1) to the variables, we obtain the transformed variables (identified with the subscript p):

$$\begin{aligned} \boldsymbol{\Psi}_p &= \boldsymbol{T}_0 \boldsymbol{\Psi}_f \\ \boldsymbol{i}_p &= \boldsymbol{T}_0 \boldsymbol{i}_f \end{aligned} \tag{2.8}$$

From the second (2.8) identity, we can also state $\boldsymbol{i}_f = \boldsymbol{T}_0^t \boldsymbol{i}_p$, from which, by replacing the latter in (2.7) and replacing the result in the first identity of (2.8), we obtain

$$\boldsymbol{\Psi}_p = \boldsymbol{L}_p \boldsymbol{i}_p \tag{2.9}$$

in which $\boldsymbol{L}_p = \boldsymbol{T}_0 \boldsymbol{L}_f \boldsymbol{T}_0^t$ is the matrix of transformed inductances. Note that, by virtue of the properties set out above, the reverse relationship also applies, that is, $\boldsymbol{L}_f = \boldsymbol{T}_0^t \boldsymbol{L}_f \boldsymbol{T}_0$.

A particularly important case is the one related to a triplet of symmetrical inductors, for which in (2.7) the self-inductances are equal to each other and all the mutual inductances are equal to each other, that is,

$$\boldsymbol{L}_f = \begin{bmatrix} L_f & L_m & L_m \\ L_m & L_f & L_m \\ L_m & L_m & L_f \end{bmatrix} \tag{2.10}$$

By applying the transformation $\boldsymbol{L}_p = \boldsymbol{T}_0 \boldsymbol{L}_f \boldsymbol{T}_0^t$, we obtain the following result:

$$\begin{aligned} \boldsymbol{L}_p = \boldsymbol{T}_0 \begin{bmatrix} L_f & L_m & L_m \\ L_m & L_f & L_m \\ L_m & L_m & L_f \end{bmatrix} \boldsymbol{T}_0^t &= \begin{bmatrix} L_f - L_m & 0 & 0 \\ 0 & L_f - L_m & 0 \\ 0 & 0 & L_f + 2L_m \end{bmatrix} \\ &= \begin{bmatrix} L_\alpha & 0 & 0 \\ 0 & L_\beta & 0 \\ 0 & 0 & L_0 \end{bmatrix} \end{aligned} \tag{2.11}$$

in which $L_\alpha = L_\beta = L_f - L_m$ and $L_0 = L_f + 2L_m$.

The transformed matrix is therefore diagonal. This result is particularly important, as it envisages three independent relationships between fluxes and currents in the Park domain, which are therefore more simple relationships with respect to those described by the phase variables. The ease of calculation for the inverse of (2.11) is also noted. The achieved results can be generalized to other systems with symmetric parameters such as resistances, capacitances, conductances, elastances, and inverse inductances. Parametric matrices of the type (2.10), that is, having the same terms on the main diagonal and equal off-diagonal terms, are defined as having three-phase symmetry.

Moreover, as a particular case of (2.10) in which the mutual terms are zero, the matrix (e.g., of equal resistors) coincides with its own transform:

$$\boldsymbol{R}_f = \begin{bmatrix} R & 0 & 0 \\ 0 & R & 0 \\ 0 & 0 & R \end{bmatrix}, \quad \boldsymbol{R}_p = \boldsymbol{R}_f \tag{2.12}$$

The information presented allows us to decouple, in the transformed variables, constructively symmetrical impedance systems, which represent the vast majority of practical applications.

In the domain of transformed scalar variables, by means of formally identical relationships, this leads to the representation of three single-phase elements that are decoupled from each other. Furthermore, the equivalence of the parameters relative to the axes α and β allows us to combine the corresponding scalar relationships in one complex relationship, in analytical form, in space vector representation. In fact, take the example:

$$\Psi_\alpha = L_\alpha i_\alpha, \quad \Psi_\beta = L_\beta i_\beta$$

If and only if $L_\alpha = L_\beta = L$ by applying (2.4), we can have the vector representation:

$$\boldsymbol{\Psi} = L\boldsymbol{i}$$

Consequently, the representation of three-phase circuit elements can be represented with single-phase equivalent circuits in the space vectors and, separately, in the zero-sequence components. By applying the Park vector transformation to all the elements of a three-phase network, we obtain a transformed network with space vectors as variables. This network will be driven by corresponding transformed generators.

As a generalization of the above, consider a generic relationship (with real coefficients) between two sets of three-phase quantities, of the type

$$\boldsymbol{a}_f = \boldsymbol{Z}_f \boldsymbol{b}_f$$

The Park transform shifts the relationship to

$$\boldsymbol{a}_p = \boldsymbol{Z}_p \boldsymbol{b}_p, \quad \text{with } \boldsymbol{Z}_p = \boldsymbol{T}_0 \boldsymbol{Z}_f \boldsymbol{T}_0^t$$

In which the matrix $\boldsymbol{Z}_p$ can always be inverse-transformed

$$\boldsymbol{Z}_f = \boldsymbol{T}_0^t \boldsymbol{Z}_f \boldsymbol{T}_0$$

EXAMPLE

To highlight the features of Park analysis, consider the example in Figure 2.5 of a load consisting of a mechanically symmetrical three-phase inductor having self-inductance and mutual inductance as in (2.10) and equivalent series resistances as in (2.12). ■

The phase variable system has the matrix form:

$$\begin{bmatrix} v_a(t) \\ v_b(t) \\ v_c(t) \end{bmatrix} = \begin{bmatrix} L_f & L_m & L_m \\ L_m & L_f & L_m \\ L_m & L_m & L_f \end{bmatrix} \frac{d}{dt} \begin{bmatrix} i_a(t) \\ i_b(t) \\ i_c(t) \end{bmatrix} + \begin{bmatrix} R & 0 & 0 \\ 0 & R & 0 \\ 0 & 0 & R \end{bmatrix} \begin{bmatrix} i_a(t) \\ i_b(t) \\ i_c(t) \end{bmatrix} \quad (2.13)$$

Now apply the stationary reference frame Park transform $\boldsymbol{T}_0$ to the voltages and currents. Note that, since the stationary reference frame transformation matrix is constant, $\frac{d}{dt}\boldsymbol{i}_f = \boldsymbol{T}_0^{-1}\frac{d}{dt}\boldsymbol{i}_p$, this leads to the expression,

$$\begin{bmatrix} v_\alpha(t) \\ v_\beta(t) \\ v_0(t) \end{bmatrix} = \begin{bmatrix} L_f - L_m & 0 & 0 \\ 0 & L_f - L_m & 0 \\ 0 & 0 & L_f + 2L_m \end{bmatrix} \frac{d}{dt} \begin{bmatrix} i_\alpha(t) \\ i_\beta(t) \\ i_0(t) \end{bmatrix} + \begin{bmatrix} R & 0 & 0 \\ 0 & R & 0 \\ 0 & 0 & R \end{bmatrix} \begin{bmatrix} i_\alpha(t) \\ i_\beta(t) \\ i_0(t) \end{bmatrix}$$

which, given the diagonality of the matrices, summarizes three independent equations.

Since the parameters for the α and β components are equal, the first two relationships can be combined into a single space vector form. With

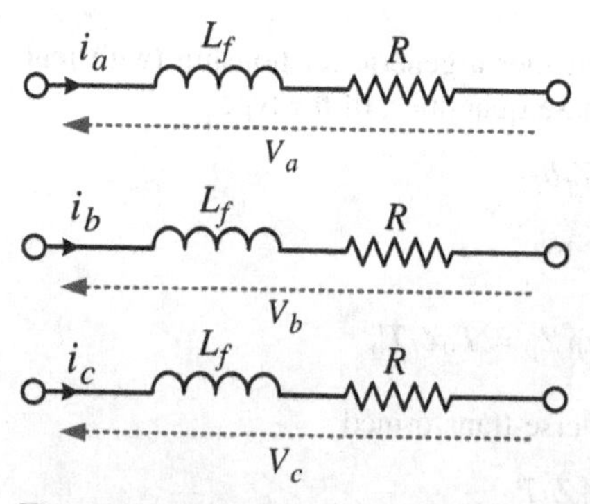

Figure 2.5 Representation of a symmetrical three-phase inductor.

$L = L_f - L_m$ and $L_0 = L_f + 2L_m$, we have

$$\bar{v} = L\frac{d\bar{i}}{dt} + R\bar{i}, \quad v_0 = L_0\frac{di_0}{dt} + Ri_0 \tag{2.14}$$

The resulting relationships, the first one complex and the second one real, constitute the transformed Park model of the inductor system (2.13).

Now consider a triplet of symmetrical sinusoidal voltages with positive phase sequence: $\bar{v} = \overline{V}e^{j\omega t}$, $v_0 = 0$. At steady state, the current will have the voltage form: $\bar{i} = \bar{I}e^{j\omega t}$, $i_0 = 0$.

Knowing also that $\frac{d\bar{i}}{dt} = j\omega\bar{I}e^{j\omega t}$, the steady-state vector relationship then becomes

$$\overline{V}e^{j\omega t} = \left(R\bar{I} + j\omega L\bar{I}\right)e^{j\omega t} \tag{2.15}$$

Apart from the rotational factors $e^{j\omega t}$, we have returned to the classical phasor diagram.

If the cyclic rotation is inverted to the negative direction, then $\bar{v} = \overline{V}e^{-j\omega t}$, $v_0 = 0$, the steady-state equation becomes

$$\overline{V}e^{-j\omega t} = \left(R\bar{I} - j\omega L\bar{I}\right)e^{-j\omega t}$$

Apart from the common rotation, the steady-state space vector presents space phase relationships that are inverted with respect to the previous case, hence its corresponding complex conjugate must be invoked to recover the relevant phasor diagram. This is due to the fact that phasors represent phase shifts in time between sinusoids, while space vectors represent spatial phase (more precisely, relationships between instantaneous magnitudes that are spatially portrayed in the complex plane).

EXAMPLE. ***RL* Transient**

The effectiveness of the space vector transformation is fully manifested in the dynamic case. As a simple example, consider the inrush transient of the three-phase symmetrical load *RL* (2.13) of Figure 2.5 driven by a set of symmetrical three-phase generators.

The system is represented by the complex linear equation given by the first of (2.14), in which the driving term is given by $\bar{v}(t) = \overline{V}e^{j\omega t}$. Since the equation is analytic, the integration can make use of the rules applicable to constant coefficient linear differential equations. The solution is the following dynamic Park current space vector (the zero-sequence components are identically zero):

$$\bar{i}(t) = \frac{\overline{V}}{R + j\omega L}\left[e^{j\omega t} - e^{-\frac{R}{L}t}\right] = \frac{\overline{V}e^{j\omega t}}{R + j\omega L}\left[1 - e^{-\left(j\omega + \frac{R}{L}\right)t}\right]$$

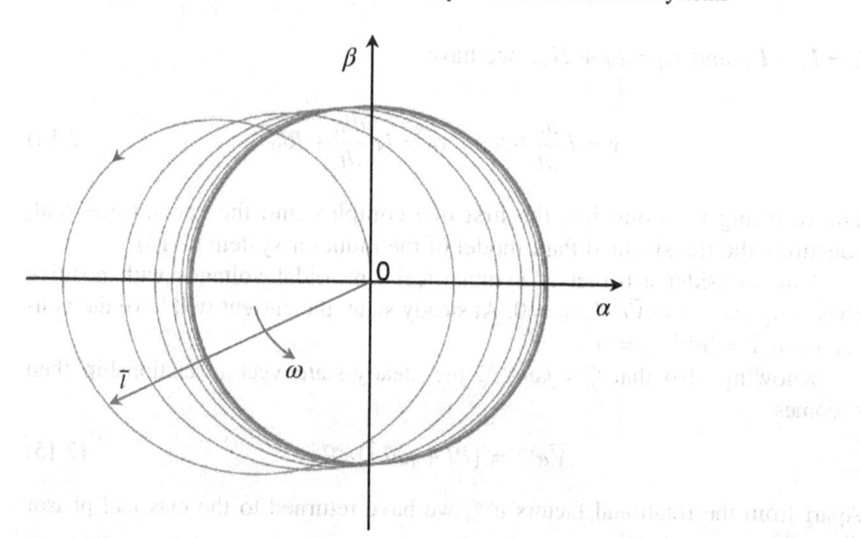

Figure 2.6 Representation of an *RL* transient by means of the stationary reference frame Park transform.

The trajectory of the vector in the Park plane depicts the dynamic of the transient (Figure 2.6). The projections on the phase axes (as in Figure 2.1) constitute the dynamics of the phase currents (excluding the coefficient $\sqrt{2/3}$). ■

2.1.3 The Park Transform and Symmetrical Components

It is known that the transformation of symmetrical components (or processing the Fortescue sequences) is an effective technique for treating sinusoidal three-phase systems in the phasor domain (complex constants having one-to-one correspondence with isofrequential sinusoidal waveforms).

The transformation of symmetrical components uses the following matrix with constant complex coefficients (shown here in rational form):

$$\overline{S} = \frac{1}{\sqrt{3}} \begin{bmatrix} 1 & \overline{\alpha} & \overline{\alpha}^2 \\ 1 & \overline{\alpha}^2 & \overline{\alpha} \\ 1 & 1 & 1 \end{bmatrix} \tag{2.16}$$

When applied to the triplet of phase phasors $\overline{V}_a\overline{V}_b\overline{V}_c$, it provides the phasors of the direct (positive), inverse (negative), and zero-sequence (zero) sequences.

$$\begin{bmatrix} \overline{V}_d \\ \overline{V}_i \\ \overline{V}_0 \end{bmatrix} = \overline{S} \begin{bmatrix} \overline{V}_a \\ \overline{V}_b \\ \overline{V}_c \end{bmatrix} \tag{2.17}$$

The inverse of (2.16) is equal to its conjugate transpose $\overline{S}^{-1} = \overline{S}^{*t}$. (Matrices with this property are called unitary matrices, an extension to the field of complex orthogonal matrices.) The inverse transformation is obtained as

$$\begin{bmatrix} \overline{V}_a \\ \overline{V}_b \\ \overline{V}_c \end{bmatrix} = \overline{S}^{*t} \begin{bmatrix} \overline{V}_d \\ \overline{V}_i \\ \overline{V}_0 \end{bmatrix} \tag{2.18}$$

The relationships between the transformations of the symmetrical components and the Park transform, as applicable to both the variables and the impedances, can now be defined.

With regard to the impedances, consider the following relationship matrix between three-phase phasors:

$$\overline{V}_f = \overline{Z}_f \overline{I}_f$$

By proceeding in a similar manner to (2.9), the application of the symmetrical components brings forth the sequence values:

$$\overline{V}_s = \overline{Z}_s \overline{I}_s, \quad \text{with} \quad \overline{Z}_s = \overline{S}\overline{Z}_f \overline{S}^{*t}$$

In the case of symmetrical phase impedances

$$\overline{Z}_f = \begin{bmatrix} \overline{Z} & \overline{Z}' & \overline{Z}' \\ \overline{Z}' & \overline{Z} & \overline{Z}' \\ \overline{Z}' & \overline{Z}' & \overline{Z} \end{bmatrix}$$

by applying the transform, we obtain

$$\overline{Z}_s = \begin{bmatrix} \overline{Z} - \overline{Z}' & 0 & 0 \\ 0 & \overline{Z} - \overline{Z}' & 0 \\ 0 & 0 & \overline{Z} + 2\overline{Z}' \end{bmatrix}$$

The result is formally identical to that obtained in (2.11) by applying the Park transform. We conclude that in the case of symmetry, the Park impedances and sequence impedances coincide.

This result has significant practical relevance. Usually the sequence and circuit parameters are known or they may be determined through the use of standard methods. The same parameter values are required for formulations involving the Park variables. Care must be taken, however, when considering these equivalences, remembering that sequence impedances or admittances, in general, are complex constants, while the Park parameters are real constants. The corresponding domain parametric relationships are summarized in Table 2.1.

In the general case, having defined the complex matrix

$$\overline{g} = T_0\overline{S}^{-1} = \frac{1}{\sqrt{2}}\begin{bmatrix} 1 & 1 & 0 \\ -j & j & 0 \\ 0 & 0 & \sqrt{2} \end{bmatrix}$$

we can establish the reciprocal relationships between sequence and Park impedances as follows:

$$\boldsymbol{Z}_p = \overline{\boldsymbol{g}}\overline{\boldsymbol{Z}}_s\overline{\boldsymbol{g}}^{-1}, \quad \overline{\boldsymbol{Z}}_s = \overline{\boldsymbol{g}}^{-1}\overline{\boldsymbol{Z}}_p\overline{\boldsymbol{g}}$$

Remember that the matrix $\boldsymbol{Z}_p$ must always resolve to real terms. Through the general relationship, we recognize that only if the transformed matrix is diagonal, having its first two terms equal, then it is the same according to both Park and sequence transforms.

For the passage procedure between the variables of two formulations, the formal identity between a sinusoidal magnitude $v(t)$ and its phasor $\overline{V}$ $(\overline{V} = Ve^{j\varphi})$ must be remembered:

$$v(t) = \sqrt{2}V\cos(\omega t + \varphi) = \sqrt{2}Re\left(\overline{V}e^{j\omega t}\right) = \frac{1}{\sqrt{2}}\left(\overline{V}e^{j\omega t} + \overline{V}^* e^{-j\omega t}\right)$$

Table 2.1 Element Correspondences Represented in the Sequence and Park Domains

Sequences	Park
$R_d = R_i$	$R_\alpha = R_\beta = R_d$
R_0	R_0
$X_d = X_i = \omega L_d = \omega L_i$	$L_\alpha = L_\beta = L_d$
$X_0 = \omega L_0$	L_0
$B_d = B_i = \omega C_d = \omega C_i$	$C_\alpha = C_\beta = C_d$
$B_0 = \omega C_0$	C_0

A triplet of sinusoidal instantaneous values $v_a(t)$, $v_b(t)$, and $v_c(t)$ can then be expressed as a function of their respective phasors $\overline{V}_a$, $\overline{V}_b$, and $\overline{V}_c$ as

$$\begin{bmatrix} v_a \\ v_b \\ v_c \end{bmatrix} = \frac{1}{\sqrt{2}} \left(\begin{bmatrix} \overline{V}_a \\ \overline{V}_b \\ \overline{V}_c \end{bmatrix} e^{j\omega t} + \begin{bmatrix} \overline{V}_a^* \\ \overline{V}_b^* \\ \overline{V}_c^* \end{bmatrix} \right) e^{-j\omega t} \tag{2.19}$$

From the first expression of (2.5) and (2.18), we have

$$\overline{v} = \sqrt{\frac{2}{3}} \begin{bmatrix} 1 & \overline{\alpha} & \overline{\alpha}^2 \end{bmatrix} \begin{bmatrix} v_a \\ v_b \\ v_c \end{bmatrix} = \frac{1}{\sqrt{3}} \begin{bmatrix} 1 & \overline{\alpha} & \overline{\alpha}^2 \end{bmatrix} \left(\begin{bmatrix} \overline{V}_a \\ \overline{V}_b \\ \overline{V}_c \end{bmatrix} e^{j\omega t} + \begin{bmatrix} \overline{V}_a^* \\ \overline{V}_b^* \\ \overline{V}_c^* \end{bmatrix} e^{-j\omega t} \right)$$

$$= \frac{1}{\sqrt{3}} \begin{bmatrix} 1 & \overline{\alpha} & \overline{\alpha}^2 \end{bmatrix} \left(\overline{S}^{*t} \begin{bmatrix} \overline{V}_d \\ \overline{V}_i \\ \overline{V}_0 \end{bmatrix} e^{-j\omega t} + \overline{S}^t \begin{bmatrix} \overline{V}_d^* \\ \overline{V}_i^* \\ \overline{V}_0^* \end{bmatrix} e^{-j\omega t} \right)$$

that gives rise to the expression:

$$\overline{v}(t) = \overline{V}_d e^{j\omega t} + \overline{V}_i e^{-j\omega t} \tag{2.20}$$

So in the sinusoidal regime, the space vector is composed of two counter-rotating terms with an amplitude and phase equal to the direct sequence phasor and the complex conjugate of the inverse (negative) sequence phasor, respectively.

With regard to the zero-sequence element, again from (2.5) and (2.18):

$$v_0 = \frac{1}{\sqrt{3}} \begin{bmatrix} 1 & 1 & 1 \end{bmatrix} \begin{bmatrix} v_a \\ v_b \\ v_c \end{bmatrix} = \frac{1}{\sqrt{6}} \begin{bmatrix} 1 & 1 & 1 \end{bmatrix} \left(\begin{bmatrix} \overline{V}_a \\ \overline{V}_b \\ \overline{V}_c \end{bmatrix} e^{j\omega t} + \begin{bmatrix} \overline{V}_a^* \\ \overline{V}_b^* \\ \overline{V}_c^* \end{bmatrix} e^{-j\omega t} \right)$$

$$= \frac{1}{\sqrt{6}} \begin{bmatrix} 1 & 1 & 1 \end{bmatrix} \left(\overline{S}^{*t} \begin{bmatrix} \overline{V}_s \\ \overline{V}_i \\ \overline{V}_0 \end{bmatrix} e^{-j\omega t} + \overline{S}^t \begin{bmatrix} \overline{V}_s^* \\ \overline{V}_i^* \\ \overline{V}_0^* \end{bmatrix} e^{-j\omega t} \right)$$

from which is obtained:

$$v_0(t) = \frac{1}{\sqrt{2}} \left(\overline{V}_0 e^{j\omega t} + \overline{V}_0^* e^{-j\omega t} \right) = \sqrt{2} Re \left(\overline{V}_0 e^{j\omega t} \right)$$

By comparison with (2.19), we recognized that under sinusoidal conditions, the Park zero-sequence component is a sinusoid whose phasor coincides with the zero-sequence phasor. In other words, the Park and sequence zero-sequence components coincide, thus justifying their identical reference terms.

2.1.4 Powers in the Park Variables

As seen in the previous section, the Park transform is orthogonal and consequently it does not alter the values of the system powers. In fact, considering the phase voltages v_a, v_b, and v_c and currents i_a, i_b, and i_c of a three-phase system, the three-phase instantaneous power is

$$p(t) = v_a i_a + v_b i_b + v_c i_c = \mathbf{v}_f^t \mathbf{i}_f$$

Applying the Park transform to the voltages and currents and remembering that orthogonality implies $TT^t = 1$, we have

$$\mathbf{v}_f^t \mathbf{i}_f = \mathbf{v}_p^t T T^t \mathbf{i}_p = \mathbf{v}_p^t \mathbf{i}_p$$

So, the three-phase power can be expressed as

$$p(t) = v_a i_a + v_b i_b + v_c i_c = v_\alpha i_\alpha + v_\beta i_\beta + v_0 i_0 = Re\left(\overline{v}\overline{i}^*\right) + v_0 i_0$$

We assign the terms zero-sequence power to the instantaneous power associated with the zero-sequence components

$$p_0(t) = v_0 i_0$$

and Park real power to the instantaneous power of the pure three-phase system

$$p_p(t) = v_\alpha i_\alpha + v_\beta i_\beta = Re\left(\overline{v}\overline{i}^*\right) \tag{2.21}$$

Therefore, the total instantaneous power is $p(t) = p_p + p_0$.

We also define the imaginary power as the function

$$q_p(t) = v_\beta i_\alpha - v_\alpha i_\beta = Im\left(\overline{v}\overline{i}^*\right) \tag{2.22}$$

and thus the Park complex power as

$$\overline{a}_p(t) = p_p(t) + jq_p(t) = \overline{v}\overline{i}^* \tag{2.23}$$

These definitions are valid for instantaneous values and in any regime. In a symmetrical and balanced sinusoidal direct sequence system, the powers p_p, q_p, and $\overline{a}_p$ are constant and coincide, respectively, with the active, reactive, and complex powers. The instantaneous energy-related quantities defined in (2.21)–(2.23) have the following properties:

- They are algebraic quantities whose sign depends on the assumed reference directions for voltages and currents.
- They are conservative, in the sense that for each of them considered separately, the algebraic sum of the absorbed powers extended to all the elements of a three-phase isolated network is null.[4]

[4] Assuming that the three-phase voltages outside of the three-phase elements separately satisfy the same loop equations, for the linearity of the transformation to hold, the same equations must separately satisfy the Park voltage components. Similarly, if the three line currents separately satisfy the same equations at the nodes of the same graph of the voltages, the same equations will satisfy the Park current components. By Tellegen's theorem, we have a null result for the sum extended to all sides of the graph of the products between arbitrary pairs of voltage and current components. The conservation of the powers as sums of these products can be implied from this.

- They are measurable by means of constant coefficient linear combinations of products between phase voltages and currents.
- The sign q_p depends on the cyclic rotation assigned to the phases.

Under periodic conditions, the averages of the instantaneous powers P_p, P_0, Q_p, and $\overline{A}_p$ taken over the entire period obviously manifest the same properties established for the instantaneous values. The average P_p of $p_p(t)$ is the pure three-phase active power.

2.1.5 Stationary Reference Frame Three-Phase Components

Summarizing the above, we can apply the expressions listed in Table 2.2 to the following three-phase elements.

Note that if the zero-sequence terms are absent, as happens in many cases, the expressions become formally similar to the corresponding single-phase versions as long as the space vectors are used as substitutes for the instantaneous or average quantities.

2.1.6 Rotary Reference Frame Rotating Park Transform

The stationary reference frame Park transform (2.1) is a special case of a more general transformation with a rotating reference frame. The (orthogonal) matrix

Table 2.2 Primary Expressions for the Application of the Stationary Reference Frame Park Transform Applied to Three-Phase Systems

	Symmetrical resistors	Symmetric inductors	Symmetrical capacitors	Symmetrical sinusoidal generators (direct sequence)
Ohm's law for stationary reference frame	$\overline{v} = R\overline{i}$ $v_0 = R_0 i_0$	$\overline{v} = L\frac{d\overline{i}}{dt}$ $v_0 = L_0\frac{di_0}{dt}$	$\overline{i} = C\frac{d\overline{v}}{dt}$ $i_0 = C_0\frac{dv_0}{dt}$	$\overline{v} = \overline{V}e^{j\omega t}$ $\overline{a}_s = \overline{A}_s e^{j\omega t}$
Instantaneous power	$Ri^2 + R_0 i_0^2$	—	—	—
Average power (active)	$RI^2 + R_0 I_0^2$	—	—	—
Instantaneous energy	—	$\frac{1}{2}Li^2 + \frac{1}{2}L_0 i_0^2$	$\frac{1}{2}Cv^2 + \frac{1}{2}C_0 v_0^2$	—
Average energy	—	$\frac{1}{2}LI^2 + \frac{1}{2}L_0 I_0^2$	$\frac{1}{2}CV^2 + \frac{1}{2}C_0 V_0^2$	—

of the general transformation is as follows:

$$T(\theta) = \sqrt{\frac{2}{3}} \begin{bmatrix} \cos(\theta) & \cos\left(\theta - {}^{2\pi}/_{3}\right) & \cos\left(\theta + {}^{2\pi}/_{3}\right) \\ -\sin(\theta) & -\sin\left(\theta - {}^{2\pi}/_{3}\right) & -\sin\left(\theta + {}^{2\pi}/_{3}\right) \\ 1/\sqrt{2} & 1/\sqrt{2} & 1/\sqrt{2} \end{bmatrix} \tag{2.24}$$

When applied to a triple of instantaneous phase values $v_a(t)$, $v_b(t)$, and $v_c(t)$, it gives rise to the three Park components: components on the axes d, q, and the zero-sequence component:

$$\begin{bmatrix} v_d(t) \\ v_q(t) \\ v_0(t) \end{bmatrix} = T(\theta) \begin{bmatrix} v_a(t) \\ v_b(t) \\ v_c(t) \end{bmatrix}, \quad \begin{bmatrix} v_a(t) \\ v_b(t) \\ v_c(t) \end{bmatrix} = T(\theta)^t \begin{bmatrix} v_d(t) \\ v_q(t) \\ v_0(t) \end{bmatrix} \tag{2.25}$$

The angular parameter $\theta(t)$ is a generic function of time that represents, together with its first time derivative, respectively, the position and angular velocity of the Park reference axes d and q with respect to a fixed reference defined by the position $\theta(t) = 0$.

In the particular case of a stationary reference frame, with $\theta = 0$, the transformation reduces to (2.1). Note that the stationary reference frame transformation could be obtained from (2.24) for any value of θ, as long as it is constant.

As for the stationary reference frame case, the components v_d and v_q define the complex variable (rotating frame space vector)

$$\overline{v}(t) = v_d + jv_q$$

The real part v_d and the coefficient of the Park vector imaginary unit v_q can be considered as the components referenced to the direct and quadrature axes.

The identity (2.24) can also be expressed as the product of (2.1) and the rotation matrix $\boldsymbol{H}(\theta)$:

$$\boldsymbol{T}(\theta) = \boldsymbol{H}(\theta)\boldsymbol{T}_0, \quad \text{with } \boldsymbol{H}(\theta) = \boldsymbol{T}(\theta)\boldsymbol{T}_0^t = \begin{bmatrix} \cos\theta & \sin\theta & 0 \\ -\sin\theta & \cos\theta & 0 \\ 0 & 0 & 1 \end{bmatrix}$$

The transformation $\boldsymbol{H}(\theta)$ rotates the components d and q by the angle θ while leaving the zero-sequence component unaltered. This transformation has the property of geometric rotation, in particular:

$$\begin{aligned} &\boldsymbol{H}(\theta)^t = \boldsymbol{H}(\theta)^{-1} \quad \text{(orthogonality)} \\ &\boldsymbol{H}(0) = 1 \quad \text{(unit matrix)} \\ &\boldsymbol{H}(\theta + 2k\pi) = \boldsymbol{H}(\theta) \quad (\text{where } k = \text{integer}) \\ &\boldsymbol{H}(\theta_1 + \theta_2) = \boldsymbol{H}(\theta_2)\boldsymbol{H}(\theta_1) = \boldsymbol{H}(\theta_1)\boldsymbol{H}(\theta_2) \\ &\boldsymbol{H}(-\theta) = \boldsymbol{H}(\theta)^{-1} \end{aligned}$$

When $\boldsymbol{H}(\theta)$ is applied to a triple of stationary reference frame Park transformed variables, the corresponding triple of rotating reference frame variables is generated. Then, by composing the d and q terms of both triples in the space vectors, the submatrix $\begin{bmatrix} \cos\theta & \sin\theta \\ -\sin\theta & \cos\theta \end{bmatrix}$ is recognized as being equivalent to the rotation term in the complex field $e^{-j\theta}$ (note that $e^{-j\theta}$ has the same rotation properties listed above).

The following significant result therefore emerges. Given a stationary-referenced Park vector $\overline{v}^0$ with ($\theta = 0$), the same in relation to a rotating reference frame specified by a generic $\theta(t)$ is given by

$$\overline{v}(t) = \overline{v}^0(t)e^{-j\theta(t)}, \quad \overline{v}^0(t) = \overline{v}(t)e^{j\theta(t)} \tag{2.26}$$

while the zero-sequence component is unaffected by the change of reference frame. The sign of the angle in the expressions (2.26) is clarified in Figure 2.7.

Equation 2.26 is generalized to the transformation between two reference frames in arbitrary motion simply by replacing θ with the difference $\theta_2 - \theta_1$ between the angular parameters of each reference frame system.

The transformation between different reference frames alters the angles and the rotation velocity of a set of space vectors by a common value without changing the relative positions and movements between them. The energy-dependent variables are not altered, as will be seen, by the moduli of the vectors and the phase differences between them.

The rotating reference frame transformation, in many cases, allows for considerable simplifications in the structure of the equations. A particularly important case is the following. Consider a stationary frame-referenced direct sequence symmetric sinusoidal triple $\overline{v}^0 = \overline{V}e^{j\omega t}$. By applying (2.26), we obtain

$$\overline{v}(t) = \overline{V}e^{j(\omega t - \theta)} \tag{2.27}$$

If we now assume $\theta = \omega t$ (rotating reference frame with angular frequency ω), the space vector is constant (equal to the direct sequence phasor): $\overline{v}(t) = \overline{V}$.

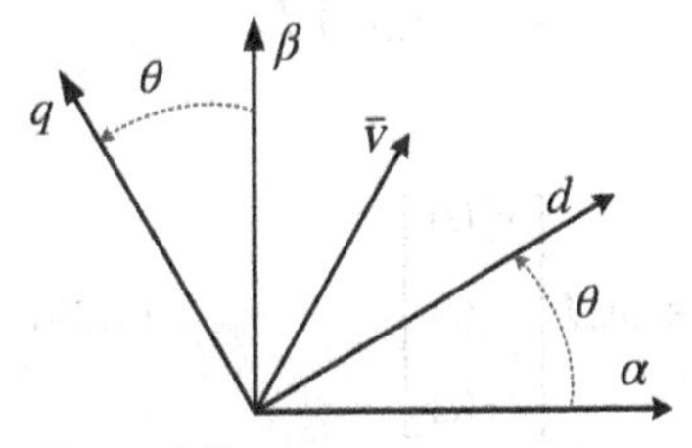

Figure 2.7 Representation of the rotating reference frame Park transform.

The major implications of this result will now be highlighted in the example of Figure 2.5. Having transformed all the rotating reference frame variables to ω the power supply frequency, the steady-state Equation 2.15 evolves to the following expression between constant magnitudes:

$$\overline{V} = R\overline{I} + j\omega L\overline{I}$$

As discussed above in relation to symmetrical components, this expression coincides with the relationship between direct sequence phasors that is formulated in the traditional analysis applied to sinusoidal conditions.

The results discussed above relative to the stationary case can be easily extended to the rotating reference frame. With regard to the space vectors, (2.26) is used. The transformation of impedances from stationary to rotating reference frames employs the matrix $\boldsymbol{H}(\theta)$. In particular, by indicating $\boldsymbol{Z}^0$ as the parameter matrix for stationary reference frames, the corresponding matrix for the rotating reference frame case $\boldsymbol{Z}$ is given by

$$\boldsymbol{Z} = \boldsymbol{H}(\theta)\boldsymbol{Z}^0\boldsymbol{H}(\theta)^{-1} \tag{2.28}$$

It is easy to verify that (2.28) does not alter the transformed matrices of symmetric matrices (2.11) and (2.12). These results are thus valid for any reference frame system.

Transformation of Derivatives The time-varying transformation $\boldsymbol{T}(\theta(t))$ is not indifferent to derivation.

Having defined the transformation with the expressions (2.25), the question arises regarding how to transform the derivatives of the phase variables. By considering the second expression in (2.25) and taking its time derivative, we obtain

$$\frac{d}{dt}\begin{bmatrix} v_a(t) \\ v_b(t) \\ v_c(t) \end{bmatrix} = \boldsymbol{T}(\theta)^t \frac{d}{dt}\begin{bmatrix} v_d(t) \\ v_q(t) \\ v_0(t) \end{bmatrix} + \frac{d\boldsymbol{T}(\theta)^t}{dt}\begin{bmatrix} v_d(t) \\ v_q(t) \\ v_0(t) \end{bmatrix}$$

$$= \boldsymbol{T}(\theta)^t \left(\frac{d}{dt}\begin{bmatrix} v_d(t) \\ v_q(t) \\ v_0(t) \end{bmatrix} + \boldsymbol{T}(\theta)\frac{d\boldsymbol{T}(\theta)^t}{dt}\begin{bmatrix} v_d(t) \\ v_q(t) \\ v_0(t) \end{bmatrix} \right)$$

which can be reformulated as

$$\boldsymbol{T}(\theta)\frac{d}{dt}\begin{bmatrix} v_a(t) \\ v_b(t) \\ v_c(t) \end{bmatrix} = \frac{d}{dt}\begin{bmatrix} v_d(t) \\ v_q(t) \\ v_0(t) \end{bmatrix} + \omega_p \boldsymbol{J}\begin{bmatrix} v_d(t) \\ v_q(t) \\ v_0(t) \end{bmatrix} \tag{2.29}$$

where

$$J = T(\theta)\frac{dT(\theta)^t}{d\theta} = H(\theta)\frac{dH(\theta)^t}{d\theta} = \begin{bmatrix} 0 & -1 & 0 \\ 1 & 0 & 0 \\ 0 & 0 & 0 \end{bmatrix}, \quad \omega_p = \frac{d\theta}{dt}$$

The result appears more significant in the complex field. Deriving the second expression from (2.26) with respect to time:

$$\frac{d\bar{v}^0}{dt} = \left(\frac{d\bar{v}}{dt} + j\omega_p\bar{v}\right)e^{j\theta} \tag{2.30}$$

This result gives rise to additional terms proportional to ω_p ("motion-related" terms) every time the derivative of a vector appears in the equations.

The consequences of expressions (2.29) and (2.30) are that in the transformation from stationary to rotating reference frames, the substitution rules indicated in Table 2.3 can be applied.Again observing the expressions (2.27) and (2.30), the formal analogy with phasors is clear: taking $\omega_p = \omega$ equal to the frequency of the symmetric sinusoidal power supply, under symmetrical and balanced sinusoidal conditions (direct sequence), the space vectors reduce to constants and we revert to the usual sequence phasor diagrams.

EXAMPLE. *RL* Transient on Rotating Reference Frames

Consider again the *RL* symmetrical three-phase load inrush transient (2.13) of Figure 2.5 powered by a three-phase direct symmetrical generator.

Now consider the transformation to a rotating reference frame at the power supply frequency ω. Applying the developed expressions, we obtain the vector equation:

$$\overline{V} = L\frac{d\bar{i}(t)}{dt} + j\omega L\bar{i}(t) + R\bar{i}(t) \tag{2.31}$$

Table 2.3 Applicable Replacement Identities between Stationary and Rotating Reference Frames

Stationary reference frame	Rotating reference frame
v_α	v_d
v_β	v_q
$\frac{dv_\alpha}{dt}$	$\frac{dv_d}{dt} - \omega_p v_q$
$\frac{dv_\beta}{dt}$	$\frac{dv_q}{dt} - \omega_p v_d$
$\overline{v}^0$	$\overline{v}$
$\frac{d\overline{v}^0}{dt}$	$\frac{d\overline{v}}{dt} + j\omega_p\overline{v}$

with constant driving term $\overline{V}$. Integrated as a linear differential equation with constant parameters, we obtain an explicit expression for the Park current:

$$\overline{i}(t) = \frac{\overline{V}}{R + j\omega L}\left[1 - e^{-\left(j\omega + \frac{R}{L}\right)t}\right]$$

Compared with the result obtained earlier, we recognize it as the same but with a change of reference frame. Now the diagram in the Park rotating plane is even more significant: the rotating and damped transient converges to the constant value of the steady-state phasor (Figure 2.8). ■

2.1.6.1 RMS powers and values in rotating reference frames

All the expressions for power and RMS values shown in the preceding paragraphs are independent of the position or the velocity of the reference frame. In the rotating frame case, it is sufficient to replace the subscripts d and q for the subscripts α and β in all expressions and consider, when applied, the general transformation T in place of T_0.

For a proof of this, simply apply (2.26) to the two terms of the product $\overline{v}\overline{i}^*$ and rewrite the main expressions.

$$p(t) = v_d i_d + v_q i_q + v_0 i_0 = Re\left(\overline{v}\overline{i}^*\right) + v_0 i_0$$

$$p_p(t) = v_d i_d + v_q i_q = Re\left(\overline{v}\overline{i}^*\right) \tag{2.32}$$

$$q_p(t) = v_q i_d + v_d i_q = Im\left(\overline{v}\overline{i}^*\right) \tag{2.33}$$

$$\overline{a}_p(t) = p_p(t) + jq_p(t) = \overline{v}\overline{i}^* \tag{2.34}$$

$$v_a^2 + v_b^2 + v_c^2 = v_d^2 + v_q^2 + v_0^2 = v^2 + v_0^2, \quad \text{with } v^2 = v_d^2 + v_q^2 = \overline{v}\overline{v}^* \tag{2.35}$$

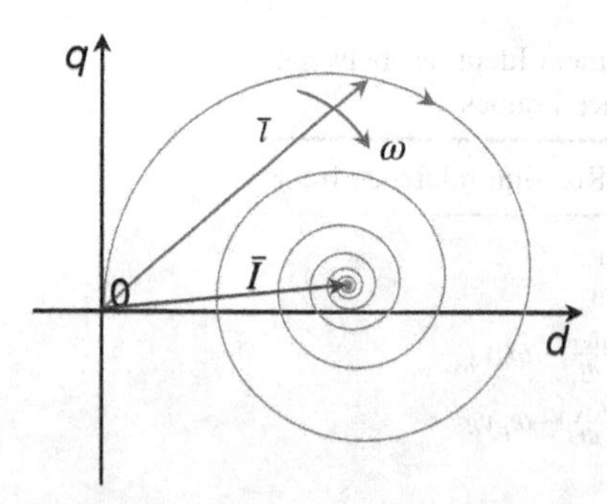

Figure 2.8 Representation of an *RL* transient by means of the space vector with the rotating reference frame Park transform.

$$V_a^2 + V_b^2 + V_c^2 = V_d^2 + V_q^2 + V_0^2 = V^2 + V_0^2 \quad \left(V^2 = V_d^2 + V_q^2\right) \tag{2.36}$$

$$V = \sqrt{\frac{1}{T}\int_T \overline{v}(t)\overline{v}(t)^* dt}$$

2.1.6.2 Three-phase components on a rotating reference frame

The main rules for three-phase elements following the application of the rotating reference frame Park transform are complemented by the identities listed in Table 2.4.

2.1.7 Final Considerations Regarding the Park Transform

In a general sense and in the form described so far, the Park transform may be derived in a direct manner beginning from the properties to be ascertained. This serves to justify its application, regardless of the lack of strict adherence to the historical evolution that established its mathematical formalism.

We will summarily trace this deductive process below without addressing the rather laborious analytical approach.

The conditions to be met are presented here in a logical order, albeit not a unique one, which easily enables its construction.

a. A linear third-order transformation $\boldsymbol{T}$ with real, generally not constant, coefficients. The exclusion of complex coefficients is justified by the application of the transformation to real instantaneous values and the desire to keep the transformed values real.

b. A transformation that when applied to symmetrical three-phase impedance matrices of type

$$\mathbf{Z} = \begin{bmatrix} Z & Z' & Z' \\ Z' & Z & Z' \\ Z' & Z' & Z \end{bmatrix}$$

makes the transformed impedance $\boldsymbol{Z}_p = \boldsymbol{T}\boldsymbol{Z}\boldsymbol{T}^{-1}$ diagonal.

For the desired transformation to be generally applicable, it must be independent of the values of the impedances themselves (impedance is intended, in the broad sense, as matrices with real terms of resistance, inductance, or capacitance). This property enables the decoupling in the transformed variables of impedance systems constructively characterized by three-phase symmetry, which is the great majority of cases.

Table 2.4 Primary Expressions for the Application of the Rotating Reference Frame Park Transform Applied to Three-Phase Systems

	Symmetrical resistors	Symmetric inductors	Symmetrical capacitors	Symmetrical sinusoidal generators (direct sequence)
Ohm's law for stationary reference frame	$\bar{v} = R\bar{i}$ $v_0 = R_0 i_0$	$\bar{v} = L\frac{d\bar{i}}{dt}$ $v_0 = L_0\frac{di_0}{dt}$	$\bar{i} = C\frac{d\bar{v}}{dt}$ $i_0 = C_0\frac{dv_0}{dt}$	$\bar{v} = \overline{V}e^{j\omega t}$ $\bar{a}_s = \overline{A}_s e^{j\omega t}$
Ohm's law for rotating reference frame with velocity ω_p	$\bar{v} = R\bar{i}$ $v_0 = R_0 i_0$	$\bar{v} = L\frac{d\bar{i}}{dt} + j\omega_p L\bar{i}$ $v_0 = L_0\frac{di_0}{dt}$	$\bar{i} = C\frac{d\bar{v}}{dt} + j\omega_p C\bar{v}$ $i_0 = C_0\frac{dv_0}{dt}$	$\bar{v} = \overline{V}e^{j(\omega-\omega_p)t}$ $\bar{a}_s = \overline{A}_s e^{j(\omega-\omega_p)t}$
Instantaneous power	$Ri^2 + R_0 i_0^2$	—	—	—
Average power (active)	$RI^2 + R_0 I_0^2$	—	—	—
Instantaneous energy	—	$\frac{1}{2}Li^2 + \frac{1}{2}L_0 i_0^2$	$\frac{1}{2}Cv^2 + \frac{1}{2}C_0 v_0^2$	—
Average energy	—	$\frac{1}{2}Li^2 + \frac{1}{2}L_0 i_0^2$	$\frac{1}{2}CV^2 + \frac{1}{2}C_0 V_0^2$	—

c. Orthogonal transformation: This condition, not strictly indispensable, applies to cases where the inverse and transposed transformations coincide: $\boldsymbol{T}^{-1} = \boldsymbol{T}^{t}$. The preservation of the inner products ensures the invariance of power, energy, and vector moduli.

d. The above conditions are not particularly binding and transformations with constant coefficients may also be satisfactorily implemented. The most restrictive constraint that determines the actual structure of the transformation is as follows: Considering impedance matrices with cyclic symmetry depending on an angle α, of the type

$$\mathbf{Z}(\alpha) = \mathrm{Z}\begin{bmatrix} \cos(\alpha) & \cos\left(\alpha + {}^{2\pi}/_{3}\right) & \cos\left(\alpha - {}^{2\pi}/_{3}\right) \\ \cos\left(\alpha - {}^{2\pi}/_{3}\right) & \cos(\alpha) & \cos\left(\alpha + {}^{2\pi}/_{3}\right) \\ \cos\left(\alpha + {}^{2\pi}/_{3}\right) & \cos\left(\alpha - {}^{2\pi}/_{3}\right) & \cos(\alpha) \end{bmatrix}$$

the objective is for it to be made diagonal (in the interests of decoupling the circuits) and, at the same time, rendered independent of α. This can be achieved by applying different transformations to the two sides of the matrix (in other words, in the case of inductances, different transformations for fluxes and currents).

Having searched for the two transformations as special cases of a single variant transformation $\boldsymbol{T}(\theta)$, we achieve the final structure of the desired transformation, a function only of the angular parameter θ. In such a form, the goal is achieved with the application of the transformation:

$$\mathbf{Z}_p = \boldsymbol{T}(\theta_1)\mathbf{Z}(\alpha)\boldsymbol{T}(\theta_2)^{-1}$$

in which $\theta_1 - \theta_2 = \alpha$.

The reason for the discussed condition lies in the fact that $\mathbf{Z}(\alpha)$ appears as a matrix of mutual inductances between two three-phase systems coupled on symmetric rotating structures subject to a phase shift α (the case of the stator and rotor in isotropic rotating machines). $\mathbf{Z}(\alpha)$ may still be generalized to mutual couplings of various types of three-phase systems, provided they have appropriate symmetries.

In all cases, however, it follows that coupling between the transformed systems is invariant.

Finally, it should be noted that the illustrated condition includes, as a particular case, the condition b, previously only required in order to simplify the procedure for establishing the definitive transform. By way of comment and justification regarding the required properties, we must consider that the more general forms for $\mathbf{Z}$ and $\mathbf{Z}(\alpha)$ and those characterized by a lower degree of symmetry may certainly be diagonalized; however, except in exceptional cases, this is achievable by recourse to transformations that are dependant on the impedances themselves. Thus, the general nature of the transformation would be lost, substantially undermining much of its analytical advantages.

Furthermore, the condition d requiring the dependence of the transformation on the angle θ, hence on time if the angle is varying, has the important consequence of altering the variables to which it is applied over time. In particular, it can change the frequency, for example, to the point of transforming sinusoidal quantities into constants.

Such outcomes are obviously not random. Suffice to say that mutual interactions between three-phase structures in relative motion can be seen as pairs of generators at different frequencies. The time independence of the couplings is achieved only if the transformation is able to unify, thereby altering, the frequencies of these generators by "restoring" a system to the frequency of the other.

Beyond the applications to rotating structures, the usefulness of the variant transformation lies in its capability to alter the frequencies and hence the structure as well as the solutions of the analytical models pertaining to the electrical systems to which it is applied.

2.2 GRAETZ DIODE BRIDGE RECTIFIERS

2.2.1 Six-Pulse Rectifier

The Graetz bridge three-phase rectifier is the most widely used solution for generating a DC power supply from the basic symmetrical triple sinusoidal voltages of the industrial mains network.

The circuit diagram is shown in Figure 2.9.

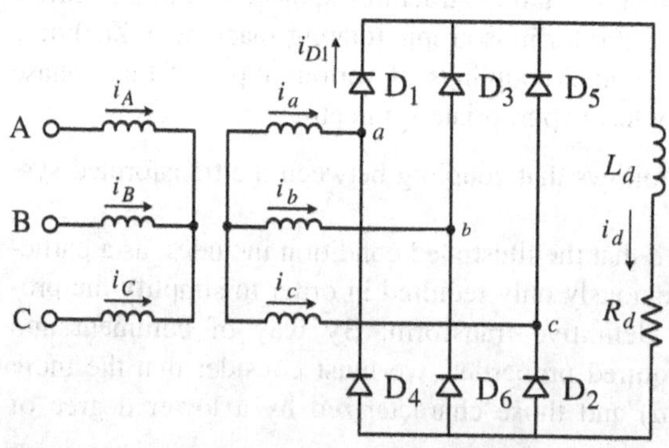

Figure 2.9 Diagram of the Graetz bridge three-phase rectifier.

2.2.1.1 Ideal operation

The bridge consists of six ideal semiconductor devices, in this case silicon diodes, arranged in three branches and grouped into two independent switching units. In reality, considering the high voltages and currents involved, each ideal diode can be composed of a number of modules connected in series to divide the applied voltage, and in parallel to distribute the current. The presence of more diodes per branch also allows for redundancy so that the rectifier, if adequately sized, can continue to function even in the event of a fault in a single diode.

A switching group is a set of diodes that conduct in succession. In the common cathode-connected group of diodes D1, D3, and D5, the conducting diode is the one with highest potential at its anode. On the contrary, in the common anode-connected group D4, D6, and D2, the diode with the lowest cathode potential will conduct.

We also define the switching index q_i as the number of current commutations that occur in the switching group during one period of the sinusoidal input voltage, in this case equal to 3, since each diode has a 120° conduction interval.

Since the switching groups are in series $s = 2$, hence (2.37), this indicates that the pulsation index of the rectified voltage q_v is equal to 6. In other words, the rectifier output voltage v_d will have six pulses per period.

$$q_v = s \cdot q_i \tag{2.37}$$

To study the ideal operation of the rectifier, consider the output current i_d to be a constant I_d. This hypothesis may be applied if the time constant of the load, that is, the ratio between the inductance L_d and the resistance R_d, is assumed to be much greater than the period of the sinusoidal input voltage.

Taking into consideration the time interval $0 \leq \omega t \leq 30°$ of the graph of the sinusoidal phase voltages applied to the input of the three-phase bridge in Figure 2.10, it can be noted that since the voltage v_c is the highest in value and v_b is the lowest, then diodes D5 and D6 will conduct. Therefore, whereas the diode voltage drops are ideally zero, Kirchhoff's voltage law indicates that

$$v_d - v_c + v_b = 0 \tag{2.38}$$

from which follows

$$v_d = v_c - v_b \tag{2.39}$$

that is, it appears that v_d is equal to the difference between the two-phase voltages v_c and v_b. As these belong to a triple of symmetrical voltages with 120° phase interval, it follows that v_d, in this interval, is equal to the line-to-line voltage.

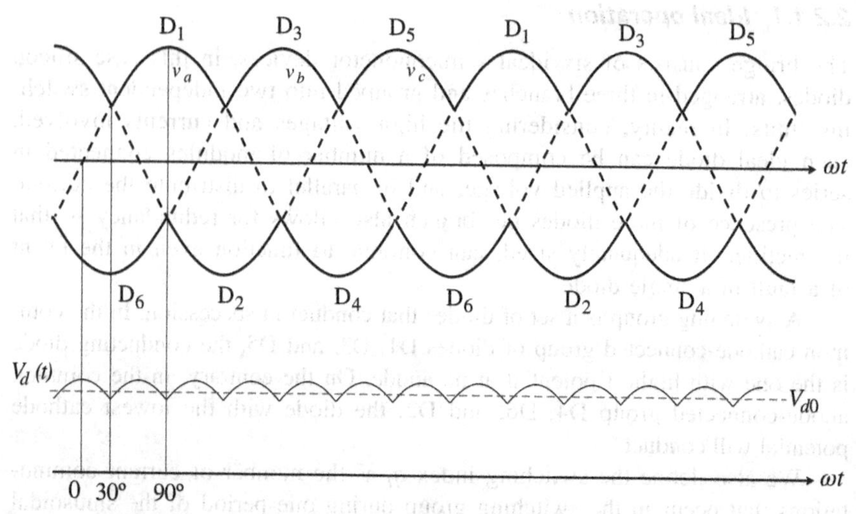

Figure 2.10 Graph of the rectified phase voltages.

Considering then the next interval $30° \leq \omega t \leq 90°$, when diodes D1 and D6 conduct, giving

$$v_d - v_a + v_b = 0 \tag{2.40}$$

from which follows

$$v_d = v_a - v_b = \sqrt{2}V_f \sin(\omega t) - \sqrt{2}V_f \sin(\omega t - 120°) \tag{2.41}$$

from which, by indicating the RMS phase voltage as V_f

$$v_d = 2\sqrt{2}V_f \cos(\omega t - 60°)\sin(60°) = \sqrt{3}\sqrt{2}V_f \cos(\omega t - 60°) \tag{2.42}$$

where v_d is still a line-to-line voltage.

The same process can be repeated for the remaining intervals, obtaining the waveform for the voltage $v_d(t)$ represented in Figure 2.10. Then, by calculating the extreme excursions of the rectified voltage, we can determine the maximum value

$$V_{dM} = v_d\,(\omega t = 60°) = \sqrt{6}V_f \tag{2.43}$$

and the minimum value

$$V_{dm} = v_d(\omega t = 30°, 90°) = \sqrt{6}V_f \cos(30°) \tag{2.44}$$

obtaining a voltage ripple amplitude of 1.15, which is quite limited.

With regard to the theoretical average value V_{d0} for the voltage v_d, we get

$$V_{d0} = \frac{q_v}{\pi} V_{dM} \sin\left(\frac{\pi}{q_v}\right) = \frac{6}{\pi}\sqrt{6}\, V_f \sin\left(\frac{\pi}{6}\right) = \frac{3\sqrt{2}}{\pi}\sqrt{3}\, V_f = 1.35\, V_c \quad (2.45)$$

where V_c is the line-to-line rectifier input voltage.

The line currents entering the three-phase bridge will have the waveforms depicted in Figure 2.11. We must take into account that the square wave represents the ideal case where the current i_d at the rectifier output is perfectly constant under the hypothesis applied to the magnitude of the load time constant *RL*. Should the ideal conditions not apply, the current i_d will have the same oscillation as the voltage v_d.

The RMS value of the bridge input line current may then be calculated as

$$I_{\text{RMS}} = \sqrt{\frac{1}{T}\int_0^T i(t)^2 dt} = \sqrt{\frac{1}{2\pi}\int_0^{2\pi} i(x)^2 dx} = \sqrt{\frac{1}{2\pi} I_d^2 \left(\frac{2\pi}{3} + \frac{2\pi}{3}\right)} = \sqrt{\frac{2}{3}} I_d \quad (2.46)$$

2.2.1.2 Harmonics of the line current

As shown above, the three-phase currents absorbed by the Graetz bridge are not sinusoidal but square waves with 120° phase shifts. In addition to the fundamental component, we have the presence of harmonics of order

$$h = k \cdot q_v + 1, \quad \text{with } k = 0, \pm 1, \pm 2, \pm 3, \ \ldots \quad (2.47)$$

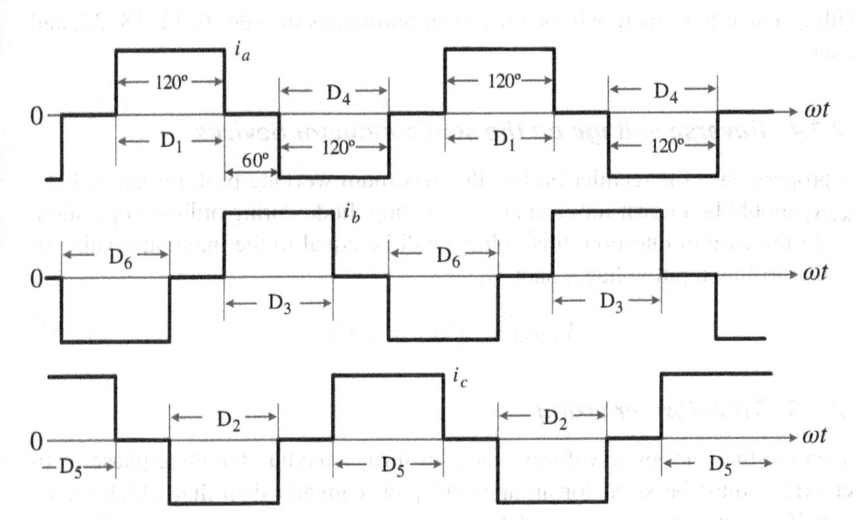

Figure 2.11 Graph of the rectifier bridge input line currents.

Since $q_v = 6$ for the case of a three-phase six-pulse bridge, the current harmonics present will be of order 1, 5, 7, 11, 13, 17, 19, and so on and will have RMS values

$$I_{1\,\mathrm{RMS}} = \frac{\sqrt{6}}{\pi} I_d \tag{2.48}$$

for the first harmonic, while for subsequent ones

$$I_{h\mathrm{RMS}} = \frac{I_{1\,\mathrm{RMS}}}{h} \tag{2.49}$$

It should be noted that the presence of low-frequency harmonics with particularly high value causes additional voltage drop, distortion of the mains voltage, and disturbances to other utility users. To avoid these drawbacks, it is normal to counteract the production of harmonics within limits set by the utility provider through the installation of appropriate filtering systems.

2.2.1.3 Harmonics of the rectified voltage

As for the harmonics present in the rectified voltage v_d, in addition to the DC component, Fourier analysis provides the further terms:

$$a_k = \frac{2}{\pi}\int_0^{\pi} v_d(\omega t)\cos(kq_v\omega t)d(q_v\omega t) = V_{d0}\cos(k\pi)\frac{2}{1-(k\,q_v)^2} \tag{2.50}$$

Hence, we have components of the order

$$h = k \cdot q_v, \quad \text{with } k = 1,\ 2,\ 3,\ \ldots \tag{2.51}$$

With q_v equal to 6, there will be only even harmonics of order 6, 12, 18, 24, and so on.

2.2.1.4 Reverse voltage on the semiconductor devices

To properly size the rectifier bridge, the maximum working peak reverse voltage V_{RWM} should be known for each nonconducting diode during ordinary operation.

In the case in question, this voltage will be equal to the maximum value of the line-to-line input voltage, namely,

$$V_{\mathrm{RWM}} = \sqrt{2}V_c = \frac{\pi}{3}V_{d0} \tag{2.52}$$

2.2.1.5 Transformer sizing

Given the line current waveforms, the transformer feeding the three-phase rectifier bridge must be sized for an apparent power greater than that which would apply if the current were sinusoidal.

Defining A_d as the rated transformer power, we have

$$A_d = 3 \cdot V_{f2} \cdot I_2 \tag{2.53}$$

where V_{f2} is the phase voltage and I_2 is the line current, both related to the secondary winding. Replacing, we get

$$A_d = 3 \, \frac{\pi}{3\sqrt{6}} V_{d0} \, \frac{\sqrt{2}}{\sqrt{3}} I_d = \frac{\pi}{3} P_{d0} = 1.05 \; P_{d0} \tag{2.54}$$

where P_d is the theoretical rectified power available at the Graetz bridge output equal to the product of V_{d0} and I_d.

It should, however, be noted the harmonic distortion in the line currents obliges the feed transformer to be oversized by 5% with respect to the power transferred to the load on the DC side, P_{d0}.

2.2.2 Twelve-Pulse Rectifiers

Figure 2.12 illustrates the circuit diagram of a 12-pulse rectifier bridge. In this case, it has been created by connecting in series two three-phase Graetz bridges fed by two secondaries of a three-winding transformer. The first bridge, identified by the output voltage v_{d1}, is fed from the star-connected secondary, while the second bridge, which supplies the voltage v_{d2} at the output, is driven by the delta-connected secondary. This implies that the voltages of the second secondary winding have a 30° phase shift with respect to those of the first secondary winding.

Consequently, the output voltages from the two rectifier bridges will also be out of phase in the same way, as shown in Figure 2.13.

Since the two line-to-line voltages V'_{ab} and V''_{ab} have equal RMS value, the following expression can be defined for the relationship between v_{d1} and v_{d2}:

$$v_{d2}(t) = v_{d1}\left(t + \frac{T}{12}\right) \tag{2.55}$$

Considering the time interval $0 \leq \omega t \leq 30°$, we have

$$v_{d1}(t) = V_{dM}\cos(\omega t) \tag{2.56}$$

$$v_{d2}(t) = V_{dM}\cos(\omega t - 30°) \tag{2.57}$$

where

$$V_{dM} = \sqrt{2}V'_{ab} = \sqrt{2}V''_{ab} \tag{2.58}$$

It follows that

$$\begin{aligned} v_d &= v_{d1} + v_{d2} = V_{dM}\cos(\omega t) + V_{dM}\cos(\omega t - 30°) \\ &= 2\, V_{dM}\cos(\omega t - 15°)\cos(15°) \end{aligned} \tag{2.59}$$

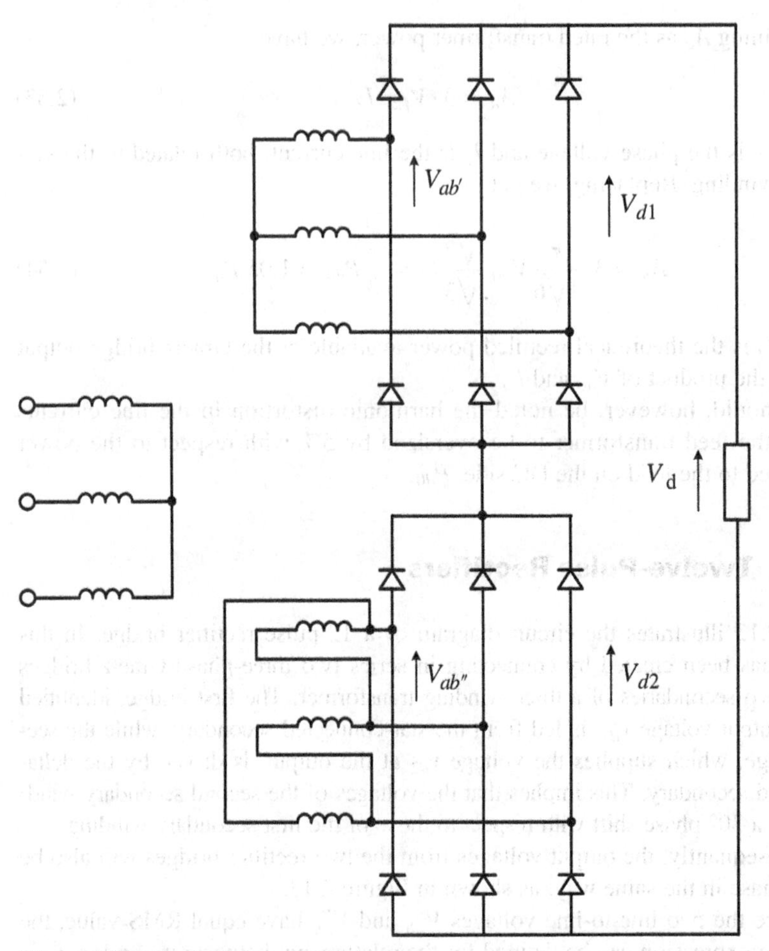

Figure 2.12 Twelve-pulse rectifier bridge.

It therefore appears that v_d is a cosine function centered at 15°. The same analysis applies to the remaining intervals, providing the waveform shown in Figure 2.13, where the waveform has 12 pulses in each fundamental period, which entails a q_v of 12.

The extreme values $V_{d\max}$ and $V_{d\min}$ can be derived as

$$V_{d\max} = 2\ V_{dM}\cos(15°) \tag{2.60}$$

$$V_{d\min} = 2\ V_{dM}(\cos(15°))^2 \tag{2.61}$$

from which it follows that the ripple amplitude of the output voltage v_d is equal to 1.035, less than that obtained with the basic three-phase Graetz bridge.

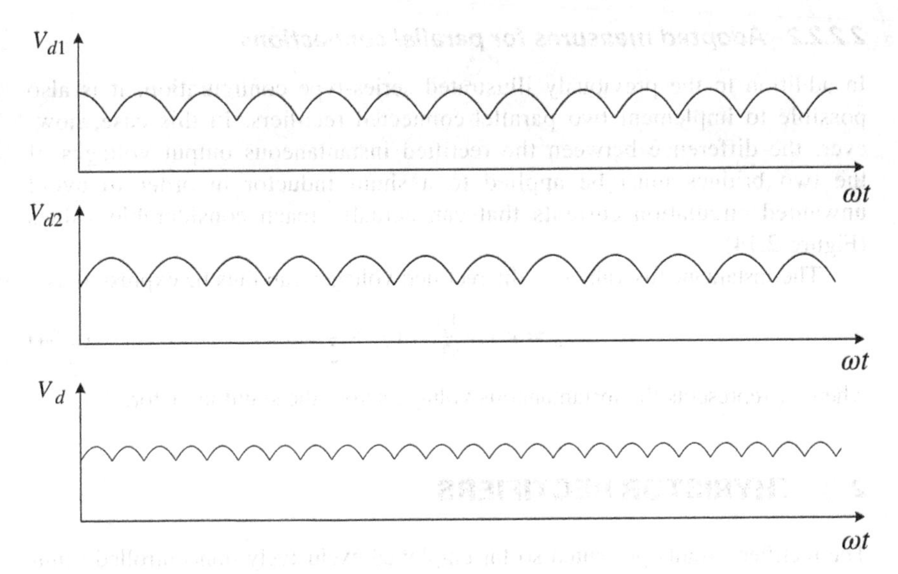

Figure 2.13 Graph of the rectified voltage of the 12-pulse bridge.

The theoretical average value V_{d0} is calculated as

$$V_{d0} = \frac{q_v}{\pi} V_{d\max} \sin\left(\frac{\pi}{q_v}\right) = V_{d0\ \text{bridge 1}} + V_{d0\ \text{bridge 2}} = 2\frac{3\sqrt{2}}{\pi} V'_{12} \qquad (2.62)$$

2.2.2.1 Harmonics

With regard to the harmonic analysis for the line currents absorbed by the primary of the transformer and the rectified voltage, the procedures applied for the three-phase six-pulse Graetz bridge can be generalized to the 12-pulse case.

For the line currents, the expression (2.47) still holds. Since in this case $q_v = 12$, the harmonics present will be of order 1, 11, 13, 23, 25, and so on and they will have RMS values

$$I_{1\ \text{RMS}} = \frac{\sqrt{6}}{\pi} \frac{I_d}{k_t} \qquad (2.63)$$

for the first harmonic, where k_t represents the transformation ratio, while for the subsequent harmonics we can still apply (2.49). Similarly, the harmonics present in the output voltage v_d, in addition to the DC component, include components of the order (2.51).

It should be emphasized that the use of a 12-pulse rectifier bridge leads to reduced harmonics with respect to a basic three-phase Graetz bridge.

2.2.2.2 Adopted measures for parallel connections

In addition to the previously illustrated series-type configuration, it is also possible to implement two parallel-connected rectifiers. In this case, however, the difference between the rectified instantaneous output voltages of the two bridges must be applied to a shunt inductor in order to avoid unwanted circulation currents that can actually reach considerable values (Figure 2.14).

The instantaneous values of the rectified voltages can thus be expressed as

$$v_d = v_{d1} - \frac{v_p}{2} = v_{d2} + \frac{v_p}{2} \tag{2.64}$$

where v_p represents the instantaneous voltage across the shunt inductor.

2.3 THYRISTOR RECTIFIERS

The rectifier circuits presented so far employed exclusively noncontrolled semiconductor devices, or models of simple silicon diodes. In practice, however, controlled semiconductor devices such as thyristors can be used, which can be driven into conduction by injecting a control current into a dedicated terminal known as the *gate*. In this way, it is possible to adjust the rectifier output voltage by acting on the phase angle α of the gate control current relative to the supply voltage.

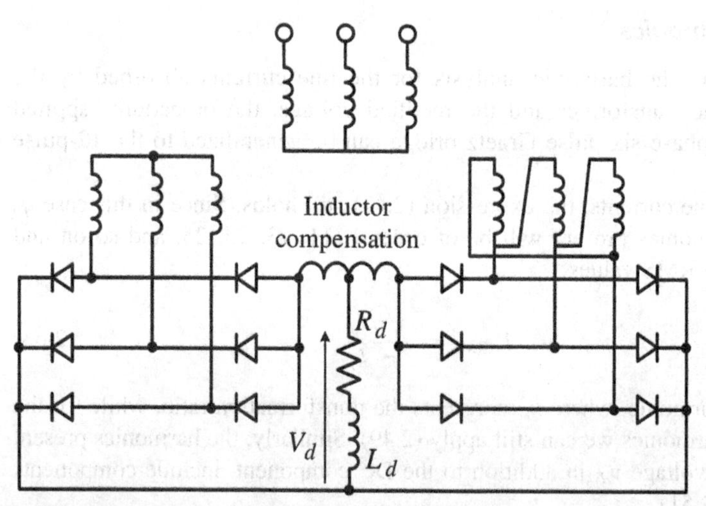

Figure 2.14 Wiring diagram of a parallel-connected bridge.

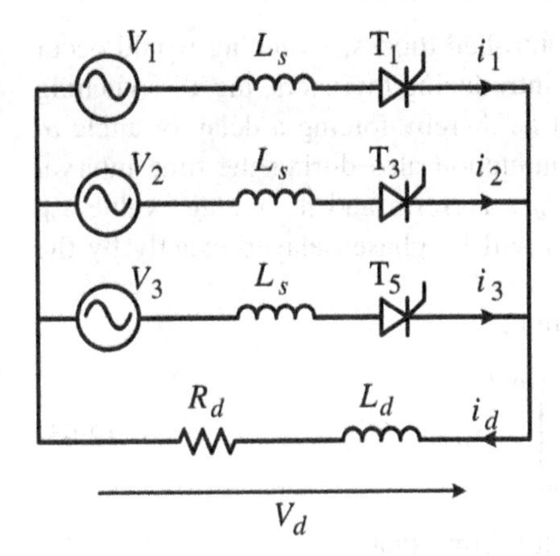

Figure 2.15 Basic star-connected three-phase rectifier circuit diagram.

2.3.1 Phase Control

To explain the operation of this control approach, we shall refer to the basic three-phase star-connected rectifier circuit consisting of only one switching group of three thyristors, depicted in Figure 2.15.

Figure 2.16 shows the voltage waveforms and the current i_d with phase control.

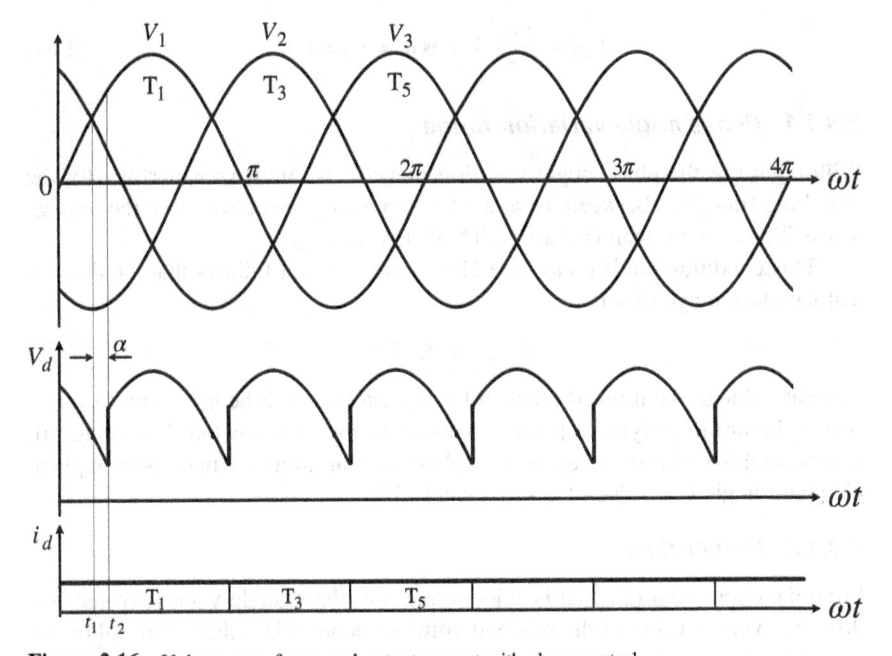

Figure 2.16 Voltage waveforms and output current with phase control.

At the instant of time t_1, with noncontrolled diodes, switching would occur with a T5-to-T1 transition. However, by introducing thyristors, the T1 switching interval may be delayed until the instant t_2, thereby forcing a delay of angle α. Hence, T5 will remain in prolonged conduction also during the time interval between t_1 and t_2, thus modifying the v_d waveform and its average value V_d. Furthermore, the square wave current i_d will be phase delayed exactly by the phase angle α.

The average value of v_d will therefore be

$$V_{d\alpha} = \frac{1}{\text{period}} \int_{\omega t_2}^{\omega t_2 + \text{period}} v_d(\omega t) d\omega t \tag{2.65}$$

since the period of $v_d(t)$ equates to 120°. It follows that

$$V_{d\alpha} = \frac{3}{2\pi} \int_{30° + \alpha}^{150° + \alpha} \sqrt{2}\, V_f \sin(\omega t) d\omega t = \frac{1}{2} \frac{3\sqrt{2}}{\pi} V_c \cos\alpha \tag{2.66}$$

where V_c is the line-to-line voltage.

It should be noted that the term $\frac{1}{2}\frac{3\sqrt{2}}{\pi} V_c$ represents half of the theoretical average value of the rectified voltage of a three-phase Graetz bridge. This is because the examine circuit has only one switching group. In the case of the three-phase bridge, we will then have that

$$V_{d\alpha} = \frac{3\sqrt{2}}{\pi} V_c \cos\alpha = V_{d0} \cos\alpha \tag{2.67}$$

2.3.1.1 Phase angle variation range

With regard to the phase angle variation range α, we must consider that for the switching transition between T5 and T1 to occur, it is necessary that the voltage across T1 is greater than that across T5, that is, $v_1 > v_3$.

This condition holds up to $\omega t = 210°$, from which it follows that the theoretical variation range of α is

$$0° \leq \alpha \leq 180° \tag{2.68}$$

In reality, due to the turn-off times of the thyristors, it will be necessary to wait a time t_q before reapplying a positive voltage to the thyristor that is turning off, otherwise this will turn on again. This phenomenon limits the maximum applicable phase angle to a value of approximately 155°.

2.3.1.2 Power flow

From the expression (2.67) it is evident that, for phase angle values greater than 90°, the average value of the rectified voltage becomes less than zero, following a cosine function of α.

On the other hand, the first harmonic of the line current, for $\alpha \neq 0°$, undergoes a phase shift that actually coincides with the applied delay angle. It follows that the power flow will be

$$P = P_{d0} \cos \alpha = 3\ V_f I_1 \cos \alpha \tag{2.69}$$

This implies that absorption of a reactive power Q will be present equal to

$$Q = 3\ V_f I_1 \sin \alpha \tag{2.70}$$

Ultimately, phase control with angle $\alpha < 90°$ allows for the adjustment of the average rectified voltage and the active power transferred from the three-phase network to the direct current load, but it introduces a reactive power absorption.

On the other hand, operating with $\alpha > 90°$, the voltage V_d will become negative and consequently, given that the current I_d is constant and positive, the active power flow is reversed and passes from the DC section to the AC side. Thus, the operation of the so-called *naturally commutated inverter* is achieved. The reactive power, also in this case however, will be absorbed by the rectifier.

2.3.2 Noninstantaneous Switching

In the discussion undertaken so far, device switching has always been considered as instantaneous. In reality, a certain time interval is required for the current to be transferred between two consecutive semiconductor devices in the switching group. To illustrate this phenomenon, we will again refer to a simple star-connected three-phase rectifier circuit, as shown in Figure 2.15.

First of all, we assumed that the current i_d is constant, with value equal to I_d, and that the line inductance on the AC side L_S has a nonzero value.

We also define two cyclic currents i_c and i_{c1}. The first passes through the loop that contains the T1 and T3 thyristors in a counterclockwise direction, while the second passes through the loop that includes the T1 thyristor and the load in a clockwise direction. Referring to Figure 2.17, at the instant t_0 the phase control triggers with a delay angle α and lasts until the moment t_1 when the switchover between T1 and T3 begins.

Applying Kirchhoff's voltage law to the loop crossed by the cyclic current i_c, we obtain

$$v_1 - L_s \frac{di_1}{dt} + L_s \frac{di_2}{dt} - v_2 = 0 \tag{2.71}$$

whereas for the currents we can apply the following:

$$\begin{cases} i_1 = i_{c1} - i_c = I_d - i_c \\ i_2 = i_c \end{cases} \tag{2.72}$$

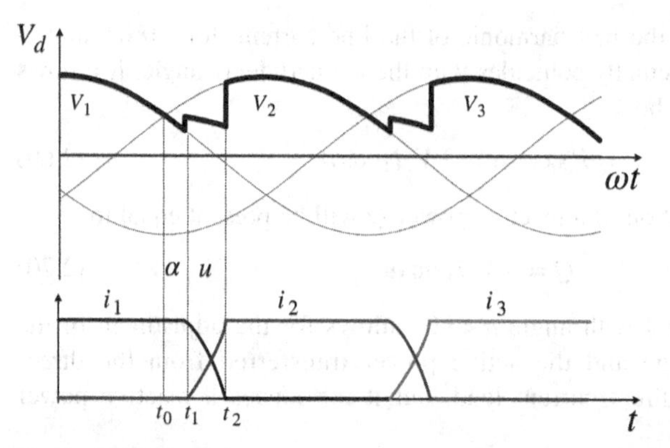

Figure 2.17 Voltages and currents in the presence of phase control and switching.

Substituting (2.72) with (2.71), we obtain

$$L_s \frac{di_c}{dt} = \frac{v_2 - v_1}{2} \tag{2.73}$$

Setting a reference instant t_1, we can redefine the phase voltages v_1 and v_2 as

$$\begin{cases} v_1(t) = \sqrt{2}\, V_S \cos(\omega t + 60°) \\ v_2(t) = \sqrt{2}\, V_S \cos(\omega t - 60°) \end{cases} \tag{2.74}$$

where V_S is the RMS value of the phase voltage. It follows that

$$v_2(t) - v_1(t) = \sqrt{3}\sqrt{2}\, V_S \cos(\omega t - 90°) = \sqrt{3}\sqrt{2}\, V_S \sin(\omega t) \tag{2.75}$$

and is therefore a line-to-line voltage. Resolving (2.73) we have an expression for the current i_c as

$$i_c = \frac{1}{2\, L_S} \int_{t_1}^{t} \sqrt{6} V_S \sin(\omega\tau) d\tau + \text{const} = \frac{\sqrt{6} V_S}{2\, L_S} \left[-\frac{\cos(\omega\tau)}{\omega} \right]_{t_1}^{t} + \text{const} =$$
$$= \frac{\sqrt{6} V_S}{2\, X_S} [\cos(\omega t_1) - \cos(\omega t)] + \text{const} = \frac{\sqrt{6} V_S}{2\, X_S} [\cos(\alpha) - \cos(\omega t)] + \text{const} \tag{2.76}$$

By imposing the conditions $i_c(t_1) = i_2(t_1) = 0$ and $i_c(t_2) = I_d$, then const = 0 and

$$I_d = \frac{\sqrt{6} V_S}{2\, X_S} [\cos(\alpha) - \cos(\alpha + u)] \tag{2.77}$$

where u is the switching angle equal to $\omega(t_2 - t_1)$. Thus, we obtain the expression

$$\frac{X_S\, I_d}{\sqrt{6}V_S} = \frac{\cos(\alpha) - \cos(\alpha + u)}{2} \tag{2.78}$$

Considering the case in which $\alpha = 0°$, we have

$$\frac{X_S\, I_d}{\sqrt{6}V_S} = \frac{1 - \cos(u)}{2} \tag{2.79}$$

Observing this last expression, it is easy to see that the switching angle primarily depends on the AC side line reactance, on the current required by the load and on the maximum value of the line-to-line voltage that feeds the rectifier circuit.

The next step is to calculate the average value V_d of the rectified voltage by considering the instant at which the switchover occurs between T1 and T3, during which both thyristors will be simultaneously conducting.

Applying Kirchhoff's voltage law to the two loops consisting of the concerned branches and the load, and assuming zero voltage drop across the conducting thyristors, we have

$$\begin{cases} v_1 - L_s \dfrac{di_1}{dt} - v_d & = 0 \\ v_2 - L_s \dfrac{di_2}{dt} - v_d & = 0 \end{cases} \tag{2.80}$$

Substituting (2.72) with (2.80), we obtain

$$\begin{cases} v_1 + L_s \dfrac{di_c}{dt} - v_d & = 0 \\ v_2 - L_s \dfrac{di_c}{dt} - v_d & = 0 \end{cases} \tag{2.81}$$

By means of term summation of the (2.81) system of equations, we obtain the expression for the voltage v_d that emerges as

$$v_d = \frac{v_1 + v_2}{2} \tag{2.82}$$

It can be noticed how the rectified voltage during the switching transient is equal to the instantaneous average value of the two relevant phase voltages, that is, less than the ideal level.

This leads to the conclusion that in the case of noninstantaneous switching the average value V_d of the rectified voltage undergoes a quantifiable drop with a decrease in the average voltage V_x that may be calculated as

$$V_x = \frac{1}{\frac{2\pi}{3}} \int_{\alpha}^{\alpha+\frac{2\pi}{3}} (v_{d\,\text{ideal}} - v_{d\,\text{real}})d(\omega t) = \frac{3}{2\pi} \int_{\alpha}^{\alpha+u} \left(v_2 - \frac{v_1 - v_2}{2}\right) d(\omega t) \tag{2.83}$$

$$= \frac{3}{2\pi} \int_{\alpha}^{\alpha+u} \left(\frac{\sqrt{6}V_S \sin(\omega t)}{2}\right) d(\omega t) = \frac{3}{2\pi}\frac{\sqrt{6}\,V_S}{2}[-\cos(\omega t)]_{\alpha}^{\alpha+u}$$

Finally, we have

$$V_x = V_{d0}\frac{\cos\alpha \; - \; \cos(\alpha+u)}{2} \tag{2.84}$$

Expressing the voltage drop as a relative value with respect to the theoretical average voltage V_{d0} we have

$$\frac{V_x}{V_{d0}} = \frac{\cos\alpha \; - \; \cos(\alpha+u)}{2} = \frac{X_S I_d}{\sqrt{6}V_S} \tag{2.85}$$

In the case where there is no phase control, the applicable expression is

$$\frac{V_x}{V_{d0}} = \frac{1 \; - \; \cos(u)}{2} = \frac{X_S I_d}{\sqrt{6}V_S} \tag{2.86}$$

from which it can be easily deduced that the voltage drop due to noninstantaneous switching depends on the same factors that affect the switching angle u.

Finally, it is possible to calculate the effective average value of the rectified voltage V_d with both phase control and noninstantaneous switching

$$V_d = V_{d0} - V_\alpha - V_x = V_{d0} - V_{d0}(1 - \cos\alpha) - \frac{V_{d0}}{2}[\cos\alpha - \cos(\alpha+u)] \tag{2.87}$$

from which we finally obtain

$$V_d = V_{d0}\frac{\cos\alpha + \cos(\alpha+u)}{2} \tag{2.88}$$

In the case of a three-phase Graetz bridge, for which the results obtained for the basic star-connected circuit are valid, if the angle u should exceed 60°, there would be interference between the two switching groups. In practice however, the switching angle u is generally less than 20°, which limits the voltage drop V_x as a relative value with respect to V_{d0} well below 10%, allowing for the switching groups to always remain independent.

Furthermore, it should be noted how the switching angle is dependent on the phase control angle. In fact, if α increases, then angle u will tend to decline. To highlight this interaction, refer to the expression (2.73).

In cases where there is phase control with $\alpha > 0°$, at the instant that switching starts, the difference between the two-phase voltages will be greater compared with the case in which the angle α is zero. Consequently, the derivative of the current will also be greater, which will reduce the time taken to perform the switching, and hence the angle u will be smaller.

2.4 FORCED SWITCHING CONVERTERS

In this section we analyze the different types of forced switching converters currently available and the related control techniques. The purpose is to identify their possible applications in DC traction systems, in particular for harnessing the energy related to regenerative braking by returning it to the AC network.

2.4.1 Sinusoidal PWM Modulation

Pulse-width modulation or PWM is a sinusoidal control technique widely used in power electronics. It is primarily used to adjust the output amplitude and frequency of an *inverter* to obtain an approximately sinusoidal waveform while maintaining a low harmonic content.

To achieve PWM modulation, the following two control voltage waveforms are required:

- a triangular wave v_t, called the *carrier*, with amplitude V_t and *switching frequency* f_s
- a sine wave v_c, called *the modulating signal*, with amplitude V_c and frequency f_c

We can now introduce the following two fundamental parameters:

- m_a the amplitude modulation ratio defined as

$$m_a = \frac{V_c}{V_t} \tag{2.89}$$

- m_f the frequency modulation ratio defined as

$$m_f = \frac{f_s}{f_c} \tag{2.90}$$

By means of a comparator device that continuously compares the two waves, the output voltage v_o can take on either of two levels:

$$\begin{cases} v_c > v_t \Rightarrow v_o = +V_d \\ v_c < v_t \Rightarrow v_o = -V_d \end{cases} \tag{2.91}$$

Figure 2.18 illustrates the control voltage waveforms v_t and v_c and the output voltage v_o. Also, the first harmonic component of the output $v_{o(1)}$ is shown.

To calculate the harmonic content of the output voltage $v_o(t)$, consider the case in which $m_f \gg 1$, that is, $T_c \gg T_s$, where the latter are, respectively, the periods of v_c and v_t. Under these conditions, in a time interval equal to T_s, the control voltage can be considered approximately constant. From this, it follows that the average value $v_o(t)$ during T_s is

$$v_{o(av)}(t) = V_d \frac{v_c(t)}{V_t} = V_d \frac{V_c}{V_t} \sin(\omega_c t) = m_a V_d \sin(\omega_c t) \tag{2.92}$$

It can therefore be concluded that in the case of $m_f \gg 1$, the "instantaneous average value" of the output voltage is a sinusoidal function representing the fundamental component of $v_o(t)$. It has a frequency equal to that of the modulating wave and an amplitude equal to $m_a V_d$ while m_a has a value between 0 and 1, that is, as long as $V_c \leq V_t$. Thus by varying V_c and f_c, we can vary the amplitude and frequency of the output voltage.

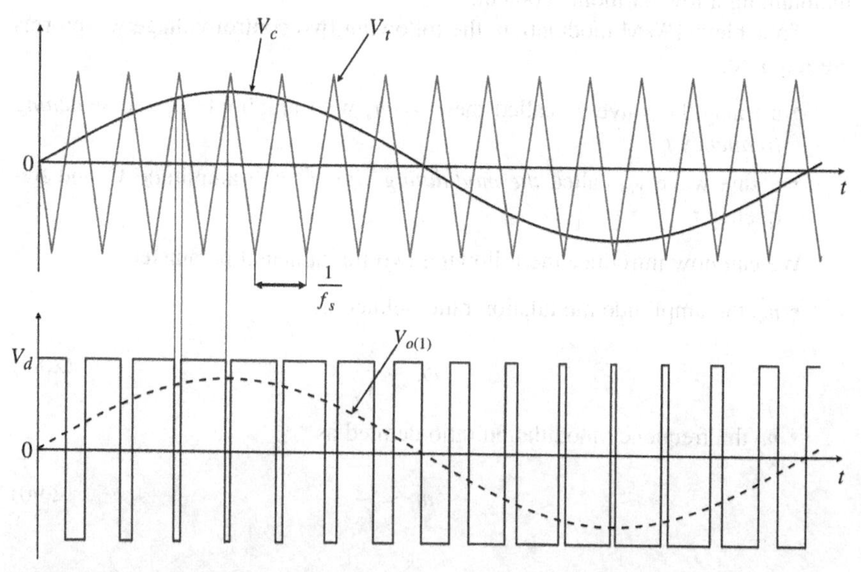

Figure 2.18 PWM modulation waveforms.

2.4.1.1 Harmonics

In actuality, even with $m_f \gg 1$, the control voltage will never be perfectly constant during any period of T_s. For this reason, $v_o(t)$ will also contain harmonics of order higher than the fundamental, whose order h is given by the expression

$$h = j\,m_f \pm k \tag{2.93}$$

with j and k belonging to the set of natural numbers, where if k is odd, then j is even and vice versa.

Figure 2.19 illustrates the frequency spectrum of the harmonics of $v_o(t)$ for the case with $m_a = 0.8$.

In general, if m_f is even, then $v_o(t)$ will contain both even and odd harmonics. On the other hand, if m_f is integer and odd, then $v_o(t)$ will have mirror symmetry and consequently it will have only higher order odd harmonics.

Ultimately, we must note that the sinusoidal PWM modulation technique shifts the voltage harmonics to high frequencies, as a function of the frequency modulation ratio, but does not reduce their amplitudes. For the current harmonics, since their amplitudes are inversely related to the order h, also the lowest orders are generally attenuated to 0.

2.4.1.2 Asynchronous modulation

It has been shown that, to vary the frequency of the output voltage, it is necessary to act on the modulating wave frequency. This may be performed with a simultaneous variation of the *switching frequency* f_s, thus maintaining an integer ratio for m_f corresponding to a *synchronous modulation*, or by leaving f_s constant and ensuring that m_f can take on fractional values. The latter case

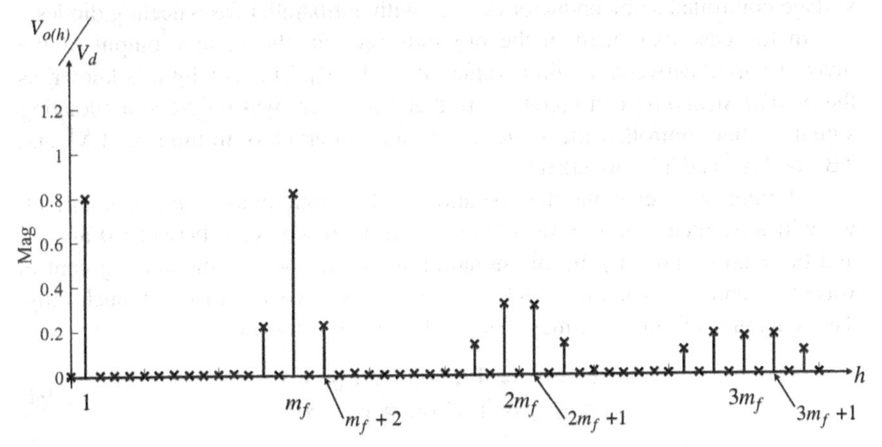

Figure 2.19 The frequencies spectrum of the harmonics of $v_o(t)$ with $m_a = 0.8$.

takes the name of *asynchronous modulation* and it may give rise to subharmonic components in the output voltage with frequencies lower than the fundamental.

Even if small in amplitude, these components can generate significant currents in predominantly inductive loads, for example, electric motors, with dangerous consequences such as parasitic and undesirable heating effects.

2.4.1.3 Overmodulation and square wave operation

The term overmodulation applies when the modulating signal amplitude exceeds that of the carrier ($m_a > 1$). In such circumstances, the amplitude of the first harmonic of the output $v_o(t)$ ceases to be directly proportional to the amplitude modulation ratio and the converter no longer functions in a linear manner. With this type of modulation, the amplitude of the fundamental component $v_o(t)$ increases, but lower order harmonics are also generated at frequencies that are not multiples of the fundamental.

If m_a continues to grow further, until the modulating signal no longer intersects the carrier, then the output voltage $v_o(t)$ will no longer be sinusoidal but becomes an alternating square wave. In this case, it will no longer be possible to adjust the amplitude $v_o(t)$, but only its frequency and all the successive odd order harmonics of the fundamental will be present.

Square wave operation is generally adopted in inverter-driven AC machines in order to achieve the highest motor rotation speeds.

2.4.2 Complete Single-Phase Full-Bridge Inverter

Figure 2.20 shows the circuit diagram of a single-phase full-bridge inverter based on IGBT devices (acronym for *insulated-gate bipolar transistor*), that is, voltage controlled semiconductor devices with antiparallel freewheeling diodes.

In the case examined in the previous section, the voltage output of the inverter varied between the limit values V_d and $-V_d$. This is why it is known as the *bipolar switching* configuration. In that case, there was only one modulating signal v_c that controlled the switches in pairs, alternately turning on TA^+ and TB^- or TA^- and TB^+ together.

Moving on to examine the operation of the single-phase full-bridge circuit, we will introduce *unipolar switching*, where $v_o(t)$ will vary between 0 and V_d and between 0 and $-V_d$. In this situation, there will be two modulating control waves, v_c and $-v_c$, each of which will trigger the switches of one branch only. The dynamics of branch A may then be described as follows:

$$\begin{cases} v_c > v_t \Rightarrow \mathrm{TA}^+ \text{ on } \Rightarrow v_{\mathrm{AN}} = V_d \\ v_c \leq v_t \Rightarrow \mathrm{TA}^- \text{ on} \Rightarrow v_{\mathrm{AN}} = 0 \end{cases} \tag{2.94}$$

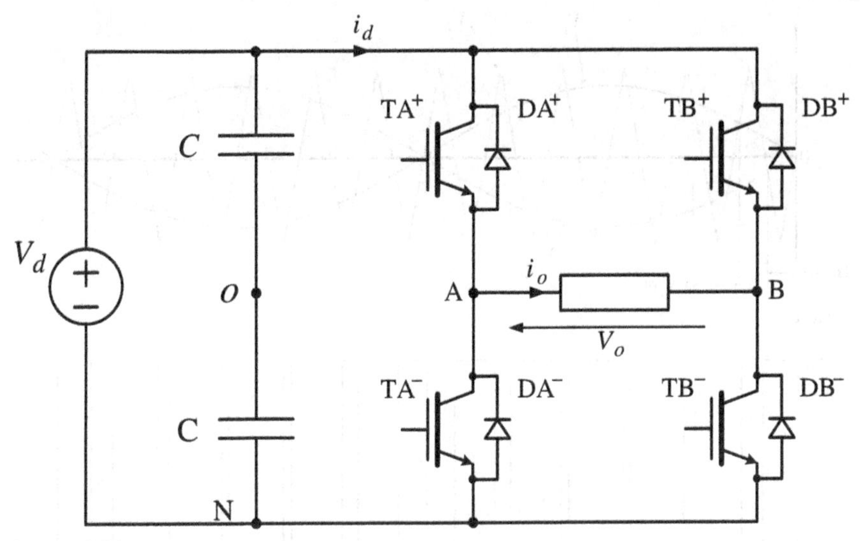

Figure 2.20 Single-phase full-bridge inverter.

while for branch B

$$\begin{cases} -v_c > v_t \Rightarrow \text{TB}^+ \text{ on} \Rightarrow v_{\text{BN}} = V_d \\ -v_c \leq v_t \Rightarrow \text{TB}^- \text{ on} \Rightarrow v_{\text{BN}} = 0 \end{cases} \tag{2.95}$$

Since the output voltage $v_o(t)$ is equal to

$$v_o = v_{\text{AN}} - v_{\text{BN}} \tag{2.96}$$

the resulting waveforms will be as illustrated in Figure 2.21.

The presence of the antiparallel recirculation diodes allows the voltages v_{AN} and v_{BN} to remain independent of the sign of the output current i_o. It should also be noted that there are time intervals during which the voltage $v_o(t)$ takes on a zero value due to the simultaneous triggering of the upper and lower switches in the two branches. These correspond to momentary load short-circuit conditions during which the current i_d will be null, with no circulation through DC source V_d. At these moments the current i_o can flow through the load, thanks to the presence of the recirculation diodes.

A great advantage in the use of unipolar switching, compared to bipolar, is related to the harmonic content of the output voltage $v_o(t)$, despite both voltages v_{AN} and v_{BN} having the harmonic components described by the expression (2.5). By way of demonstration, let us assume a frequency modulation ratio $m_f = 6$.In this case, for $t > 0$, the relative position of v_c and v_t is equal to that between $-v_c$ and v_t for $t > T_c/2$. Therefore,

$$v_{\text{BN}}(t) = v_{\text{AN}}\left(t - \frac{T_c}{2}\right) \tag{2.97}$$

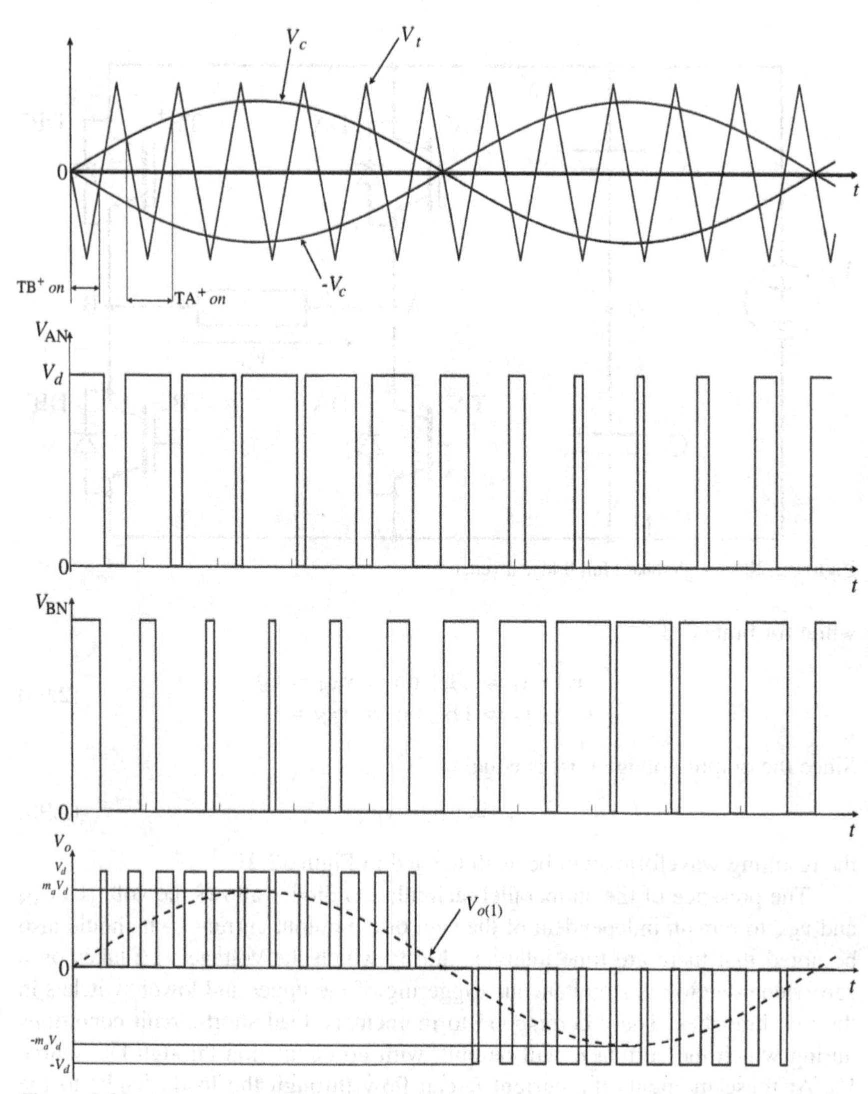

Figure 2.21 Voltage waveforms with unipolar switching.

For the harmonic component of order $h = 6$, we have

$$\begin{cases} v_{\mathrm{AN}(6)}(t) = V_{(6)}\sin(6\omega_c t) \\ v_{\mathrm{BN}(6)}(t) = V_{(6)}\sin\left(6\omega_c\left(t - \dfrac{T_c}{2}\right)\right) \end{cases} \tag{2.98}$$

Expanding the second expression in (2.98), we obtain

$$v_{\mathrm{BN}(6)}(t) = V_{(6)}\sin(6\omega_c t - 3\omega_c T_c) = V_{(6)}\sin(6\omega_c t) \tag{2.99}$$

from which it follows that, by applying (2.96), the sixth harmonic of the output voltage $v_o(t)$ is null.

Conversely, considering a generic odd order harmonic with $h = 2k + 1$, then

$$\begin{aligned} v_{BN(h)}(t) &= V_{(h)}\sin\left(h\omega_c\left(t - \frac{T_c}{2}\right)\right) \\ &= V_{(h)}\sin\left(h\omega_c t - \frac{(2k+1)\omega_c T_c}{2}\right) \\ &= V_{(h)}\sin(h\omega_c t - \pi) = -V_{(h)}\sin(h\omega_c t) \end{aligned} \tag{2.100}$$

In this case the two components are added together and are present in $v_o(t)$ with amplitude equal to $2V_{(h)}$.

This leads to the conclusion that, unlike the bipolar switching case, selecting an even integer value for m_f makes it possible to eliminate even lower order harmonics in the output voltage and thus obtains the spectrum, as shown in Figure 2.22.

2.4.3 The Three-Phase Inverter

Three-phase inverters are widely used when a DC network must be interfaced with the industrial mains or with a three-phase load.

To demonstrate the relevant operation and characteristics, we will refer to the circuit diagram, as shown in Figure 2.23.

In order to obtain a set of three symmetrical output voltages, three modulating control voltages v_{cA}, v_{cB}, v_{cC} are compared with the triangular carrier wave v_t. The modulating waveforms are sinusoidal and are related as follows:

$$v_{cB}(t) = v_{cA}\left(t - \frac{T_c}{3}\right) \quad \text{and} \quad v_{cC}(t) = v_{cA}\left(t - \frac{2T_c}{3}\right) \tag{2.101}$$

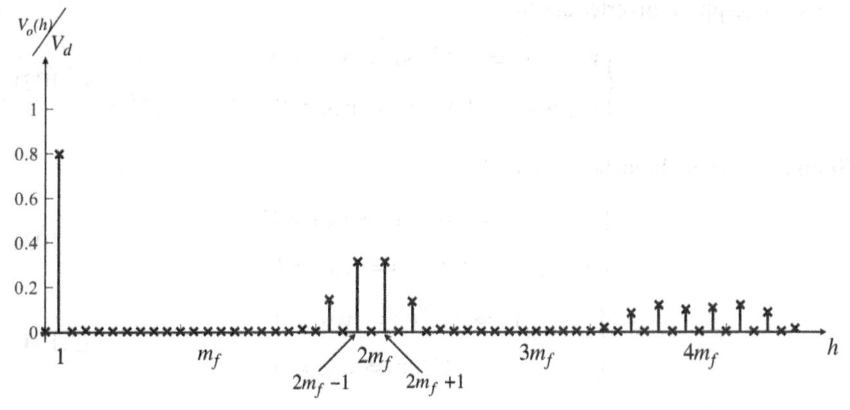

Figure 2.22 The frequency spectrum of the harmonics in $v_o(t)$ with unipolar switching.

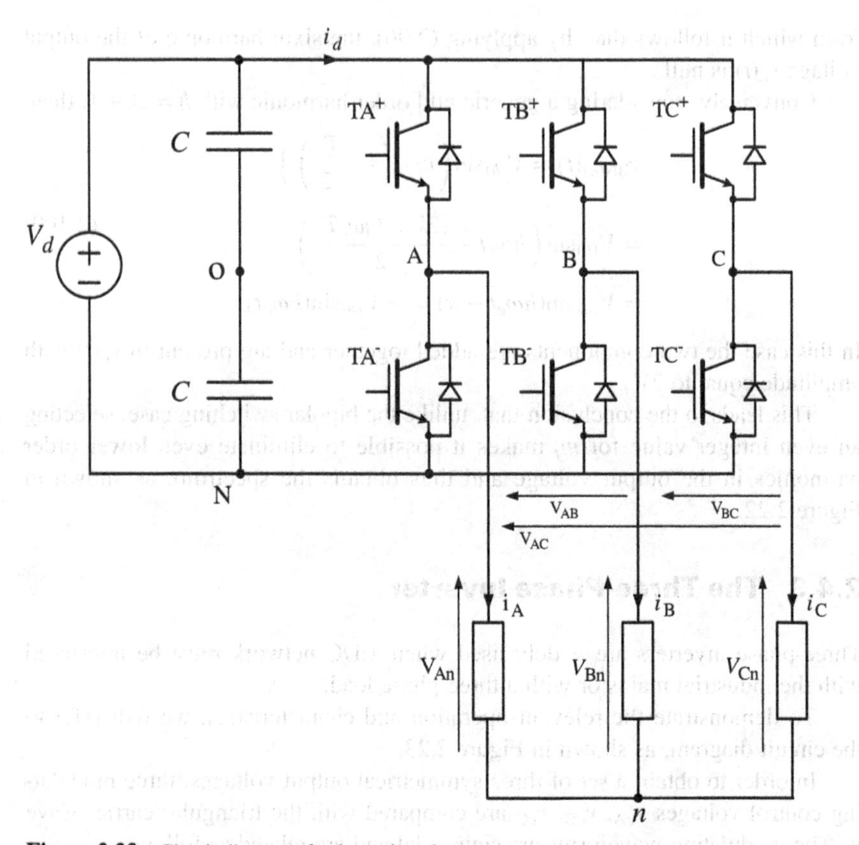

Figure 2.23 Circuit diagram of a three-phase inverter.

The switching conditions of the controlled semiconductor devices for branch A of the three-phase inverter are then

$$\begin{cases} v_{cA} > v_t \Rightarrow \text{TA}^+ \text{ on } \Rightarrow v_{AN} = V_d \\ v_{cA} \leq v_t \Rightarrow \text{TA}^- \text{ on} \Rightarrow v_{AN} = 0 \end{cases} \tag{2.102}$$

Similarly, for the branches B and C,

$$\begin{cases} v_{cB} > v_t \Rightarrow \text{TB}^+ \text{ on} \Rightarrow v_{BN} = V_d \\ v_{cB} \leq v_t \Rightarrow \text{TB}^- \text{ on} \Rightarrow v_{BN} = 0 \end{cases} \tag{2.103}$$

$$\begin{cases} v_{cC} > v_t \Rightarrow \text{TC}^+\text{on} \Rightarrow v_{CN} = V_d \\ v_{cC} \leq v_t \Rightarrow \text{TC}^-\text{on} \Rightarrow v_{CN} = 0 \end{cases} \tag{2.104}$$

The output line-to-line voltages v_{AB}, v_{BC}, and v_{CA} are then derived as

$$\begin{cases} v_{AB} = v_{AN} - v_{BN} \\ v_{BC} = v_{BN} - v_{CN} \\ v_{CA} = v_{CN} - v_{AN} \end{cases} \tag{2.105}$$

The waveforms of the control and output voltages in relation to branch A are shown in Figure 2.24.

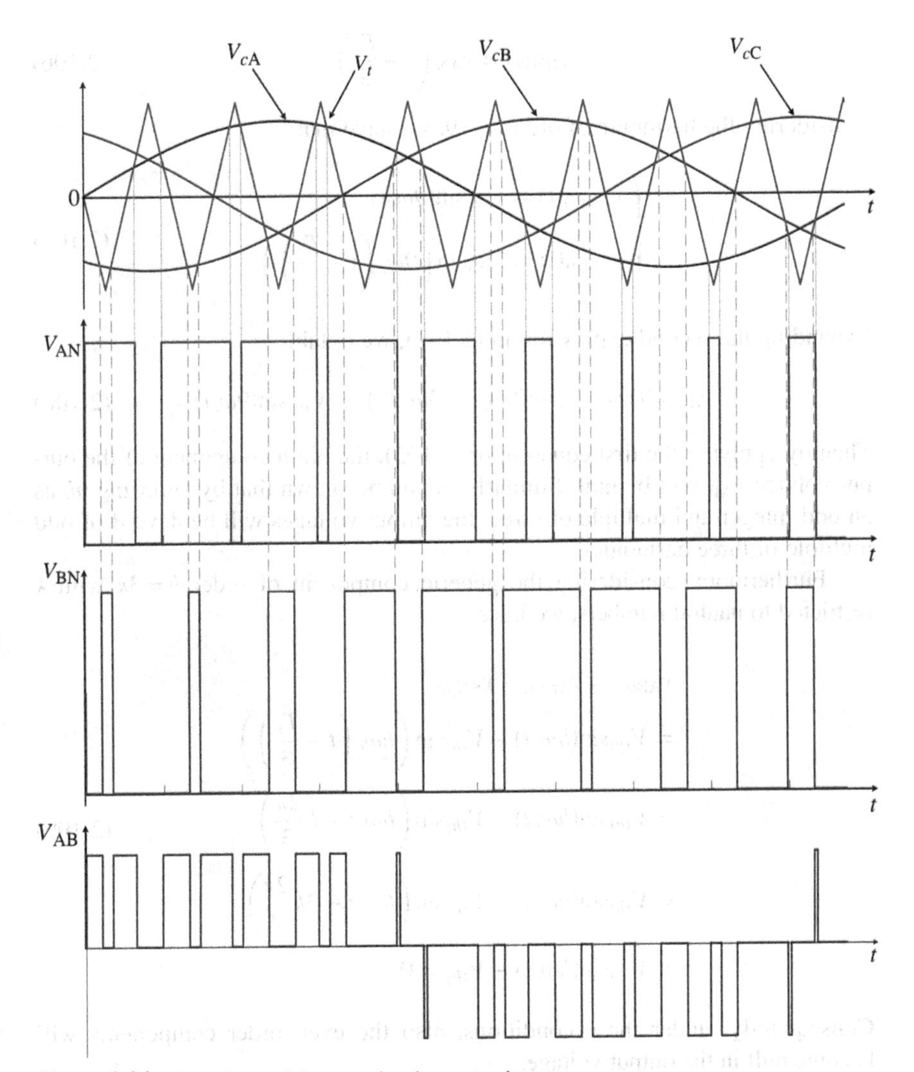

Figure 2.24 Waveforms of the control and output voltages.

To perform a harmonic analysis of the output voltages, a frequency modulation ratio of m_f equal to 9 was assumed.

As in the case of the single-phase full-bridge inverter, the orders of the harmonic voltage components v_{AN}, v_{BN}, and v_{CN} are defined by the expression (2.5). For simplicity, we carry out only the calculation for the v_{AB}, but the results obtained are also perfectly valid for v_{BC} and v_{CA}.

First, note that choosing a value of m_f that is an integer, odd, and a multiple of three, as in this case, then the relative position of v_{CA} with respect to v_t for $t>0$ is the same as v_{CB} with respect to v_t for $t>T_c/3$. Therefore, it follows that

$$v_{BN}(t) = v_{AN}\left(t - \frac{T_c}{3}\right) \tag{2.106}$$

Considering the harmonic of order $h=9$, we have that

$$\begin{cases} v_{AN(9)}(t) = V_{(9)}\sin(9\omega_c t) \\ v_{BN(9)}(t) = V_{(9)}\sin\left(9\omega_c\left(t - \frac{T_c}{3}\right)\right) \end{cases} \tag{2.107}$$

Expanding the second expression in (2.107), we obtain

$$v_{BN(9)}(t) = V_{(9)}\sin(9\omega_c t - 3\omega_c T_c) = V_{(9)}\sin(9\omega_c t) \tag{2.108}$$

Then by applying the first equation of (2.105), the ninth component of the output voltage v_{AB} will be null. Similarly, it can be shown that by choosing m_f as an odd integer and multiple of three, the output voltages will be devoid of odd multiple of three harmonics.

Furthermore, considering the generic component of order $h=3k$ with k restricted to natural numbers, we have

$$\begin{aligned} v_{AB(h)} &= v_{AN(h)} - v_{BN(h)} \\ &= V_{(h)}\sin(h\omega_c t) - V_{(h)}\sin\left(h\omega_c\left(t - \frac{T_c}{3}\right)\right) \\ &= V_{(h)}\sin(h\omega_c t) - V_{(h)}\sin\left(h\omega_c t - h\frac{2\pi}{3}\right) \\ &= V_{(h)}\sin(h\omega_c t) - V_{(h)}\sin\left(h\omega_c t - 3k\frac{2\pi}{3}\right) \\ &= V_{(h)}\sin(h\omega_c t) - V_{(h)} = 0 \end{aligned} \tag{2.109}$$

Consequently, under these conditions, also the even order components will become null in the output voltage.

The RMS value of the first harmonic of v_{AB} can be calculated as

$$V_{AB(1)RMS} = \frac{V_{AB(1)}}{\sqrt{2}} = \frac{V_{AN(1)}}{\sqrt{2}}\sqrt{3} = m_a \frac{V_d}{2}\frac{\sqrt{3}}{\sqrt{2}} = 0.612\ m_a V_d \qquad (2.110)$$

and it is significantly less than the value of V_d supply level.

2.4.3.1 Square wave operation

By setting $m_a \gg 1$, as already seen previously, the sinusoidal modulating voltages no longer intersect the carrier waveform and the voltages v_{AN}, v_{BN}, and v_{CN} will be positive square waves with 120° relative phase shifts. Following the expressions of (2.105), the output voltages will have the waveforms, as shown in Figure 2.25.

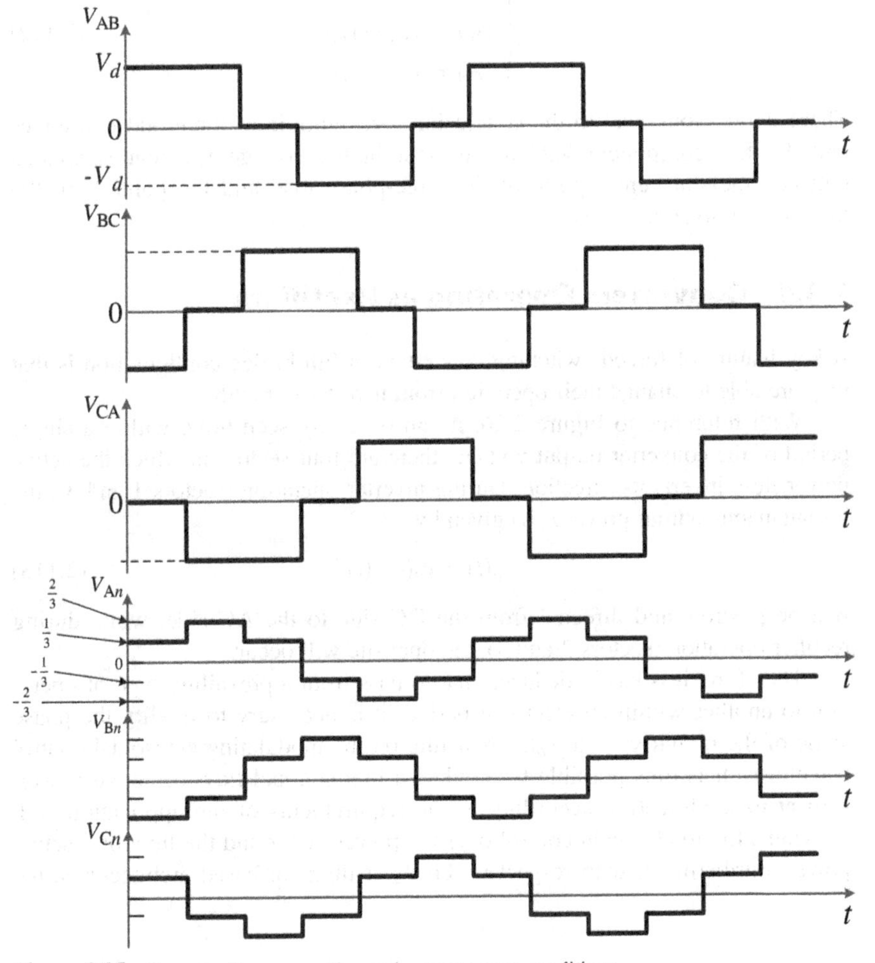

Figure 2.25 Output voltages operating under square wave conditions.

The maximum RMS value of the first harmonic of the line-to-line voltages achievable in square wave operation is

$$V_{\mathrm{AB(1)RMS}} = \frac{4}{\pi}\frac{V_d}{2}\frac{\sqrt{3}}{\sqrt{2}} = \frac{\sqrt{6}}{\pi}V_d = 0.78\ V_d \tag{2.111}$$

Figure 2.25 also shows the waveforms of the phase voltages $v_{\mathrm{A}n}$, $v_{\mathrm{B}n}$, and $v_{\mathrm{C}n}$ that are created, in this situation, by connecting the inverter output to a generic three-phase load. Harmonic analysis of these voltages shows that only the odd order components are present except for those that are multiples of three. Because of their square shape, voltages v_{AN}, v_{BN}, and v_{CN} do include all the odd harmonics and the DC component $V_d/2$.

From Kirchhoff's voltage law, we have

$$\begin{cases} v_{\mathrm{AN}} = v_{\mathrm{A}n} + v_{n\mathrm{N}} \\ v_{\mathrm{BN}} = v_{\mathrm{B}n} + v_{n\mathrm{N}} \\ v_{\mathrm{CN}} = v_{\mathrm{C}n} + v_{n\mathrm{N}} \end{cases} \tag{2.112}$$

These expressions lead to the conclusion that multiple of three odd harmonics and the DC component $V_d/2$ are present in the voltage $v_{n\mathrm{N}}$ that is created between the star center point of the three-phase load and the point N of the three-phase inverter.

2.4.4 Converters Operating as Rectifiers

A key feature of forced switching converters in full-bridge configuration is that they are able to change their operation from inverter to rectifier.

With reference to Figure 2.26, it can be easily seen how, within a single period of the converter output voltage, there are four sectors in which the active power flow inverts its direction. During inverter operation (sectors 1 and 3), the instantaneous output power $p_o(t)$ given by

$$p_o(t) = v_o(t)i_o(t) \tag{2.113}$$

will be positive and directed from the DC side to the AC side, while during rectifier operation (sectors 2 and 4), the opposite will occur.

Therefore, it is easily deduced that to pass from a prevailing type of operation to another within an electrical period, it is necessary to modify the phase angle of the output voltage $v_o(t)$ by acting on the modulating sinusoidal control waveform. It is thus possible to regulate a bidirectional flow of active power, subject to modification according to the requirements of the application and, consequently, to also gain control over the power factor and the flow of reactive power. Furthermore, with respect to diode- or thyristor-based architectures, the

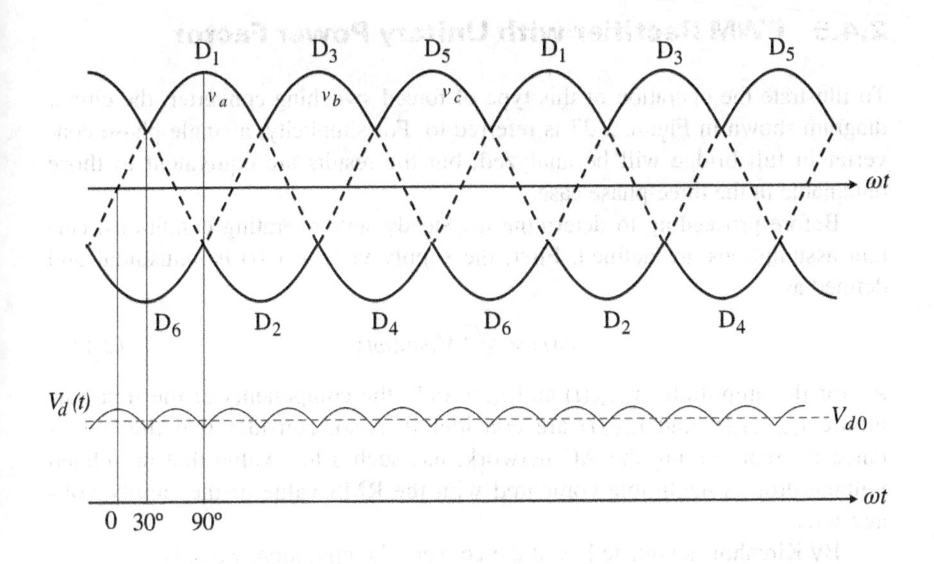

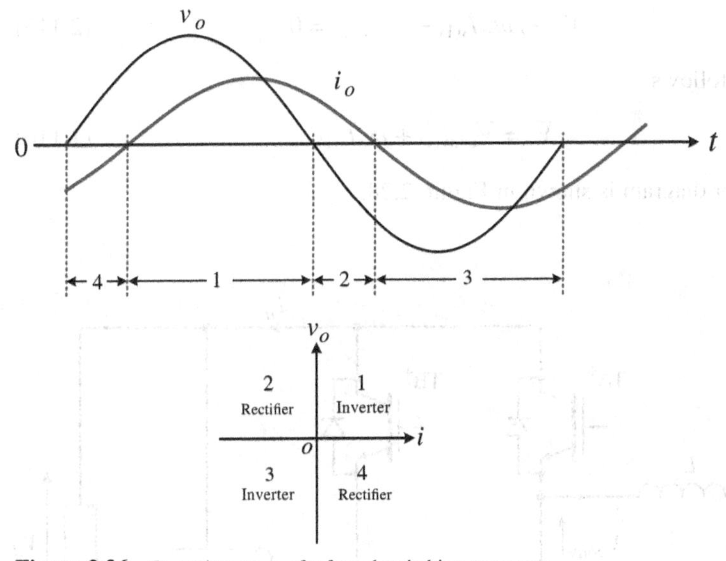

Figure 2.26 Operating range of a forced switching converter.

AC current at the input will be essentially sinusoidal, leading to a significant reduction in the DC side harmonic content.

In contrast, however, this type of rectifier requires complex control systems to adjust the output voltage and involves greater power losses due to the high *switching* frequencies for the semiconductor devices.

2.4.5 PWM Rectifier with Unitary Power Factor

To illustrate the operation of this type of forced switching converter, the circuit diagram shown in Figure 2.27 is referred to. For simplicity, a single-phase converter in full bridge will be analyzed, but the results are equivalent to those obtainable in the three-phase case.

Before proceeding to determine the steady-state operating conditions, certain assumptions are defined. First, the supply voltage $v_s(t)$ is sinusoidal and defined as

$$v_s(t) = \sqrt{2}\, V_s \sin(\omega t) \tag{2.114}$$

As for the amplitudes $v_{\text{conv}}(t)$ and $i_s(t)$, only the components of the first harmonic $v_{\text{conv}(1)}(t)$ and $i_{s(1)}(t)$ are considered. Also, consider that the inductance L_s, representing the AC network, has such a low value that its related voltage drop is negligible compared with the RMS value of the supply voltage $v_s(t)$.

By Kirchhoff's voltage law at the converter's input loop, we have

$$\overline{V}_s - j\omega L_s \overline{I}_{s(1)} - \overline{V}_{\text{conv}(1)} = 0 \tag{2.115}$$

from which follows

$$\overline{V}_s = \overline{V}_{\text{conv}(1)} + jX_s \overline{I}_{s(1)} \tag{2.116}$$

whose phasor diagram is shown in Figure 2.28.

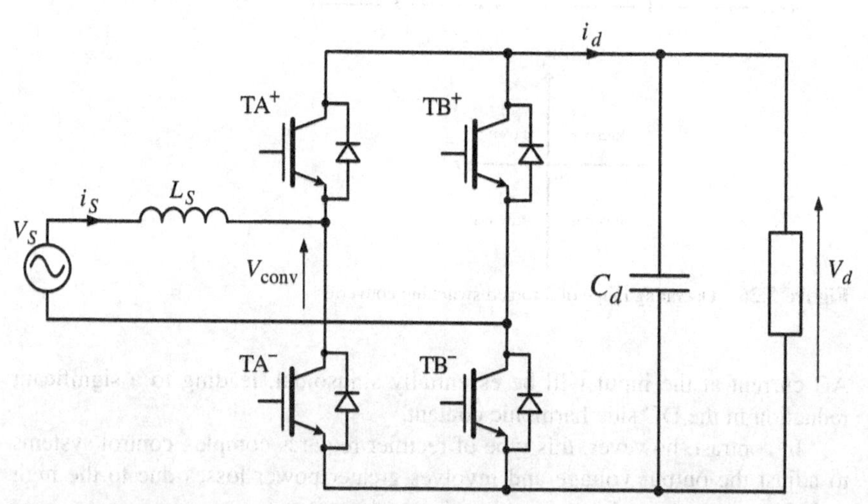

Figure 2.27 Single-phase PWM rectifier.

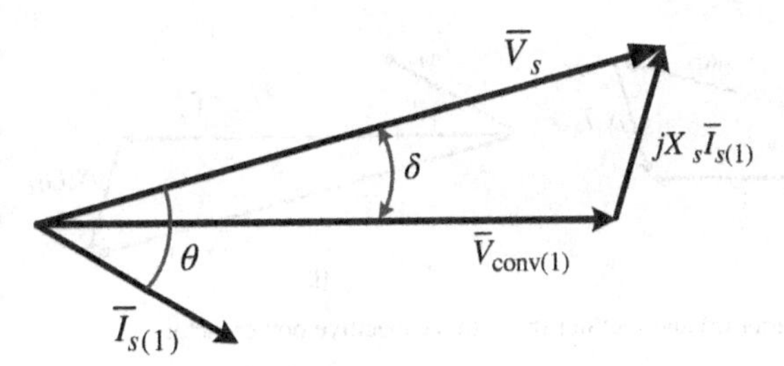

Figure 2.28 Phasor diagram of the voltages at the converter input.

The active power P delivered by the AC source represented by $v_s(t)$ will thus be equal to

$$P = V_s I_{s(1)} \cos\theta \tag{2.117}$$

By applying the sine theorem to the two right-angled triangles formed by the relative phasor representation of (2.116),

$$\frac{X_s\, I_{s(1)}}{\sin\delta} = \frac{V_{conv(1)}}{\sin(90° - \theta)} \tag{2.118}$$

resulting in

$$X_s I_{s(1)} \cos\theta = V_{conv(1)} \sin\delta \tag{2.119}$$

Substituting (2.119) with (2.117) finally provides an expression for the transmitted active power as

$$P = \frac{V_s V_{conv(1)}}{X_s} \sin\delta \tag{2.120}$$

The reactive power Q, also delivered by the AC network, can be defined as

$$Q = V_s I_{s(1)} \sin\theta \tag{2.121}$$

Rewriting (2.116), we have

$$V_s = V_{conv(1)} \cos\delta + X_s I_{s(1)} \cos(90° - \theta) \tag{2.122}$$

$$V_s - V_{conv(1)} \cos\delta = X_s I_{s(1)} \sin\theta \tag{2.123}$$

The expression for the reactive power Q is then

$$Q = \frac{V_s\left(V_s - V_{conv(1)}\cos\delta\right)}{X_s} \tag{2.124}$$

From the P and Q expressions thus obtained, it is possible to identify the different types of operation according to the relative positions of the voltage phasors of the converter's input loop.

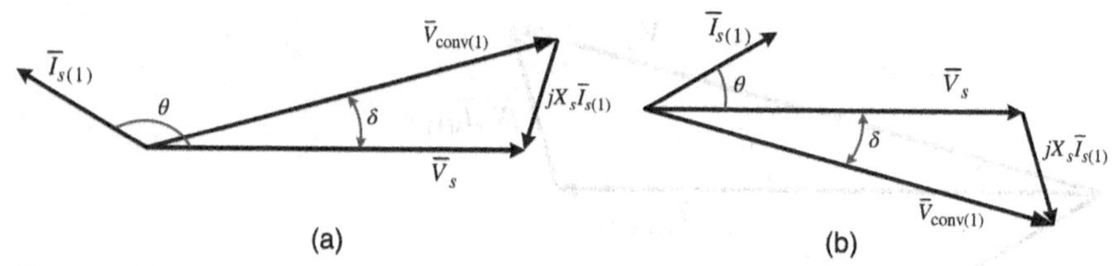

Figure 2.29 Operation with inverter (a) and rectifier (b) with a capacitive power factor.

Considering, for example, the case shown in Figure 2.28, it has been shown that the AC network transfers both active and reactive power to the DC side. By ideally rotating the voltage phasor of $v_{conv(1)}(t)$, while holding the supply voltage phasor $v_s(t)$ stationary, the other converter operating conditions can be extrapolated.

In Figure 2.29a, for example, the inverter operation is illustrated where both powers flow from the DC side to the AC source.

Finally, Figure 2.29b illustrates the case where the active power is absorbed by the DC network and the reactive power is supplied to the AC grid, effectively establishing a rectifier with capacitive power factor.

The flexibility of this type of converter is therefore advantageous in addressing different requirements besides electrical energy conversion. The main one is the compensation of the reactive power in the AC network it is connected to, with relative correction of the power factor.

With regard to operation as a rectifier, the working condition that has the widest application is at unity power factor, that is, with voltage $v_s(t)$ and current $i_{s(1)}(t)$ in phase.

2.4.5.1 Rectifier operation conditions

As previously stated, a very important feature of this type of rectifier is that input current remains sinusoidal and with low harmonic distortion. This is achievable, thanks to the use of the sinusoidal PWM modulation technique that, with a high-frequency modulation ratio ($m_f \gg 1$), allows for the generation of a relatively undistorted sinusoidal voltage $v_{conv}(t)$, thanks to the presence of only the high-order harmonic components.

Assuming a low inductive voltage drop on L_s, and a consequent approximate equivalence of RMS values between $v_{conv(1)}(t)$ and $v_s(t)$, within the linear modulation range of the converter ($0 < m_a < 1$), we have

$$\hat{V}_{conv(1)} = m_a V_d \tag{2.125}$$

from which follows

$$\hat{V}_{conv(1)} \cong \sqrt{2}\, V_s \cong m_a V_d \tag{2.126}$$

Given that the amplitude modulation ratio is less than 1, we have

$$V_d > \sqrt{2}\, V_s \tag{2.127}$$

This result represents the fact that the inductance L_s at the rectifier input can transfer energy to the capacitor C_d on the DC side with a higher voltage level. For this reason, this type of converter is also known as the *PWM boost rectifier.*

At initial power up, the semiconductor devices are not driven. This allows the capacitor C_d to be charged through the recirculating diodes that behave as a single-phase noncontrolled rectifier bridge. Because of the considerable value of the inrush current, it is possible to install a damping resistor, in series on the DC side, in order to limit the charging transient of C_d. Once the condition has been reached such that

$$v_d = \sqrt{2}\, V_s \tag{2.128}$$

it is possible to start triggering the switches to close, thus raising the voltage V_d. The PWM rectifier voltage output will then become adjustable only after the condition expressed by (2.127) has been established.

2.4.5.2 High-frequency oscillations

Figure 2.30 and Figure 2.31, respectively, illustrate the circuit diagram of a three-phase PWM boost rectifier and the graph of the relative input voltages and current waveforms under unity power factor conditions. Of course, the operation is the same as the single-phase case studied so far.

There is an evident high-frequency ripple superimposed on the sinusoidal line current, which is caused by the power switch's *switching* action and has a

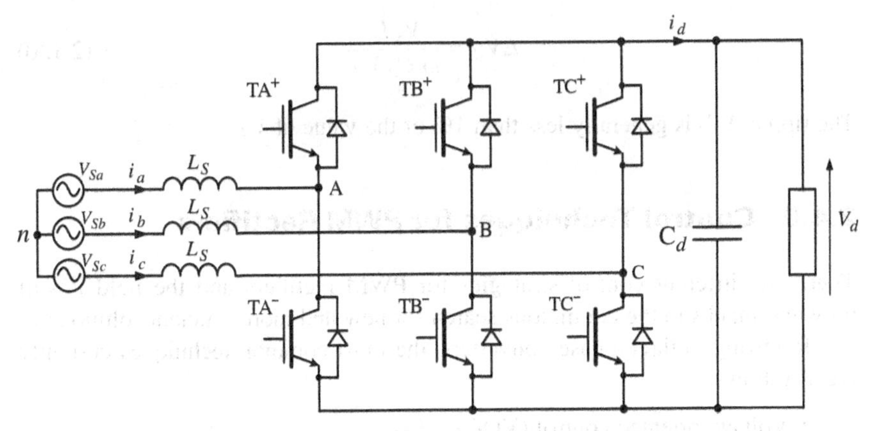

Figure 2.30 Three-phase PWM boost rectifier.

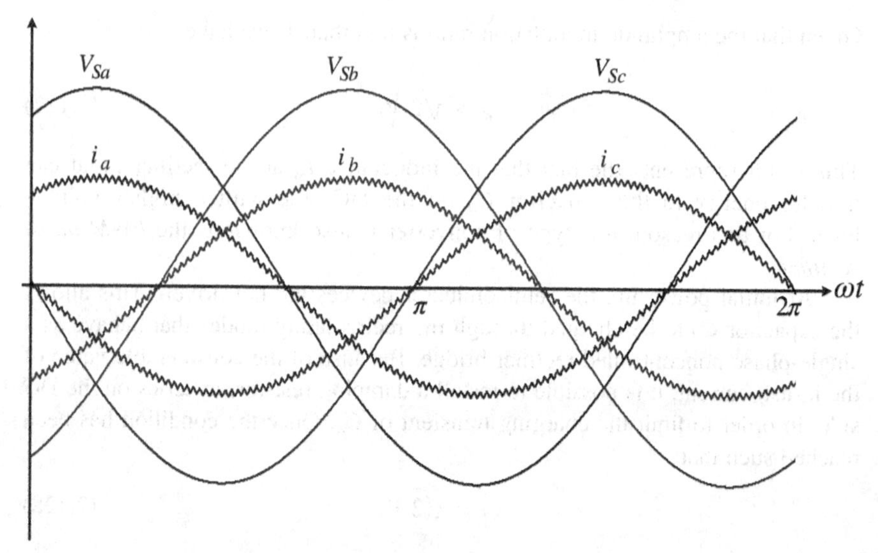

Figure 2.31 Input voltages and currents in a three-phase PWM rectifier.

frequency equal to f_s. The amplitude ΔI_s of this ripple is defined by the expression

$$\Delta I_s = \frac{V_d}{2\,L_s f_s}\left[1 - \left(\frac{v_s}{V_d}\right)^2\right] \tag{2.129}$$

It is dependent on the line inductance L_s and is maximum when the phase voltage v_s takes on zero value.

As for the output DC voltage v_d, it also fluctuates around its average value V_d with a peak-to-peak excursion that depends on the capacitance C_d and has value

$$\Delta V_d = \frac{V_s\,I_s}{\omega\,C_d\,V_d} \tag{2.130}$$

The ripple ΔV_d is generally less than 1% of the value of V_d.

2.4.6 Control Techniques for PWM Rectifiers

There are different control strategies for PWM rectifiers and the field is still growing, thanks to the continuous search for new and more efficient solutions.

Referring to three-phase converters, the most common techniques currently are as follows:

- Voltage-oriented control (VOC)

- Virtual flux-oriented control (VFOC)
- Direct power control (DPC)

The first two are based on the control of voltage space vectors and virtual flux, while the third is based on the control of active and reactive power.

2.4.6.1 Voltage-oriented control

Consider only branch A of a three-phase PWM rectifier, as shown in Figure 2.32.

Due to the presence of the two controlled switches, the branch can assume a total of four theoretical states.

In reality, the state in which both conduct simultaneously is forbidden because it would constitute a dangerous short-circuit capacity for the capacitance C_d. The same applies to the possibility of TA^+ and TA^- being simultaneously off because in that case the voltage v_a at node A would be determined by the direction of the current i_a. There are therefore only two possible valid states, namely, those in which one switch is turned off and the other is on.

The state of the branch A may be described by introducing variable a, defined as

$$\begin{cases} a = 0, & \text{if } TA^+ \text{ off and } TA^- \text{ on} \\ a = 1, & \text{if } TA^+ \text{ on and } TA^- \text{ off} \end{cases} \tag{2.131}$$

The same can be done for the branches B and C.

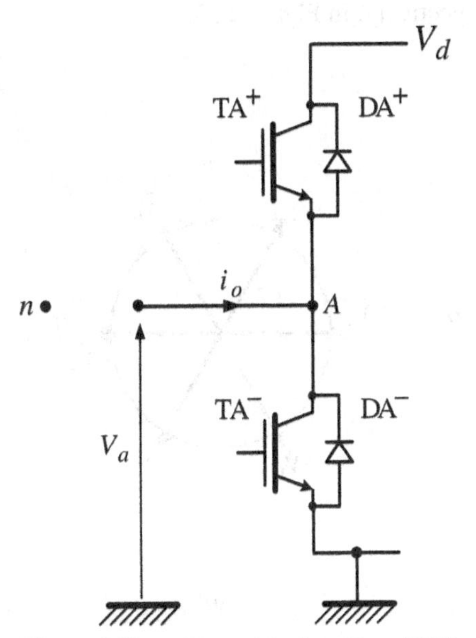

Figure 2.32 Branch A of a three-phase PWM rectifier.

The following expressions can be written for the voltages at nodes A, B, and C and the output voltage V_d:

$$\begin{bmatrix} v_a \\ v_b \\ v_c \end{bmatrix} = V_d \begin{bmatrix} a \\ b \\ c \end{bmatrix} \tag{2.132}$$

Considering the triple sets of line-to-line voltages v_{ab}, v_{bc}, and v_{ca} and phase voltages v_{an}, v_{bn}, and v_{cn} at the converter input, we obtain the other expressions

$$\begin{bmatrix} v_{ab} \\ v_{bc} \\ v_{ca} \end{bmatrix} = V_d \begin{bmatrix} 1 & -1 & 0 \\ 0 & 1 & -1 \\ -1 & 0 & 1 \end{bmatrix} \begin{bmatrix} a \\ b \\ c \end{bmatrix} \tag{2.133}$$

$$\begin{bmatrix} v_{an} \\ v_{bn} \\ v_{cn} \end{bmatrix} = \frac{V_d}{3} \begin{bmatrix} 2 & -1 & -1 \\ -1 & 2 & -1 \\ -1 & -1 & 2 \end{bmatrix} \begin{bmatrix} a \\ b \\ c \end{bmatrix} \tag{2.134}$$

The three switching variables *abc* imply the possibility of having a total of eight rectifier states.

The states 000 (0) and 111 (7) are defined null because they would set the input voltages to zero, since all of the rectifier input terminals would be connected together either to the positive or to the negative pole of the DC output.

Considering the states 1–6, termed as "active," these are obtained by applying the stationary Park transform onto axes α and β. This generates the line-to-line voltage space vectors illustrated, respectively, in Figure 2.33.

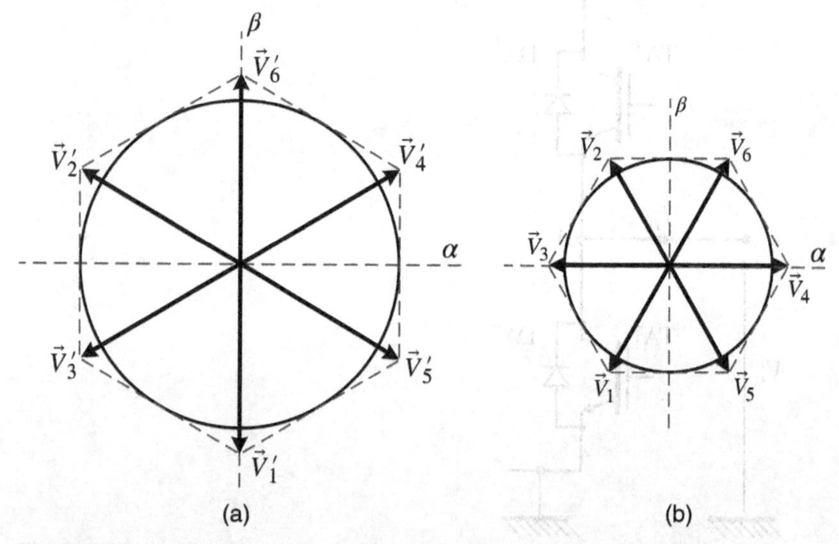

Figure 2.33 Space vectors of the line-to-line voltages (a) and phase voltages (b).

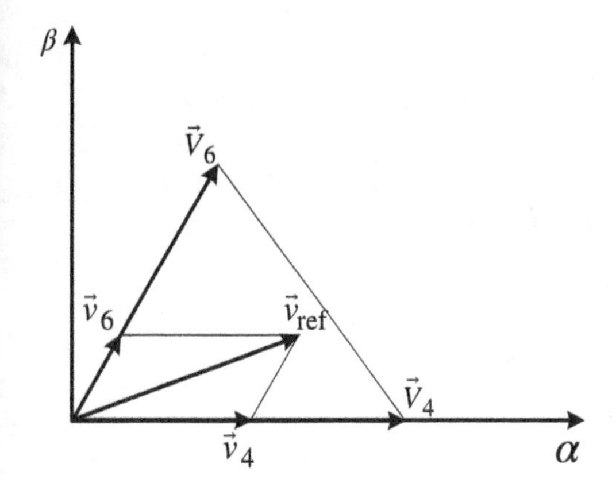

Figure 2.34 Defining the average value of the reference space vector.

The α, β plane is thus divided into six sectors representing the active state space vectors. The reference voltage space vector will be constant within the switching period T_s, but it is not likely to coincide with that of one of the six active states. To establish the mean value during T_s, two vectors are used that define the sector in which it is contained, as shown in Figure 2.34.

In the illustrated case, the vectors 4 and 6 will be alternately applied for a time proportional to the projections that the reference vector creates on them. Vector 4 will be selected for a time interval t_4 and vector 6 for t_6, both within the switching period. The switching times therefore must satisfy the condition

$$T_s = t_4 + t_6 + t_0 \tag{2.135}$$

where t_0 is the time interval remaining within T_s in which either a 0 or 7 null state is applied.

In the *voltage-oriented control* technique, a rotating reference frame with d, q axes is also used.

Figure 2.35 shows the voltage and current space vectors in both reference systems. To achieve the condition of operation at unity power factor, the two carriers must be in phase. In this manner, aligning the voltage vector with the d rotating axis will cancel both its q-axis component and that of the current vector.

The rectifier control architecture is illustrated in Figure 2.36.

The three-phase voltages and currents to the converter are transformed into their respective space vectors, initially into the stationary reference frame with α, β axes and subsequently into the rotating reference frame with d, q axes.

For the transition between these reference systems, it is necessary to identify the v_α and v_β voltage space vectors as they provide the angle ωt required for the transformation. In fact, the two systems are linked by

$$\vec{v}^{dq} = \vec{v}^{\alpha\beta} e^{-j\omega t} \tag{2.136}$$

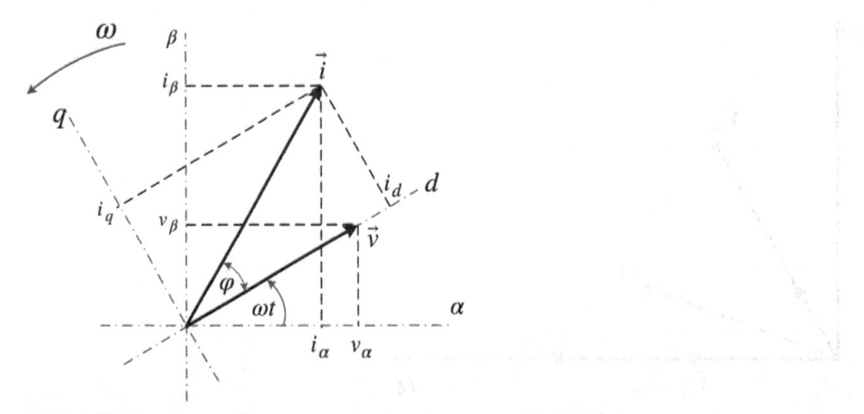

Figure 2.35 Reference systems for fixed and rotating reference frames.

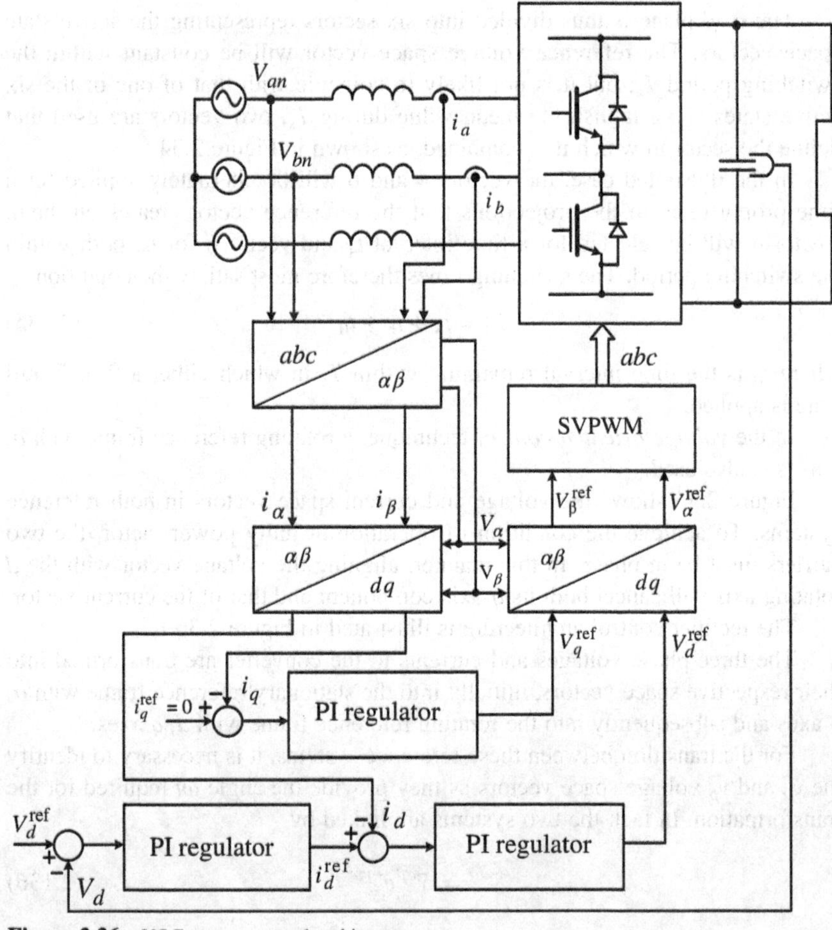

Figure 2.36 VOC system control architecture.

The reference value of the current vector q-axis component, namely, i_q^{ref}, is set to zero, while i_d^{ref} is derived from the output of a proportional-integral controller that processes the error in the output voltage V_d with respect to its reference V_d^{ref}. Two other PI controllers provide the references for the voltage space vector v_d^{ref} and v_q^{ref} that are subsequently transferred to the stationary reference domain as v_α^{ref} e v_β^{ref}. These latter components are finally used for regulating the PWM modulation by means of the space vector, also known as SVPWM, thus driving the switching action of the converter semiconductor devices.

2.4.6.2 Virtual flux-oriented control

To explain the operation of this control technique, it is assumed that the input of the PWM rectifier is connected to the stator of a virtual AC machine, as shown in Figure 2.37.

Consider the resistive and inductive parameters as equal across all phases with values R and L.

The phase voltages v_a, v_b, and v_c are induced by the action of a virtual magnetizing flux ψ, whose space vector has components in the stationary reference frame axes α, β, namely, ψ_α and ψ_β. In reality, the space vector ψ represents the virtual three-phase network flux supplying the converter from the AC side.

Other space vectors are defined as follows:

- $\vec{v}_s$, vector of the voltages at the converter input terminals
- $\vec{i}_s$, vector of the line currents at the converter input terminals

Neglecting the resistive voltage drops, we can establish the expression

$$\frac{d\vec{\psi}}{dt} = \vec{v}_s + L\frac{d\vec{i}_s}{dt} \tag{2.137}$$

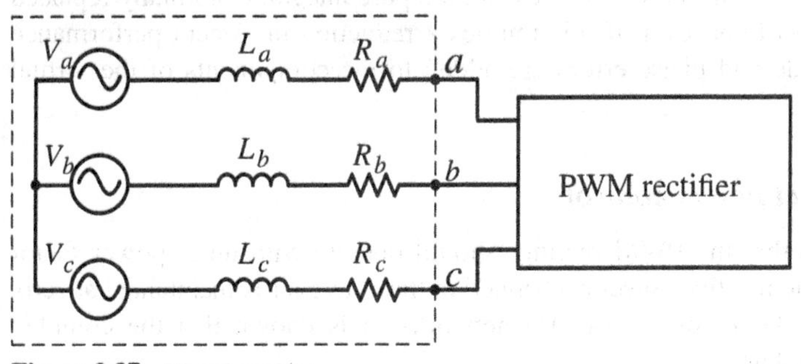

Figure 2.37 Virtual machine.

By invoking the stationary reference frame axes, we can expand (2.137) according to its components:

$$\begin{cases} \dfrac{d\psi_\alpha}{dt} = v_{s\alpha} + L\dfrac{di_{s\alpha}}{dt} \\ \dfrac{d\psi_\beta}{dt} = v_{s\beta} + L\dfrac{di_{s\beta}}{dt} \end{cases} \tag{2.138}$$

Integration then allows the α and β virtual flux components to be expressed as follows:

$$\psi_\alpha = \int \left(v_{s\alpha} + L\frac{di_{s\alpha}}{dt} \right) dt \tag{2.139}$$

$$\psi_\beta = \int \left(v_{s\beta} + L\frac{di_{s\beta}}{dt} \right) dt \tag{2.140}$$

When implementing the relevant control structure, the VFOC technique also incorporates the concepts introduced when examining the *voltage-oriented control* approach. Hence, we now introduce the rotating reference frame in such a way that the virtual flux space vector is aligned with the direct axis d and the relative angle with respect to the stationary reference frame is defined as

$$\theta = \arctan\left(\frac{\psi_\beta}{\psi_\alpha}\right) \tag{2.141}$$

The error signal for the rectified output voltage V_d is processed by a PI controller that employs an estimation algorithm to provide the reference values of the d and q components of the phase current space vectors. The control system will use the latter to generate the components of the voltage space vector at the rectifier input, which, in turn, are finally converted into the three alternating voltages by means of sinusoidal PWM.

The pure integral used to calculate the components of the virtual flux vector in Equations 2.139 and 2.140 could saturate due to noise or offsets present in the current or voltage sensors. For this reason, the pure integral is normally replaced with a low-pass filter, even if this implies a reduction in system performance since magnitude and phase errors are added to the components of the virtual flux vector.

2.4.6.3 Direct power control

The condition that the PWM rectifier should operate with unity power factor requires that the reactive power exchanged in the converter is maintained at zero.

From the theory of three-phase networks, it is known that the complex power is defined as

$$\overline{S} = 3\,\overline{V}_f \overline{I}^* = P + jQ \tag{2.142}$$

where $\overline{V}_f$ is the phasor of the phase voltage and $\overline{I}^*$ is the complex conjugate of the line current phasor. An analogous expression applies to the space vectors

$$\bar{s} = \vec{v}\vec{i}^{\,*} = p + jq \tag{2.143}$$

where p and q represent the instantaneous values of the active and reactive power, which may be expressed separately as

$$p(t) = v_\alpha i_\alpha + v_\beta i_\beta = Re\left(\vec{v}\vec{i}^*\right) \tag{2.144}$$

$$q(t) = v_\beta i_\alpha - v_\alpha i_\beta = \mathrm{Im}\left(\vec{v}\vec{i}^*\right) \tag{2.145}$$

Applying the inverse Park transform, it is possible to calculate the instantaneous values of the powers as a function of the line currents and the phase voltages. It follows that, by measuring the instantaneous levels at the converter input, we can calculate the power exchanged with the AC network.

We thus proceed by introducing three control variables x, y, and z that will determine the operational state of the PWM rectifier. The variable x indicates an arbitrary 30° sector that contains the space vector of the converter's input voltages $\vec{v}$ in the α, β plane. It takes on integer values between 1 and 12 as defined by

$$x = \mathrm{int}\left(\frac{\delta}{30^\circ}\right) + 1 \tag{2.146}$$

The angle δ defines the phase of the space vector $\vec{v}$ in the stationary reference frame and is determined by its components v_α and v_β.

To better describe the operation of the DPC system, we have introduced the control scheme, as illustrated in Figure 2.38.

The control variables y and z are obtained from the binary outputs of the two hysteresis controllers that are, in turn, driven by the instantaneous error signals related to the active and reactive powers. The amplitude of the hysteresis loop is determined by the tolerance band on the Δp and Δq errors and is adjusted so as to remain within the converter's desired average switching frequency.

The instantaneous active power p is compared with its reference value p^{ref}, which is, in turn, derived from the response V'_d generated by a PI controller that operates on the error of the DC voltage output V_d.

Since the active power at the rectifier input is obviously linked to the output DC side, the latter being proportional to the square of the voltage V_d, the reference signal p^{ref} will then be obtained as the product of the signals V'_d and V_d.

As for the reactive power, its reference value q^{ref} is set to zero to maintain operation at unity power factor, and it is continuously compared with the instantaneous value q calculated by the system.

The values of the state variables x, y, and z obtained from the control system are then used to determine which status to assign to the converter in order to minimize power errors, allowing for fine space vector-driven PWM control applied to the input voltages.

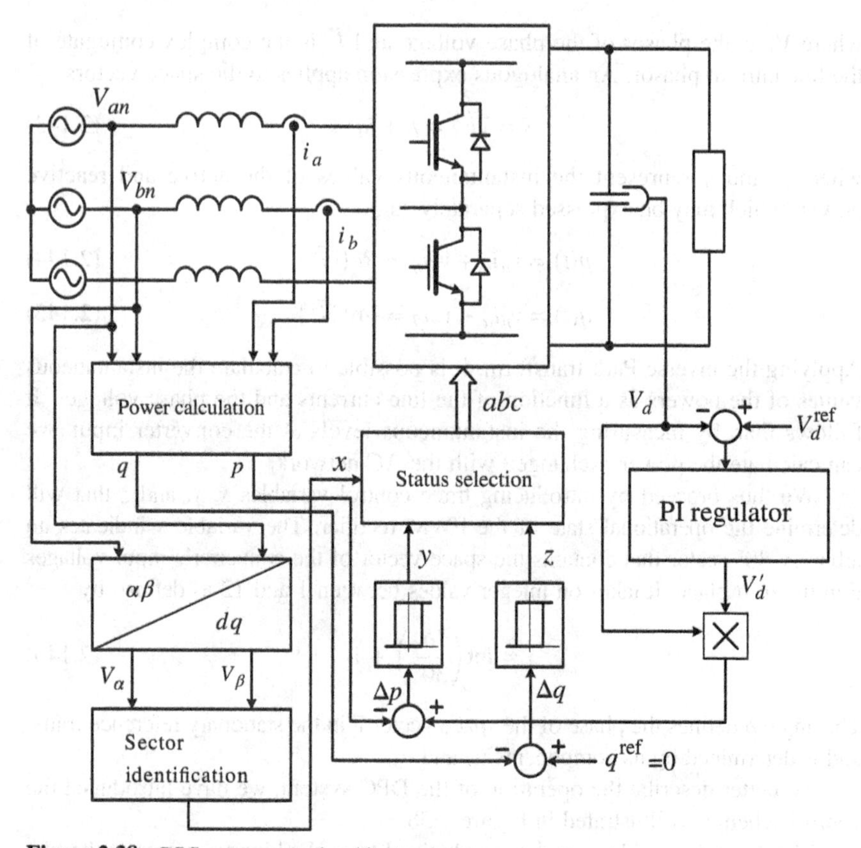

Figure 2.38 DPC system control architecture.

In conclusion, it must be emphasized that the DPC control technique is capable of providing a better power factor at the converter input compared with the other previously illustrated systems. It also has good dynamic characteristics and does not require the transformation of the magnitudes to a rotating reference frame, thus significantly simplifying the control system. However, contrary to the VOC and VFOC solutions, the *direct power control* approach does not allow for the controlled semiconductor devices to be switched at a fixed frequency and this introduces harmonic filtering issues in the design of the devices.

2.4.7 Multilevel Converters

Multilevel converters are becoming more widely used in high-power applications such as electrical traction.

The main advantages inherent in this approach can be summarized as follows:

- Low harmonic distortion

- Reduced stress on the individual control components
- High energy efficiency

On the other hand, their characteristics also lead to the need for a greater number of power modules for their operation, thus making the construction of the relevant control circuits more complicated and increasing installation costs, especially for the high-voltage components.

In the field of power electronics, the tendency is toward increasingly higher switching frequencies, especially in converters that operate with considerable powers. This is because of the consequent reduction in low-frequency harmonic distortion and less bulky passive components such as transformers. In contrast, however, increasing the *switching* frequency involves greater switching losses, which can reach considerable levels, especially in high-voltage applications.

To mitigate the voltage stresses affecting the semiconductor devices, one solution is to replace the single switch with a series of two or more elements, thus dividing the applied voltage among a number of devices. Figure 2.39 depicts a single branch of a three-phase converter implementing a series switch combination.

However, this configuration introduces a number of problems. First of all, the switches belonging to the same pair must switch simultaneously, otherwise one of them will be momentarily subjected to the entire applied voltage. Furthermore, the overall *dv*/*dt* transient that occurs when the two switches in series are switched is equal to the sum of those of the individual devices, which can create undesirable effects in adjacent low-voltage circuitry and additional electromagnetic interference.

To cope with these problems, instead of simultaneously switching the semiconductor devices, the circuit topology of the converter circuit is arranged so as

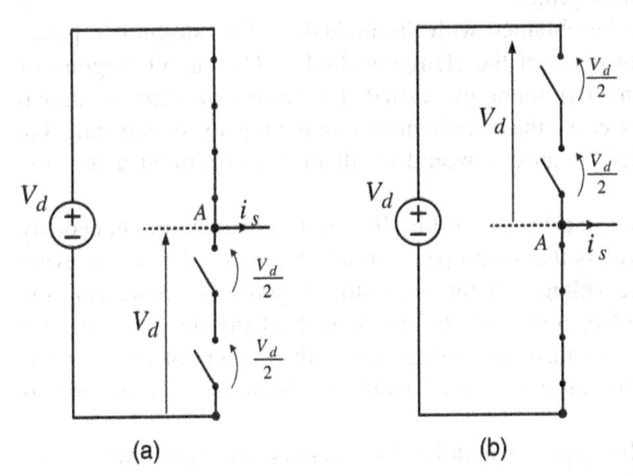

Figure 2.39 Branch of a converter with switches in series.

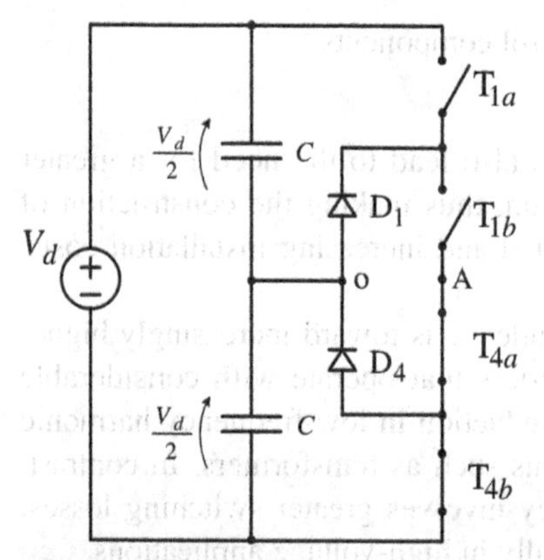

Figure 2.40 Branch with switches in series, clamping diodes, and capacitive voltage divider.

to distribute their relative stresses in a balanced manner. Figure 2.40 illustrates the case in which clamping diodes and a capacitive voltage divider are employed for this purpose.

In this configuration, depending on the state of the switches, the number of voltage levels at point A increases by 1. If the upper pair of switches T_{1a} and T_{1b} are turned on, the output voltage v_{AO} will have a value equal to $V_d/2$, while when the lower switch pair conducts, v_{AO} will be equal to $-V_d/2$. Finally, there is also the possibility of turning on only the internal switches T_{1b} and T_{4a}, thus creating a short circuit through the diodes D1 and D4 and nulling the output voltage.

This converter configuration is referred to as *diode-clamped* and will thus have three levels. Its output voltage v_{AO} may therefore assume three different values and only three acceptable operating conditions exist for which adjacent switch pairs will conduct together.

Similar results can be obtained with the inclusion of a capacitor in parallel with the switches in place of the clamping diodes. The circuit diagram of one branch of this converter topology, called the *flying capacitor*, is shown in Figure 2.41. In this case, the simultaneous switching of switch pair T_{1b} and T_{4a} is not permitted because it would result in a short circuit across the capacitor.

Similarly, T_{1a} and T_{4b} also cannot be allowed to conduct simultaneously because of the high currents that could pass through these switches in the event of an unbalance of the voltage on the capacitor. Figure 2.42 represents the acceptable conditions of operation of the one branch of the floating capacitor converter. Also, in this case there are three levels of the output voltage v_A; however, they are achievable through four possible combinations, of which one is redundant.

Compared with other types of multilevel converters, the *diode-clamped* (or *neutral point-clamped*) configuration is very widespread for its high energy

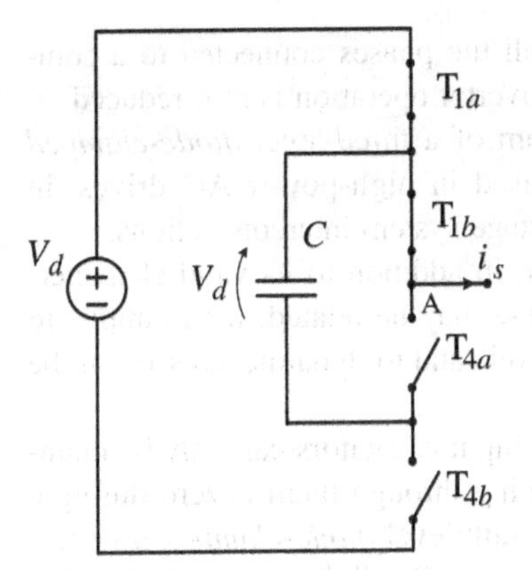

Figure 2.41 Branch of a converter with a floating capacitor.

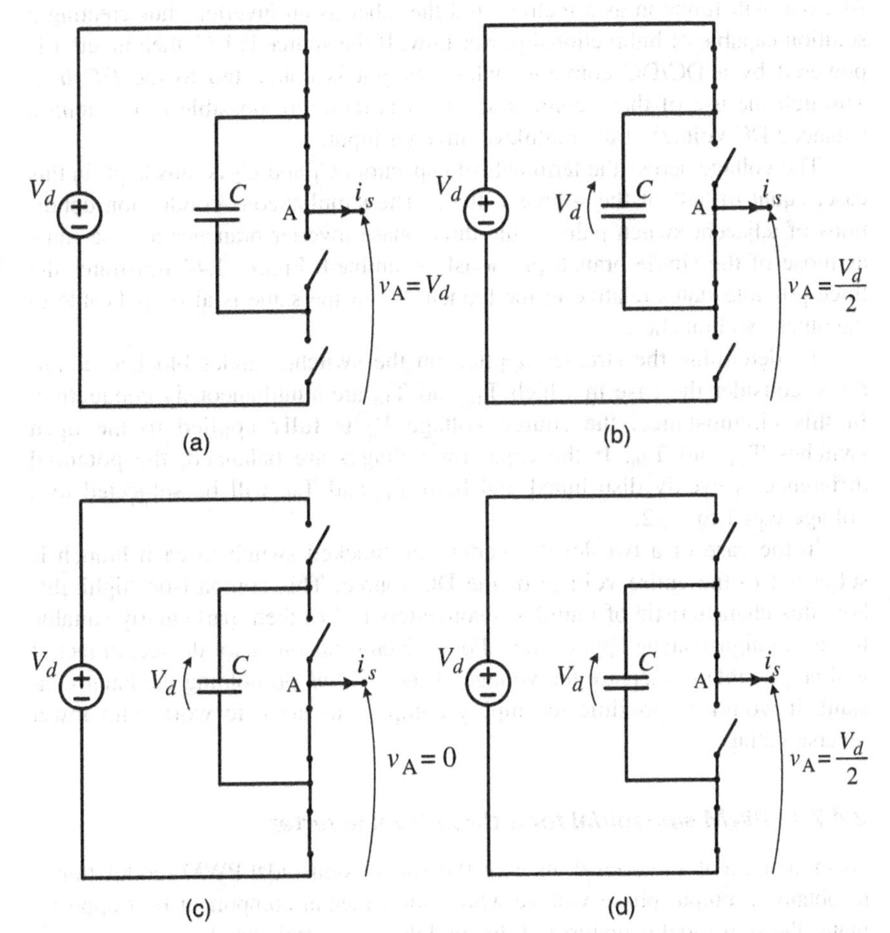

Figure 2.42 Operating conditions of a converter with a floating capacitor. Three-phase three-level diode-clamped inverter.

efficiency and economy. Furthermore, with all the phases connected to a common *DC link*, the capacitance required for converter operation is also reduced.

Figure 2.43 illustrates the circuit diagram of a three-level *diode-clamped* inverter. This type of converter is widely used in high-power AC drives, in reactive power compensation, and in high-voltage system interconnections.

The use of this type of inverter, however, in addition to its valid characteristics introduces some inherent problems. These may be related, for example, to the balancing of the voltage at intermediate levels and to dynamic stresses on the semiconductor devices.

A state of voltage balance on the inverter input capacitors can only be maintained if the average value of the current flowing through them is zero during a period of the output voltage. For this reason, multilevel *diode-clamped* inverters are often used in pairs and connected to the same *DC link*, thus implementing the so-called *back-to-back* converter configuration. If both source and load are AC, one will function as a rectifier and the other as an inverter, thus creating a solution capable of bidirectional power flow. If the source is DC, then inverter is powered by a DC/DC converter whose output is connected to the *DC link*. Through the use of these configurations, it is therefore possible to maintain a balanced DC voltage at the multilevel inverter input.

The voltage across the terminals of capacitors C_1 and C_2 is thus kept, in this case, equal to half of the source voltage. The simultaneous conduction conditions of adjacent switch pairs in the three-phase inverter branches are the same as those of the single branch previously examined. Figure 2.44 illustrates the three possible states relative to the branch A, but the same is also applicable to the other two branches.

To determine the stresses applied on the switches under blocked conditions, consider the case in which T_{1a} and T_{1b} are simultaneously conducting. In this circumstance, the source voltage V_d is fully applied to the open switches T_{4a} and T_{4b}. If the capacitor voltages are balanced, the potential difference is evenly distributed and both T_{4a} and T_{4b} will be subjected to a voltage equal to $V_d/2$.

In the case of a two-level inverter, the blocked switch in each branch is subjected to the entire voltage of the DC source. This comparison highlights how this characteristic of multilevel converters makes them particularly suitable for use in high-voltage applications. For a given semiconductor device, in fact, it will be possible to increase the voltage of the *DC link* or, holding the latter constant, it would be possible to employ components sized to work with lower reverse voltages.

2.4.7.1 PWM sinusoidal for a three-level inverter

As in the case of a two-level inverter, the goal of sinusoidal PWM modulation is to obtain an output phase voltage whose fundamental component best approximates the sinusoidal waveform of the modulating control signal.

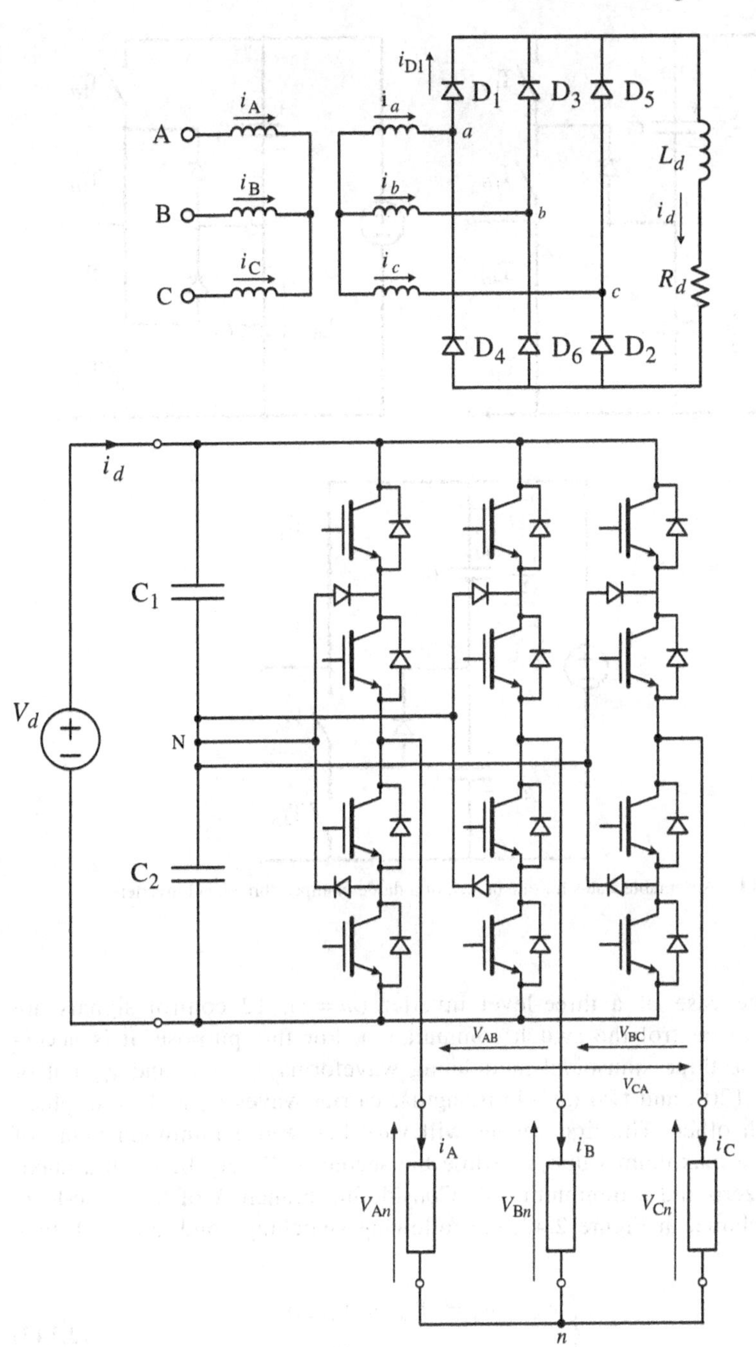

Figure 2.43 Three-phase three-level diode-clamped inverter.

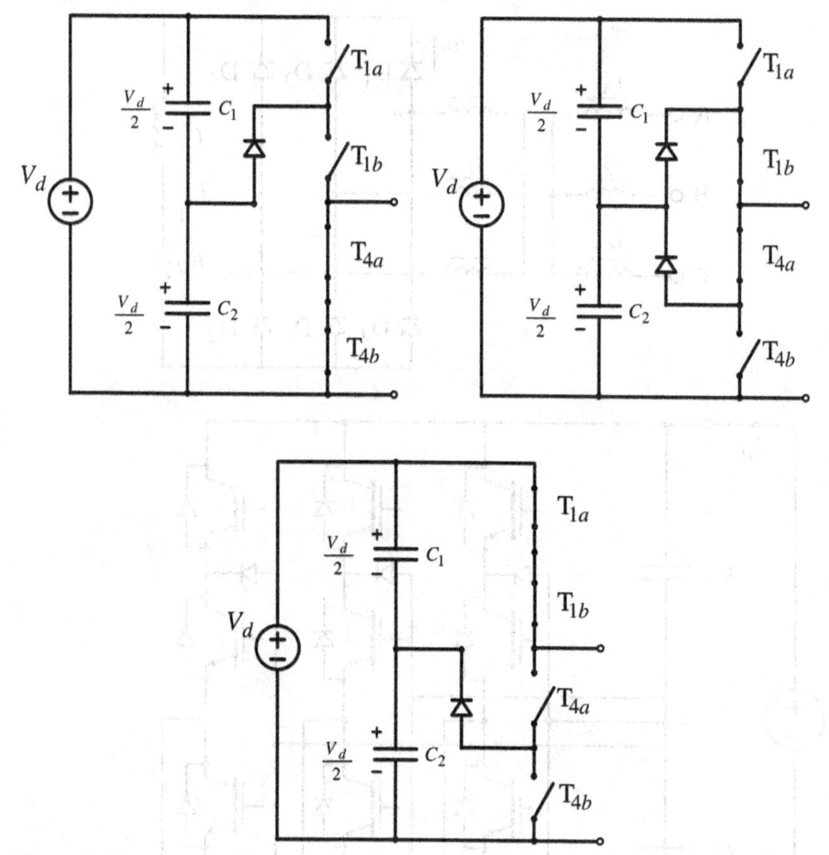

Figure 2.44 Acceptable states for one branch of a diode-clamped three-level inverter.

In the case of a three-level inverter ($m = 3$), 12 control signals are required to control the switch commutation. For this purpose, it is necessary to use three sinusoidal modulating waveforms v_{ca}, v_{cb}, and v_{cc} out of phase by 120°, and two ($m - 1$) triangular carrier waves v_{t1} and v_{t2} in phase with each other. The first carrier will vary between a minimum value of zero and a maximum value V_t, while the second will vary between a maximum of zero and a minimum $-V_t$. Considering branch A of the three-level inverter shown in Figure 2.43, the following switching conditions will then apply:

$$\begin{cases} v_{ca} > v_{t1} \Rightarrow \mathrm{T_{1a}\ on, T_{4a}\ off} \\ v_{ca} < v_{t1} \Rightarrow \mathrm{T_{1a}\ off, T_{4a}\ on} \end{cases} \tag{2.147}$$

$$\begin{cases} v_{ca} > v_{t2} \Rightarrow \text{T}_{1b}\text{ on}, \text{T}_{4b}\text{ off} \\ v_{ca} < v_{t2} \Rightarrow \text{T}_{1b}\text{ off}, \text{T}_{4b}\text{ on} \end{cases} \tag{2.148}$$

In order to use the same group of carrier signals to generate the control pulses for the switches of the branches B and C, a frequency modulation ratio m_f is chosen as odd and a multiple of three.

Figure 2.45 shows the control and inverter output voltage waveforms at three levels related to the branch A. It can easily be seen that the phase voltage at the converter output takes on three different values in a fundamental period, namely, zero, $V_d/2$, and $-V_d/2$.

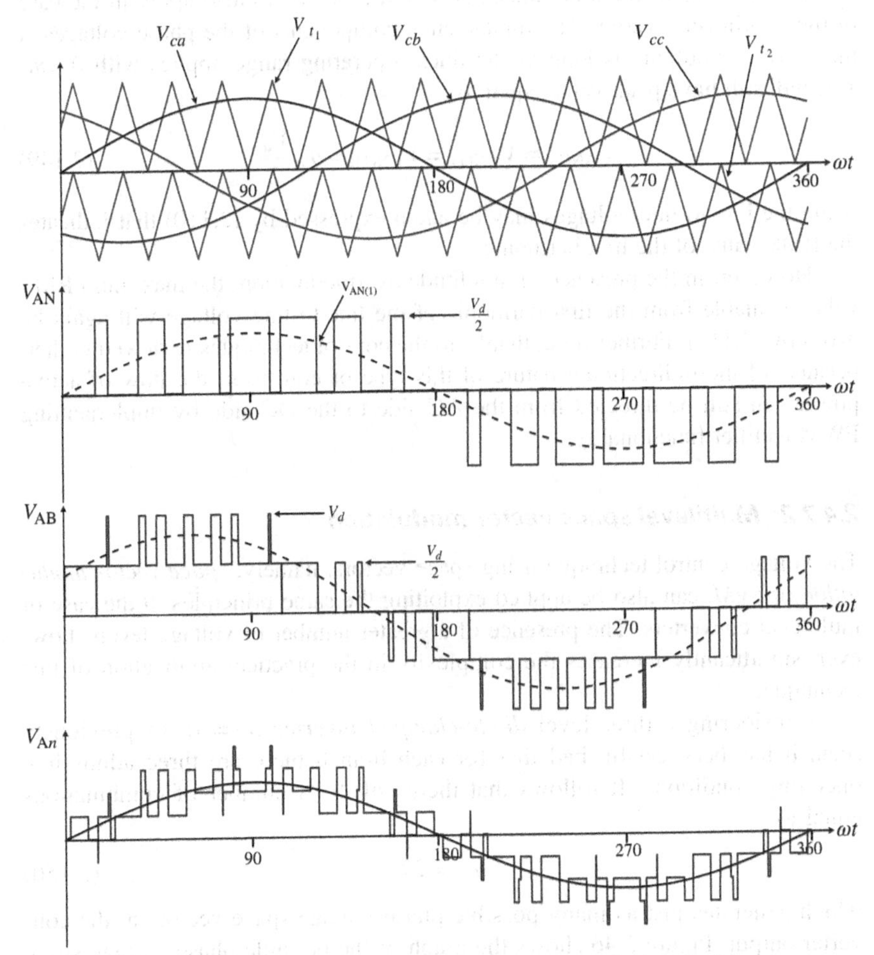

Figure 2.45 Waveforms of the control signal and output voltages of a three-level inverter.

Concerning the expected order of the output voltage harmonic content, the considerations made for the three-phase inverter with two levels are still valid. The changes can be seen in the amplitudes of those components which, in the case of the three-level inverter, will be significantly reduced. This is due to the fact that the inclusion of an additional level in the output voltage allows its waveform to better approximate that of a sinusoid and the greater the number of converter levels, the lower will be the amplitude of the higher harmonics.

Ultimately, the reduced harmonic distortion in the voltages produced by multilevel inverters is another great benefit that characterizes the use of this technology, especially when the goal is to limit the harmonic contamination in electrical networks.

The primary features of sinusoidal PWM continue to also apply in the case of the multilevel inverter. The fundamental components of the phase voltages at the converter output, as long as the linear operating range applies with $0 \leq m_a \leq 1$, will still have peak value equal to

$$V_{AN(1)} = V_{BN(1)} = V_{CN(1)} = m_a \frac{V_d}{2} \tag{2.149}$$

while the line-to-line voltages may be again expressed by (2.110) that indicates the RMS value of the first harmonic.

However, in the presence of amplitude overmodulation, the maximum RMS value available from the first harmonic of the line-to-line voltage will again be given by (2.111). Furthermore, thanks to the control techniques reviewed earlier, because of the bidirectional nature of this type of converter, the flow of active power can still be directed from the AC side to the DC side by implementing PWM rectifier functionality.

2.4.7.2 Multilevel space vector modulation

The voltage control technique using space vectors, namely, *space vector modulation* or *SVM*, can also be applied exploiting the same principles in the case of multilevel converters. The presence of a greater number of voltage levels, however, significantly increases the complexity in the practical application of this technique.

Considering a three-level *diode-clamped inverter* ($m = 3$), as previously seen, it has been established that for each branch there are three admissible operating conditions. It follows that there will be a number of combinations equal to

$$m^3 = 27 \tag{2.150}$$

which generates just as many possible phase voltage space vectors at the converter output. Figure 2.46 shows the graph of the possible phase voltage space vectors for a three-level inverter.

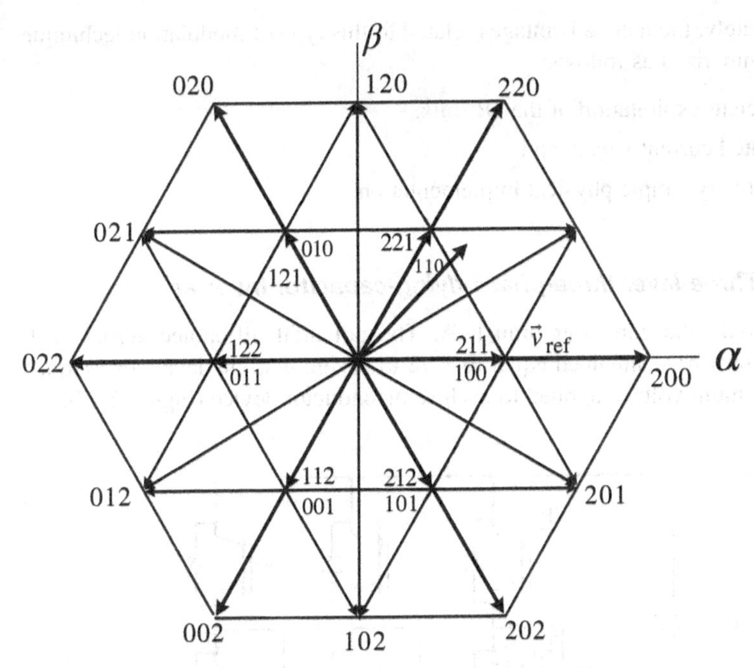

Figure 2.46 Space vectors of the phase voltages of a three-level inverter.

Establishing a reference $\vec{v}$ref, the vectors that may be used to create the appropriate space vector are those that form a triangle where $\vec{v}$ref is at the tip. The control sequence is structured in order to be able to achieve the minimum possible number of switching operations to transit between successive states.

Although the principle of operation is the same as that for a two-level inverter, the space vector selection algorithm must work with a greater number of possible states, with considerable complication. Also, since it is possible to have two different states producing the same space vector (which can be seen in Figure 2.46 in those that form the internal hexagon), it is necessary to implement additional criteria in order to choose which of the two to apply, possibly minimizing the number of switching operations.

Clearly, as the number of levels of the converter increases, the space vector selection algorithm will become more and more complex. However, the benefits of this control technique do not significantly increase with increasing voltage levels. In fact, relying on a reasonable compromise between the complexity of the algorithm and the quality of the output voltage waveforms, there is no significant advantage in using the *SVM* with converters having more than five levels.

Ultimately, the main advantages related to this type of modulation technique can be summarized as follows:

- efficient exploitation of the *DC link*,
- limited current *ripple*, and
- relatively simple physical implementation.

2.4.7.3 Three-level three-phase flying-capacitor inverter

Consider only the converter branch A. The potential difference across each capacitor must be maintained equal to $V_d/2$ to ensure that this level corresponds to the maximum voltage applied to each semiconductor device (Figure 2.47).

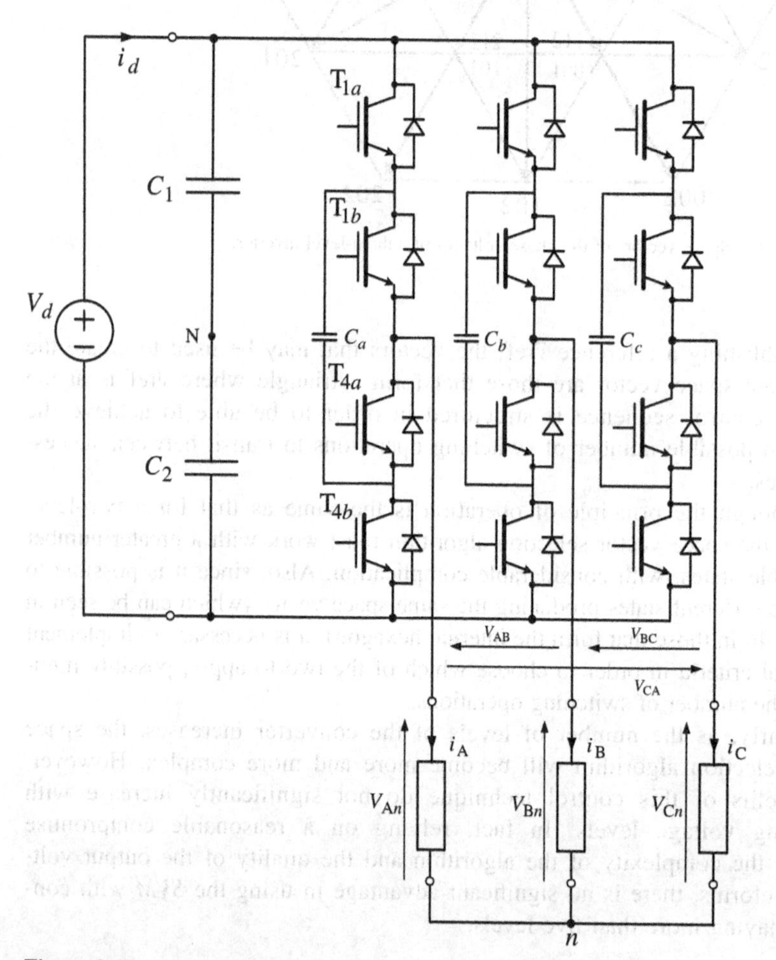

Figure 2.47 Three-level three-phase flying-capacitor inverter.

The output voltage between point A and point N will have three levels defined by the following states of the controlled switches:

$$
\begin{aligned}
&1100 \Rightarrow T_{1a}, T_{1b} \text{ on} \Rightarrow v_{AN} = \frac{V_d}{2} \\
&1010 \Rightarrow T_{1a}, T_{4a} \text{ on} \Rightarrow v_{AN} = 0 \\
&0101 \Rightarrow T_{1b}, T_{4b} \text{ on} \Rightarrow v_{AN} = 0 \\
&0011 \Rightarrow T_{4a}, T_{4b} \text{ on} \Rightarrow v_{AN} = -\frac{V_d}{2}
\end{aligned}
\tag{2.151}
$$

There are evidently two states for which the output voltage becomes zero. One of these can be defined as redundant.

The voltages across the capacitors C_a, C_b, and C_c may vary depending on the prevailing state of the converter switches and the sign of the currents i_a, i_b, and i_c.

Considering only the branch A, it can be observed that in the states 1100 and 0011, the current i_a, depending on its sign, passes only through the conducting pair of controlled switches or only through the associated recirculating diodes. In these situations, without passing current, the voltage across C_a remains unchanged.

In contrast, in the states 0101 and 1010 with zero output voltage, the current path i_a includes the capacitor C_a, thus changing its charge. When the switches T_{1b} and T_{4b} are turned on, i_a is positive, then C_a will discharge, while if i_a is negative, it will participate in charging the capacitor. The same occurs in a complementary manner when switches T_{1a} and T_{4a} conduct.

Obviously, the same behavior applies to the other two phases of the converter. This leads to the conclusion that, in order to maintain a stable balance of voltages across the capacitors C_a, C_b, and C_c, it is necessary to make appropriate use of the redundant states of the converter as a function of the signs of the phase currents.

As the number of voltage levels increases, the balancing of the capacitor voltages becomes increasingly complex due to mutual influences that arise in certain switching states. In a five-level inverter, for example, there are three groups of redundant states. Depending on the signs of the currents, the first group charges and discharges only one capacitor, while the second acts on two capacitors at each instant by alternately charging one and discharging the other. Finally, the third group simultaneously modifies the charge of three capacitors, in a manner such that two are charged while a third is discharged or vice versa, depending on the sign of the load current.

Because of these phenomena, *flying-capacitor* inverters with more than three levels cannot operate at zero power factor, that is, as static reactive power compensators.

Ultimately, compared with *diode-clamped* inverters, the floating capacitor category has the advantage of having redundant conditions for the output voltages, but the control and maintenance of the capacitor voltage level, in practice, proves to be a complex operation. Moreover, the high number of capacitors required to implement this type of structure is demanding both in terms of dimensions and costs compared with the use of clamping diodes in multilevel *diode-clamped* converters.

2.4.7.4 Cascaded multilevel inverter

A multilevel converter with a cascaded structure is formed by connecting in series the outputs of a number of single-phase full-bridge inverters (also known as the H-bridge configuration), each powered by an independent DC source. Figure 2.48 shows the diagram of three groups of p units to construct an inverter having m levels with $m = 2p + 1$.

From what has been analyzed previously, it is known that the single-phase full bridge can provide three values for the output voltage, that is, V_d, $-V_d$, and zero. By connecting in series all the outputs of the units of each branch of the converter, it is evident that the total output voltage will be

$$v_o = v_{o1} + v_{o2} + \cdots + v_{op} - 1 + v_{op} \tag{2.152}$$

Consider the case of a three-phase inverter with 11 levels achieved with this type of structure. For the implementation of such a converter, five full-bridge inverter units are required for each phase. The waveforms of the output voltages of the five bridges and the entire branch are shown in Figure 2.49.

The output voltage of each single cell can be expressed as

$$v_{ok} = \begin{cases} V_d \rightarrow \theta_k < \omega t < \pi - \theta_k \\ 0 \rightarrow 0 < \omega t < \theta_k, \quad \pi - \theta_k < \omega t < \pi + \vartheta\theta_k, \quad 2\pi - \theta_k < \omega t < 2\pi \\ -V_d \rightarrow \pi + \theta_k < \omega t < 2\pi + \theta_k \end{cases} \tag{2.153}$$

From which the total branch output voltage may be derived as

$$v_o = v_{o1} + v_{o2} + v_{o3} + v_{o4} + v_{o5} \tag{2.154}$$

which can be expressed as the sum of its harmonic components:

$$v_o(\omega t) = \sum_{n=1,5,7,\ldots} \frac{4V_d}{n\pi}[\cos(n_1) + \cos(n_2) + \cdots + \cos(n_5)]\sin(n\omega t) \tag{2.155}$$

Thanks to the presence of many voltage levels, it is possible to achieve a good approximation of the sine waveform in the output voltage v_o. Moreover, by choosing appropriate values for the angles θ_k, it is possible to eliminate the higher level harmonics that have a greater influence on voltage distortion.

The square wave operation of the individual cascaded modules implies a limited number of switching operations, since each semiconductor device is switched on and off only once during the fundamental period, thus limiting switching losses. On the other hand, this mode of operation of the individual

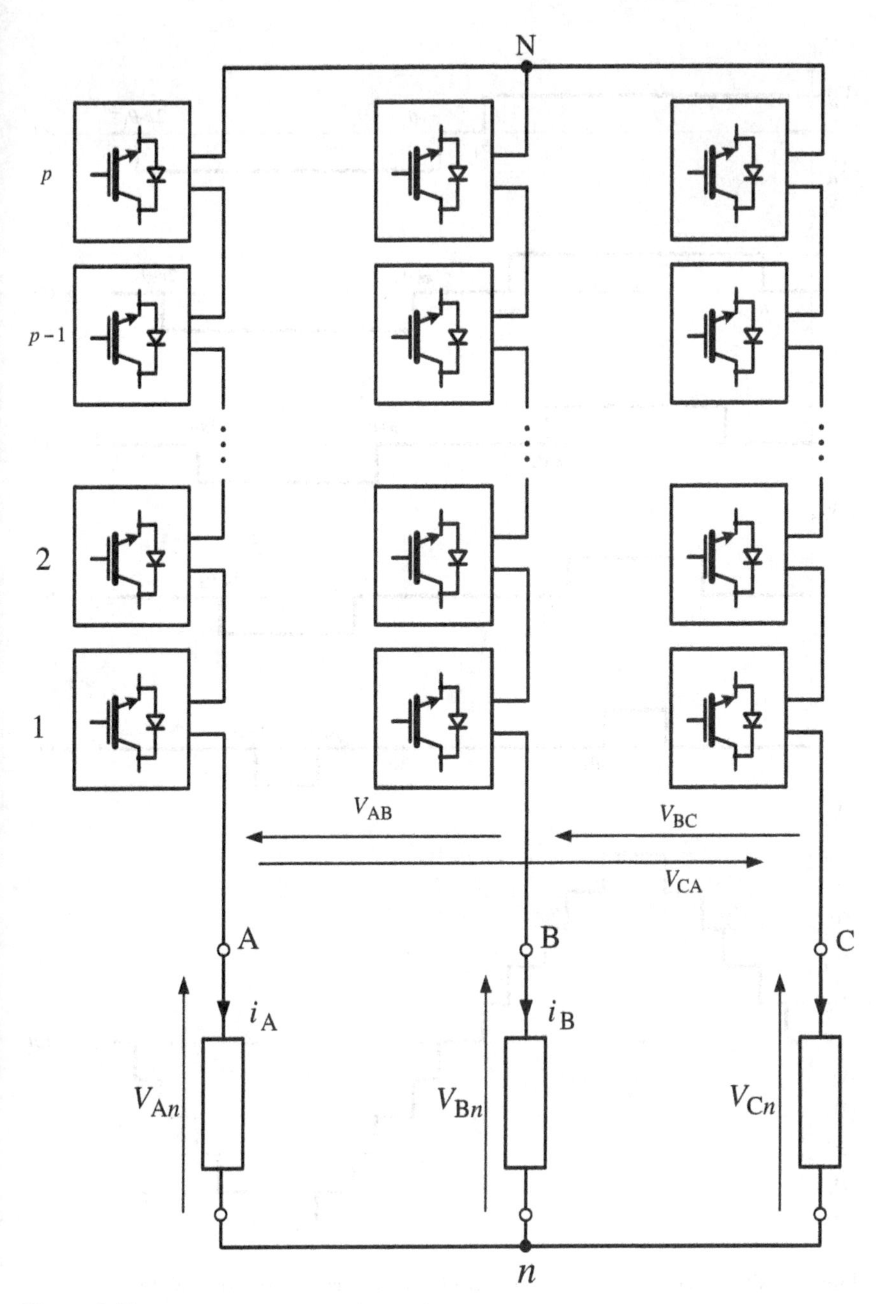

Figure 2.48 Three-phase inverter with cascade structure.

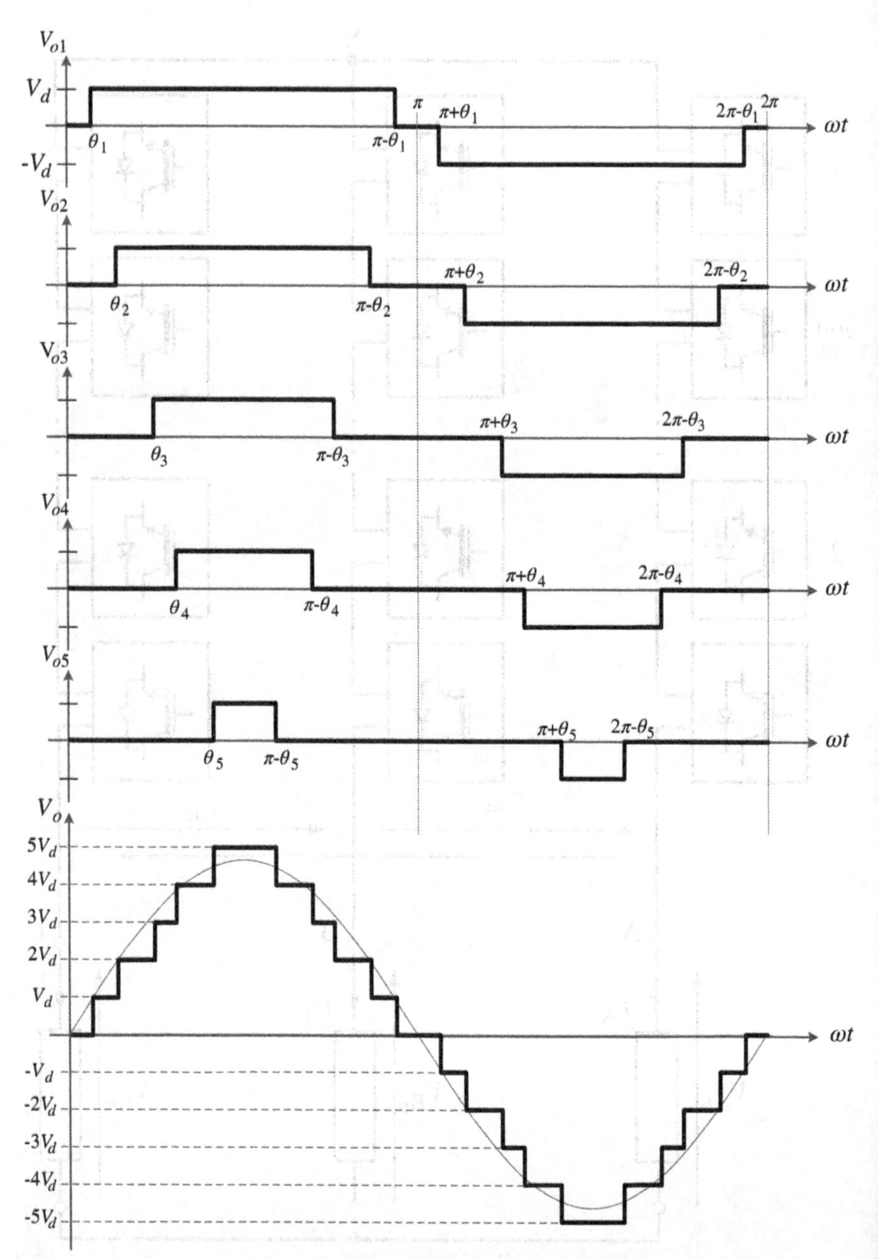

Figure 2.49 Output voltages of one inverter's branch with an 11-level cascaded structure.

bridges does not allow for precise output voltage amplitude adjustment. If there are p units, the voltage v_o will have a peak value of pV_d.

For this reason, square wave operation is used only in situations where amplitude adjustment of the output voltage is not required, or in which it is possible to change the voltages of the sources on the DC side.

If output voltage amplitude adjustment is required, it is possible to implement a PWM scheme with the same amplitude modulation ratio m_a, but for each bridge of the multilevel converter. Obviously, in the three-phase case, the voltages produced by the B and C branches will be out of phase, respectively, by 120° and 240° with respect to that generated by branch A.

Control techniques based on space vectors can also be used in the case of multilevel cascaded structures, although, as previously noted, they become complex to implement when the voltage levels become too numerous.

This type of converter is generally used in high-power and high-voltage applications, since it is possible to obtain high performance at relatively lower switching frequencies compared with those used in classic two-level inverters. A strong limitation to their use, however, is the need to have separate DC sources for each unit, which is not always easy to implement.

Chapter 3

DC Railway Electrification Systems

In urban, suburban, and regional electric transport networks that normally employ DC traction, the energy needed to power the electrical drive systems is taken from a three-phase distribution network at medium or high voltage and then rectified to DC by means of rectifier bridges at conversion traction power substations (TPSSs).

Rail and metro lines are powered by TPSSs located within an appropriate proximity in order to ensure the best power performance and safety for the transportation system. However, the choice of a final location must take the proximity to primary grid transmission or distribution stations into account in order to ensure that sufficient power is available for the traction plant. In the case of urban surface transport such as trams and trolleybuses, given the urban context and the absence of adequate power supply lines, it is difficult to locate substations adjacent to the transport lines themselves. Therefore, it is preferable to locate the substations at the actual existing primary grid stations with a provision for adequately protected connection lines.

The following are the main criteria for establishing a grid connection strategy:

- Size of the plant
- Location of the plant with respect to the network and the presence, within the interested area, of installation or production facilities and lines as well as primary and secondary stations.
- Network operational factors related to the connected plant.
- Possibility to extend stations, primary and secondary stations, and, more generally, the possibility for network expansion.
- The presence of protection and automation equipment on the network.
- Continuity and quality of service requirements.

Electrical Railway Transportation Systems, First Edition. Morris Brenna, Federica Foiadelli, and Dario Zaninelli.

In general, the choice of power system must meet criteria related to the electrical characteristics of the vehicles in circulation, operating frequencies, plant service availability, and reliability standards. System reliability is of paramount importance, not only in relation to powering the actual trains, but also for all power supplies serving the security, safety, and emergency systems, such as signaling, remote control, tunnel lighting, and ventilation, as well as the entire telecommunication infrastructure.

3.1 CONNECTION OF ELECTRICAL SUBSTATIONS

Electrical substation connectivity strategies and their relative distances can differ according to the nature of the transport service in question. Metro lines and trams in an urban context typically have supply voltages between 750 and 1500 V, although, for historical reasons, it is also possible to find different levels. The substations are positioned with several kilometers between each one, generally in the range of 2–10 km, depending on the vehicle types and traffic intensity. In the railways, which operate at higher voltages, usually between 1500 and 3000 V, the average distance between two TPSSs can vary between 15 and 20 km. For regional transport networks characterized by rather intense traffic, or in the case of higher power feeds such as those required for high-speed lines, these distances may actually be significantly reduced.

Figure 3.1 illustrates a possible generic substation connection scheme for DC traction lines, highlighting typical components that may be installed downstream of the network interface points:

- primary three-phase lines;
- conversion substations;
- traction circuit, whose positive conductor, in the case of rail transport, is implemented by an overhead contact line or a third rail.

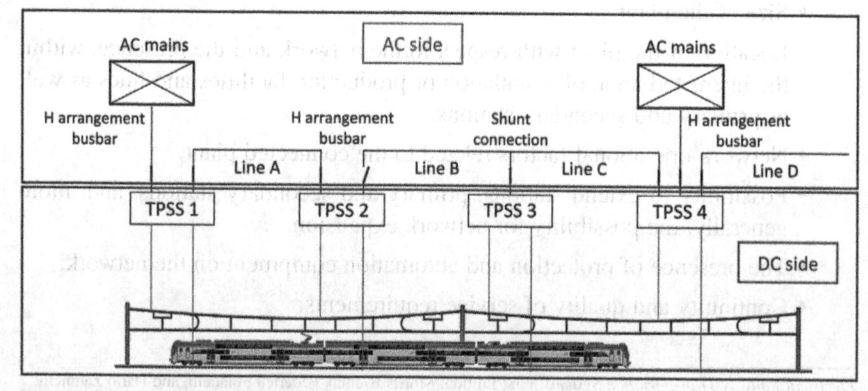

Figure 3.1 Architecture of a distribution system for the supply of DC traction networks.

The connection of the conversion substations to the three-phase industrial network can take place at both medium and high voltage with various possible interface solutions, depending on the service requirements and local conditions. The following are the main cases:

- Direct TPSS supply from a nearby grid station at high or medium voltage through a traction-dedicated primary line (primary traction line).
- Power feed through primary traction lines at high or medium voltage from nearby TPSSs.
- Mixed supply feeds via primary traction lines at high or medium voltage derived from both a nearby TPSS and a grid station.

There are essentially two types of possible connectivity solutions: branch connection or cascade.

In the case of a branch connection, the conversion substation is connected to a grid station via a primary line that is part of the railway network, or the branch connection may be implemented by tapping into a passing primary line. In both cases, an input disconnector is envisaged for isolating the TPSS from the grid for maintenance purposes.

On the other hand, in the cascade or "H arrangement busbar" configuration, the high-voltage primary line is connected to high-voltage busbars that serve as supplementary output terminals to feed another downstream TPSS. Both ends of the primary line terminating on the substation HV busbars are equipped with a circuit breaker in series with a disconnector.

The power derived from the grid is of fundamental importance, since it supplies a public utility service and therefore must ensure excellent quality of service.

The primary supply voltage levels may vary generally in the range of 15–30 kV for medium-voltage connections and 132–150 kV for connections to the high-voltage grid. However, there are countries where different voltage levels are implemented, for example, 45 and 66 kV.

Finally, to ensure service continuity, which is a fundamental characteristic of any electric rail system that serves to minimize the probability of inefficiency caused by failures and malfunctions, the principle of redundancy is often implemented in substation power supply systems. In practice and whenever possible, a dual power supply system solution is implemented, with two separate primary lines fed by grid stations that are independent of each other. Similarly, within the TPSS, dual or triple HV busbars may be installed or, along the most important railway trunk lines, a dual primary line system may be implemented consisting of two separate three-phase lines mounted on the same distribution pylons or on a separate distribution infrastructure.

The branch connection scheme envisages the connection of the conversion plant to one or two primary lines by means of circuit breakers and disconnectors. Furthermore, dedicated arrangements can provide for the alternative connection

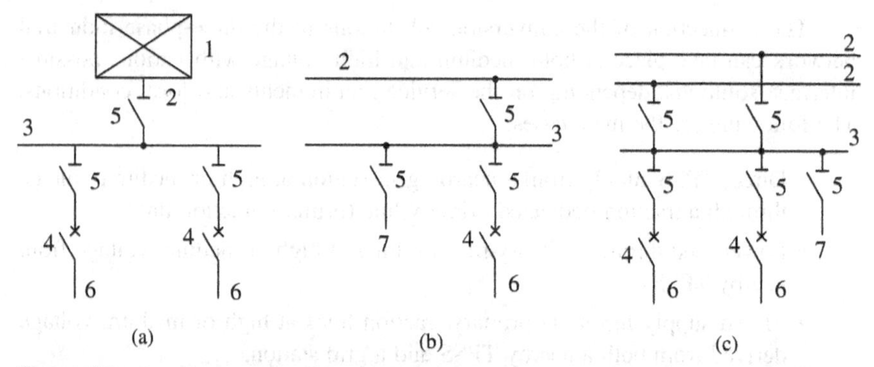

Figure 3.2 Branch connection schemes. (a) From the grid. (b) From a railway primary. (c) From a dual primary. 1: grid station; 2: HV primary three-phase lines; 3: substation HV three-phase busbar; 4: HV three-pole group circuit breakers; 5: three-pole disconnectors; 6: conversion group power supply; 7: mobile substation connection point.

of mobile substations during maintenance operations. The possible connection schemes are shown in Figure 3.2.

The H arrangement busbar configuration is more complex and envisages circuit breakers also on the primary line connections and a greater number of disconnectors compared to the branch solution. The system usually requires connection to a primary line and a dual busbar system, as shown in Figure 3.3.

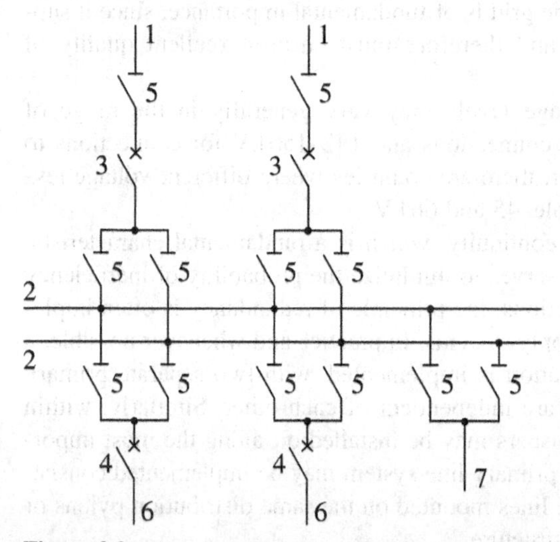

Figure 3.3 "In series" substation connection scheme with a dual HV busbar system. 1: HV primary three-phase lines; 2: HV three-phase busbars; 3: three-pole line circuit breakers; 4: three-pole group circuit breakers; 5: three-pole disconnectors; 6: conversion group power supplies; 7: mobile substation connection point.

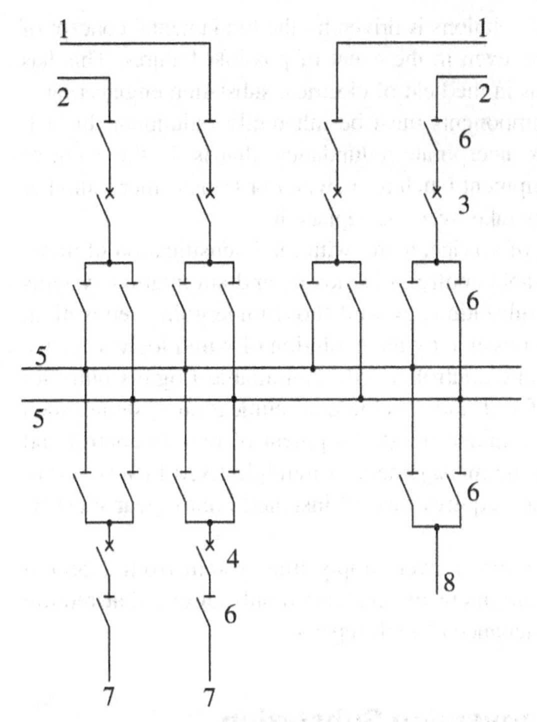

Figure 3.4 Branch connection of a conversion substation with HV dual busbar system on two independent primary lines. 1: Odd primary line; 2: even primary line; 3: three-pole line circuit breakers; 4: switches group pole; 5: HV busbars; 6: three-pole disconnectors; 7: conversion group power supply; 8: mobile substation connection point.

The more important power substation nodes may have multiple primary lines that supply two or more busbar systems, as represented in Figure 3.4. In this case, the complexity of the system requires the installation of a number of switching and protection devices such as disconnectors, circuit breakers, measurement transformers, and relays.

3.2 STRUCTURE OF TRACTION POWER SUBSTATION

Today's electrical substations for the power supply of metro and railway lines have been built over the years, in line with the construction of new lines or the extension of existing lines and in the context of the development of increasingly innovative technologies. For this reason, a wide variety of solutions have been implemented and they can have profound structural differences.

TPSSs are composed of various types of equipment, each performing a well-defined task related to functionality and protection to ensure a reliable power supply to the contact line.

The development of all new solutions is driven by the fundamental concept of continuity and quality of service, even in the event of possible failures. This has resulted in significant innovations in the field of electrical substation engineering.

First, all key substation components must be inherently redundant. In fact, modern substation architectures incorporate redundancy, that is, in the event of failure of a critical piece of equipment (such as a switch or transformer), another similar device will automatically take over and replace it.

The demand for continuity of service, along with the intensification of transport activity, mandates that reliable control, diagnostic, and automation systems be installed in all newly built substations as well those undergoing renovation. These systems have been made possible by the evolution of wired logic systems, modern computers, and electronic controllers (Programmable Logic Controller (PLC)). The implementation of fast and reliable communication systems, such as optical fiber networks, has sustained the development of remote control and command systems, centralizing the management of multiple fixed facility installations in a single location. This requires that all installed controlgear must be motorized.

The ability to control the entire power supply line system from a central location has sustained the development of unattended substations that require only personnel for routine maintenance or fault repairs.

3.2.1 Diagram of a Conversion Substation

Conversion TPSSs are composed of different parts, which, albeit with differing voltage and power levels, are common in all transport applications (railway, tram, and metro systems).

A typical TPSS power supply architecture is shown in Figure 3.5.

The substation is divided into two macrosections:

1. The AC section, which houses the controlgear, protection, and measurement equipment for connection to the grid, and the controlgear, protection, and measurement equipment for feeding the group converters and transformers.
2. The DC section, which includes the AC/DC converters, line filters, and the controlgear, protection, and measurement equipment for the traction lines.

3.2.1.1 Input line

The TPSS input power is carried by HV or MV cable or overhead lines. The input line is associated with incoming protection gear, line, and earth breakers and disconnectors as well as remote control equipment.

The input lines, both HV and MV, may be either overhead lines or underground cable. Substations with overhead input lines will have visible

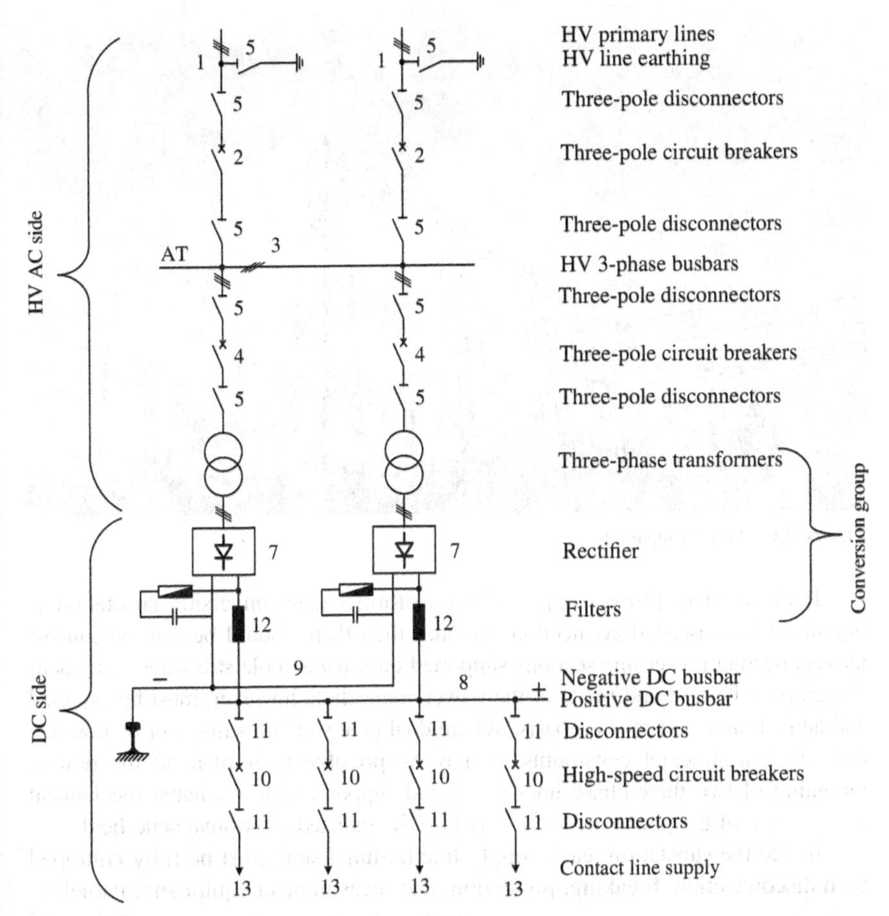

Figure 3.5 Functional diagram of a railway TPSS.

tower pylons at the entrance with spark gap anchor points for the conductors (Figure 3.6).

The sequence of elements at the input line is generally identical in all substations and consists of the following:

- VT: voltage transformer
- S: switch (disconnector) with earthing blade
- CT: current transformer
- I: circuit breaker
- S: pantograph or rotating disconnector
- SA: surge arrester (for the transformer)

The chain is repeated on each of the three phases.

Figure 3.6 Primary input line.

If the primary power supply of a transformer and conversion substation is organized for cascaded connection (in–out), then there should be a provision for interconnecting power line sections supported on separate pole structures. The identification of the type and mode of the power connection, however, must be assessed and agreed on a case-by-case basis. Where dual pole structures may not be possible due to environmental constraints, it may be possible to implement the double mounting of two three-phase lines on shared supports with a suitable mechanical arrangement of the phases in order to reduce the radiated electromagnetic field.

Inside the substation, each supply line terminal stall must be fully equipped with disconnection, breaking, protection, and measurement equipment, namely:

- input line spark gap anchor points;
- three voltage transformers installed upstream of the line disconnector, with appropriate turns ratio and performance parameters matched to the protection system;
- three voltage transformers installed upstream of the line disconnector, (i.e., dual secondary VT), with appropriate turns ratio and performance parameters also for the measurement of power and energy;
- a three-pole line disconnector with interlocked earthing blades;
- a three-pole line circuit breaker with adequate breaking characteristics determined by the worst-case short circuit conditions, complete with current transformers with transformation ratio and performance characteristics matched to the protection system;
- a system of three current transformers, distinct from the previous ones (i.e., dual secondary CTs), with transformation ratio and performance characteristics suitable for the measurement of power and energy;

Figure 3.7 Substation busbar.

- a three-pole busbar disconnector;
- a set of remote-controlled selective distance protectors.

If a substation is equipped with multiple input lines, then pantograph disconnectors are installed downstream of the surge arrester (Figure 3.7). Each input line is equipped with its pantograph disconnectors, one for each row of busbars. This busbar system constitutes the central node of the substation, acting as a hub for the connection of the group transformers. It is normal in all substations to install two rows of busbars in order to ensure adequate system redundancy. Normally they do not operate in parallel to avoid a fault on one row propagating to the other and to contain short-circuit currents. The busbars can be interconnected to each other via a dedicated disconnector called a bus sectionalizer that allows, for example, live voltage to be applied to the second set of three bars with the input line closed only on the first one (Figure 3.8).

Figure 3.8 Triple set of (a) closed and (b) open pantograph disconnectors.

Figure 3.9 A detail of the gravel under the high-voltage busbars.

Figure 3.10 132 kV voltage transformer.

Gravel is laid beneath high-voltage equipment (Figure 3.9). Its function is important because it assists water drainage and prevents its accumulation over time. All high-voltage equipment also contains oil, both for its operation and for the lubrication of movable parts. This oil is particularly flammable; hence, gravel is useful for the drainage of any spills and extinguishing fire in the event of ignition.

The first measurement transformers are installed immediately after each primary line anchor point. The first types are single-phase voltage transformers (Figure 3.10).

CTs are employed for measuring current and one is installed on each phase. They are normally of conventional construction with toroidal cores.

VTs and CTs, besides providing voltage and current measurements to the input lines, are necessary for giving various indications and status recordings as also required for metering purposes.

For each incoming phase, a rotary-type disconnector is installed to provide air insulation to the high-voltage section, enabling completely safe access (Figure 3.11). Unlike breakers, they cannot be operated under loaded conditions because they are not equipped with arc suppressors; hence, their activation must be interlocked with breaker switches so that they operate only with an open circuit. The central column has 90° rotation and drags the switching blade, while the fixed contacts are mounted on the lateral columns.

Figure 3.11 Rotary disconnector switch in open and closed positions.

The opening procedure for these devices is rather complex. Once the power contacts are open, the switch is grounded through a special blade, thus making it safe (Figure 3.12). Three-phase disconnectors are mounted on sturdy steel tripods and they can be controlled either manually or by means of an electromechanical servo.

The circuit breaker and the current transformer are located in a single multi-functional unit (Figure 3.13).

The circuit breaker is the only element able to open the circuit under loads; for this reason, it is equipped with an electric arc extinguishing system.

Finally, the earthed arrester ensures protection against possible surges on the primary lines that would damage the various components of the substation equipment. Proximity to the protected equipment is critical; if the arresters are positioned at too great a distance, they will not be effective in ensuring adequate protection.

The active part of the arrester is composed of a suitable number of variable resistance metal oxide resistors. Under normal operating conditions, the arrester presents high impedance to the ground. If there is a voltage surge, air ionization will ignite an electrical arc across the electrodes (starting voltage). The electric arc is an almost-pure short circuit between phase and earth, and causes a strong current through the arrester that trips the overcurrent protection.

Figure 3.12 Rotary disconnector switch with earthing blade.

Figure 3.13 Unit consisting of current transformer, circuit breaker, and rotary disconnector.

3.2.1.2 Transformer group

Power supply to the conversion group takes place via the switching and protection equipment together with the measurement devices necessary for their operation.

The transformer group is that section of the TPSS that has the task of transforming the HV or MV input line voltage to levels that can be used by the various loads.

Transformers are categorized into three types according to their function:

1. *Traction transformers*

   They typically have two secondary windings (one connected in delta and the other star) for feeding groups of 12-pulse rectifiers, which, in turn, feed the rail traction loads (Figure 3.14).

   In the substations, resin construction traction transformers are installed indoors, while they are mounted outdoors if oil-filled. The neutral terminal may be earthed depending on the management policy of the grid to which they are connected.

   The specific nature of traction loads, characterized by frequent peaks of short duration, requires implementation design criteria that differ with respect to industrial systems.

Figure 3.14 Power transformer with dual secondary.

In particular, they must be able to withstand rail power overload conditions without suffering any damage, according to the following rating cycles:

- 200% of the rated power supply for an interval of 2 h
- Delivery of 233% of the rated power for an interval of 5 min
- Combined overload cycles interspersed with intervals of less than 6 h

Furthermore, the transformer must be capable of tolerating the absorption of harmonic currents, particularly from the ferromagnetic core.

In higher power substations connected to the high-voltage grid, it is useful to compensate for line voltage variations by means of an on-load tap changer (OLTC).

The OLTC is a rotary mechanical device in an oil bath fitted inside the transformer enclosure and provides the connections for the terminals of a number of primary windings. An important aspect to be emphasized is that the choice of switching the primary coil is due to the lower current with respect to the MV secondary; the switchover between winding occurs without interrupting the primary current; this avoids the formation of electric arcs and ensures continuity of service. The location of the changeover switch operating under loaded conditions is governed by an electronic system that coordinates several transformers equipped with switches in order to spread the energy load across the primary lines.

In substations connected to the MV network, a vacuum variator may be installed and has the task, during the installation stage, of adapting the output voltage as a function of the actual input level.

2. *Auxiliary service transformers*

These are normally MV–LV transformers with a single secondary with the task of powering the services of the substation itself. The auxiliary services transformers are normally installed inside the substation building, near the rectifier units. This choice is driven by factors of logistic convenience. In fact, these transformers are normally fed from the star secondary of the traction transformer and provide a secondary voltage of 400 V. For safety reasons, the transformers are protected by appropriately sized metal barriers that ensure adequate distances and safety margins.

These transformers feed fundamental loads such as the substation control system, the 230–400 V facility mains, and the lighting system.

3. *Secondary line transformers*

Dedicated transformers may supply secondary railway infrastructure power lines. If present, they are generally HV–MV transformers with a single secondary (Figure 3.15).

The outputs from the medium-voltage transformers are connected to the substation MV control room. The relevant control panels house the measurement systems and lines' switch controls.

Two types of transformers may be installed in the TPSS: oil-filled and dry.

Oil-filled transformers are housed in metal casings that contain the mineral oil that performs two functions:

- Insulation between the windings and with respect to the chassis.
- Dissipation of the heat produced in the windings by the Joule effect and generated in the iron core by hysteresis and eddy current effects.

Figure 3.15 HV–MV secondary line transformer.

They are often equipped with an oil reservoir, a raised cylindrical container with a capacity of about 1/10th of that of the chassis, to allow expansion of the liquid with increasing temperature. A silica-gel crystal filter allows the reservoir to ventilate, preventing any buildup of moisture that could compromise the oil's dielectric strength; this filter is rather delicate, however, and requires regular replacement. An alternative solution is to seal the chassis while allowing for a cushion of air or nitrogen in the upper section, which enables the expansion of the liquid without undue mechanical stress on the chassis itself.

Dry transformers can be air-filled or with windings cast in resin. They require virtually no maintenance and they are the preferred solution in environments where there is a high risk of fire.

In this regard, the installation of oil transformers is more critical because measures must be taken to prevent the leakage of oil and propagation of a possible fire. Transformers with high amounts of oil must have a proper collection sump, while the higher power transformers require appropriate fire-resistant insulating walls. Oil-filled transformers are also more critical in their maintenance requirements and end-of-life disposal. On the other hand, until a few years ago, oil-filled transformers were achieving higher powers, although they were bulkier and required higher installation and maintenance costs.

Cast-resin transformers are becoming more widespread due to their reliability and reduced environmental impact. They are manufactured for voltage levels up to 36 kV and for power ratings that may reach around 30 MVA. Frequently, the windings of cast-resin transformers are made with aluminum coil tape with a thermal expansion coefficient similar to that of the resin in which they are encapsulated. Low-voltage windings are made of aluminum sheets that have the same height as the coil.

Whereas oil-filled transformers reach their operating temperature within a few minutes due to the fast circulation of the oil, the dry types take a much longer time, approximately an hour and a half, so they can withstand greater overloads. With appropriate axial fans, manufacturers also allow for overloads equal to one and a half times the rated power, also on a permanent basis.

The current density of the windings in oil is greater than that permissible in dry transformers; furthermore, the distances between the windings and the core are lower in oil-filled transformers, given the high dielectric strength of the mineral oil. The core cross section is also lower for oil-filled transformers. The consequences of these structural features have an impact on losses in the ferromagnetic core and in the conductors; under nominal operating conditions, oil-filled transformers have lower no-load losses, but more significant losses when supplying higher loads.

Transformers are classified according to their cooling solutions. For oil-filled transformers, there are two types of refrigerant: the oil contained within the inner housing, in contact with the windings, and the fluid external to the housing, which is normally demineralized water. The circulation of the cooling fluids can be natural or forced. For cast-resin transformers, the only refrigerant is the external fluid and, in most cases, it is air (ONAN).

3.2.1.3 Conversion group

Over the decades, substation conversion groups have seen considerable progress in terms of performance, efficiency, maintenance requirements, and reliability.

In the earliest electrification systems, rotating converters were used; for example, the 3 kV railways built in the United States in the second decade of the twentieth century were powered by groups consisting of five machines: a synchronous three-phase motor, activated as in asynchronous mode, two dynamos connected in series, each with a terminal voltage of 1500 V, and two exciters.

Similar rotating units were employed in Italy for the electrification of the 3 kV Benevento–Foggia line, but on that occasion experimental trials were also carried out at the Apice substation, with a static mercury vapor rectifier group (Figure 3.16).

The excellent results achieved favored the adoption of this type of rectifier in all subsequent substations, which were greatly simplified. It was thus possible to implement, where necessary, unmanned substations that were remotely supervised and controlled from the main substations.

A further milestone was the introduction, starting from the 1960s, of silicon rectifiers, not only in all new substation sites, but gradually as total replacements for the mercury devices. This type of rectifier played a significant role in the development of electric traction, both for DC and grid frequency AC systems, thanks to the possibility of installing rectifiers directly on board the vehicles.

As seen in Chapter 2, conversion systems without switching control are based solely on diodes, for which the conduction state depends on the value of the anode–cathode voltage. Obviously, such systems only allow for the conversion from alternating voltage to direct voltage (AC/DC conversion).

The appearance of controlled semiconductors has greatly expanded the scope of electronic conversion, offering multiple possibilities for the high-performance regulation of voltage and current.

For the greater part, substations are characterized by conversion systems based on diode bridges. The reason for this choice lies in the simplicity and reliability of the system, as external controls are unnecessary and the cooling requirements are satisfied by natural air convection, given the very low losses involved. Conversely, there is no possibility of regulating the system voltages in order to control power flow and the quality of power supplied to the load.

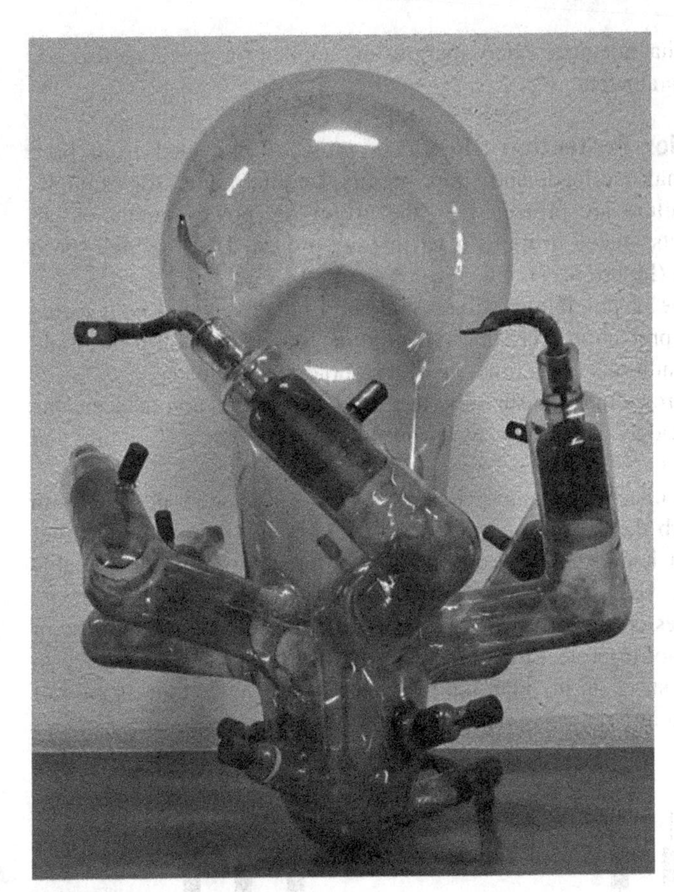

Figure 3.16 Mercury arc rectifier.

3.2.1.3.1 Use of Diodes in Substations The use of silicon diodes allows the implementation of high-power converters characterized by excellent performance, high reliability, and reduced dimensions. This type of device also allows very short start-up times and operation with ambient temperatures between −50 and 50 °C. Even in the case of high-power 3 kV substations, the converters may be cooled by natural ventilation, reducing energy consumption and increasing overall substation reliability. The combination of these factors also makes the implementation of remote-controlled substations possible, which can therefore remain unmanned.

With regard to their application limits in conversion substations, we must consider that diodes have poor overload ratings and a high sensitivity to overvoltages. In traction substations, rectifiers must be sized according to the expected maximum peak currents and they must be adequately protected against short circuits. Moreover, overvoltage surges that may arise during normal operation require that adequate safety margins are applied for the diode reverse-working voltage parameters.

3.2.1.3.2 Rectifier Protection Rectifiers must be protected from both internal and external overloads and short circuits. Internal faults, for example, due to a diode failure are protected by the group AC power supply circuit breaker, while faults related to the contact line are protected by the high-speed line circuit breaker (Figure 3.17).

The occurrence of possible overloads and short circuits should be detected by an overcurrent protection relay that relies on an instantaneous trip threshold for short-circuit conditions and a time-based threshold for overloads.

Overvoltage protection is ensured by means of two attenuation devices, one for each 6-pulse bridge. These devices are installed in both the AC section and the DC rectifier unit.

The AC side of each attenuation device consists of capacitors in series with resistors that absorb the voltage spikes caused by switching within the rectifier itself. On the other hand, the DC side is fitted with a surge protector (variable resistance arrester), which protects the rectifier from external overvoltages.

There is a series-connected current limit fuse for each diode. The purpose of the power fuse is not to protect the diodes against overloads caused by the passage of the rolling stock on the line, but rather to protect the rectifier in case of internal faults, such as a defective diode in the rectifier bridge.

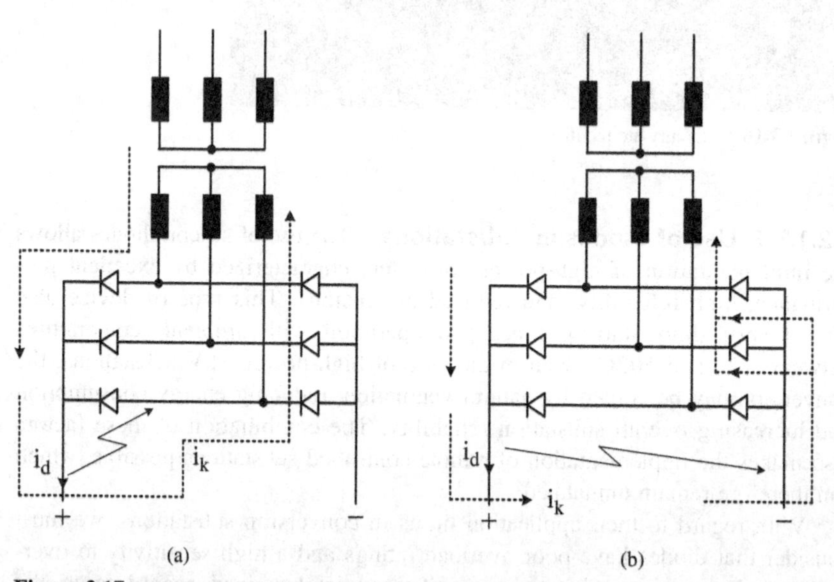

Figure 3.17 Internal and external faults in an AC/DC converter. (a) Internal short circuit due to a diode failure. (b) External short circuit due to a line fault.

The peak inverse voltage indicates the maximum voltage that a diode can withstand during its switching action. As an example, consider a diode used in a substation rectifier where the peak inverse voltage is equal to 2400 V.

Given that during normal operation the diodes are subjected to repetitive overvoltage spikes due to switching between the branches, the semiconductor modules must be sized so that the peak inverse voltage parameter is related to the nominal working voltage with a safety factor of 1.5–2.5. A safety factor of less than 1.5 is used where the overvoltage levels are generally well known, while safety factors up to 2.5 are generally preferred in the construction of plants where the diodes are connected to high-power networks and where the maximum overvoltage levels are mostly unknown.

In practice, one device alone per branch is never used, but a number of diodes in series are connected so that each of them is subject to a lower reverse voltage. It is also necessary to provide for possible dissymmetry in the values of the reverse voltages, on the order of 10–15%, due to structural differences between the individual diodes or different working temperatures. It is therefore normal practice to fit equal high-value resistors in parallel with each series diode in order to ensure that the reverse voltage is partitioned as evenly as possible.

Finally, it is essential to take into account that possible failures can cause an individual diode to short circuit and lose its unidirectional properties; dedicated alarms are arranged to signal such events. Under this condition, the total reverse voltage is no longer partitioned over n diodes but over $n - 1$ diodes; thus, to avoid the entire group's failure, the healthy devices must be able to withstand the maximum reverse voltage in order to ensure continued operation of the rectifier bridge.

For example, in new-generation substations supplying 3 kV lines, there are four series diodes for each branch (for a total of 24 diodes for each 6-pulse bridge). Diversely, in older generation substations, there are 5 diodes per branch (for a total of 30 diodes in a 6-pulse bridge), each with a lower rated reverse voltage but with greater overall conduction losses. It is clear that the choice should always be to fit diodes with the highest possible maximum peak repetitive reverse voltage rating (VRRM) so as to limit the number of semiconductors in series and the associated losses. By reducing losses, in addition to increasing the rectifier's efficiency, it is possible to reduce the weight and dimensions of the associated heat sinks.

An overheating protection system is installed in each rectifier; the cabinet temperature sensors are positioned where the greatest accumulation of heat is expected. If a first temperature limit is exceeded, the system generates an alarm, while a second alarm will cause the rectifier group output to disconnect.

3.2.1.3.3 Sizing of the Converter Group and Overload Capacity Depending on the location of a substation and the power that it must supply to the line, one or two converter groups may be installed, which, in the case of the 12-pulse scheme, can have the rectifier bridges connected in series or in parallel.

While it has been seen that, for high voltages, a number of diodes are normally connected in series, it is not advisable for diodes within the same group to be connected in parallel since even small structural differences or different operational temperatures could cause a very uneven distribution of currents between the different semiconductors. However, given the high current performance of modern diodes, this need does not arise. If the output current must be increased, as is the case for 750 V metro lines, it is sufficient to connect two or more converters in parallel. The uniformity of the current distribution between converters is ensured by the droop related to the series impedance of the transformers in each group.

For the case where the bridges are connected in series, the following advantages apply:

- Less sensitivity to unbalances between the secondary windings of the converter transformer, which in 12-pulse systems can arise due to different transformation ratios in the star and delta primary–secondary couplings.
- No need to use interphase coils, conversely required for parallel connections.
- Reduced short-circuit current in the case of decoupled transformers.

On the other hand, the series topology does involve one notable disadvantage: It is not possible to increase the output current and, therefore, substation overload capacity is compromised.

With regard to the choice of the number of series diodes, the formula may be used:

$$n_s = \frac{K_V V_{MAX}}{K_S V_{RRM}} \tag{3.1}$$

where

- V_{MAX} is the maximum permissible continuous voltage on the DC side (3600 V in 3 kV systems) system;
- K_V is a factor greater than 1, which takes the operating overvoltages into account;
- K_S is a voltage distribution coefficient;
- V_{RRM} is the diode maximum peak repetitive reverse voltage (VRRM) and is supplied by the device manufacturer.

Regarding a substation's required overload capacity, regulations prescribe the admissible percentages with respect to the rated continuous current as a function of the overload duration.

For example, for a 3 kV system with 3600 V output voltage and current equal to 1000 A, the rated rectifier power is then 3600 kW with all diodes in operation, and 40 °C is the reference operational ambient temperature. This is the power that the equipment can provide under continuous operation without overloads. Typical overload cycles are as follows:

- For 3.6 MW groups (3600 V and 1000 A):
  - 100% (2000 A) for 2 h
  - 200% (3000 A) for 5 min
- 5.4 MW groups (3600 V and 1500 A):
  - 100% (3000 A) for 2 h
  - 133% (3500 A) for 5 min

Given the low thermal inertia of the semiconductor modules, the rectifiers must be sized for the maximum load, while the transformer must allow for the overloads to occur in succession, starting from steady-state conditions at rated load and then three times in 24 h, provided that consecutive cycles are appropriately spaced by at least 6 h and that the transformer generates nominal power between consecutive cycles.

In case of failure of a single diode, considering the worst-case scenario of one diode breakdown for each branch of the bridge, the rectifier is able to supply a maximum power of 3400 kW (always considering a working temperature of 40 °C). Under these conditions, the rectifier does not need to be shut down and it can continue to function, albeit with lower than rated power transfer.

Only in the event of a fault in a second diode in the same branch must the rectifier be taken out of service. However, in case of simultaneous failure of one diode in each branch, the protection parameters of the power system must be automatically readjusted to lower values in order to avoid thermal damage to the power unit.

3.2.1.3.4 Thyristor-Based Controlled and Reversible Substations In order to adjust a substation's rectifier output voltage or even to reverse its power flow, a possible solution is to replace the diodes with thyristors without altering the basic circuit.

The active power that the rectifier provides to the DC network is a function of the control angle α according to (3.2):

$$P_d = V_d \cdot I_d = V_{d0} \cdot \cos\alpha \cdot I_d \tag{3.2}$$

In any operational condition, the current I_d always circulates in the same direction since, following diode behavior, the devices do not allow reverse current flow.

For $\alpha \leq \pi/2$, the output voltage remains positive. In such conditions, the thyristor bridge is used to adjust the output voltage as a function of load demands and thus provides a better quality of power supply with respect to the diode bridge.

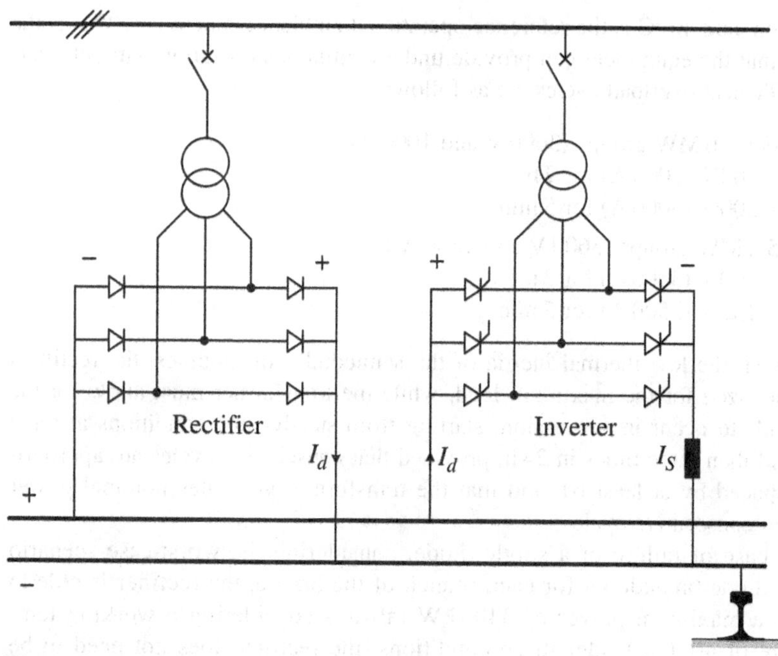

Figure 3.18 Diagram of a reversible substation.

Conversely, as already seen, the mean value of the rectified voltage becomes negative for $\pi/2 \leq \alpha \leq \pi$ and thus the active power also changes direction and flows from the DC side to the AC side. In other words, the rectifier becomes reversible and for control angles greater than 90°, it behaves like an inverter. To achieve this function, the output terminals to the contact line must be inverted. In a reversible substation, this allows for regenerative braking power to be restored to the grid when there are no other traction loads on the contact line capable of absorbing it.

In reversible substations, the rectifier and inverter groups are implemented separately to avoid having to frequently reverse the DC busbars according to the output voltage sign change. Furthermore, since the traction load power demand is typically substantially greater than any recovered regenerative braking power, a reversible substation is equipped with only one inverter group as opposed to the two or more rectifier groups present.

Figure 3.18 shows the functional diagram of a reversible substation that uses thyristor-based converters. Since a controlled bridge has greater rectified voltage harmonic content than that of a diode bridge, it is necessary to include a series smoothing inductance L_S between the bridge and the contact line.

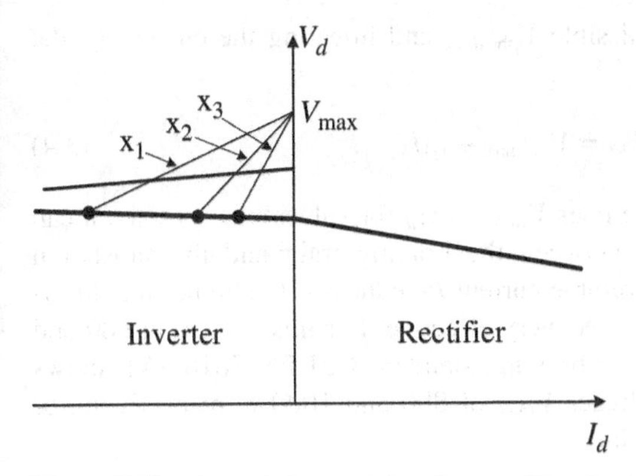

Figure 3.19 External characteristics of a reversible substation.

3.2.1.3.5 Recovering Regenerative Braking Energy Consider a substation that includes diode rectifier bridges for AC–DC conversion and one thyristor-based bridge for DC–AC conversion. The voltage–current characteristic (*V–I*) is represented in Figure 3.19.

During operation, if the thyristor bridge should generate a voltage lower than that of the diode bridge, $V_{d\alpha} < V_{d0}$, this would result in a circulating current between the two bridges that would cause a loss of power without any practical benefit. For this reason, the thyristor bridge should always generate a voltage higher than that of the unloaded diode bridge. Under these conditions, the diode bridges are reverse-biased and the thyristor bridge can function only in the presence of a train subject to electrical braking.

In any case, the generated voltage cannot exceed the stability limit value corresponding to the angle α_{lim} given by π minus a reasonable safety margin needed to ensure complete switching between two thyristors.

The amount of recoverable energy, which depends on the current I_d injected by the train, is a function of the distance of the train from the substation according to the following expression:

$$I_d = \frac{V_{ph} - V_{d\alpha}}{R} \tag{3.3}$$

where V_{ph} represents the voltage generated at the train's pantograph during the braking phase and R is the electrical resistance of the line section between the train and the substation.

As will be seen, knowing that $R = R(x) = rx$, where r is the resistance of the line circuit per unit length and x is the distance of the train from the TPSS, in the case of only one train undergoing regenerative braking that is generating the maximum permissible $V_{\text{ph max}}$ and injecting the current I_d, the substation voltage will be

$$V_{da} = V_{\text{ph max}} - rxI_d \tag{3.4}$$

Figure 3.19 shows the slope lines $V_{\max} - rxI_d$ for values $x_3 > x_2 > x_1$. It can be noted that as the distance between the braking train and the substation increases, the value of the recoverable current I_d reduces – the higher the difference $V_{\text{ph max}} - V_{da}$, the more the recovery effective. For this reason, in 600 and 750 V networks with regenerative braking, standard CEI EN 50163/A1 allows for peak instantaneous line voltages $V_{\max}$ of 800 and 1000 V, respectively, or 33% higher than the nominal value.

Displacement Power Factor As already mentioned in Section 2.3, the introduction of phase control α generates a phase shift between the first current harmonic I_1 and the supply voltage, which causes an absorption of reactive power and a reduction of the power factor. Taking into account also that the switching is noninstantaneous, the phase shift increases by a further angle u. By approximating the current waveform as trapezoidal, the DPF (displacement power factor) turns out to be (Equation 2.84)

$$\text{DPF} = \frac{\cos\alpha + \cos(\alpha + u)}{2}$$

From this the PF (power factor) is derived as a function of the rms current I:

$$\text{PF} = \text{DPF} \cdot \frac{I_1}{I}$$

Note that the greatest influence on the power factor is borne by the control angle α. A wide regulation range can cause considerable phase shift and, therefore, a significant drop in DPF (e.g., assuming $u = 10°$ and $\alpha = 45°$, DPF = 0.64). This requires the installation of an efficient AC side power factor correction system that, for the high power levels typical of railway substations, can be very onerous.

Harmonics and Disturbances As seen, the filtering system consists of a single DC side inductance in practice, and in reality, with noninstantaneous switching and possibly with phase control, this implies that the amplitude of the line current harmonics is less than ideal, since the shape of the current waveform is more rounded. Since the analytical calculation is very complex and time consuming, it is preferable to use the graphs and tables published in the norms that indicate the multiplication factors for each harmonic. Table 3.1 shows the amplitudes of the current harmonics in the ideal and real cases.

Table 3.1 Harmonic Levels in the Ideal Case and in the Case of Instantaneous Real-Switching

H	5	7	11	13
Ideal	0.20	0.14	0.09	0.07
Real	0.17	0.10	0.04	0.03

Table 3.2 Ideal Amplitude of the Rectified Voltage First Harmonic

q	2	4	6	8	10	12	14
6	0.5	0.25	0.167	0.125	0.1	0.083	0.071
12			0.167			0.083	

The same analysis applies to 12-pulse converters where, however, only the harmonics of order $h = 12n \pm 1$ appear, namely, the orders 1, 11, 13, 23, 25, . . ., with the advantage that the lower order harmonics (fifth and seventh), which are those with the greater amplitude, ideally should not appear. In reality, minor constructional dissymmetries can generate them, but with negligible amplitudes.

The harmonics present in the rectified voltage are of even order and are indicated in Table 3.2 for ideal converters.

The actual voltage harmonics deviate from the ideal values. Unlike the case for current harmonics, the contribution of noninstantaneous switching worsens the harmonic content of the rectified voltage, which is further degraded again by phase control. As in the previous case, costly and time-consuming analytical calculation is not feasible for the derivation of harmonics levels in the presence of noninstantaneous switching and phase control; it is preferable to reference the amplitude of the α parametric curves as a function of the switching angle u.

3.2.1.3.6 Comparison Between Diode and Thyristor Rectifiers The first railway applications of voltage-controlled thyristor converters date back to the late twentieth century. By way of example, the first applications include the following:

- The power system of the light rail "Nockeby-line" in Stockholm that was upgraded with thyristor rectifiers in the first months of 1984.
- The first metro line in Busan, Republic of Korea; it entered service in July 1985 and was completed in June 1994, the entire line is powered by thyristor converters.
- The second Busan metro line has nine substations equipped with thyristor rectifiers. The first section with six substations has been in operation since

mid-1999, while the second was commissioned in early 2002. An extension with a new substation and the upgrading of another were completed in August 2007.

- The power system of the "Roslagsbanan" commuter line in Stockholm, which was upgraded with a thyristor-based plant in the first months of 1995.
- The first Dallas light rail line, whose power system was constructed with 16 substations based on thyristor converters. The line was opened in the summer of 1996.

The thyristor bridge possesses a number of advantages compared to traditional systems based on diode rectifiers, but it also introduces new problems.

Compared to diode bridges, the use of thyristors in converter substations implies lower line losses, thanks to the possibility of stabilizing the output voltage under varying current demand. Consequently, for a given delivered power, there is a reduction in overall current consumption and lower Joule effect losses. Furthermore, when upgrading from diode to thyristor technology, the improved voltage drop compensation allows for more intense network traffic, or, for a given traffic level, greater distances between the substation locations.

On the other hand, as discussed previously, railway regulations provide for a very wide variation in the DC line voltage, from −33 to 20% of the rated rectified voltage. In order to regulate voltage excursions exceeding a 50% range, a wide control angle variation range is required. The control angle is also responsible for the phase shift of the absorbed current fundamental harmonic with respect to the phase voltage of the bridge power supply. Consequently, there is a significant absorption of reactive power that requires the installation of large power factor correction equipment. Phase control also has an impact on the harmonic content, both in the AC side absorbed current and in the rectified voltage. In particular, increasing angle α greatly increases the amplitude of the rectified voltage harmonics, which results in a more "jagged" superimposed ripple. This requires the installation of higher performance DC side filters in order to limit the harmonics to acceptable levels.

Compared to conventional diode rectifiers, the compensation of the voltage drop also requires a further oversizing of the power transformers in terms of rated voltage and power. This implies that the process of revamping a substation mandates the replacement of the entire transformer-rectifier group.

A further consideration is the need to install a control system and drivers to trigger the thyristors, which degrades the overall reliability and maintainability of the system. The control system implemented on these types of converters is composed of a feedback loop that measures the bridge output voltage and compares it with a reference level, which in turn is a function of the current delivered by the bridge.

Another difference is the lower short-circuit current delivered by thyristor bridges, which could jeopardize the proper operation of the protection circuits and oblige the installation of forced air cooling, again impacting the reliability of the system.

Finally, some considerations on the possible energy savings could be achieved, thanks to the recovery of vehicle braking energy.

With a normal diode bridge, it is not possible to direct power flow from the DC side to the AC side. This implies that to apply regenerative electric braking, there must necessarily be another locomotive motor that absorbs the energy injected by the braking train. If this condition is not verified, the voltage on the contact line surges to the maximum limit, tripping the traction protection systems and interrupting the regenerative braking, thus causing the excess power to be dissipated in dedicated onboard resistors.

With regard to power reversibility, the thyristor bridge allows for operation as an inverter but only through the inversion of the DC side voltage. In the field of electric traction, this mode of operation is not feasible with the original bridge configured as a rectifier, given that in no case may the voltage on the contact line be reversed.

It would be possible to feasibly reverse power flow only by arranging two separate thyristor bridges in antiparallel, one functioning as a rectifier and the other as an inverter. It is clear that this would entail a considerable increase in costs and system complexity.

3.2.1.3.7 Use of Voltage Source Converters (VSCs) Currently, there is need to find innovative systems that allow for the regulation of the output voltage and the bidirectional power flow to take advantage of the train regenerative braking.

These functions can be obtained using VSC converters that are also able to reduce the harmonics content and the reactive power absorption.

The VSC configuration has many advantages:

- Possibility of maintaining a constant output voltage up to the substation maximum load.
- Possibility of adjusting the output line voltage to compensate for voltage drops or for balancing the load across various substations.
- Absorption of AC side power with almost unitary power factor.
- Reduction of the DC side harmonic content.
- Intrinsic converter power flow reversibility achieved by reversing the current rather than the voltage.

As is known, the use of these converters has solved most of the problems related to diode bridges and has ensured the benefits of voltage regulation that has caused significant problems for thyristor bridge implementations.

The ability to reverse power flow during train braking without the need for dedicated bridges or having to reverse the connections is particularly suitable for high traffic systems, for which the energy savings can be significant.

Yet inevitably, VSC converters also have some disadvantages compared to the traditional Graetz bridge topology:

- Higher cost
- Need for a control system
- Lower component reliability with respect to diodes
- Greater conduction losses that imply a lower conversion efficiency and the need for liquid cooling systems

When upgrading traditional substations to VSC technology, it is necessary to replace the entire converter–transformer assembly. In fact, as seen in Chapter 2, the DC side voltage must always be higher than the peak of the AC side voltage. This implies the need to precharge the DC side capacitors that prevent the absorption of current peaks (inrush currents) during substation start-up.

Since the IGBT modules used in VSCs are less resistant to current peaks compared to diodes or thyristors, it is necessary to provide a high-speed substation circuit breaker to protect against external faults.

3.2.1.3.8 Power Supply to the DC Busbars The DC busbars are fed downstream of the conversion groups via servo-driven disconnectors. The busbar system then feeds the various contact line sections. Each power supply is protected by one high-speed circuit breaker, which can be entirely redundant or only partially redundant, by means of a backup busbar that incorporates multiple high-speed breakers. Since the entire system is on the same voltage, the various line sections may be connected in parallel and fed by both adjacent TPSSs (bilateral power feed).

This particular connection scheme allows for the exchange of part of the train braking energy with other trains in traction mode, even with nonreversible diode-based substations, since the power flows only in the DC section. It follows that the high-speed protection circuit breakers and relays must be bidirectional. Although this behavior is taken for granted in current electromechanical systems, when upgrading to power electronics solutions the employed switching devices must be able to conduct and interrupt current bidirectionally.

The contact lines fitted with overvoltage and overcurrent protection systems are fed from the DC busbars. The chain of switching and protection equipment therefore consists of

- bipolar disconnector,
- high-speed circuit breakers,
- surge arrester,
- horn circuit breaker, and
- connection to the contact line.

The output of the rectifier is connected to the main busbars, downstream of which the measurements and power units that house the high-speed circuit

Figure 3.20 Positive and negative busbar distribution system.

breakers are installed. The system is composed of two bars, that is, the anode (positive) and the cathode (negative) bars, which are generally supported by appropriate insulators for obvious safety reasons, and placed at a height of about 3 m above ground. The busbar system is usually divided into two or more sections so as to prevent a line failure causing an entire substation shutdown. The sections can then be linked by means of busbar couplers. For easy identification and to avoid installation errors, the bars are painted in two different colors: *red* for the anode bar (positive pole) and *blue* for the cathode bar (negative pole). There is also a yellow or bare copper bar that constitutes the ground collector (Figure 3.20).

Bipolar Disconnector The power supply unit houses one or two bipolar disconnectors. The bipolar disconnectors can be either manually operated or activated by an electric servomotor as part of a general equipment remote control system.

High-Speed Circuit Breaker High-speed circuit breakers are the most delicate element of a DC power supply system and the safety of people and preservation of equipment both on land and on board vehicles depend on their efficiency.

The opening of a direct current circuit with voltages on the order of 750–3000 V and where higher powers are at play cannot be carried out with the means used to interrupt AC currents of the same voltages. In fact, in the case of

AC, the interruption takes place after a zero-crossing of the current, through the interposition of a more or less deionized insulating area between the open switch contacts (an area that prevents the arc from forming again).

Instead, in the opening of a DC circuit, an arc is formed that tends to remain in continuity, and thus the means for interruption mentioned above would have no effect. Furthermore, where the opening occurs automatically due to a short circuit, the current increases with a gradient on the order of MA/s; that is, in brief instants, the current would reach the value of the short circuit, damaging generators or converters that power the circuit. It is therefore necessary for the switch to be operational within a few thousandths of a second from the beginning of the fault between positive and negative poles that tends to allow the current to increase beyond the system's current nominal value. The breaker must also be able to extinguish the arc within a period on the order of two or three hundredths of a second. For extinguishing the arc, the high-speed breaker uses the continuity of the current in generating an "electromagnetic blowout." Forcing the arc to lengthen its path increases the voltage drop until it reaches the normal value of the voltage across the open circuit and therefore forces the current to zero. The electric arc is in fact pushed toward an indented chimney (Figure 3.21) due to the Lorentz force created between the arc current itself and the magnetic field generated by the blowout chokes.

Figure 3.21 High-speed breaker for direct current applications.

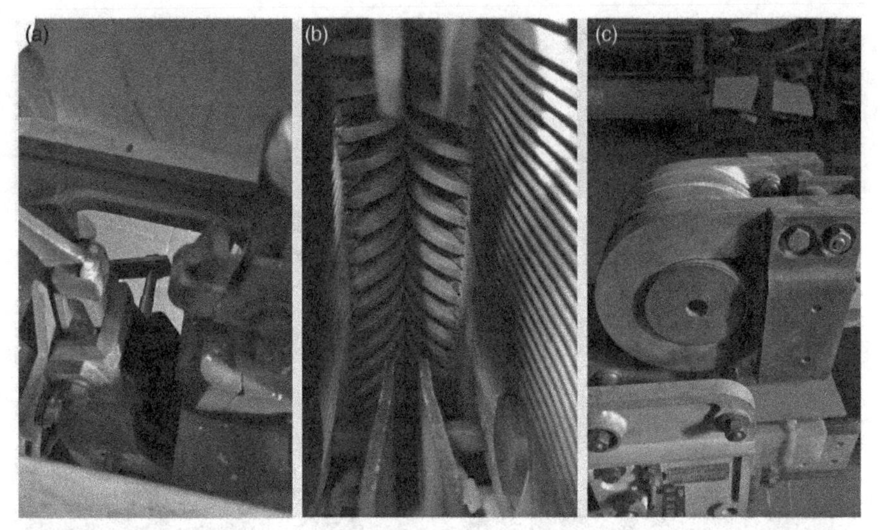

Figure 3.22 Main parts of a high-speed breaker. (a) Main contacts. (b) Chimney. (c) Blow-out chokes.

From Figure 3.22, it can be observed how the three components, current, magnetic field, and force, are perpendicular to each other; therefore, the ratio that links the module of the three magnitudes according to the Lorentz force is

$$F = B \cdot l \cdot I$$

where l is the electric arc length within the magnetic field B generated by the blowout coils.

The most common high-speed breakers are of a reclosing type; therefore, the substations are equipped with devices arranged so that after each overload or short-circuit surge, the conditions for insulation of the lines (test line circuit) are automatically checked and once the fault has been resolved, the switch closes, reactivating the power supply once again. These devices consist of a resistance that is inserted between the switch and the line: If the fault persists, it detects a short-circuit current (limited by the same resistance), and then the power supply is switched on again. If, however, the fault is extinguished, then the line can be supplied once again by excluding the test resistance.

Given the frequency with which an abnormal condition of an overload or short circuit occurs, which causes a high number of trips from the high-speed circuits, it is normal to establish a reclosing system coordinated to the grounding connection. The reclosing only allows one attempt, then the switch will remain open until the intervention by a special team of operators.

Surge Arresters for the DC Side These surge arresters are generally of the variable resistance type and their task is to protect the line in the event of

Figure 3.23 Surge arrester for DC contact line.

excessive overvoltages due to circuit breaker activations caused by internal operations or external events of atmospheric origin (Figure 3.23).

Horned Disconnector The boundary between the substation and the contact line is the disconnector, known as the "horned" type in jargon. This name comes from the shape of the fitted spark gap. These disconnectors are connected immediately downstream of the high-speed circuit breaker and the surge arrester (Figure 3.24). In addition to these switches, known as first row disconnectors, usually other devices known as second row disconnectors are also installed. Their role as alternative connections is to ensure the continuity of power supply in the event of faults and to provide safe access for maintenance. In other words, in the event of a total substation outage, the second row disconnectors should close to ensure continuity of supply to the contact line from an adjacent substation.

As in the case of the power supply disconnectors, these can also be controlled either manually or by an electromechanical servo. On opening, the

Figure 3.24 Horned disconnectors.

operating lever and the relative counterweight transmit the clockwise movement to the intermediate insulator, causing the main contact blade of the left-hand insulator contact clamps to come out. Until the maneuver is completed, any current flowing through the main disconnector contact will be deviated through the auxiliary contacts of the arc-suppression lever, which remains in the lowered position. On completing the maneuver, the above-mentioned lever rotates clockwise, the auxiliary contacts open, and an arc forms between them, considerably heating the surrounding air, which, as it rises, pushes the arc itself upward into the dedicated horns until it extinguishes.

3.3 BRAKING ENERGY RECOVERY SYSTEMS FOR DC RAILWAY APPLICATIONS

Transportation systems are one of the areas in which there are many opportunities for energy conservation, while obtaining significant operating cost reductions at the same time. In particular, electric railway systems are immediately eligible for energy efficiency improvement, since they use electric energy that can be generated through renewable sources or that can be stored for better exploitation.

A way to save energy in electrified transportation systems is the recovery of the braking energy coming from the trains equipped with electric regenerative braking.

At present, in DC systems this energy is mainly dissipated by dedicated rheostats, while a small amount is delivered to other trains in an accelerating phase.

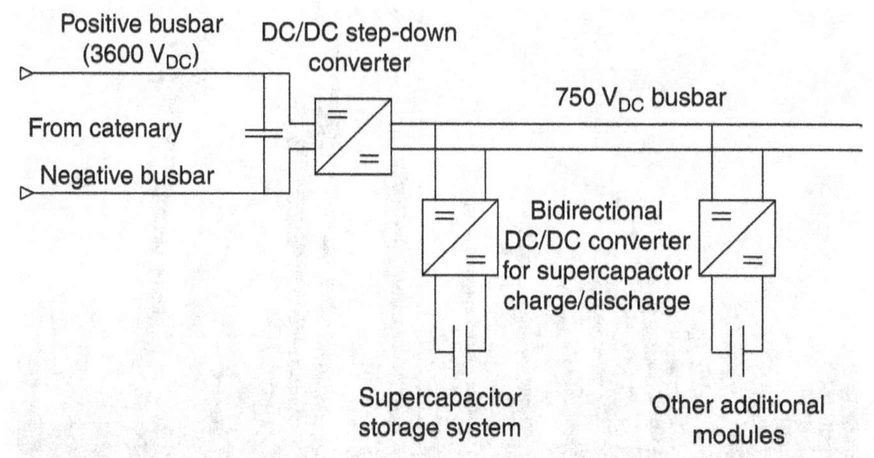

Figure 3.25 Connection of the supercapacitor storage system to high-voltage DC lines.

In fact, various statistical analyses show that there is a low probability of having contemporarily one braking and another accelerating train on the same line.

A solution for the braking energy saving is its recovery in storage devices installed near the railway line. Since the braking energy presents high-power peaks for a short time, one of the best solutions is the installation of supercapacitor banks. In fact, supercapacitors have all the characteristics needed for this purpose in terms of maximum power and number of charge–discharge cycles. Nevertheless, due to their poor energy density, the space needed for their installation suggests their application in fixed power substations.

Another feature is that supercapacitors, like the other storage systems, cannot be connected directly to the railway supply electric line. In particular, for heavy railway systems or subways, the high voltage value of the catenary requires a dedicated DC/DC stepdown converter to adapt the system voltage to the storage voltage. A possible scheme for the connection of supercapacitor modules to high-voltage DC railway lines is depicted in Figure 3.25. The same solution can be used for lower voltage subway lines avoiding the installation of the first DC/DC stepdown converter.

3.3.1 Braking Energy Recovery Systems in Subway Lines

In a subway line, TPSSs are generally distributed in sequence in such a way that the voltage drop does not exceed 33% of the nominal value.

At the same time, it is obviously necessary to consider the topography of the line, as well as the environmental conditions. The result is that the TPSSs can also not be equally spaced.

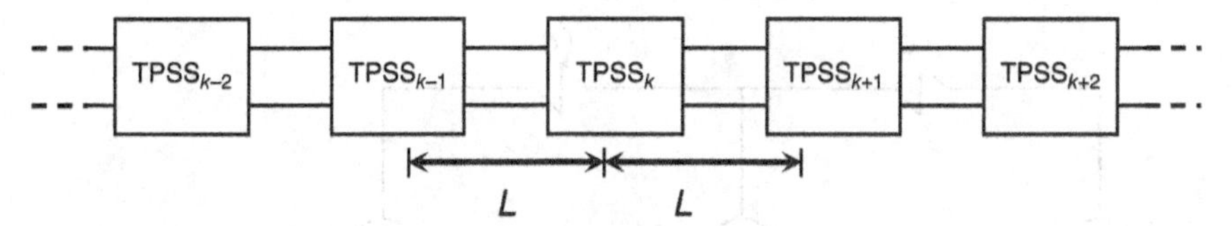

Figure 3.26 Distribution of the TPSSs along a subway line.

In order to create a probabilistic sizing criteria, one has to first suppose that the TPSSs are equally distributed, as may be seen in Figure 3.26, and that the TPSS is modeled as a real voltage generator, with its own internal resistance.

In this case, there is no discontinuity between the TPSSs, so each TPSS does not have its own supply area, but every train in the subway line is supplied with the contribution of the all TPSSs. However, each train is predominantly supplied from the nearest TPSSs, while the farthermost ones do not contribute in any significant way.

With these hypotheses, the subway line can be thought of as being constituted by various independent sections. In this case, the TPSSs are modeled as ideal voltage generators because the internal resistance can be neglected. In this situation, the load between two TPSSs is supplied only by the nearest TPSSs because the ideal generators decouple the space between the two TPSSs from the other circuital loop constituting the subway line, as represented in Figure 3.27.

Using a probabilistic approach, it is possible to assume that there is a probability that TPSS$_k$, and TPSS$_{k+1}$, emit the same power and that each TPSS, in the line under consideration, then has a probabilistic supply area that is equal to half the considered line ($L/2$).

The TPSSs have to be sized according to $N-1$ security criteria, that is, if a TPSS is not available, the other $N-1$ TPSSs have to give more power in order to guarantee normal traffic conditions, even if the trains are working at reduced performance levels caused by increased voltage drops.

Considering the case that all the TPSSs are modeled as ideal voltage generators, it is possible to assume that only the two TPSSs nearest to the one that is out of order are overloaded, while the others are working normally.

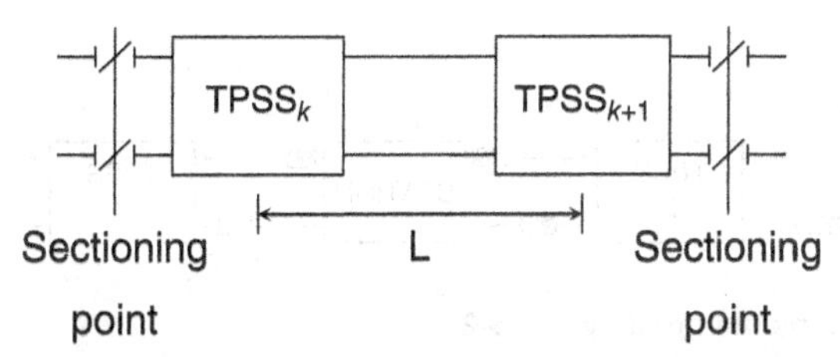

Figure 3.27 Equivalent circuit of the decoupling introduced by ideal TPSSs.

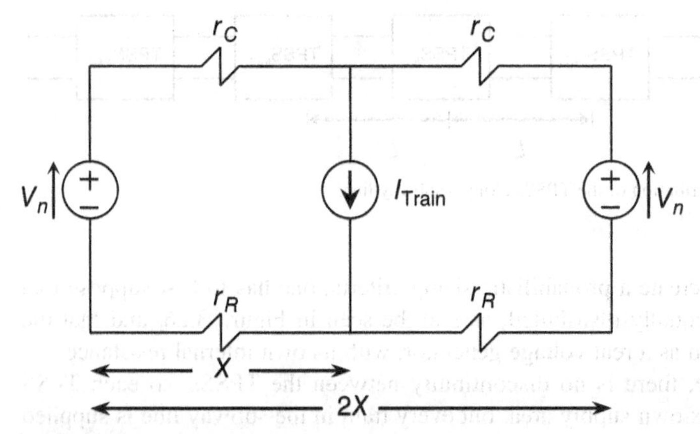

Figure 3.28 Electrical scheme of the line between two ESSs.

Given the length of the route and the daily traffic volume, it is possible to determine the number of TPSSs in such a way that each TPSS can give power with a voltage value of between +20% and −33% of the nominal one.

It is important to determine the number of TPSSs in order to establish the total size of the storage system needed for the energy recovery of the whole subway line.

Figure 3.28 shows what happens between the two TPSSs, from an electric standpoint, and considers the TPSS as an ideal voltage generator, and the train as an ideal current generator.

In Figure 3.28, the two TPSSs supply the subway line with its nominal voltage, the distance between two TPSSs is unknown and equal to 2X; the rail and contact line linear resistances are r_R and r_C, respectively.

Under the hypothesis that the subway line between two stops is occupied at the same time by only one train, in order to guarantee a safe distance, each TPSS supplies, in the probabilistic meaning, four trains (Figure 3.29). This is the case chosen for this analysis, which considers maximum train traffic. During other periods, each TPSS can supply less than four trains, relative to traffic conditions.

In order to apply a probabilistic algorithm, the choice has been made to divide the traction diagram of each train into 11 intervals, each with the same time duration.

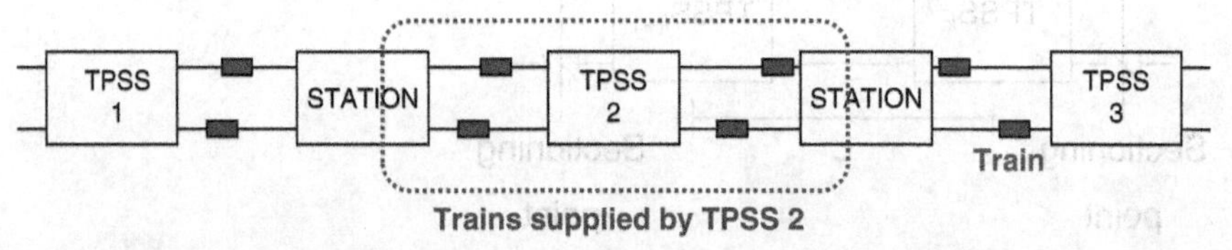

Figure 3.29 Supply area of TPSS 2.

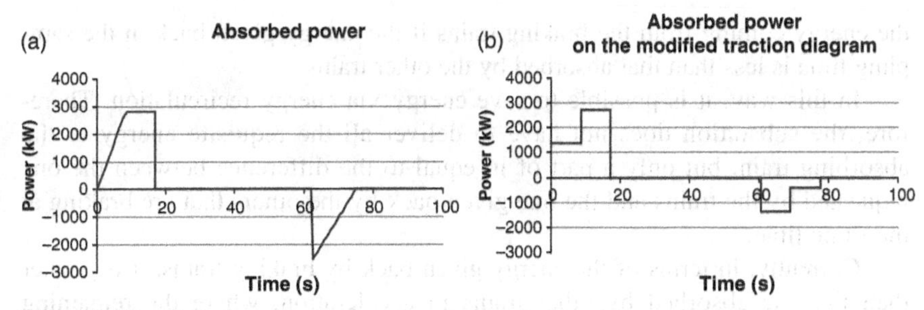

Figure 3.30 (a) Real traction diagram of a subway train (b) and the modified one.

The mean value of the absorbed and given power for each interval has been chosen to keep the same value of the total absorbed energy in the traction phase and the same energy given back in braking phase.

Each mean power value for each interval has the same probability to manifest itself because the sample time is constant. The absorbed power diagram changes as shown in Figure 3.30.

The possible combination of the various intervals among the four trains are described by a rectangular matrix 114×4, where 11 is the number of the intervals in which the traction diagram has been divided and 4 is the number of trains supplied by the considered TPSS.

Summing up in all the possible ways the powers of the four trains supplied by the TPSS at the same time and multiplying by their probabilities, it is possible to obtain the mean powers column vector and the corresponding probabilities one. The diagram of the average powers, and their probabilities obtained in this manner, is reported in Figure 3.31.

In the sampling time, it can happen that some trains are in the absorbing phase, while others are in a braking phase. Trains in the starting phase absorb all

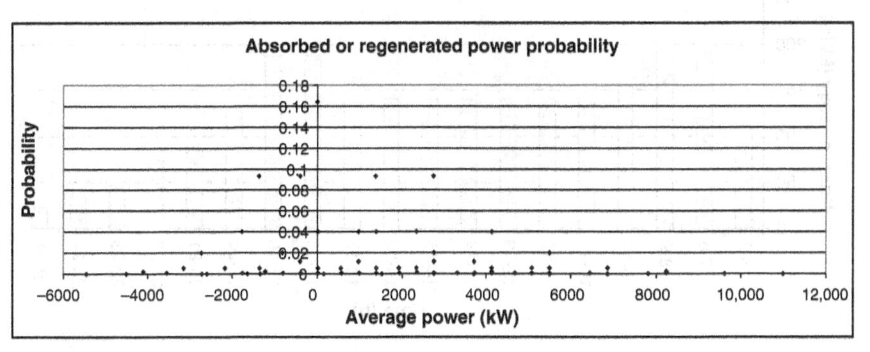

Figure 3.31 Absorbed or given back powers probability spectrum during maximum traffic conditions.

the energy coming from the braking trains if the energy given back in the sampling time is less than that absorbed by the other trains.

In this way, it is possible to save energy via energy recirculation. Therefore, the substation does not have to deliver all the requisite energy to the absorbing train, but only a part of it, equal to the difference between the one requested by the trains and the one given back by the others that are braking at the same time.

Currently, in terms of the energy given back by braking trains, it is greater than the one absorbed by other trains in acceleration, where the remaining energy is dissipated.

It is possible to observe that the most probable situation, where it is possible to have this condition, corresponds to the case of a braking train and the other three either in a coasting phase or stopped at a station.

In order to determine the energy that, with a probabilistic concept, is dissipated in a determined time, it is necessary to multiply the mean negative powers previously obtained for the corresponding probabilities and then to sum up these obtained values.

By applying this algorithm to the energy lost in 1 h without storage systems, and considering that the traction diagram of Figure 3.32 is equal to 213.6 kWh, that is the 57% of the total braking energy. The remaining 43% is absorbed by other trains in a starting phase. Moreover, the total braking energy corresponds to the 42% of the energy delivered by the TPSSs, without the regenerative braking.

The same procedure can be applied in the other subway service periods. The only variable that has to be changed for each hour is the train frequency on the line, therefore the duration for which each power value is absorbed from the TPSS in that hour.

Represented in Figure 3.32 is the average lost energies in the various periods of a working day without storage systems, characterized by different traffic

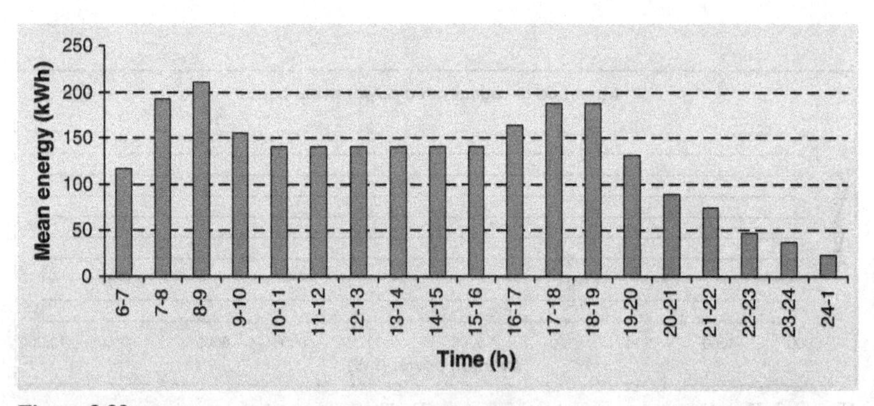

Figure 3.32 On average lost energy per hour without storage system for a subway line, considering different traffic conditions during a working day.

conditions. These energies are not linearly proportional to the traffic conditions, because an increase in train frequency also increases the probability of having energy recirculation among the various trains.

The storage system has to be fast charged because the power given back by the trains behaves impulsively, but it has also to be discharged quickly because the starting power of the trains is also impulsive.

Supercapacitors are the storage systems that feature such behavior. They are suitable for energy demands during high power peaks, ranging from a few milliseconds to a few minutes. Indeed, they are characterized by high power density. They can be fast charged with high currents and with every charge profile. Another characteristic is their high charge and discharge cycle numbers, and if they work in accordance with the indicated prescriptions, they do not require any maintenance.

During the recovery phase, the energy coming from the trains is stored in the supercapacitors and given back to the trains in the starting phase as soon as possible, so the supercapacitors are continuously charged and discharged.

The energy E_{sc} that has to be exchanged in every cycle is around 8–10 kWh for a traditional subway. However, not all the stored energy can be used since a supercapacitor can work between its nominal voltage V_{sc_n} and $V_{sc_n}/2$. Therefore, the total capacity has to be oversized according to the following relation:

$$E_{sc} = \frac{1}{2} \cdot C \cdot V_{sc_n}^2 - \frac{1}{2} \cdot C \cdot \left(V_{sc_n}/2\right)^2 = \frac{3}{8} \cdot C \cdot V_{sc_n}^2$$

3.4 CONTACT LINES

The contact and power supply lines for electric railway traction, generally situated outdoors like other overhead electrical lines, must bear both its own mechanical stresses of the traction system, typically the pantograph–catenary interaction, and those due to atmospheric agents, climate changes, temperature changes, and chemical aggression from the outside environment.

The specific performance of the contact lines has evolved considerably over time due to the needs related to the drive system's overall performance, and due to the growing increase in the speed and energy absorption of trains, all in accordance with the constraints imposed by the original structuring of systems.

In the past, contact lines were generally composed of a copper messenger wire of 120 mm^2 section, with fixed anchoring, calculated for a mechanical tension of 1075 N at 15 °C, and by two contact wires adjusted from 100 mm^2 regulated to 750 N (Figure 3.33). The initial total cross section was thus 320 mm^2.

Exceeding speeds of 150 km/h, it became necessary to increase the quality of the wire tap and to improve the interaction between the pantograph and

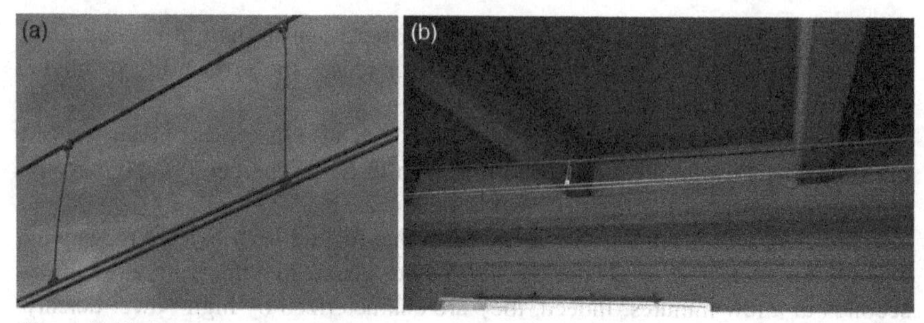

Figure 3.33 Representation of a contact line. (a) Railway. (b) Subway.

catenary, therefore, it was decided to switch to the messenger wire and increase the mechanical tension with equal mass of the contact line. This situation, which immediately proved to be very valid, was then also extended to lines with speeds below 150 km/h, for which the standard classic of 320 mm^2 required the adjustment of the messenger wire to a mechanical tension of 1375 N and the contact wires to 1000 N.

With the intensification of traffic and the increase in nominal rating value, it was then decided, in conjunction with the renovation of the contact lines, to increase the total section up to 440 mm^2 with the addition of a second messenger wire of 120 mm^2 and by modification of the mechanical tension of the same, which was changed to 1125 N for each. Consequently, the contact line thus far considered technologically appropriate to the needs of a 3 kV power supply system, also used below 150 km/h, became the 440 mm^2 section, formed by two messenger wires of 120 mm^2 and two contact lines of 100 mm^2. Values of standard 3 kV$_{DC}$ contact lines are reported in Table 3.3.

In lines at the highest speed up to 250 km/h with direct current, to further improve the needs of current uptake to the pantograph, two contact lines of

Table 3.3 Standard 3 kV$_{DC}$ Contact Lines

Characteristics of railway lines		Contact line	Messenger wires			Contact wires		
Maximum speed (km/h)	Intensity of traffic	Copper wire section (mm^2)	Number	Section (mm^2)	Mechanical tension (N)	Number	Section (mm^2)	Mechanical tension (N)
200	Low	320	1	120	1375	2	100	1000
200	Mean	440	2	120	1125	2	100	1000
200	High	610	2	155	1000	2	150	1125
250	Mean	540	2	120	1500	2	150	1875
250	High	610	2	155	1625	2	150	1875

150 mm^2 and a copper–cadmium messenger wire of 160 mm^2 were introduced. On such contact line, of a total 460 mm^2 (equal however to 440 mm^2 copper equivalent), it was possible to considerably raise the mechanical tension of the conductors, which were changed to 2750 N for the messenger wire and 1500 N for the wires.

To remedy the difficulties arising from the use of cadmium and further develop the performance of the contact line, based on the experience gained and the tests performed, the standard section for this type of line (high-speed direct current) was raised to 540 mm^2 with the use of

- two messenger wires of 120 mm^2 regulated to power line of 1500 N;
- two contact wires of 150 mm^2 regulated to power line of 1875 N.

A new standard of 540 mm^2 is being implemented for the upgrading of traditional lines in continuation or alongside the high-speed lines. The main reason is the use of the same materials of new contact lines in the 25 kV system and, in particular, the contact wire of 150 mm^2 in place of the 100 mm^2 one, which also allows operating time to be increased by reducing the frequency of replacements due to reaching maximum wear.

In this regard, studies are underway to define the characteristics of the new contact lines so that the same, while achieving the required quality of planned uptake for modern high-speed lines, may be implemented with the renewal of only the contact conductor, minimizing trips to masts and related foundation blocks. Finally, to cope with the need for greater absorption of current on some routes with steep slopes and/or with heavy traffic, the use of messenger wires of 155 mm^2, in place of those of 120 mm^2, has been made. Therefore, in some cases a total section of 610 mm^2 can be reached.

In addition to the measures up to this point, the following interventions can also be carried out:

- Addition of a branch feeder (or reinforcement feeder) of 120 mm^2, only if mechanically supported by the existing piling. Compared to the catenary, the feeder is not subject to the interaction with the pantograph, therefore, its installation and maintenance is simpler.
- Installation, some on double-track lines, split between two TPSS, of a special booth containing the equipment for the parallel connection between the conductors of the two tracks, so as to advantageously exploit the probable nonsimultaneity of high loads in both directions, or the exchange of energy between a train in traction and one braking on steep lines (with a track traversing uphill and the other downhill).
- Increase of the installed power in the substation, with the addition of further conversion groups.
- Automatic voltage adjustment through the adjustment of the transformation ratio or via controlled converters, so as to keep the voltage at the output from the substations constant.

3.4.1 Constructive Aspects of the Line

The fundamental problem is to obtain regular current uptake through a contact that, from an electrical point of view, is certainly not perfect. In fact, the relative speed between the pantograph and the contact wire allows for satisfactory service as long as the pressure from the tap and the elastic characteristics of the catenary are appropriate. The line must also have vertical elasticity $c = \frac{\Delta y}{F}$, understood as the ratio between the lifting of the wire and the applied force, directed upward as uniformly as possible.

A perfect contact line should have the following characteristics:

- Have the contact conductor(s) at a constant height.
- Maintain constant conductor height also at pantographs transit.
- Maintain the contact conductors according to a symmetrical zigzag in respect to the axis of the road (staggering).
- Not be affected by any appreciable action of side winds.

Respect of the above points is essential to ensure good uptake current in the pantograph–catenary interaction. The system used to ensure the above is called *catenary suspension*, in which a top messenger wire supports one or two contact wires (Figure 3.34a). For lower speed applications, an urban tramway cross suspension with only one contact wire can be allowed (Figure 3.34b).

3.4.2 Catenary Suspension

This system is typically used most often in railways and sometimes in metros. The contact wire (single or double) is suspended from a messenger wire, which in turn is supported by suitable brackets anchored to the support masts whose

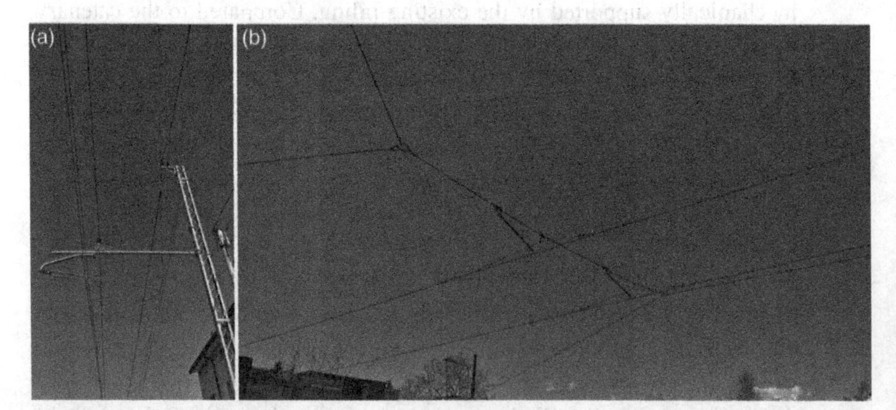

Figure 3.34 Catenary suspension for railway application with reinforcement feeder (a) and cross suspension for tramway application (b).

distance, defined as span, is typically between 50 and 65 m. The messenger wire, which supports the contact wire by droppers located at a short distance, is placed according to a catenary with a maximum deflection on the order of 1 m. Under resting conditions, the contact wire of linear mass m (kg/m), subjected to an axial pull T (N), is placed between two consecutive supports, according to a catenary.

If the origin of the axes is allowed to coincide with the vertex, the suspension equation is the following:

$$y = \frac{m \cdot g \cdot x^2}{2T} \tag{3.5}$$

where

- m = the linear mass of the wire (kg/m)
- g = the acceleration of gravity (m/s^2)
- x = the distance between the two points of support (m)
- T = the mechanical tension of the wire (N)

The maximum deflection d is obtained by placing $x = l/2$ (Figure 3.35):

$$d = \frac{m \cdot g \cdot l^2}{8T} \tag{3.6}$$

The length of the droppers is calculated so that the wire or contact wires are horizontal with respect to the rail level. The contact wires are grooved to be fixed to the droppers as reported in Figure 3.36a). The electrical continuity between the contact wire and the messenger wire, which is necessary for sharing the current between the two conductors because the contact strip only absorbs current from the contact wire, is ensured by special continuity bypasses (Figure 3.36b) mounted at distances varying between 120 and 180 m.

To allow even wear of the pantographs' contact strips, the contact wire is arranged in a zigzag pattern on the straight sections with respect to the track's

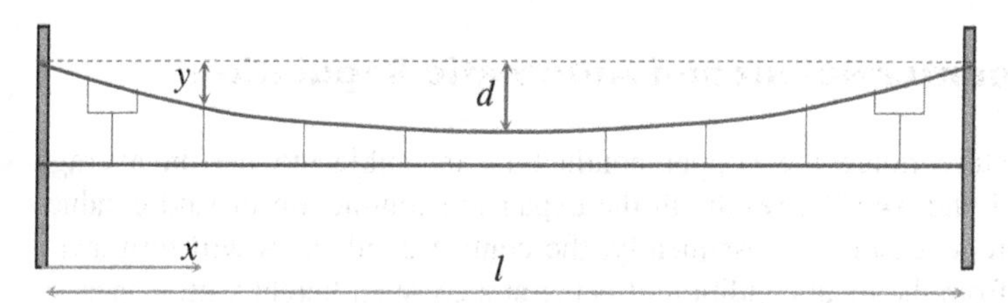

Figure 3.35 Deflection of the messenger wire.

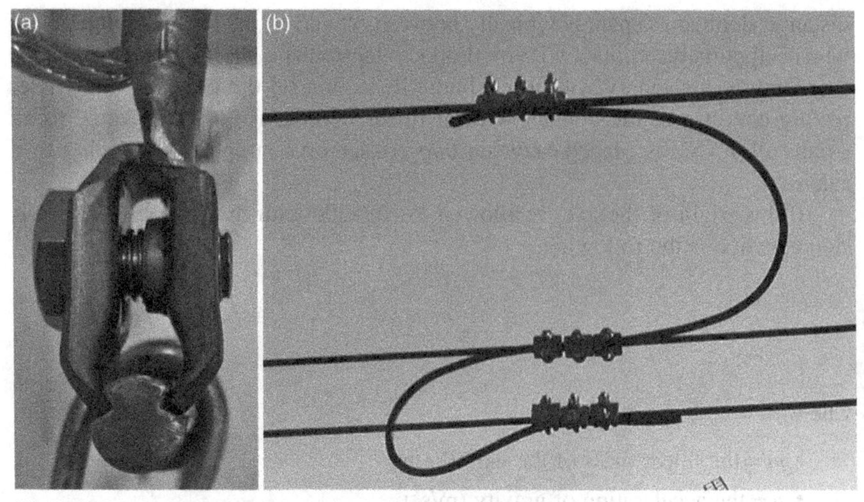

Figure 3.36 (a) Grooved contact wire. (b) Continuity bypasses.

centerline. Such zigzag pattern is defined staggering p of the contact line, and is obtained by suitable arms and tie-rods that position the wire alternately at ±20 cm with respect to the axis of the track. In practice, the contact wire is contained in a 40 cm strip that coincides with the pantograph contact area (Figure 3.37a). In curves of a radius greater than 3000 m, the staggering is created in a straight line; in those of a radius lower than 3000 m, it is obtained by mounting the suspensions externally to the curve and creating bays of length L satisfying the following ratio (Figure 3.37b):

$$L \leq 4 \cdot \sqrt{(a \cdot p)} \tag{3.7}$$

where

- a = the radius of curvature;
- p = the staggering.

3.4.3 Counterweight and Automatic Regulation

The temperature range the copper conductors are subjected to (on average between −15 and +45 °C) results in the expansion/contraction of said conductors in different seasons. Consequently, the contact conductors will turn away excessively from the ideal condition of being at a constant height with respect to

Figure 3.37 Staggering of the contact line. (a) Straight line. (b) Curve.

the track level. To avoid this, a conductor counterweight is used (only contact line or contact line + messenger wire), which is aimed at maintaining the tension almost constant by a system of masses and pulleys, allowing the necessary freedom of longitudinal sliding (Figure 3.38).

The expansions of counterweighted conductors are made compatible with the constraints of support and staggering by

- the rotation, on the horizontal plane, of the staggering tie-rods, when only the contact wires have a counterweight;
- the rotation of the entire bracket and thus the entire suspension, even when both the messenger wire and the contact wire have a counterweight.

These rotations are to be contained, due to construction constraints of the components within ±35 cm or within a total excursion of 70 cm. The

Figure 3.38 Counterweight and automatic regulation.

thermal excursions produce a maximum expansion of 1 cm every 10 m in the copper conductors; therefore, to contain any tie-rod and bracket rotations within the mentioned 70 cm, the fixed point and the counterweight point of the conductor must not exceed 700 m in distance. Hence, there is need to resort to counterweight spans of 1400 m with a fixed central point and two counterweights, one for each end.

The conductor is supported while outstretched, as shown in Figure 3.39, by masts spaced apart for up to a maximum of 60 m. All line devices (strut support, tie-rod, staggering arms) are live and are supported by the mast with post insulators.

3.4.4 Electrical Calculations of the Traction Lines

The electrical calculation of the DC lines serves to determine the copper section of the same line and the distance between substations, starting from the following elements:

- power supply voltage
- planimetric trend of each segment

Figure 3.39 Support mast of a catenary DC contact line, with ground bracket and catenary tensioning system.

- maximum absorption by trains
- maximum expected rail traffic
- permissible continuous average and maximum voltage drops for short periods

3.4.4.1 Line kilometric resistance

The kilometric resistance of a railway line power supply circuit r includes the resistance of the catenary r_c and that of the track r_r:

$$r = r_c + r_r \left[\frac{\Omega}{\text{km}}\right] \tag{3.8}$$

3.4.4.1.1 Contact Line Resistance The catenary resistance (messenger wire + contact wires) is evaluated using the following ratio:

$$r_c = \rho_{\text{Cu}} \left[\frac{\Omega \cdot \text{mm}^2}{\text{m}}\right] \cdot \frac{1000\,[\text{m}/km]}{S_c[\text{mm}^2]} \tag{3.9}$$

where S_c is the overall cross section of all the copper messenger wires composing the aerial line (considering the total nominal section minus 15–20%, in order to take account of wear).

In practice, all the wires (and therefore also the messenger wire) are made of copper; therefore, all contribute to defining the overall cross section of the conductor S_c. Knowing that the resistivity of copper has the value $\rho_{Cu} = 0.018 \left[\frac{\Omega \cdot mm^2}{m}\right]$, the final ratio is

$$r_c = 0.018 \cdot \frac{1000}{S_c} = \frac{18}{S_c} \left[\frac{\Omega}{km}\right] \tag{3.10}$$

3.4.4.1.2 Track Resistance The railway track is characterized by the linear mass m (kg/m) of the rails and of the length l of the individual sections of the rail; the latter are connected to one another by joints, although currently the trend is to create very long sections by welding the same (Long-Welded-*Rail* LWR). If the resistivity and the density of the steel with which the rails are constructed are known, it is possible to express their resistance per unit of length as a function of the linear mass,[1] through the following ratio:

$$r_r = \frac{0.75}{m} \left[\frac{\Omega}{km}\right]$$

In order to take the wear of the rails and the additional resistance introduced by the connection between the different sections into account, the equivalent resistance of the rail is approximated with the following ratio:

$$r_r = \frac{0.9}{m} \left[\frac{\Omega}{km}\right] \tag{3.11}$$

For example, for a standard rail of 60 kg/m, the resistance of the rail is equal to 0.015 Ω/km.

The resistance of the track is therefore approximately of a lower magnitude than that of the contact line.

To reduce the kilometric resistance, we must therefore intervene on the contact line by increasing the conductor section or by adding a branch feeder.

3.4.5 Voltage Drops

The evaluation of voltage drops in traction systems is important both in rating stages of the system and in planning stages of the railway operation to verify that even under the worst conditions, the voltage at the pantograph is always within the permissible limits.

[1] In the case of rails, it is preferred to express the resistance as a function of the linear mass m, which is the most characteristic and known parameter, unlike the cross section.

In direct current systems, voltage drops are determined only by the resistive component of the power supply line, neglecting the transient phenomena that require a more detailed study.

An accurate assessment of voltage drops can only be carried out through dedicated software that simulates the actual operating rail traffic, considering the spatial and time variation of the absorption of every single train in the line.

However, in the rating phases of the contact line, where there is no definitive knowledge of the actual rail traffic, simple ratios may be used to estimate the voltage drops for different configurations of the traction system assuming the current absorbed by the trains to be constant. In addition, a further possibility is introduced of considering the substations as ideal, that is, modeled through a constant voltage generator.

It is necessary to first define the type of power supply, which can be unilateral (one end only) or bilateral.

In unilateral power supply, trains are powered by a single substation as shown in Figure 3.40a, while in bilateral power supply the segment is powered from both ends by two substations connected in parallel, as shown in Figure 3.40b.

Normally, in railway and metro traction bilateral power supply is used, leaving the end sections and deposits with unilateral power supply.

The calculations of voltage drops for the most common examples are summarized in the following sections.

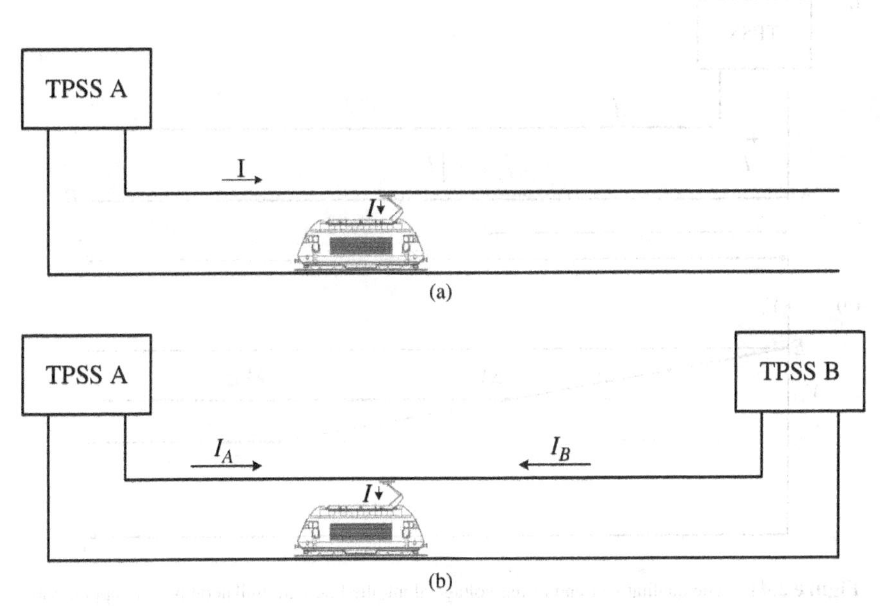

Figure 3.40 Unilateral (a) and bilateral (b) power supply.

3.4.5.1 Voltage drops with unilateral power supply

3.4.5.1.1 Concentrated Load In the case of unilateral power supply, with reference to Figure 3.41a, it is assumed that in the trunk A–B, of length L, supplied only from substation A, there is one train that absorbs current I. Given x (km) is the distance of the train from the substation A and r (Ω/km) is the line resistance of the line (including contact lines and rails), the voltage drop between A and the generic point at a distance x from substation A is

$$\Delta V_x = r \cdot x \cdot I \tag{3.12}$$

The trend of the voltage drop is, therefore, a function of the position x of the train and is represented by a straight line with gradient rI.

The voltage drop ΔV_x increases when the train moves away from substation A and reaches the maximum value when it is at line end B:

$$\Delta V_{\max} = r \cdot L \cdot I \tag{3.13}$$

In the A–B path, the voltage drop then varies linearly from 0 ($x = A$) to $V_{\max}$ ($x = B$), and its average value is

$$\Delta V_{av} = \frac{r \cdot L \cdot I}{2} \tag{3.14}$$

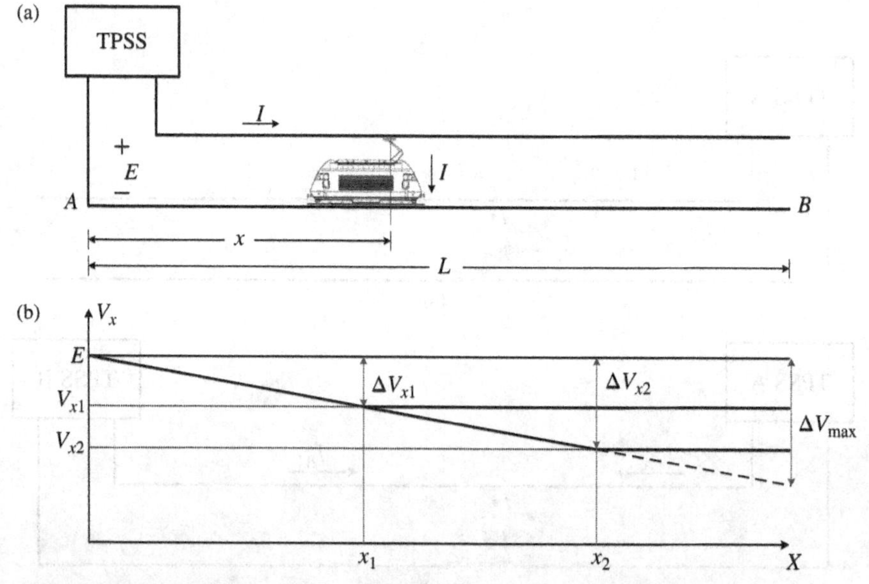

Figure 3.41 The qualitative trend of the voltage along the line with unilateral power supply, with a concentrated load.

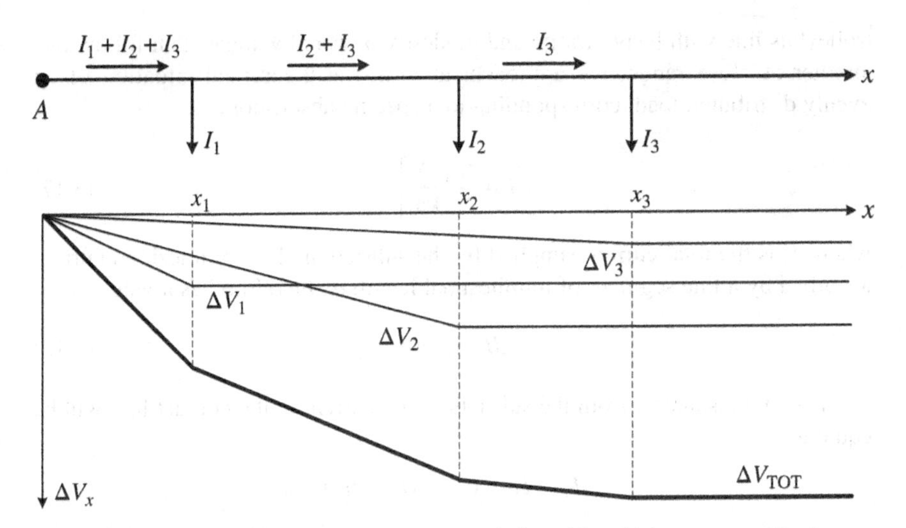

Figure 3.42 Voltage drop in a unilateral power supply line with multiple concentrated loads.

3.4.5.1.2 Multiple Concentrated Loads If there are several trains on the line that absorb different current I_1, I_2, . . . , I_n in the distances $x_1 < x_2 < \cdots < x_n$ from substation A, it is possible to determine the diagram of the total voltage drop by the principle of superimposition of the effects, as shown in Figure 3.42.

The voltage drop at point x_3, namely, that farthest from the train and thus representative of the worst case, is determined as

$$\Delta V_{x3} = r \cdot [x_1 \cdot I_1 + x_2 \cdot I_2 + x_3 \cdot I_3] = r \cdot \sum_{i=1}^{n} x_i \cdot I_i \rightarrow \Delta V_{\max} = r \cdot \sum_{i=1}^{n} x_i \cdot I_i \tag{3.15}$$

If there are n trains uniformly distributed along the segment length L that absorb all of the same current I, the total current absorbed by the line will be $I_S = n \cdot I$. Therefore,

$$\begin{aligned} \Delta V_{\max} &= \Delta V_n = r \cdot \sum_{i=1}^{n} x_i \cdot I_i = r \cdot I \cdot \sum_{i=1}^{n} x_i = r \cdot I \cdot \sum_{i=1}^{n} L \cdot \left(\frac{i}{n}\right) \\ &= r \cdot I \cdot L \left(\frac{1}{n} + \frac{2}{n} + \frac{3}{n} + \cdots + \frac{n}{n}\right) = r \cdot I \cdot L \cdot \left(\frac{1+n}{2}\right) \end{aligned} \tag{3.16}$$

3.4.5.1.3 Uniformly Distributed Load If the number of trains in segment L becomes very large, representing the typical case of an urban tram or trolleybus line with heavy traffic and modestly powered vehicles that follow one another at close range, we approach, at best, the theoretical situation of an evenly distributed load, corresponding to a specific absorption:

$$i = \frac{I_s}{L}\left[\frac{\text{A}}{\text{km}}\right] \tag{3.17}$$

where I_S is the total current supplied by the substation. The infinitesimal current absorbed by a line segment of infinitesimal length dx therefore has a value of

$$dI = i \cdot dx \tag{3.18}$$

and at a distance x from the substation, the current in the contact line will be equal to

$$I_x = I_s - x \cdot i = (L - x) \cdot i \tag{3.19}$$

Given $d\Delta V_x$, the infinitesimal voltage drop of segment dx is therefore:

$$d \cdot V_x = r \cdot I_x \cdot dx = r \cdot (L - x) \cdot i \cdot dx \tag{3.20}$$

Thus the voltage drop in point x will be equal to its integral and has a parabolic curve according to the following ratio:

$$\Delta V_x = \int_0^x d\Delta V_x = \int_0^x r \cdot (L - x) \cdot i \cdot dx = r \cdot i \cdot x \cdot \left(L - \frac{x}{2}\right) \tag{3.21}$$

The voltage drop and the current along a line of length L with unilateral power supply and evenly distributed load is shown in Figure 3.43.

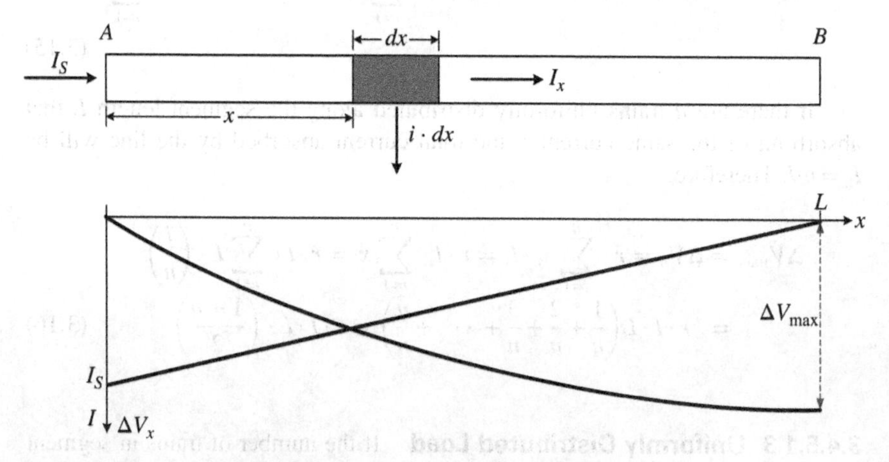

Figure 3.43 Line with unilateral power supply, uniformly distributed load.

The maximum value of the voltage drop is obtained at end B of the line ($x = L$):

$$\Delta V_{\max} = r \cdot i \cdot \frac{L^2}{2} = \frac{r \cdot L \cdot I_s}{2} \tag{3.22}$$

It can be noted that the maximum value of the voltage drop in the case of uniformly distributed load is half that would occur with concentrated load.

The average voltage drop in this case assumes the value

$$\Delta V_{av} = \frac{2}{3} \cdot V_{\max} = \frac{r \cdot L \cdot I_s}{3} \tag{3.23}$$

3.4.5.2 Voltage drops with bilateral power supply

In the evaluation of the voltage drops in the case of bilateral power supply, it is first necessary to consider the substation's operating voltage.

3.4.5.2.1 Substations Operating at the Same Voltage Assuming that both substations A and B, located at the ends of the line segment, are delivering the same voltage E.

Concentrated Load Assuming then that $E_A = E_B = E$ is constant for each supplied current, as shown in Figure 3.44, it is possible to calculate the voltage at x, starting respectively from the E_A or the E_B, as follows:

$$V = E - r \cdot x \cdot I_A = E - r \cdot (L - x) \cdot I_B \tag{3.24}$$

where I_A and I_B are the currents supplied respectively from the substations A and B.

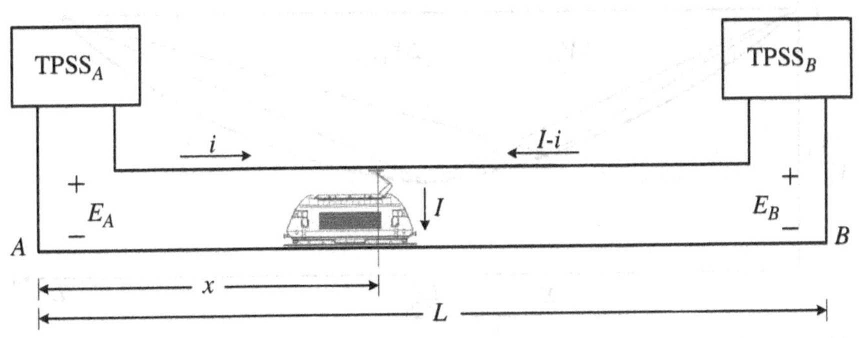

Figure 3.44 Bilateral power supply, concentrated load at a distance x.

From Equation 3.24, knowing that $I_A + I_B = I$, it follows that

$$x \cdot I_A = (L - x) \cdot I_B \rightarrow x \cdot I_A = (L - x) \cdot (I - I_A) \tag{3.25}$$

From this, it can be derived that the voltage drop

$$\Delta V_x = r \cdot x \cdot I_A = r \cdot x \cdot \frac{L - x}{L} \cdot I \tag{3.26}$$

It should be noted that

$$\begin{aligned} x = 0 &\rightarrow I_A = I \quad \text{current supplied entirely from TPSS}_A \\ x = L &\rightarrow I_A = 0 \quad \text{current supplied entirely from TPSS}_B \end{aligned}$$

The voltage drop ΔV_x varies with square law in abscissa x. The point with the minimum voltage depends on the position of the train. From the graph shown in Figure 3.45, the parabolic trend of the voltage drop along the line can be noted.

Evaluating the value ΔV_x between 0 and L, it can be noted that the voltage drop reaches its maximum value for $x = L/2$, in which

$$\Delta V_{\max} = \Delta V_{x=L/2} = r \cdot \frac{L}{2} \cdot \frac{\left(L - \frac{L}{2}\right)}{L} \cdot I = \frac{r \cdot L \cdot I}{4} \tag{3.27}$$

Note that $\Delta V_{\max}$ is reduced to 1/4 of that that was found in the case of a concentrated load and unilateral line power supply, which is why the bilateral power supply, where applicable, is always preferable over the unilateral supply.

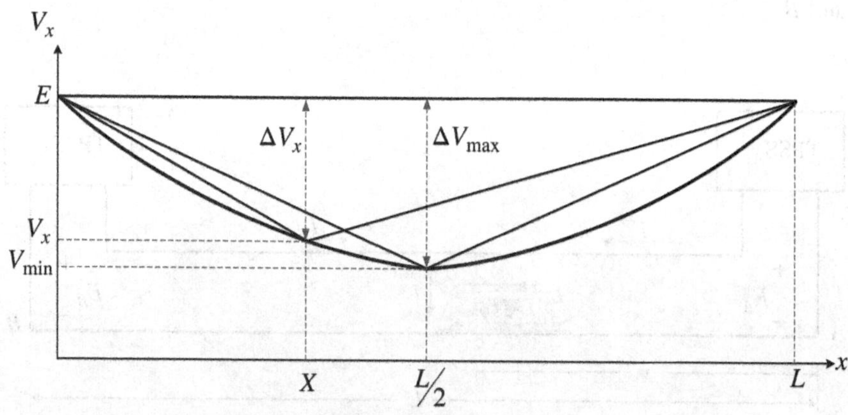

Figure 3.45 Voltage along the line with bilateral power supply and concentrated load.

Already introduced under the assumption $I = \text{const}$, the average voltage drop is

$$\Delta V_{av} = \frac{1}{L} \cdot \int_0^L r \cdot I \frac{(L-x) \cdot x}{L} dx = \frac{r \cdot L \cdot I}{6} = \frac{2}{3} V_{max} \tag{3.28}$$

In the case of several concentrated loads in the segment considered, the currents supplied by the substations must be determined and a diagram of the total drops due to the principle of superposition of the effects can be created.

Uniformly Distributed Load Similar to the case of unilateral power supply, it is assumed that per unit length a current is absorbed:

$$i = {}^{I_s}/_{L} \left[\frac{A}{km}\right] \tag{3.29}$$

Since the load is distributed evenly between the two substations, said substations will deliver the same current value equal to $I_s/2$.

As shown in Figure 3.46, the load absorbed by a line segment of length dx is

$$dI = i \cdot dx \tag{3.30}$$

In the generic segment x, we have

$$I_x = \frac{I_s}{2} - x \cdot i = \frac{I_s}{2} - \frac{I_s}{L} \cdot x = \frac{(L - 2 \cdot x)}{2 \cdot L} \cdot I_s \tag{3.31}$$

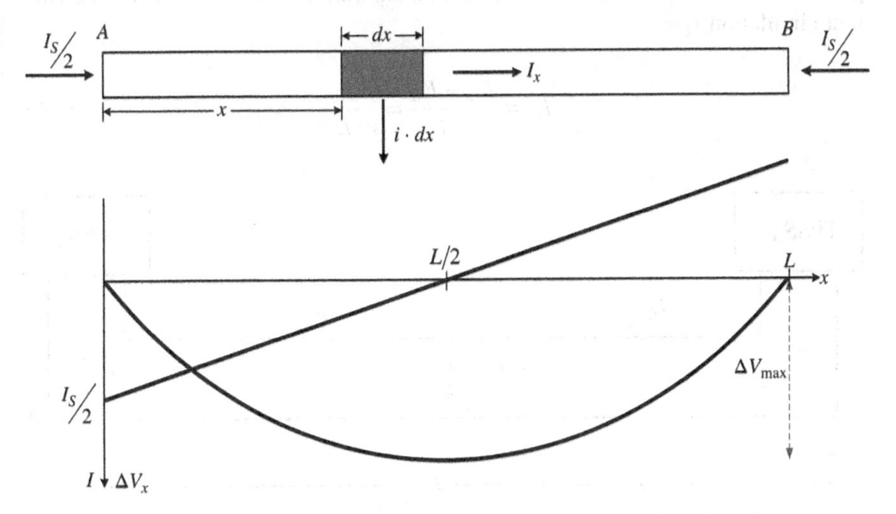

Figure 3.46 Bilateral power supply, load evenly distributed.

Therefore, the infinitesimal voltage drop on the infinitesimal segment length dx is

$$d \cdot V_x = r \cdot I_x \cdot dx = r \cdot \frac{(L - 2 \cdot x)}{2 \cdot L} \cdot I_s \cdot dx \tag{3.32}$$

$$\Delta V_x = \int_0^x d \cdot V_x = \frac{r \cdot I}{2 \cdot L} \int_0^x (L - 2 \cdot x) dx = r \cdot \frac{x \cdot (L - x)}{2 \cdot L} \cdot I_s \tag{3.33}$$

The maximum voltage drop is found for $x = {}^{L}/_{2}$, from which

$$\Delta V_{max} = \frac{r \cdot L \cdot I_s}{8} \tag{3.34}$$

It can be noted that, also in this case, with evenly distributed load, ΔV_{max} is equal to half of that that would occur with concentrated load in the same line configuration conditions.

The average voltage drop is still

$$\Delta V_{av} = \frac{2}{3} \cdot V_{max} \tag{3.35}$$

3.4.5.2.2 Substations Operating at Different Voltages Assume that substation A delivers a voltage E_A higher than the voltage E_B supplied from substation B, as reported in Figure 3.47.

Assuming initially that the substations are ideal and reversible, the circuit may be fixed by applying the principle of superposition. Considering only the presence of the substation and the effect of the train as null, there would be current circulation I_C:

$$I_C = \frac{E_A - E_B}{r \cdot L} = \frac{\Delta E}{r \cdot L} \tag{3.36}$$

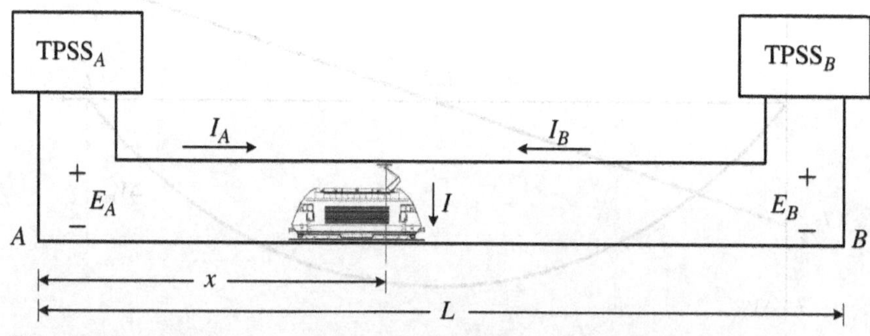

Figure 3.47 Bilateral power supply, concentrated load.

such that the two partial currents supplied by the substation would be

$$\begin{cases} I'_A = I_C \\ I'_B = -I_C \end{cases} \tag{3.37}$$

Considering now only the train effect, there is a division of the currents as in the case of substations delivering the same voltage:

$$\begin{cases} I''_A = I \cdot \dfrac{(L-x)}{L} \\ I''_B = I \cdot \dfrac{x}{L} \end{cases} \tag{3.38}$$

Combining the effects, therefore

$$\begin{cases} I_A = I'_A + I''_A = I \cdot \dfrac{(L-x)}{L} + I_C = I \cdot \dfrac{(L-x)}{L} + \dfrac{\Delta E}{r \cdot L} \\ I_B = I'_B + I''_B = I \cdot \dfrac{x}{L} - I_C = I \cdot \dfrac{x}{L} - \dfrac{\Delta E}{r \cdot L} \end{cases} \tag{3.39}$$

where it is possible to find the value of the voltage drop in a generic point x of the segment:

$$\Delta V_x = r \cdot x \cdot \frac{(L-x)}{L} \cdot I + r \cdot x \cdot I_C = \mathrm{r} \cdot x \cdot \frac{(L-x)}{L} \cdot I + r \cdot x \cdot \frac{\Delta E}{r \cdot L} \tag{3.40}$$

To find the point with the maximum voltage drop, the derivative of ΔV_x must be equal to zero with respect to x:

$$\frac{dV_x}{dx} = r \cdot I - \frac{2 \cdot I \cdot r \cdot x}{L} + \frac{\Delta E}{L} = 0 \tag{3.41}$$

which shows the x-coordinate with the maximum voltage drop:

$$x_{\Delta V_{\max}} = \left(I \cdot r + \frac{\Delta E}{r \cdot L} \right) \cdot \frac{L}{2 \cdot I \cdot x} = \frac{L}{2} \cdot \left(1 + \frac{\Delta E}{x \cdot I \cdot L} \right) = \frac{L}{2} \cdot \left(1 + \frac{I_C}{L} \right) \tag{3.42}$$

$\Delta V_{\max}$ is located in proximity to the lower voltage substation.

In the previous calculations, the fictitious current I_c was also considered as circulating in the line in the absence of a load in the case where the lower voltage substation B would be able to receive it, that is, a reversible one.

In the real case where the substations are not reversible, the theory shown thus far is valid only when the output current is positive. This is always true for the substation A at higher voltage, while it is not always true for the substation B at lower voltage, for which the term I_c is subtracted. Therefore, as long as the current I_B tends to be negative, that is, for $x < x_0$, substation B remains off due to the reverse bias of the diode bridges:

$$I_B < 0 \rightarrow \frac{x}{L} \cdot I - \frac{\Delta E}{r \cdot L} < 0 \rightarrow x < \frac{\Delta E}{r \cdot I} = x_0 \tag{3.43}$$

Under these conditions, the train has a unilateral power supply only from substation A. For $x > x_0$, both substations contribute to the train's power supply.

Differences in voltages can be unintentional or intentional. In the first case, it is due to a different power supply voltage from the two substations, while different output voltages from the substation may be intentionally created, for example, through the change in the transformation ratio of the transformers to control the distribution of the load between the same.

3.4.5.3 Parallel connection of the two tracks

In the double-track lines, the two rails may be connected in parallel, for example, at the end of the segment, when the power supply is unilateral, or at half segment when the power supply is bilateral. This practice is used only in cases where it leads to a real advantage, for example, with the crossing of lines, because putting them in parallel requires the installation of different switchgear to maintain the division of the line into multiple segments that can be individually isolated in case of failure.

3.4.5.3.1 Double-Track Unilateral Line with Parallel End Examining Figure 3.48, it can be noted that this situation is equivalent to having a line

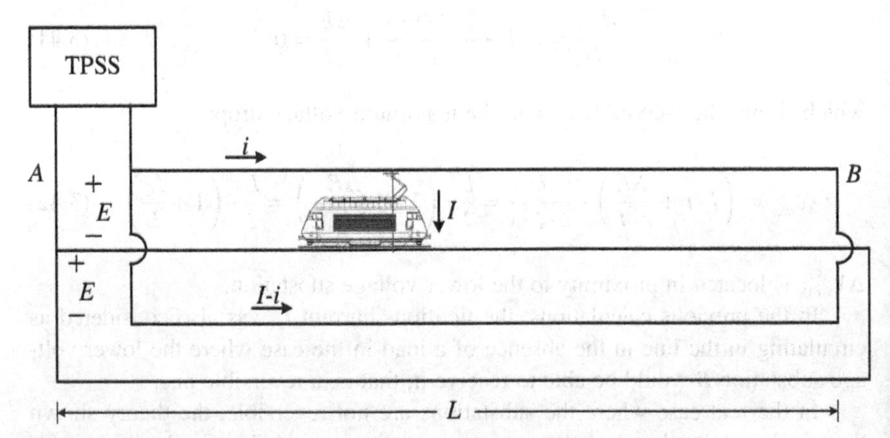

Figure 3.48 Double unilateral line with the parallel point at the end of the line.

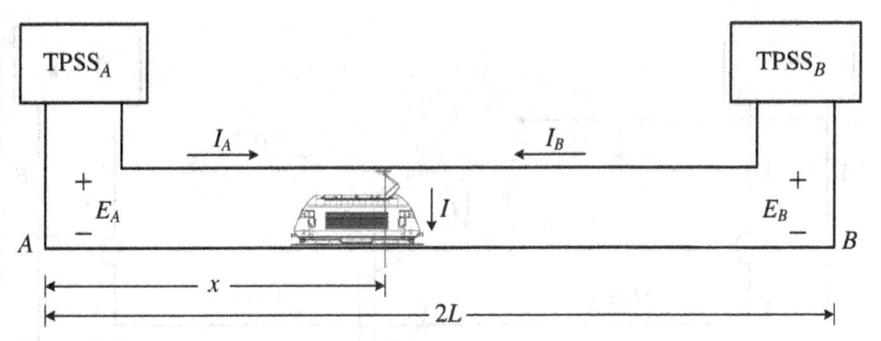

Figure 3.49 Diagram equivalent to double unilateral line with the parallel point at the end of the line.

with bilateral power supply substations operating at the same voltage, but with a distance between the two TPSS equal to $2L$ rather than L, as shown in Figure 3.49.

Placing the parallel at the end line, there is a maximum voltage drop equal at half line equal to $x = 2 \cdot L/2 = L$, that is,

$$\Delta V_{\max} = \frac{r \cdot L \cdot I}{2} \tag{3.44}$$

Without the parallel node, Figure 3.49, there would be a double maximum voltage drop:

$$\Delta V_{\max} = r \cdot L \cdot I \tag{3.45}$$

in the case of a train at the end of the line ($x = L$). This is true only if the other track is at no-load, that is, in the absence of trains, a condition that occurs in low traffic lines or for crossing of lines in which a train is in drive and the other in braking. In metro lines, characterized by high traffic and reduced distance between substations, the parallel between the lines at half segment is not generally applied.

3.4.5.3.2 Double-Track Line with Bilateral Power Supply Parallel at the Central Point We can consider two substations that power two tracks and the presence of an additional parallel connection halfway between them in order to decrease the voltage drop, as shown in Figure 3.50.

Assuming that the train is near substation A, that is, $0 \leq x \leq L/2$, this configuration is equivalent to having a bilateral central power supply without parallel in which the resistance of the o–B segment is equal to 1/3 compared to the referenced case.

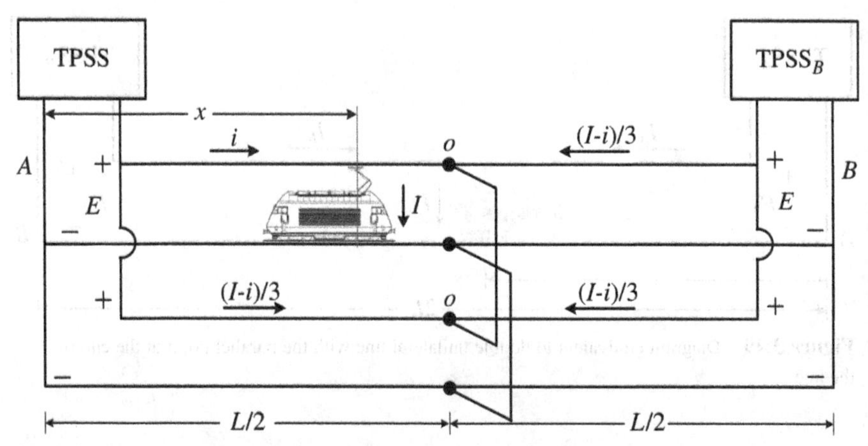

Figure 3.50 Double line with bilateral power supply and parallel node centerline.

Given voltage V at the head of the train, the voltage drop as a function of x provides:

$$\Delta V_x = E - V = r \cdot x \cdot i = (I - i) \cdot r \cdot \left(\frac{L}{2} - x\right) + \frac{(I - i)}{3} \cdot r \cdot \frac{L}{2} \tag{3.46}$$

From Equation 3.46, it follows that

$$i = (I \cdot (2L - 3x))/2L \tag{3.47}$$

Therefore, the voltage drop as a function of position x of the train is

$$\Delta V_x = r \cdot x \cdot i = r \cdot x \cdot I \cdot \frac{(2L - 3x)}{2L} \tag{3.48}$$

It can be noted that this value corresponds to the voltage drop that would occur along a single line powered bilaterally lengthwise 2/3·L (Figure 3.51).

Symmetrically applying the reasoning outlined above, for a train near the substation B, that is, $L/2 \leq x \leq L$, a specular trend is obtained compared to the previous case. Therefore, the voltage drop along the entire route assumes the trend of Figure 3.52.

3.4.5.4 Power supply system efficiency

To evaluate the power supply system efficiency, we can initially consider a train that it travels in time T on a line of length L at a constant speed $s = dx/dt$, and that absorbs a constant current I. The power absorbed by the train when in a

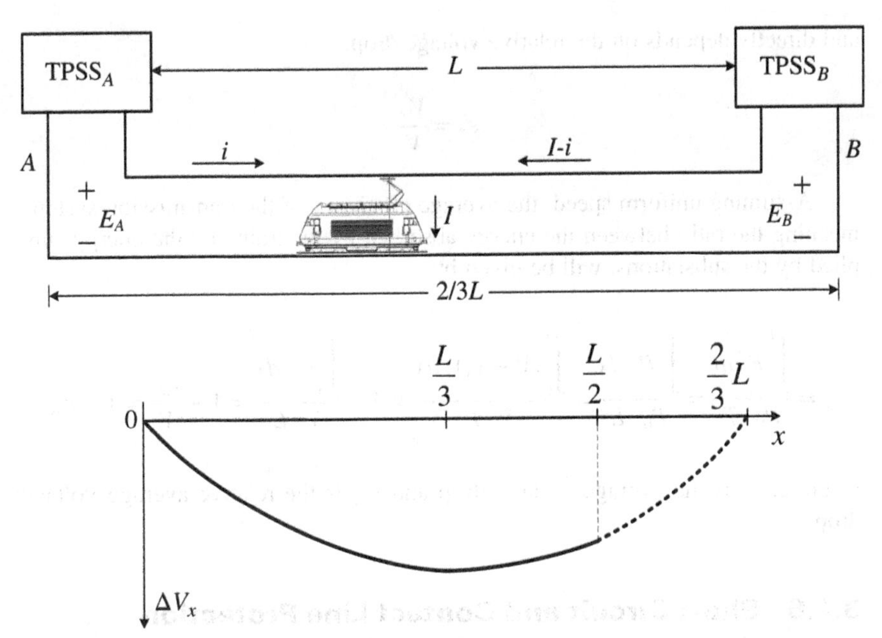

Figure 3.51 Equivalent diagram and relative performance of ΔV_x.

generic position x is equal to

$$P = V_x \cdot I = (V - V_x) \cdot I$$

Assuming that the substations are delivering the same voltage V in the bilateral power supply, the power supplied by the system is, therefore, $P_0 = V \cdot I$. Therefore, the instantaneous system efficiency of the line, given by the ratio between the effective power absorbed by the train and the power output from the substations, will be

$$\eta_x = \frac{P}{P_0} = \frac{V_x}{V} = 1 - \frac{V_x}{V} = 1 - \varepsilon_x$$

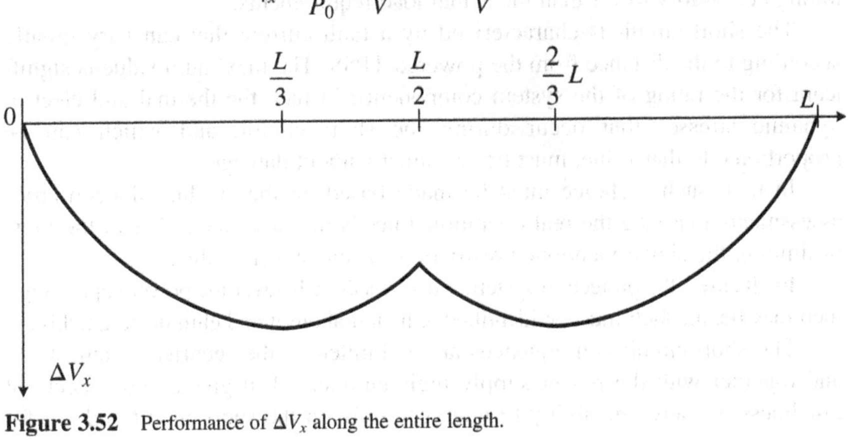

Figure 3.52 Performance of ΔV_x along the entire length.

and directly depends on the relative voltage drop:

$$\varepsilon_x = \frac{V_x}{V}$$

Assuming uniform speed, the average efficiency of the transmission system, meaning the ratio between the energy absorbed by the train and the energy supplied by the substations, will be given by

$$\eta = \frac{\int_0^T P \cdot dt}{P_0 \cdot T} = \frac{\int_0^L P \cdot dx}{P_0 \cdot L} = \frac{\int_0^L (V - v_x) \cdot dx}{V \cdot L} = 1 - \frac{\int_0^L v_x \cdot dx}{V \cdot L} = 1 - \frac{v_{av}}{V} = 1 - \varepsilon_{av}$$

where Δv_{av} is the average voltage drop and ε_{av} is the relative average voltage drop.

3.4.6 Short Circuit and Contact Line Protection

To avoid harmful consequences to the contact line (thermal and/or mechanical), any short circuit on said line must be able to be detected and eliminated in the shortest possible time by the opening of the high-speed breakers and automatic disconnectors powering the segment affected by the fault.

It is necessary, therefore, that the operation of electric traction systems and calibrations of various protective bodies are such as to ensure the timely and coordinated trip of the protections in case of a short circuit located anywhere. These criteria cannot and must not admit exceptions, even when compliance would restrict train movements due to abnormal system situations such as the decommissioning of one or more feeders of an entire TPSS or maximum current tuning necessarily lower than the actual load requirements.

The short circuit is characterized by a fault current that can vary greatly according to the distance from the powered TPSS. The maximum value is significant for the rating of the system components; in fact, the thermal and electrodynamic stresses that occur during the short circuit, and which can be proportional to that value, must be sustained without damage.

In fact, such a choice must be made based on the "technical–economic" assessments that take the real operational needs into account and with the view of limiting the above-mentioned restrictions as much as possible.

In all cases, the protection system must be safe; whatever the power supply regimen may be, the fault must be identified at its initial onset and eliminated quickly.

The short-circuit consequences are as limited as the security system itself, and together with the power supply regimen adopted, it guarantees aspects of timeliness, namely, the ability to intervene only on the short circuit and not for

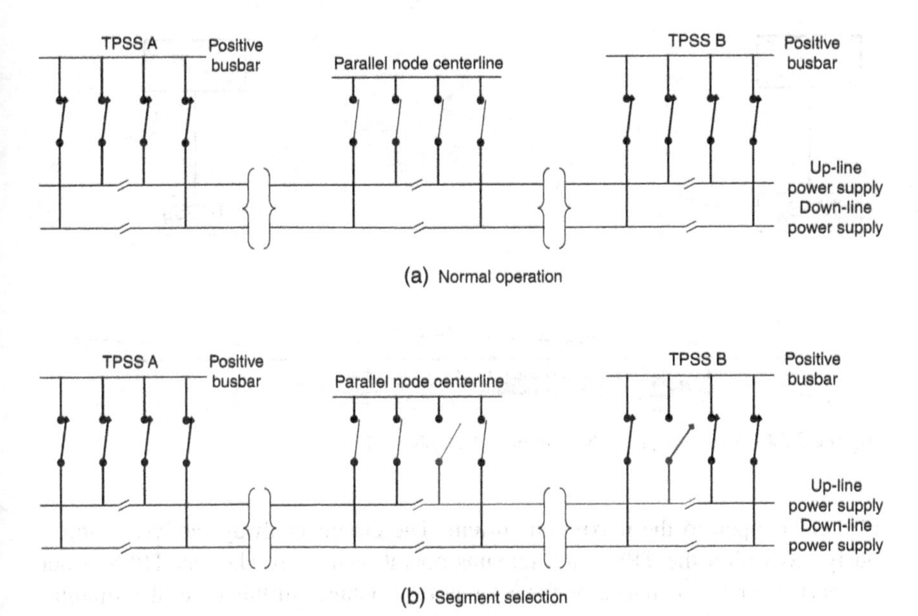

Figure 3.53 Reconfiguration of the contact line after a line fault.

operating currents, and selectivity, that is, the ability to locate the fault section of the trunk in order to power down only limited portions of the line.

In order to manage the DC traction network in order to select a segment in case of fault, the direct current line of contact is divided into sections, each of which is protected by a high-speed breaker. To allow a minor unavailability of the system, in midroute between two TPSSs, the line is sectioned so as to reduce the sudden switching off in case of fault. A disconnector system, however, proceeds to maintain the various sections in parallel with each other according to the diagram in Figure 3.53 showing the positive conductors, while the negative that corresponds to the track is connected to ground.

The protection must be able to clearly distinguish the maximum load current that is available and the minimum short-circuit current that generally occurs when the fault is found at the point farthest from the TPSS. Such discernment is more critical on a metro line where the currents at play, with the same power metro cars, are on the order of a few thousand amperes and fault currents away from the TPSS are lower than the calibration of the surge for the maximum current of the switches. In these fault situations and because of the almost always bilateral type of power supply of the contact line, it may happen that, after opening the circuit breaker of the TPSS that initially detects the fault, the short circuit is powered by the TPSS farthest away (Figure 3.54).

This dangerous event is excluded from an interlock circuit, referred to as simultaneous release, applied to the switches that feed the contact line in parallel and that causes the simultaneous opening of these devices in the event that one

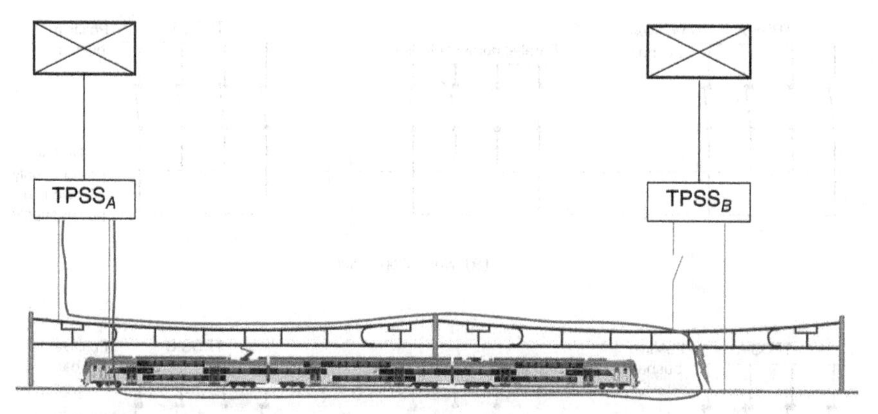

Figure 3.54 Power supply of the fault from the farthest TPSS.

of them is open to the maximum current. The circuit is always active, continuously fed within the TPSS, and remains operational even when an TPSS is out of service for maintenance or for a temporary outage. In this case, the simultaneous opening function is transferred to the TPSS breakers adjacent to the one excluded.

The case of a fault in the current section in the vicinity of the TPSS induces currents with a very high rising edge such as to have the breaker intervene, thanks to a safeguard called "current gradient."

The high-speed breaker can open even in the case of a "nonelectrical" event, that is, when dangerous current situations occur along the continuous line, signaled by the service staff or by the passengers, allowing the powering down of the entire contact line (electrical emergency).

In other cases, the fault may occur on the power supply lines that carry the energy from the TPSS to the contact line, for example, in the case of a cable insulation failure (ground fault) that causes the opening of all the high-speed breakers and the disconnection of the conversion group.

3.4.6.1 Protective relay of DC traction lines

The protection system of the lines at 3 kV DC must intervene in case of a fault, triggering alarm signals or opening the high-speed breaker command of the power supply for the protected line, all in relation to the modes set in the calibration phase. The calibration of the trip parameters is specific for each line in relation to its electrical and geometrical characteristics determining the values of the maximum and minimum short-circuit currents, as well as the variation of the current gradients (di/dt).

The trip must be subsequent to the passing of at least a threshold calibration and must remain active, even once the monitored magnitudes have returned to normal values.

The circuit breaker opening must follow the de-exciting of the relative restraint coil, whose power supply must therefore be ensured through a dedicated contact system.

The system must provide protection for

- maximum voltage,
- minimum voltage,
- directional maximum current,
- nondirectional maximum current,
- maximum temperature of the line conductors,
- directional overcurrent conditioned to the gradient, and
- minimum apparent resistance conditioned to the current gradient.

3.4.6.1.1 Maximum Voltage Trip The protection must operate by implementing a moving average of the samples detected in the interval voltage ΔT (adjustable as a parameter).

When the average value of the voltage detected is greater than the set calibration value, the protection should trigger and establish the trip (alarm signal or breaker-opening command) when the average voltage remains above the dropout value for a time that can be preset in the calibration phase.

3.4.6.1.2 Undervoltage Trip In the same criteria described in the preceding section for detecting an excess over the maximum allowable voltage, when the detected value in the ΔT interval (adjustable as a parameter) is less than the set calibrated value, the protection must start and establish the trip (signaling alarm or breaker-opening command) if the average voltage below the dropout value persists for a time that has been preset during calibration.

3.4.6.1.3 Trip for Maximum Directional Line Current The protection must intervene, normally causing the opening of the high-speed breaker, if the line output current is greater in the line than certain preset thresholds, both in regular power supply conditions and in the degradation of power supply conditions.

The same protection is used to protect the substation from possible reverse currents.

3.4.6.1.4 Trip for Maximum Nondirectional Line Current The protection must normally intervene, causing the opening of the high-speed breaker, if the current flowing through the same, regardless of the direction, exceeds a threshold value that has been preset in the calibration phase. The trip must also be implemented with an intentional adjustable delay to allow short overloads.

3.4.6.1.5 Trip for Maximum Temperature of the Contact Line The protection must normally intervene, causing an alarm signal or the opening of the high-speed breaker, if the current affecting the contact line causes a rise in temperature of the most critical conductors from a thermal point of view. Not being able to normally measure the temperature of the contact wire, this is estimated based on the value of the average current delivered by the respective power supply.

3.4.6.1.6 Trip for Maximum Current Conditioned to the Gradient The protection must intervene, causing the opening of the high-speed breaker, if the outgoing current to the contact line is affected by increases, in addition to preset values with gradients higher than those set in the calibration phase. Said trip must also be implemented with an intentional adjustable delay.

3.4.6.1.7 Trip for Minimum Apparent Resistance Conditioned to the Current Gradient The protection must intervene, thus opening the high-speed breaker, where with the outgoing current from the TPSS apparent resistance, calculated as the ratio of voltage and line current, suffers decreases characterized by current gradients higher than those set in the calibration phase to below the established values. Said trip must also be implemented with an intentional adjustable delay.

3.5 PROBABILISTIC METHODS FOR RATING THE TPSS

To rate the substations that feed an electric drive system, different criteria may be followed.

Trains' graphic timetable can be established, and knowing the characteristics of the convoys, the calculations of the current absorption in the different substations in each instant of time allows for obtaining the average current, the effective current, and the peak values.

Another method may be to assess the annual energy consumption based on the specific fuel consumption and at the scheduled time; then set, according to experience, the annual hours of use and therefore the nominal power to be installed.

Still another method is that proposed by De Koranyi in his article, which originally planned to evaluate the performance of the instantaneous current absorbed by a train traversing a section between two stations and obtain the average and effective current. Subsequently, knowing the location of the TPSSs (which are marked *A*, *B*, *C*, . . .), it is assumed that there is a constant average number of trains between the substations *A* and *C*, and it is assumed that TPSS_B

provides half the current required by the present convoys between *A* and *C*. By repeating this reasoning for each TPSS, it is possible to evaluate the average and effective current of each substation and then carry out the rating. Downstream of these considerations, it is necessary to verify that the formerly rated TPSS can tolerate all possible current peak combinations absorbed by the trains as overloading.

Finally, it is also possible to proceed as explained below: examine the absorption of convoys for different load conditions and obtain the average and effective current for the convoy; probabilistically determine, depending on the time of year, the number of trains available on the line and their composition; calculate the total current absorbed by the line by the drive system; then, knowing the location of the TPSS, evaluate the absorption of current in the various substations and proceed with the rating of the same.

Currently, the state-of-the-art method is to proceed with the rating of the TPSS according to the last of the newly exposed methods, which will be examined in more detail in the following sections.

3.5.1 The Probabilistic Method: General Information and Conditions

This method was developed in 1967 by Giorgio Meregalli (in fact, it is called the "probabilistic method" or "Meregalli" method), with particular reference to the operation of a metro line.

In this method, any mutual position in time of the trains themselves is deemed equally probable.

The occurrence or lack of contemporaneous absorptions (thus the superimposition or lack thereof of said absorptions in the substation) in this case may be considered as a matter of pure probability. This criterion evaluates the calculation procedure described below, which is mostly adopted in metro lines.

3.5.2 Representation of Absorption in a Train

3.5.2.1 Real absorption

The absorption of a train, without considering the possibility of the electric regenerative braking, can be fixed only by knowing the characteristic curves of the vehicle employed, the planoaltimetric trend of the line, the distance between the stops, and the average behavior of the drivers.

Having all the necessary data, the absorption may be set as a function of time; therefore, both the average current i_m and the effective current i_q may be derived, as well as the form factor f (ratio between the effective current and the average current).

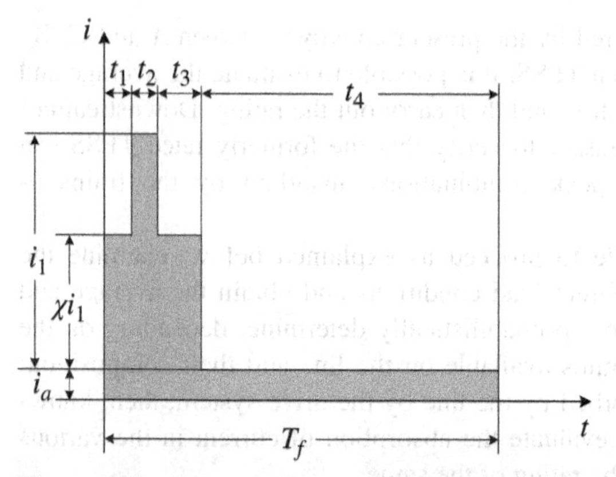

Figure 3.55 Discretization of the absorbed current over time.

The current trend over time is summarized for a preset segment (traversed in T_f: the time that includes idling time) as reported in Figure 3.55, which comprises three current significant values:

- a peak intensity $i_1 + i_a$ of duration $t_2 = \gamma \cdot T_f$ for each segment;
- an intermediate intensity value $\chi \cdot i_1 + i_a$ of duration $t_1 + t_3 = \delta \cdot T_f$;
- the current absorption of the auxiliary services, considered to be constant, with intensity $i_a = \alpha \cdot i_1$ and a duration of the idling time.

If i_m is the real average current and i_q is the actual rms current, using the previous notations and defining f as the actual form factor and ζ as the ratio between the real average current and peak current,

$$\zeta = \frac{i_m}{i_1}$$

Therefore,

$$\begin{cases} \gamma \cdot (1+\alpha) \cdot i_1 + \delta \cdot (\chi + \alpha) \cdot i_1 + (1 - \gamma - \delta) \cdot \alpha \cdot i_1 = i_m \\ \gamma \cdot (1+\alpha)^2 \cdot i_1^2 + \delta \cdot (\chi + \alpha)^2 \cdot i_1^2 + (1 - \gamma - \delta) \cdot \alpha^2 \cdot i_1^2 = i_q^2 \end{cases} \tag{3.49}$$

From the foregoing, the desired values of δ and of χ are obtained, namely,

$$\begin{cases} \delta = \dfrac{(\zeta - \alpha - \gamma)^2}{f^2\zeta^2 - \alpha \cdot (2\zeta - \alpha) - \gamma} \\ \chi = \dfrac{\zeta - \alpha - \gamma}{\delta} \end{cases} \tag{3.50}$$

3.5.3 Supply of a Substation

Assuming a double-track line fed by a single substation and traveled on by trains with time intervals between them equal to T_i; still assuming T_f as the average time it takes to travel a segment, including the time to stop at a station and T_s as the time necessary for a train to travel the line.

The number of trains N simultaneously in operation on the line will be equal, for each track, to the ratio between T_s and T_i, and thus for both tracks, at

$$N = 2 \cdot \frac{T_s}{T_i} \tag{3.51}$$

It is assumed that we can consider equally probable every mutual spacing between the maximum peaks of the train absorption, and be able to predict the loads purely considering the likelihood of connections of said peaks.

This is justified by the following considerations: first, the maximum absorption peak of a train, in a metro, lasts a few seconds and regardless is a modest fraction of the entire interval, which can be less than 1/15 of the total time.

The total duration of the traction phase is modest and less than 1/5 of the total time.

It must still be taken into account that, particularly at peak times (which are crucial to secure the rated power of the power supply groups), the interval T_i between trains is usually not more than twice or at most three times the T_f travel time of a segment. In such conditions, it is sufficient to have delays or advances on the order of a few tens of seconds to match the peaks of two trains that are generally not simultaneous.

3.5.3.1 Absence of superposition of the peaks

Now, we will consider a line and first examine the case where there is no superposition of peaks.

If D is the duration of time considered (e.g., the peak period of 1 h and a half) and N is the number of trains in line, there is absorption by the motors for the fraction of time β of each train; therefore, $\beta \cdot N$ for all trains and as a total for the duration $\beta \cdot N \cdot D$; on the other hand, there is only absorption of auxiliary components during the remaining time $(1 - \beta \cdot N) \cdot D$.

Under these conditions, the following absorptions would occur, according to Figure 3.55.

- $i_1 + i_1 - N \cdot i_a$ during $\gamma \cdot N \cdot D$ with probability $\gamma \cdot N$
- $\chi \cdot i_1 + N \cdot i_a$ during $\delta \cdot N \cdot D$ with probability $\delta \cdot N$
- $N \cdot i_a$ during $[1 - (\gamma + \delta) \cdot N] \cdot D$ with probability $1 - (\gamma + \delta) \cdot N$

3.5.3.2 Presence of superposition of the peaks

In reality, there may be simultaneous absorption of two or more of the N trains present on the line.

To calculate the probability that there is superposition between the two peaks, each of which lasts in the range considered D, $\beta \cdot D$ seconds, meaning β is probable, N all possible combinations of two superpositions is considered for the trains, each of which will have the probability β^2. Similarly, if all the possible combinations of three superpositions are considered, each combination will have the probability β^3 and so on until the one combination of N trains with superposition of all peaks simultaneously, with probability β^N.

When evaluating the number of possible combinations, it must be considered that when all the possible superimposition combinations m are taken into account, with $m < N$, all existing trains are not available for calculation, since those that have been already considered for the higher order combinations will be subtracted from them.

For this reason, it would be advantageous to resume the argument starting from the maximum combination.

The sole combination of N trains with probability β^N, on average, will last the seconds in the interval $DT = \beta^N \cdot D$.

If we consider the combination of N trains with superimposition of $(N-1)$ peaks, referencing the expression of the binomial coefficient:

$$\binom{n}{k} = \frac{n!}{k! \cdot (n-k)!}$$

will yield a number of possible combinations equal to

$$\frac{N!}{(N-1)! \cdot (N-N+1)!} = N$$

The overall probability is therefore equal to $N \cdot \beta^{N-1}$. It must, however, be inferred from these combinations already considered in the combination of order N, each having a probability β^N, and comprehensively equal to $N \cdot \beta^N$.

In total, therefore, the order of probability peaks $(N-1)$ will be

$$P_{N-1} = N \cdot \beta^{N-1} - N \cdot \beta^N = N \cdot \beta^{N-1} \cdot (1-\beta)$$

and therefore the probable duration:

$$T_{N-1} = P_{N-1} \cdot D = N \cdot \beta^{N-1} \cdot (1-\beta) \cdot D$$

Similarly, for the order combinations $(N - 2)$, the total unadjusted probability is

$$\beta^{N-2} \cdot \frac{N!}{(N-2)! \cdot (N-N+2)!} = \frac{N \cdot (N-1)}{2} \cdot \beta^{N-2}$$

From this, however, the already considered order combinations $(N - 2)$ can be both of the order N and the order $(N - 1)$. The first are

$$\frac{N!}{(N-2)! \cdot (N-N+2)!} = \frac{N \cdot (N-1)}{2}$$

each with probability β^N; therefore,

$$\frac{N \cdot (N-1)}{2} \cdot \beta^N$$

The latter, as already seen, are N; in each of them, there is a number of combinations equal to

$$\frac{(N-1)!}{(N-2)! \cdot (N-1-N+2)!} = (N-1)$$

in total, therefore, $N \cdot (N - 1)$, each with probability$\beta^{N-1} \cdot (1 - \beta)$.

Ultimately the correct probability of superimposition of $(N - 2)$ absorption is therefore

$$P_{N-2} = \frac{N \cdot (N-1)}{2} \cdot \beta^{N-2} - \frac{N \cdot (N-1)}{2} \cdot \beta^N - N \cdot (N-1) \cdot \beta^{N-1} \cdot (1-\beta)$$
$$= \frac{N \cdot (N-1)}{2} \cdot \beta^{N-2} \cdot \left[1 - \beta^2 - 2 \cdot \beta \cdot (1-\beta)\right] = \frac{N \cdot (N-1)}{2} \cdot \beta^{N-2} \cdot (1-\beta)^2$$

with respective probable duration equal to

$$T_{N-2} = P_{N-2} \cdot D$$

Proceeding in a completely analogous way, it is found that for any integer m between 0 and N, the correct probability of superimposition of $(N - m)$ absorptions is

$$P_{N-m} = \frac{N!}{(N-m)! \cdot m!} \cdot \beta^{N-m} \cdot (1-\beta)^m$$

To sum up, for each single value, then

$$\begin{cases} P_N &= \beta^N \\ P_{N-1} &= N \cdot (1-\beta) \cdot \beta^{N-1} \\ P_{N-2} &= \dfrac{N \cdot (N-1)}{2} \cdot (1-\beta)^2 \cdot \beta^{N-2} \\ P_{N-3} &= \dfrac{N \cdot (N-1) \cdot (N-2)}{6} \cdot (1-\beta)^3 \cdot \beta^{N-3} \\ P_{N-m} &= \dfrac{N!}{(N-m)! \cdot m!} \cdot (1-\beta)^m \cdot \beta^{N-m} \\ P_3 &= \dfrac{N \cdot (N-1) \cdot (N-2)}{6} \cdot (1-\beta)^{N-3} \cdot \beta^3 \\ P_2 &= \dfrac{N \cdot (N-1)}{2} \cdot (1-\beta)^{N-2} \cdot \beta^2 \\ P_1 &= N \cdot (1-\beta)^{N-1} \cdot \beta \\ P_0 &= (1-\beta)^N \end{cases} \tag{3.52}$$

For this, the duration of any condition k is given by

$$T_k = P_k \cdot D$$

It can be noted that $T_0 = P_0 \cdot D$ is the time during which no trains are in the traction phase, and therefore absorption occurs only by auxiliary components.

Those that happens in the superimpositions is best to study in more detail, starting with the first of only two currents, each with probability β and duration for the single peak equal to $\beta \cdot T_f$.

Indicating the generic time with the letter T, the duration of the superimposition can apparently vary between 0 and $\beta \cdot T$ with instant starting of the second peak included between $-\beta \cdot T_f$ and $+\beta \cdot T_f$, compared to the first peak. Having assumed that the probability does not vary during the period considered, the average length of a superimposition will be $\beta \cdot T_f/2$. The probability that in any case there is superimposition is given by the probability that the start of the second peak drops, as has been said, $-\beta \cdot T_f$ and $+\beta \cdot T_f$ (i.e., within a time period $2\beta \cdot T_f$), which is therefore equal to 2β. Thus, the total duration of the double peaks is equal to

$$2 \cdot \beta \cdot \beta \cdot \frac{T}{2} = \beta^2 \cdot T$$

that is, the duration value already considered previously. In a similar manner, it can be shown that even with higher order peaks, the total duration of each peak is equal to that that was considered previously.

Below, in general, we will not take the duration of the division in a certain number of peaks into account, but will simply consider the probability and duration of each peak according to (3.52).

The average duration of each single peak is equal to $\beta \cdot T_f$ for the first order, and $\beta \cdot T_f/2$ for subsequent orders.

The number of probable peaks n of various orders is given by the ratio between the probable duration of all the peaks of each order and the single peak duration; therefore, for the first peak it will be

$$n_1 = \frac{P_1 \cdot D}{\beta \cdot T_f} = N \cdot (1 - \beta)^{N-1} \cdot \frac{D}{T_f}$$

and for subsequent peaks,

$$n_j = \frac{2 \cdot P_j \cdot D}{\beta \cdot T_f}$$

where P_j is given by (3.52).

3.5.4 Power Supply by a Single Substation

In the previous discussion, the fraction of time in which the drive motors of a train absorb was generally indicated with β, or are equal to $\gamma + \delta$ if all the trains have the same composition.

Instead, in the case where the compositions are different from train to train, the number of cars that make up a train is indicated with v, and the percentage of trains having the composition required is indicated with τ. Index 1 is used (or 2, 3, etc.) to indicate the larger and then gradually smaller compositions, respectively.

The average composition will be

$$v_m = v_1 \cdot \tau_1 + v_2 \cdot \tau_2 + \cdots \tag{3.53}$$

The absorption of a train, if i_v is that of a car, is obviously $v \cdot i_v$. When there is superimposition in the absorption of two trains in the composition v_1 and v_2, the total power consumption will be equal to $(v_1 + v_2) \cdot i_v$. If P_2, there is a probability of having double peaks given by (3.52), because the number of trains in the composition is v_1 or v_2, respectively equal to $N_1 = \tau_1 \cdot N$ or $N_2 = \tau_2 \cdot N$; the probability that there are connections of the peaks of two trains with the same or different composition will be given by

$$\begin{cases} P_{11} = \tau_1^2 \cdot P_2 \\ P_{12} = \tau_1 \cdot \tau_2 \cdot P_2 \\ P_{21} = \tau_1 \cdot \tau_2 \cdot P_2 = P_{12} \\ P_{22} = \tau_2^2 \cdot P_2 \end{cases} \tag{3.54}$$

At this point, for a given type of metro operation, it is necessary to verify that the rectifier to be installed is able to withstand all the probable current for the probable duration of time that may occur. After passing this test, for rating substations, it must be ensured that the rated current of the group or groups is not lower than the effective current for anticipated loads.

3.5.5 Form Factor for Substation

The form factor F of a substation plays a role of particular importance, because its knowledge and that of the average current is often sufficient to define the power to be installed.

The calculation of the form factors of the current absorbed by the train and of the current delivered by the TPSS can be achieved using relations (3.55).

In general, given P probability of a current i, the average value I_m, the rms value I_q, and the form factor F can be expressed as

$$\begin{cases} I_m = \sum P \cdot i \\ I_q = \sqrt{\sum P \cdot i^2} \\ F = \dfrac{I_q}{I_m} \end{cases} \tag{3.55}$$

3.5.6 Power Supply with Several Substations

When a line is fed from different substations in parallel, they influence each other reciprocally, so that each gives part of its load to the other and vice versa, the other takes some of the load from the first.

The calculation of these mutual influences is rather laborious and we will discuss some of the most significant cases here.

It is interesting, however, to note that the substations tend to balance the loads in general.

Below, it will be generally assumed that

a. the resistance per unit of length of the contact line (including that of the circuit) is constant and equal to r,

b. the internal voltage drop at the substations is proportional to the current supplied I and equal to $R \cdot I$,

c. the value of R is equal for all substations, and

d. the load voltages of all substations are equal.

The introduced hypotheses may be considered acceptable because in the practical implementation of metro lines, substations have a uniform construction and the same components are used.

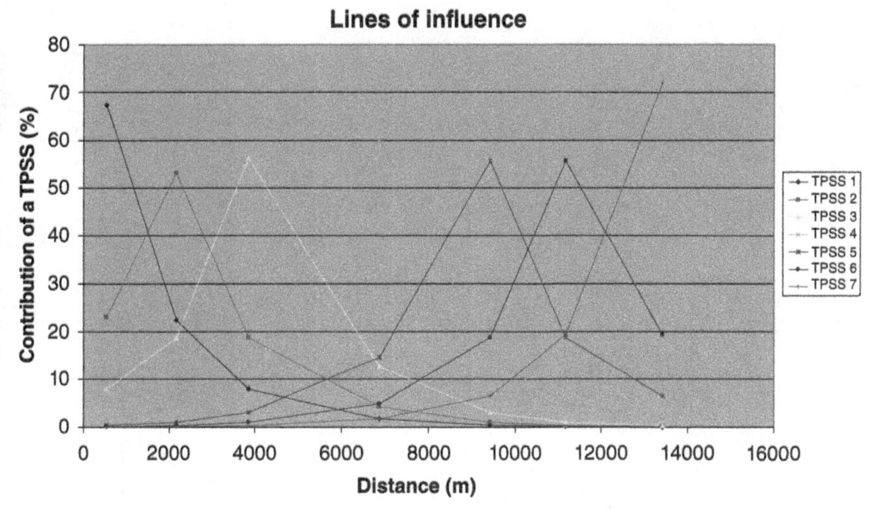

Figure 3.56 Lines of influence of the contributions of each substation for a unit load.

If the three conditions (b), (c), and (d) are met, then the reciprocity theorem applies, which states that given a line with several substations in parallel, any substation χ contributes to the power supply of a load I found at the substation γ with a current $I_{\chi\gamma}$ that is equal to the current $I_{\gamma\chi}$ with which the substation γ would contribute to feed the same load I found at the substation χ.

As a corollary, it follows that if a load I moves from a substation χ to another adjacent one γ, assuming that all the conditions previously introduced are met, the current with which any of the line substation contributes to the load power supply varies linearly from the values that each of them takes in χ to those that it will take in γ.

As a result of the two previous theorems, we see how it is possible to find the contribution of a particular substation to a load I placed along the line, to assume a load I in the same TPSS, calculate the contributions of each substation, reflect them on a graph diagram, and connect them with straight segments. In this manner, the line of influence of a substation for a unit load that is found in any point of the line can be found.

Furthermore, assuming load I in different cases, initially in the first TPSS, then in the second, and so on up to the last TPSS and repeating the above-stated procedure for each case, we obtain the lines of influence of the contributions of each substation for a unit load that is found in any point of the line.

In Figure 3.56 the contributions are expressed as a percentage and the load I is clearly equal to 100%.

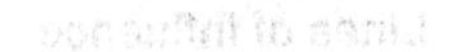

Chapter 4

AC Systems at Mains Frequency

Meeting trains' high power demand requires satisfying the uptake demands, minimizing the resistive losses and a sufficiently high voltage that implies, for the same power, a contained value of current.

The use of a single-phase alternating current allows some of the fundamental objectives for electric traction to be achieved:

- The possibility of raising the voltage level of the contact lines, thereby reducing the current values to be taken up, and at the same time ensuring the most appropriate values for the power supply of the actuators and onboard converters through the simple use of transformers
- The possibility of maintaining the single-wire contact line
- The direct connection to the industrial network
- The increase in the line's surface power density

The need to be able to further exploit the economy of high-voltage transmission without having to introduce an excessive voltage level on the contact lines has led to the study and development of the 2 × 25 kV system that, by using a system with three voltage levels (contact line, track, and a "negative" power supply), allows power to be transmitted at 50 kV and makes it available to trains at a voltage of 25 kV.

These power supply systems are used for lines that require high surface density powers and this involves the need to tap into the energy of very powerful high-voltage networks. Consequently, they are extremely costly, not only for the entire unit of HV substations, but also and especially for primary lines. The cost of the latter at the voltage levels usually employed in the industrial HV network considerably affects the total cost of the systems, which is why attempts are made to minimize its development. The choice of the industrial AC system has the advantage of being able to connect the substations (which have simple transformers) directly to the national transmission network and increase their spacing.

Electrical Railway Transportation Systems, First Edition. Morris Brenna, Federica Foiadelli, and Dario Zaninelli.

4.1 CONFIGURATION OF THE POWER SUPPLY SYSTEM

The power supply system provides the transportation, processing, and distribution of electricity from HV delivery points up to loads that are mobile (train) and fixed peripheral posts (FPP).

The structure of the power supply system is basically conditioned by the need to limit the unbalances it creates on the high-voltage network from the withdrawal of energy in typical single-phase electric traction of 25 kV–50 Hz, and by the need to give maximum operational continuity to traction power substations (TPSSs). The disadvantage of the single-phase power supply at a power frequency derived directly from the industrial HV network is that unbalance can be introduced into the national transmission network due to the high single-phase drawing power, especially when compared to the powers involved in the early industrial electrification systems. This unbalance can be very damaging to generators, motors, and other three-phase loads due to the asymmetry of the voltages that result from them.

With regard to currents, the ratio

$$K_i = {}^{I_i}/_{I_n}$$

between the negative sequence component I_i of the current and the rated value I_n must not exceed the limits conditioned by the type of construction, the rated power, and the cooling system for three-phase generators; these limits are generally understood to be between 5 and 10%.

It also harms other civil and industrial loads supplied by the three-phase network and the asymmetry of voltages, as a direct consequence of the unbalance in the currents. Considering, in this respect, the voltage unbalance factor

$$u = {}^{V_i}/_{V_d}$$

the ratio between the negative sequence component of line voltages V_i and the positive sequence component V_d; this can be expressed with sufficient approximation by the coefficient of unbalance:

$$u \cong K = {}^{P_{sp}}/_{P_{sc}}$$

where P_{sc} is the short-circuit power of the three-phase network when the single-phase power P_{sp} requested is taken.

The coefficient K is a vector quantity that represents the maximum percentage value within which the differences in individual line voltages are contained, with respect to their average value. The presence of three-phase industrial motors determines the following limits:

- $K < 1\%$ in continuous operation
- $K < 1.5\%$ for a few minutes

The following two approaches can be taken to respect these limits:

a. To connect the TPSSs to nodes or industrial HV network lines with sufficiently high values of short-circuit power P_{sc}. This solution has been facilitated by the strengthening of the general network. The highest powered networks P_{sc}, however, are those at a voltage of 220 kV or higher; this signifies a considerable increase in the connection, HV equipment, and transformer costs: in comparison to the 60–130 kV systems, there is approximately a 30–50% increase. For French and Spanish HS lines, for example, the normal power supply voltage is 220 kV and exceptionally reaches 400 kV.

b. To distribute the single-phase traction loads onto the three phases of the network. This measure involves considerable complications in the TPSS circuits and on the railway lines, as will be shown later.

In general, in order to evenly distribute the single-phase loads on the three-phase network having line voltage V_P, it is necessary to insert three transformers as shown in Figure 4.1; taking into account that one of the secondary terminals

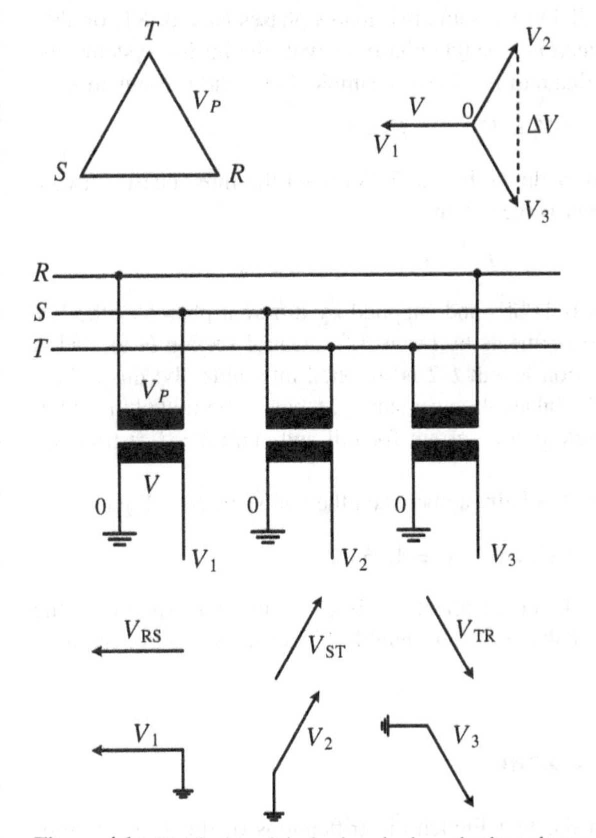

Figure 4.1 Distribution of single-phase loads on the three-phase network, with a three-phase sequence. *Vp*: primary line voltage; *V*: line voltage; ΔV: potential difference between the phases, in line.

of each transformer must be connected to the track, the three voltages available for the line circuit's power supply would form a star, with the effective value: $V = {}^{V_p}/_h$ (h = transformation ratio).

It is evident that balance would only be guaranteed when the three transformers have identical instantaneous loads and with an equal power factor.

Since two transformers are generally installed in each TPSS, other similar solutions need to be adopted: It can be observed in any case that, as the TPSSs positioned at the average distance L are connected to different phases of the three-phase HV network, the need arises to divide the railway line into segments that are V voltage powered, not into phases, with each other; the segments must, therefore, be separated by sectioning with neutral parts, indicated with N in Figure 4.2.

4.1.1 Substations with Transformers in Parallel

The power supply diagram is shown in Figure 4.2a: The two transformers of each TPSS are connected in parallel to the same two mains phases (RS, or ST, or TR), therefore, the primary connection is single phase, as with the busbar systems and HV equipment. The TPSS diagram is relatively simple, but clearly results in

$$V_i = V_d; \qquad K = 1$$

By indicating P_1 as the power drawn by the TPSS from the three-phase network nodes, the single-phase power is picked up:

$$P_{sp} = P_1$$

The line section between two TPSSs and supplied by different phases of the HV network is divided into two segments by means of a neutral section N located at the halfway point. Each section length $L/2$ is supplied intermittently: this is how the advantage is lost from the bilateral power supply, which is permitted in continuous and in alternating single-phase current for rail and centralized distribution use.

The potential difference lies halfway between the two segments (Figure 4.1):

$$\Delta V = \sqrt{3} \cdot V = 42.5\,\text{kV}$$

Even though the insulation level for all of the line circuits corresponds to the rated voltage of 25 kV, the voltage is considerably higher at the end of the neutral segments.

4.1.2 The Scott Diagram

A brilliant idea adopted in the first French electrifications of the 1950s is provided by the Scott diagram, which is comprised of two transformers that have a

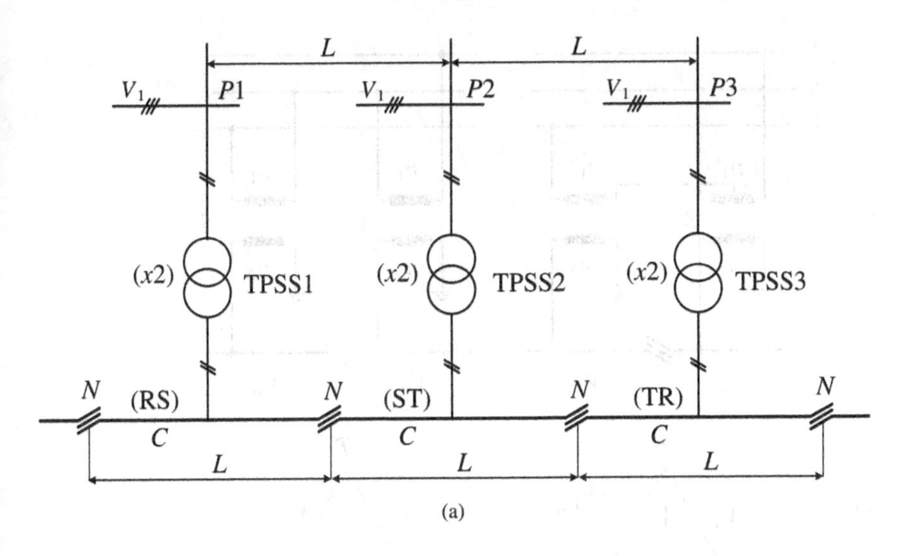

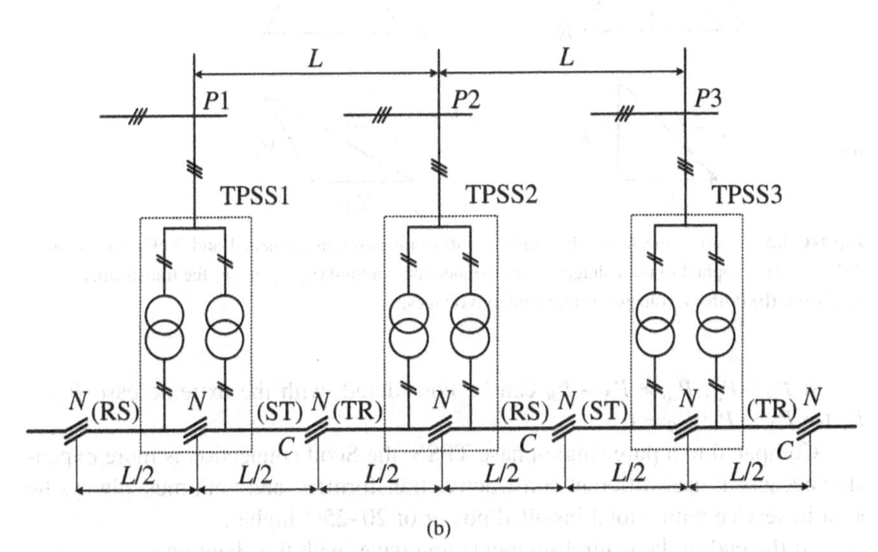

Figure 4.2 Single-phase AC power supply 25 kV–50 (60) Hz. (a) Substations with transformers in parallel. (b) Substations with V connected transformers. P: three-phase industrial HV network nodes; voltage V_p; TPSS: traction substations; C: contact line; N: sectioning with a neutral section; L: wheelbase of the substations.

primary side with different numbers of coils (Figure 4.3); the output voltages V_A and V_B are out of phase by $\pi/2$.

If the loads P_A and P_B of the two segments are equal and have identical power factors, the loads on the three-phase network are perfectly balanced, which is $P_m = 0$.

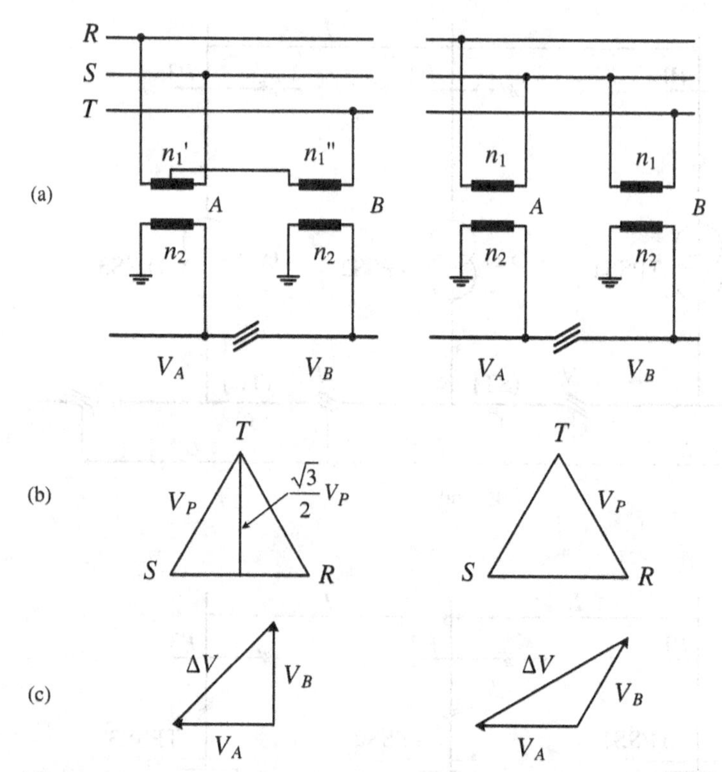

Figure 4.3 Scott connection (left) and V (right) of the two transformers A and B of a substation 25 kV–50 Hz supplied to the voltage V_p. (a) Connection diagram (n_1', n_1'', n_1, n_2 the transformer turn numbers). (b) Primary voltages. (c) Secondary voltages.

If $P_A > P_B$, $P_{sp} \approx P_A - P_B$ can be considered, with the extreme case $P_{sp} = P_1$ for $P_A = P_1$; $P_B = 0$.

Compared to a pure single-phase TPSS, the Scott connection is more expensive because it uses different constructive transformers, and both must always be kept in service with a total installed power of 20–25% higher.

At the end of the neutral segments you have, with this diagram

$$\Delta V = \sqrt{2} \cdot V = 35.4\,\text{kV}$$

4.1.3 The *V* Diagram

In order to reduce the equivalent single-phase P_m load of the three-phase network and to use identical transformers, the two transformers of each TPSS can be connected to different phases of the HV network, as in Figures 4.2b and 4.3.

In this case, the primary power supply for the TPSS, the busbars, and the HV equipment must be three phase. The whole line is then divided into long $L/2$

segments, separated by neutral N sections, both in the TPSS and halfway through the section.

The secondary voltages U_A and U_B are mutually out of phase by $2 \cdot {}^{\pi}/_{3}$ (Figure 4.3, on the right).

The equivalent single-phase P_m load corresponds approximately to the highest powers P_A or P_B; we have $P_{sp} \approx {}^{P_1}/_{1}$ for $P_A = P_B = {}^{P_1}/_{2}$.

In order to optimize the distribution of the tensile load on the three RS–ST–TR phases of the primary network (Figure 4.1), the transformer connections are established so that on segments of the line, the voltages will follow each other in order sequence 3:

$$V_1 - V_2 - V_3 - V_1 - V_2 \cdots$$

The potential difference at the ends of each neutral N section is still $\Delta V = \sqrt{3} \cdot V$.

4.1.4 Order Sequence 6

In order to reduce the potential difference between the sections of line, it is possible to carry out a three-phase star of the V voltages, where the line circuit can be divided into six or more consecutive segments (Figure 4.4): For each primary line voltage, the secondary terminal connected to the track varies; in the star of tension $V_1 - V_6$ center 0 is, as usual, the potential of the track.

In this case, we have

$$\Delta V = V \tag{4.1}$$

4.1.5 Evolution of Solutions

During the first French single-phase electrification at 50 Hz in Savoy at the beginning of the 1950s, concerns were focused on the issue of unbalance and the Scott connection of the two transformers for the substation in Annecy was implemented; the abovementioned TPSS was, in fact, connected to a "weak" node of the industrial HV network at 42 kV.

In subsequent electrification processes, the Scott connection was used in further applications, but it was noted how the actual substance of tensile loads and the high power of generating stations made coefficients for the K unbalanced, corresponding to the pure single-phase withdrawals, acceptable for other industrial users, which greatly simplified the TPSSs by connecting the two identical transformers in parallel.

In order to avoid excluding the possibility of supplying power with the same group phases of more TPSSs, the phases were normally rotated, therefore, the neutral sections came to be located halfway along section L in the middle of

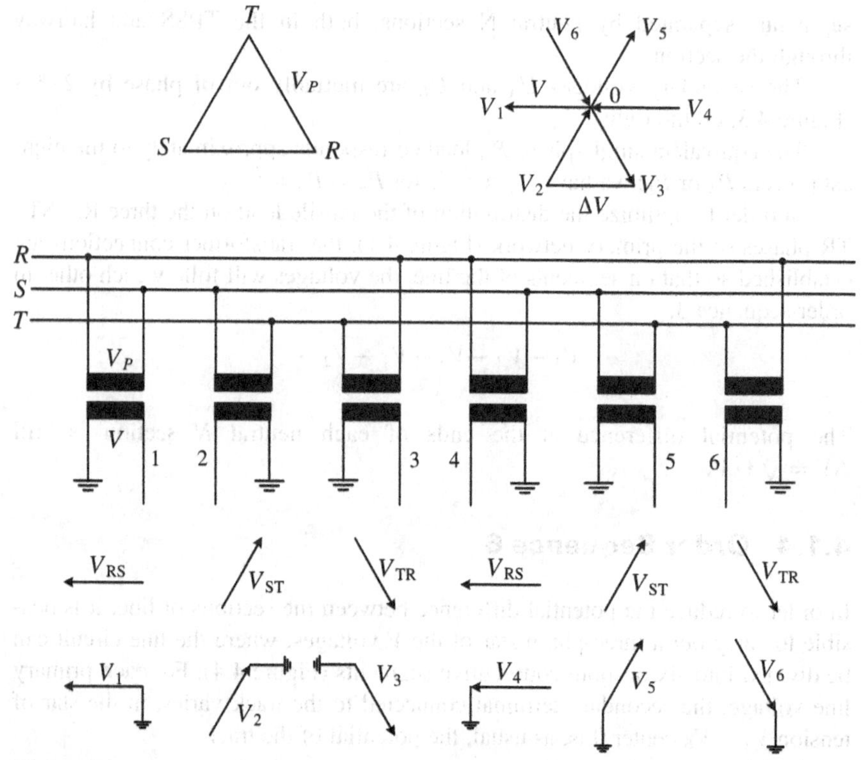

Figure 4.4 Distribution of single-phase loads on the three-phase network, with a six-phase sequence.

the countryside. Connection V was only necessary in regions with a high concentration of traffic, with the drawback of having the neutral sections also in correspondence with the TPSSs, often located near stations.

In more recent years, the considerable increase in the power supplied to each TPSS, such as to require the use of 40 MVA and also 60 MVA transformers, and more stringent limits of the coefficient K imposed by energy suppliers, forced engineers to more frequently resort to V connections and to supplying power to the TPSSs at 220 and 380 kV voltages, with a heavy increase in costs for all of the HV equipment and machinery.

In such circumstances it may be difficult to reconcile the location of the TPSSs with the local presence of HV lines with suitable power: In order to avoid constructing special HV primary connection lines, which are expensive, the French engineers preferred to use the variant 2×25 kV for TGV lines, which we will come back to, to make it possible to space out the TPSSs sufficiently in order to join up the HV network links.

4.2 SUBSTATION DIAGRAM

The diagram of a substation with two transformers with *V* connections is shown in Figure 4.5, assuming that the TPSS is powered by two primary independent three phases, each protected by an automatic triple-pole switch. On the 25 kV side, in addition to the normal single-pole disconnectors, it can be seen that

a. the protective single-pole main switches that work automatically are used to interrupt the maximum number of fault currents. Over the years there has been a switch from oil breakers to those made with sulfur hexafluoride (SF6); the breaking capacity reaches 15–25 kA, with a trip time of 60 min;

b. the switch disconnector is capable of opening the normal operating currents. These devices have also changed from initially using oil and

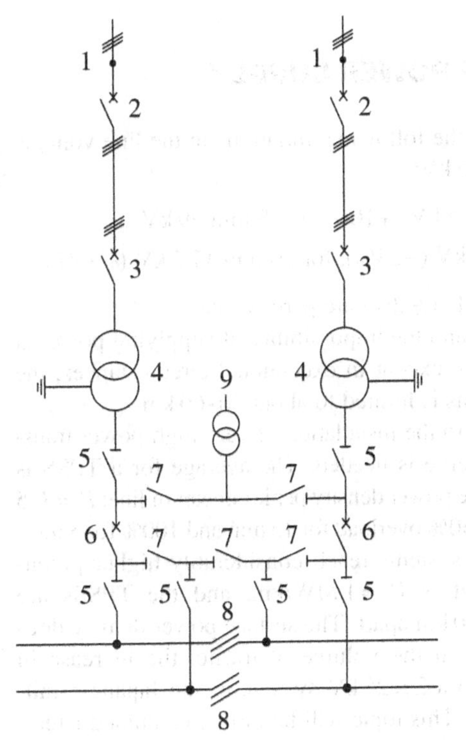

Figure 4.5 Diagram of a substation 25 kV–50 Hz with transformers connected to *V*. 1: three-phase HV lines; 2: automatic triple-pole line circuit breakers and related disconnectors; 3: automatic bipolar machinery circuit breakers; 4: single-phase transformers; 5: single-pole machinery and line circuit breakers; 6: main single-pole automatic circuit breakers; 7: single-pole disconnectors; 8: line segments with neutral sections; 9: substation power supply auxiliary services.

reduced oil to using hexafluoride; for the latter, the number of eligible openings before the replacement of the contacts is very important.

As shown in Figure 4.5, thanks to the high performance of automatic circuit breakers, the double-track lines can be provided for each direction, integrated with two feeder circuit breakers for single tracks. The diagram makes it possible to have, via disconnectors and load break switches,

- the selection of the HV arrival lines, if they are partially out of service;
- the establishment of the segments parallel upstream and downstream of the TPSS, when connection *V* of the transformers is excluded.

In the electrification of the French HS network, it was decided to no longer concentrate on a single main switch to protect both tracks for each direction: as with direct current, there is an automatic line main switch for each track and for each direction.

4.3 25 kV CONTACT LINE POWER SUPPLY

The regulations of EN 50163 permit the following variations in the line voltage, in comparison to the rated value of 25 kV:

- Maximum permanent value 27.5 kV (+10%), for 5 min 29 kV
- Permanent minimum value 19 kV (−24%), for 10 min 17.5 kV (−30%)

For frequency, variations of ±1 Hz (± 2%) are permissible.

Due to the powerful line reactor and the impossibility of supplying power at both ends of the individual segments, except in exceptional circumstances, the average *L* distance between substations is limited to about 50–60 km.

The power supply diagram leads to the installation of very high-power transformers, more so if 100% of the reserve is needed: The average for a TPSS is 2 × 15 MVA, corresponding to a surface power density per kilometer of line: $P_s = 0.5$ MW/km. Substations can withstand a 50% overload for 15 min and 100% for 5 min.

On HS lines, the power supply systems reach considerably higher potentials: On the Madrid–Seville line it is $P_s = 1$ MW/km, and the TPSSs are 2 × 20 MVA and spaced on average 40 km apart. The surface power density doubles in French high-speed lines, given the volume of traffic: the increase in power has led the SNCF to switch to a 2 × 25 kV system, as the Japanese railways did for the Shinkansen network. This topic will be further examined later.

4.3.1 Line Circuit

In the industrial system, the line circuit diagram is significantly more complex than in direct current or alternating current at railway frequency (Figure 4.6).

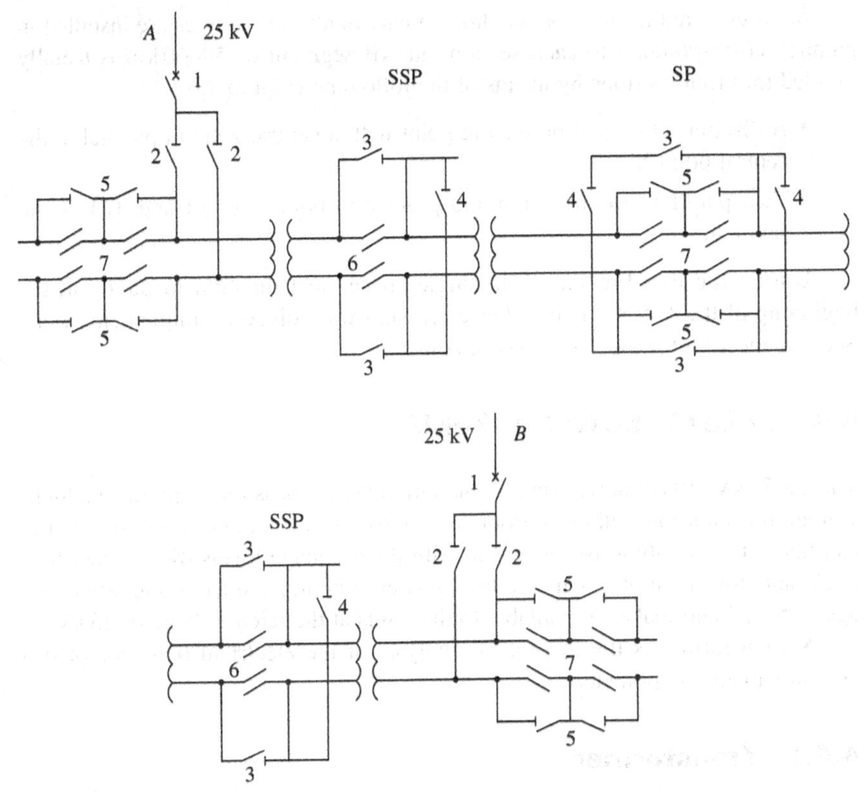

Figure 4.6 Power supply diagram of a double-track segment between substations *A* and *B*, with a sectioning and parallel point, an intermediate sectioning point, and two disconnection and parallel points. 1: the main substation circuit breakers; 2: substation line circuit breakers; 3: continuity switches (normally open in SP and closed in SSP); 4: circuit breakers in parallel (normally closed); 5: disconnectors for the emergency power supply of neutral sections; 6: sectioning; 7: sectioning with a neutral zone.

Disconnectors are quite sensitive on neutral N sections; normally they are not powered and are long enough to avoid any traction units' pantographs short-circuiting adjacent segments, across which there is the potential difference ΔV. The locomotives must, when passing in the N areas, respect the order to "lower the pantographs" or at least "open all line-side circuits"; the maneuvers are normally carried out by the train driver after seeing the appropriate signal. If the order is not followed, an automatic safety system intervenes.

Given the development of the individual line segments, there are various intermediate sections in place to limit the length of an out of service section if there is a fault. Both these simple sections (called disconnections) and those within a neutral zone are able to bypass across disconnectors, if necessary. An entire section between two substations can, in this case, be powered as necessary from one end only.

In order to reduce the voltage drops, many double-track lines are installed in parallel, corresponding to each section: an AB segment of 50–60 km is usually divided into four sections by means of the following (Figure 4.6):

- A disconnection and protection point with a neutral zone in parallel at the central point M
- Two parallel disconnection and protection points at 1/4 and 3/4 of the segment.

During the first French 50 Hz single-phase electrification in Savoy at the beginning of the 1950s, the unbalance problem was solved by implementing the Scott connection for the substation in Annecy.

4.4 2 × 25 kV–50 Hz SYSTEMS

The 2 × 25 kV–50 Hz power supply diagram makes it possible to produce a high-voltage transmission without introducing an excessive voltage level on the contact lines. This solution uses a system with three voltage levels (the contact line, track, and power supply), which allows power to be transmitted at the rated voltage of 50 kV and makes it available to the trains at the rated voltage of 25 kV.

Shown further is the theoretical analysis of the electrical behavior of this system's main components.

4.4.1 Transformer

A transformer (single phase) with three windings consists of three coils (generally coaxial with each other) wound around a single magnetic circuit (Figure 4.7).

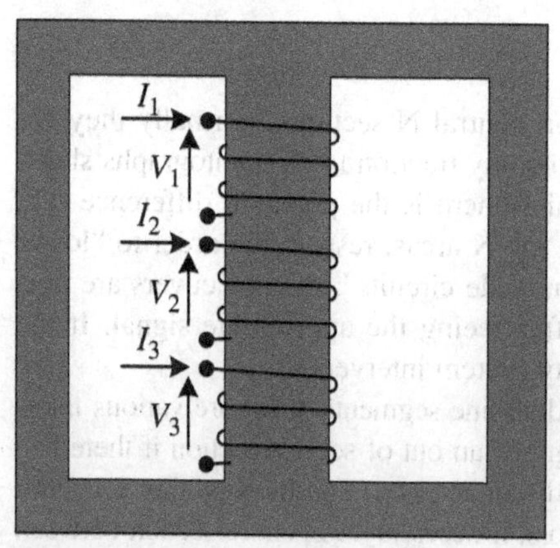

Figure 4.7 Three-winding transformer.

The equations for the transformer windings, written in phasor terms, can be borrowed from those written for the two-winding transformer, involving mutually coupled circuit breakers:

$$\overline{V}_1 = R_1 \cdot \overline{I}_1 + j\omega \cdot L_{11} \cdot \overline{I}_1 + j\omega \cdot L_{12} \cdot \overline{I}_2 + j\omega \cdot L_{13} \cdot \overline{I}_3$$
$$\overline{V}_3 = R_2 \cdot \overline{I}_2 + j\omega \cdot L_{22} \cdot \overline{I}_2 + j\omega \cdot L_{21} \cdot \overline{I}_1 + j\omega \cdot L_{23} \cdot \overline{I}_3$$
$$\overline{V}_3 = R_3 \cdot \overline{I}_3 + j\omega \cdot L_{33} \cdot \overline{I}_3 + j\omega \cdot L_{32} \cdot \overline{I}_2 + j\omega \cdot L_{31} \cdot \overline{I}$$

It will also be $L_{21} = L_{12}$ and $L_{23} = L_{32}$ and $L_{31} = L_{13}$

An equivalent circuit can now be determined for the transformer, from which the operating characteristics can be deduced. In order to simplify the model, we can accept the assumptions that lead to a substantial reduction in the number of circuit elements to be considered.

Established for the simple treatment equal to the number of turns $N_1 = N_2 = N_3 = N$ on the three windings (not a restrictive condition and one which can be removed by introducing appropriate ideal transformers), all mutual inductances are assumed as equal.

$$L_{12} = L_{23} = L_{31} = L_m$$

Expressing self-inductance in terms of mutual inductances and leakage:

$$L_{11} = L_{1d} + L_m$$
$$L_{22} = L_{2d} + L_m$$
$$L_{33} = L_{3d} + L_m$$

the previous equations can be rewritten as follows:

$$\overline{V}_1 = R_1 \cdot \overline{I}_1 + j\omega \cdot L_{1d} \cdot \overline{I}_1 + j\omega \cdot L_m \cdot \left(\overline{I}_1 + \overline{I}_2 + \overline{I}_3\right) = \overline{Z}_1 \cdot \overline{I}_1 + j\omega \cdot L_m \cdot \left(\overline{I}_1 + \overline{I}_2 + \overline{I}_3\right)$$
$$\overline{V}_2 = R_2 \cdot \overline{I}_2 + j\omega \cdot L_{2d} \cdot \overline{I}_2 + j\omega \cdot L_m \cdot \left(\overline{I}_1 + \overline{I}_2 + \overline{I}_3\right) = \overline{Z}_2 \cdot \overline{I}_2 + j\omega \cdot L_m \cdot \left(\overline{I}_1 + \overline{I}_2 + \overline{I}_3\right)$$
$$\overline{V}_3 = R_3 \cdot \overline{I}_3 + j\omega \cdot L_{3d} \cdot \overline{I}_3 + j\omega \cdot L_m \cdot \left(\overline{I}_1 + \overline{I}_2 + \overline{I}_3\right) = \overline{Z}_3 \cdot \overline{I}_3 + j\omega \cdot L_m \cdot \left(\overline{I}_1 + \overline{I}_2 + \overline{I}_3\right)$$

It is clear that the amount $L_m \cdot \left(\overline{I}_1 + \overline{I}_2 + \overline{I}_3\right)$ represents the total flux linked with the three windings ϕ_c, and that $j\omega \cdot \phi_c$ is the induced emf (equal in the three windings in relation to the equal number of turns); $X_0 = \omega \cdot L_m$ can also be identified. Interpreting the hypothesis introduced relating to the equality of the mutual inductances in physical terms, it can be noted that this leads to the deduction that each magnetomotive force acts on its own leakage circuit, and on the same magnetic circuit shared with the other two; links between pairs of windings are, therefore, not taken into consideration. The magnetic network that represents this situation is shown in Figure 4.8, and the electrical network equivalent to it, which

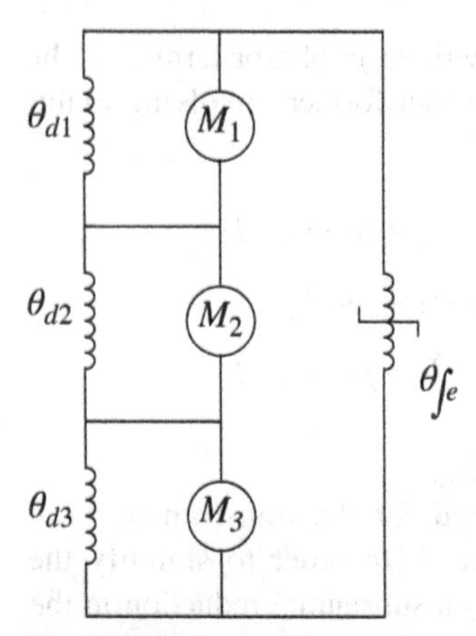

Figure 4.8 The magnetic network of a three-winding transformer, assuming that $L_{12} = L_{23} = L_{31} = L_m$.

can be obtained with the methods of duality and, on the other hand, is one that is described by the system has, of course, the configuration shown in Figure 4.9:

$$\overline{V}_1 = \overline{Z}_1 \cdot \overline{I}_1 + j\omega \cdot L_m \cdot (\overline{I}_1 + \overline{I}_2 + \overline{I}_3)$$
$$\overline{V}_2 = \overline{Z}_2 \cdot \overline{I}_2 + j\omega \cdot L_m \cdot (\overline{I}_1 + \overline{I}_2 + \overline{I}_3)$$
$$\overline{V}_3 = \overline{Z}_3 \cdot \overline{I}_3 + j\omega \cdot L_m \cdot (\overline{I}_1 + \overline{I}_2 + \overline{I}_3)$$

Finally, the introduction of two ideal transformers provides the complete equivalent circuit of the single-phase three-winding transformer (Figure 4.10).

Once the equivalent circuit of the transformer has been established, the problem lies in defining it by testing the values of the circuit elements, which are expressed in a similar way to what was seen for transformers with two

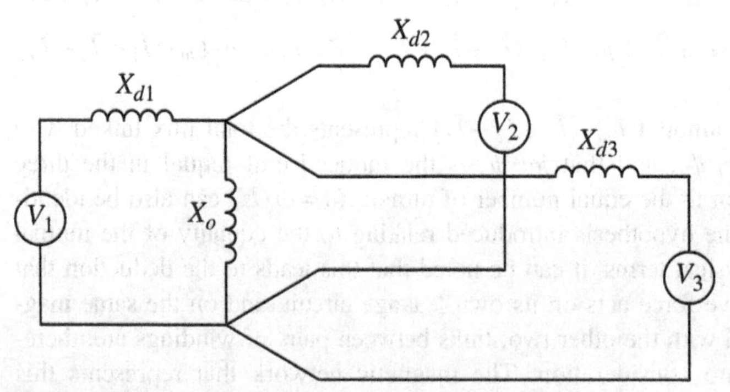

Figure 4.9 The mains supply is equivalent to the magnetic network.

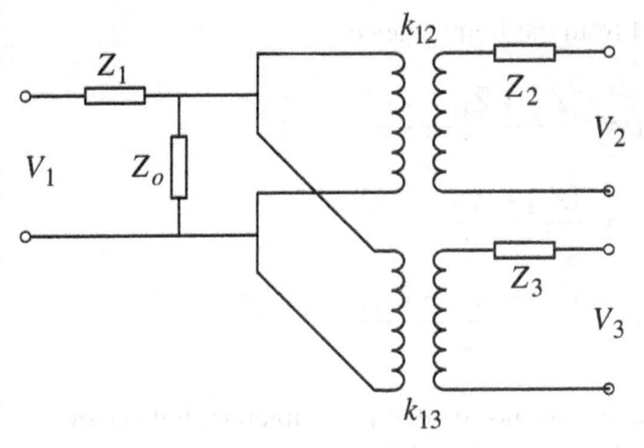

Figure 4.10 Complete equivalent circuit of the three-winding transformer.

windings, namely, through no-load losses and short circuits, and the no-load current and short circuit voltage percentage.

There is no difference between transformers with two and three windings with respect to the no-load test, from which, using the same rules, the percentages for no-load losses and currents are obtained.

As for the short circuit test, it is immediately apparent that directly determining the series of impedances $\overline{Z}_1$, $\overline{Z}_2$, $\overline{Z}_3$ is impossible.

Short-circuit tests are then carried out on the track, in which a winding is powered, the other is short-circuited, and the third has no load (Figure 4.11).

Therefore, three tests are possible, from which three track impedances are obtained, referring to the number of turns N_1:

$$
\begin{aligned}
\overline{Z}_{12} &= \overline{Z}_1 + \overline{Z}_2' \\
\overline{Z}_{13} &= \overline{Z}_1 + \overline{Z}_3' \\
\overline{Z}_{23} &= \overline{Z}_2' + \overline{Z}_3'
\end{aligned}
$$

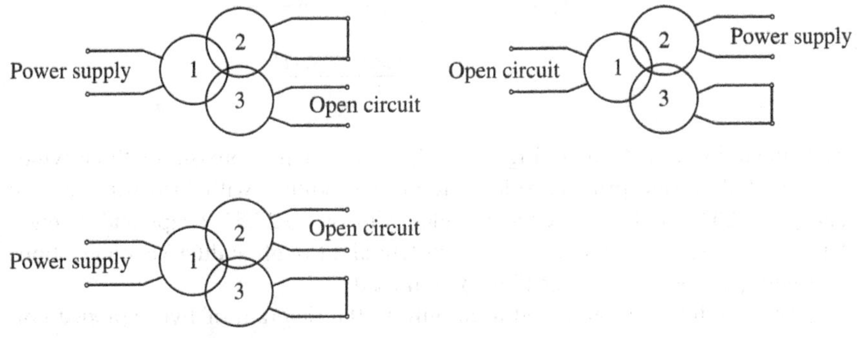

Figure 4.11 Three different short circuits on a track: 1–2 (3 no load); 1–3 (2 no load); 2–3 (1 no load).

The desired values are deduced from these impedances:

$$\overline{Z}_1 = \frac{(\overline{Z}_{12} + \overline{Z}_{13} - \overline{Z}_{23})}{2}$$

$$\overline{Z}_2' = \frac{(\overline{Z}_{12} + \overline{Z}_{23} - \overline{Z}_{13})}{2}$$

$$\overline{Z}_3' = \frac{(\overline{Z}_{13} + \overline{Z}_{23} - \overline{Z}_{12})}{2}$$

Since the impedance series values are not of definitive interest, but rather the losses in short circuits and in the percentage of short-circuit voltages, expressions similar in form to the previous ones can be written for these quantities; moreover, since the percentage of losses and the percentage of short-circuit voltages are independent from the measuring side (i.e., have a transport factor of 1), even these relations have the value $N_1 \neq N_2 \neq N_3$.

Due to the short-circuit losses at 75 °C and at the rated current, we will, therefore, have

$$P_{cc1} = \frac{(P_{12} + P_{13} - P_{23})}{2}$$

$$P_{cc2} = \frac{(P_{12} + P_{23} - P_{13})}{2}$$

$$P_{cc3} = \frac{(P_{13} + P_{23} - P_{12})}{2}$$

and for the short-circuit track voltages at 75 °C and at the rated current

$$\overline{v}_{cc1} = \frac{(\overline{v}_{cc12} + \overline{v}_{cc13} - \overline{v}_{cc23})}{2}$$

$$\overline{v}_{cc2} = \frac{(\overline{v}_{cc12} + \overline{v}_{cc23} - \overline{v}_{cc13})}{2}$$

$$\overline{v}_{cc3} = \frac{(\overline{v}_{cc13} + \overline{v}_{cc23} - \overline{v}_{cc12})}{2}$$

The substation transformer (Figure 4.12) conceptually consists of three windings, including one primary side winding that adjusts with load (on-load tap changer OLTC) and two secondary sides with two ±27.5 kV taps and a center tap connecting the track to ground potential through inductive connections depending on the type of signaling system used.

The windings are arranged according to the diagram of two wrapped columns shown in Figure 4.13; each column is wrapped with a low-voltage winding, a half-winding with high voltage, and adjustment.

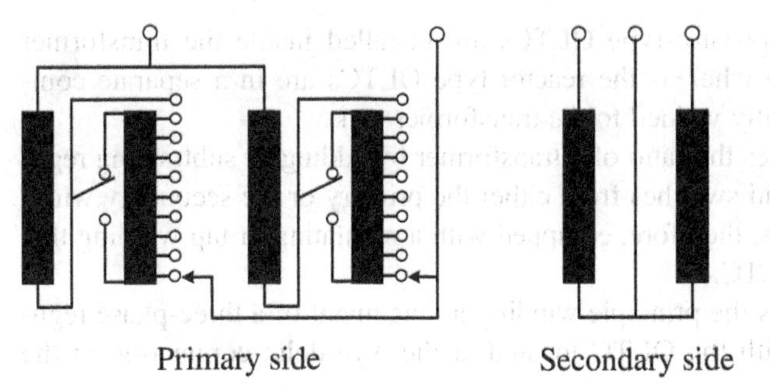

Figure 4.12 OLTC substation transformer.

4.4.1.1 On-load tap changers OLTC

Power transformers equipped with OLTCs have been the main components of electrical networks and industrial applications for nearly 90 years. OLTCs enable voltage regulation and/or phase shifting by varying the transformer ratio under load without interruption.

From the start of tap changer development, two switching principles have been used for load transfer operation – the high-speed resistor-type OLTCs and the reactor-type OLTCs.

Over the decades both principles have been developed into reliable transformer components that are available in a broad range of current and voltage applications. These components cover the needs of today's network and industrial process transformers and ensure optimal system and process control.

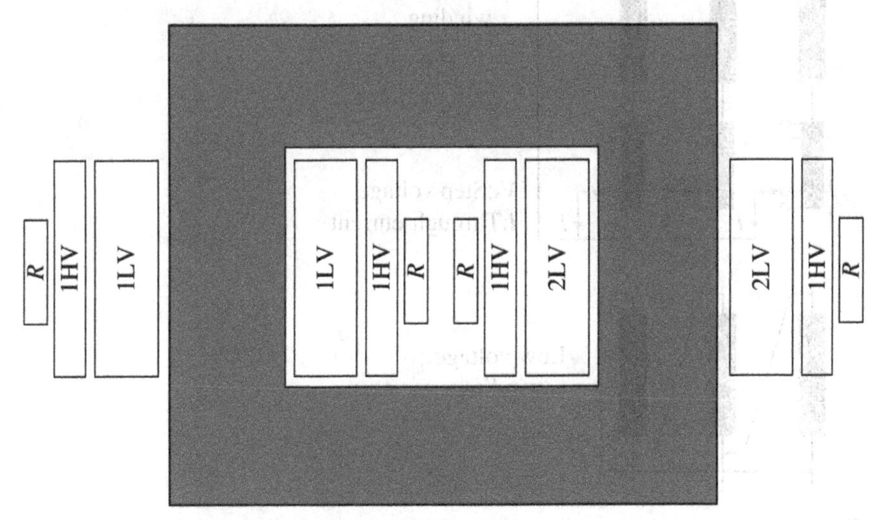

Figure 4.13 Winding arrangement of the TPSS transformer.

The majority of resistor-type OLTCs are installed inside the transformer tank (in-tank OLTCs) whereas the reactor-type OLTCs are in a separate compartment that is normally welded to the transformer tank.

The OLTC changes the ratio of a transformer by adding or subtracting regulating windings (R) and switches from either the primary or the secondary winding. The transformer is, therefore, equipped with a regulating or tap winding that is connected to the OLTC.

Figure 4.14 shows the principle winding arrangement of a three-phase regulating transformer, with the OLTC located at the wye-delta connection in the high-voltage winding.

Simple tap switching during an energized status is unacceptable due to the momentary loss of system load during the switching operation (Figure 4.15).

The "make (2) before break (1) contact concept," shown in Figure 4.16, is therefore the basic design for all OLTCs.

The transition impedance in the form of a resistor or reactor consists of one or more units that bridge adjacent taps for the purpose of transferring load from one tap to the other without interruption or appreciable change in the load current. At the same time they limit the circulating current (I_C) for the period when both taps are used. Normally, reactor-type OLTCs use the bridging position as a service position and the reactor is, therefore, designed for continuous loading.

The voltage between the abovementioned taps is the step voltage, which normally lies between 0.8 and 2.5% of the rated voltage of the transformer.

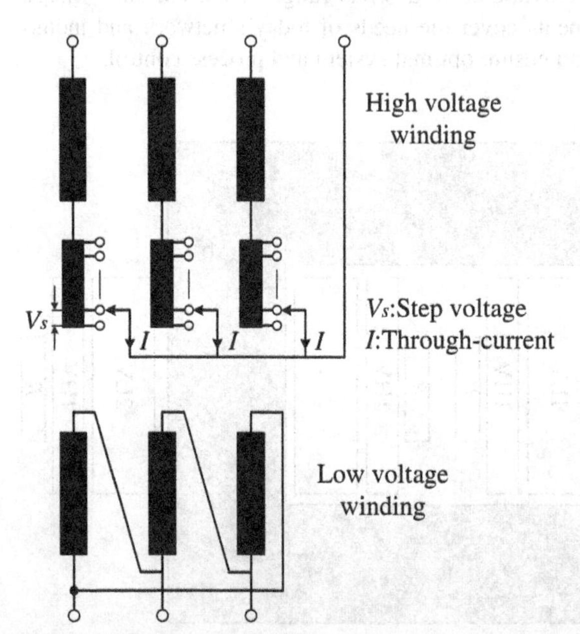

Figure 4.14 Principle winding arrangement of a regulating transformer in wye-delta connection.

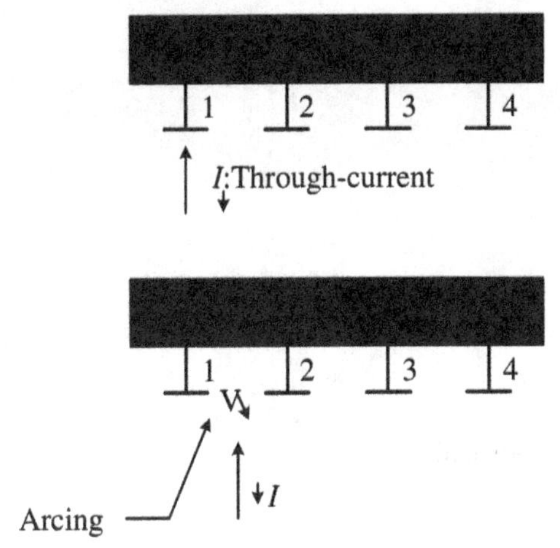

Figure 4.15 Loss of system load with single contact switching.

The main components of an OLTC are contact system transition impedances, gearings, spring energy accumulators, and a drive mechanism. Depending on the various winding arrangements and OLTC designs, separate selector switches and change-over selectors (reversing or coarse type) are also used.

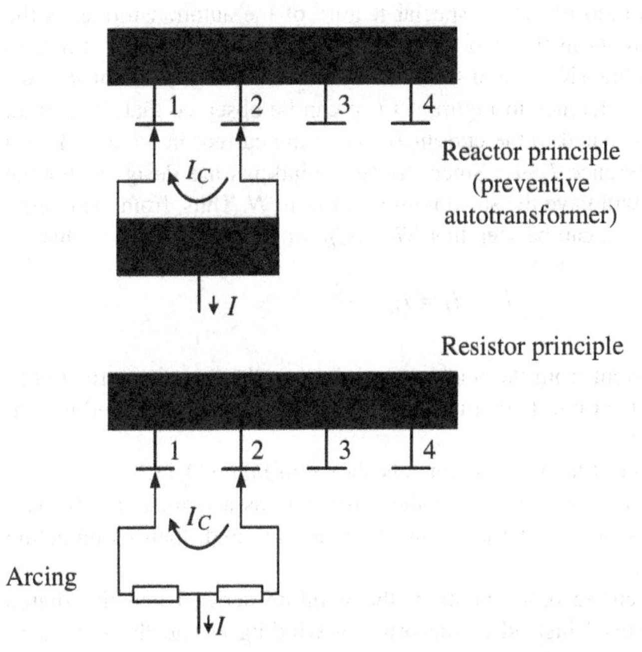

Figure 4.16 Basic switching principle "make (2) before break (1)" using transition impedances.

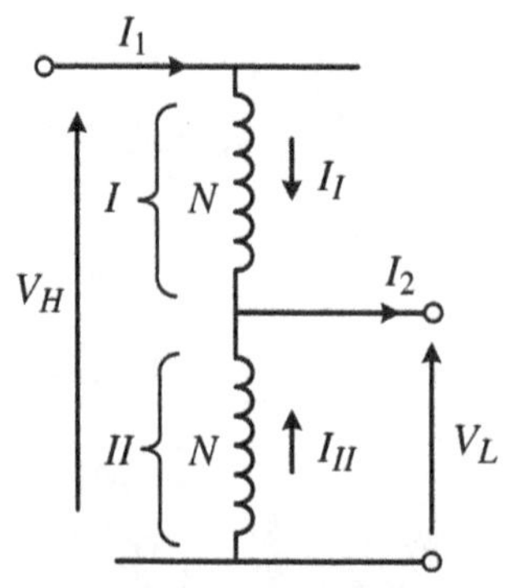

Figure 4.17 Current distribution in an autotransformer.

4.4.2 Autotransformer

Autotransformers installed in this system are necessary for redistributing the power transmitted to the train between the contact line and feeder. Normally, they have two windings designed for the same voltage as the main transformer connected between the contact line and feeder, and with a center tap connected to the track, according to the diagram in Figure 4.17.

Since the two windings are designed for the same voltage, the result is a transformation *H/L* ratio of 2:1. A special feature of the autotransformer is the division of the currents in the two windings that are subdivided according to Kirchhoff's current law (KCL) and the overall magnetomotive force of the two windings. Still with reference to Figure 4.17, it can be observed that the current in the winding (I_I) is equal to the current I_1, while the current in winding II (I_{II}) is equal to the difference $I_2 - I_1$. Since the two windings are designed for the same voltage, they will have the same number of turns *N*. Thus, from the overall magnetomotive force it can be seen that $NI_I = NI_{II}$, from which it follows that

$$I_1 = I_I = I_{II} = \frac{I_2}{2}$$

Thus, the return current from the track, once joined to the autotransformer connection, will be divided into two equal parts toward the contact line and toward the feeder.

The arrangement of the two windings is shown in Figure 4.18.

If an autotransformer is treated in the same way as a regular transformer, that is, if the details of the internal connections are ignored, then its modeling can be done as follows.

For a more accurate representation, the winding series I and the shared winding II will be used instead of the primary winding *H* and the secondary winding *L*.

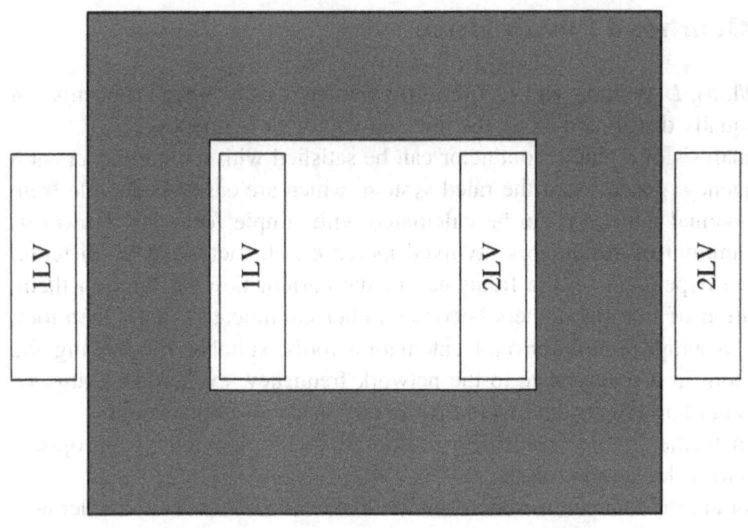

Figure 4.18 Winding arrangement of autotransformers in parallel points.

This involves a redefinition of short-circuit data for windings I and II. First, the voltage values are

$$V_I = V_H - V_L$$
$$V_{II} = V_L$$

From no-load and autotransformer short-circuit tests, the necessary parameters for windings *I* and *II* can be directly obtained. From the moment that *II* is short-circuited in a short-circuit test, the value of the short-circuit voltage, applied to *H* and, therefore, only to winding *I*, can be obtained. This test is essential to determine the short-circuit impedance value which, as will be seen later, is crucial in the functioning of the autotransformer system.

The conversion from *H* to *I* is given by

$$Z_{I,II} = Z_{HL}\left(\frac{V_H}{V_H - V_L}\right)^2 \quad \text{in p.u.} \tag{4.2}$$

Through *I* and *II* a current of 1 p.u. V_H will flow or, by changing the base and referring to V_I and V_{II} equal to $II = (V_H - V_L)/V_H$ and $III = V_L/V_H$. Considering these currents, voltages per unit become

$$V_I = Z_I \frac{V_H - V_L}{V_H} \quad \text{in p.u.}$$

$$V_{II} = Z_{II} \frac{V_L}{V_H} \quad \text{in p.u.}$$

The autotransformer can now be treated as a transformer that is simply redefining short-circuit impedances as per Equation 4.2.

4.4.3 Overhead Power Lines

The R' (Ω/km), L' (H/km), and C' (μF/km) parameters of overhead transmission lines are equally distributed along the line and vary with frequency.

The analysis of a single conductor can be satisfied with the parameters at a given frequency, generally of the rated system, which are easily deducible from technical manual tables or can be calculated with simple formulas. However, when the amount of conductors involved increases, characterized by different features and dependent on the frequency of the current flowing through them, the evaluation of line parameters becomes rather complicated. It is, therefore, necessary to adopt simulation and calculation tools, suitable for solving the problems both in a steady state to the network frequency, even considering the numerous coupling effects, and taking temporary behavior into account.

Shown further is the basic theory required for implementing appropriate algorithms for calculation tools.

First of all, the line parameters for individual conductors need to be defined, compared to the other conductors present.

The series impedance matrix is used in order to represent the voltage drop along the transmission line. For a single-phase line, as in the case of traction lines, the differential equation is

$$-\frac{\partial v}{\partial x} = R'i + L'\frac{\partial i}{\partial t} \tag{4.3}$$

The R' and L' parameters for the transmission line vary with frequency and are, therefore, not constant.

Consequently, the Equation 4.3 should be reconsidered in order to express the voltage drops in the form of phasor equations in steady state, alternating current at a given frequency:

$$-\left[\frac{dV}{dx}\right] = [Z'][I] \tag{4.4}$$

where [V] and [I], respectively, are the phasor vectors of the voltages on the ground and of the currents flowing in the conductor, assuming the earth is a node to which all voltages are referred. The matrix $[Z'] = [R'(\omega)] + j\omega[L'(\omega)]$ is the impedance matrix series, which is complex and symmetrical. The elements on the main diagonal are equal to the impedance, per unit length, of the ring formed by the ith conductor and earth is a return, $Z'_{ii} = R'_{ii} + j\omega L'_{ii}$. The elements outside the diagonal are equal to the mutual impedance per unit length between the ith and the kth conductor $Z'_{ik} = Z'_{ki} = R'_{ki} + j\omega L'_{ki}$, and determine the longitudinally induced voltage in conductor k if a current flows in conductor i and vice versa.

The resistive terms are introduced into the mutual pairs by the presence of the ground. In fact, the earth is not modeled as a conductor in itself, but is used as a reference point for measuring voltages.

The formulas used here to define the values of Z'_{ii} and Z'_{ik} are those developed by Carson. These formulas used homogeneous pieces of land that are accurate enough for studies on power systems, and are based on the following assumptions:

- The conductors are considered perfectly horizontal with the ground and long enough to reduce the problem to two dimensions.
- The air space in which they are immersed is considered leak free, with permeability μ_0 and permittivity ε_0.
- The ground is considered flat and homogeneous with uniform resistivity ρ, permeability μ_0, and permittivity ε_0.
- The conductors are considered enough far away from each other so as to disregard the proximity effects, in other words, the distance between them must be greater by at least one order of magnitude compared to the radius of the conductor.

The elements of the impedance matrix can be calculated from the geometry of the support configuration (Figure 4.19) and by the characteristics of the conductors.

The impedance is, therefore,

$$Z'_{ii} = (R'_{i-\text{int}} + \Delta R'_{ii}) + j\left(\omega\frac{\mu_o}{2\pi}\ln\frac{2h_i}{r_i} + X'_{i-\text{int}} + \Delta X'_{ii}\right) \tag{4.5}$$

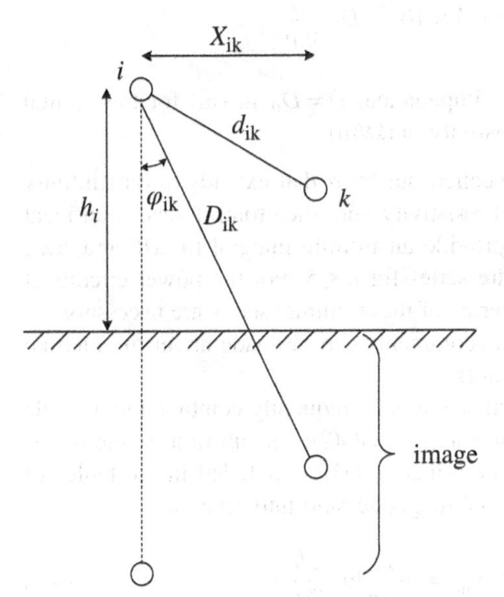

Figure 4.19 Arrangement of the conductors with their images.

while the mutual impedance is

$$Z'_{ik} = Z'_{ki} = \Delta R'_{ik} + j\left(\omega \frac{\mu_o}{2\pi} \ln \frac{D_{ik}}{d_{ik}} + \Delta X'_{ik}\right) \tag{4.6}$$

where μ_0 is the permeability of the vacuum. Using $\mu_0/2\pi = 2\times10^{-4}$ H/km, the impedances can be obtained in Ω/km. The parameters in (4.5) and (4.6) are as follows:

- $R'_{i\text{-int}}$ is the internal resistance of the ith conductor in the alternating current, in Ω/km.
- h_i is the average height of the ith conductor from the ground.
- r_i is the radius of the ith conductor.
- d_{ik} is the distance between the ith conductor and the kth conductor.
- D_{ik} is the distance between the ith conductor and the imaginary k conductor.
- $X'_{i\text{-int}}$ is the internal reactance of the ith conductor.
- $\omega = 2\pi f$ with f frequency in Hz as the pulse.
- $\Delta R', \Delta X'$ are the Carson correction factors that take the effects of the ground into account.

  The correction terms $\Delta R'$ and $\Delta X'$ take the effect of the ground into account and are functions of the angle ϕ ($\phi = 0$ for the impedance and $\phi = \phi_{ik}$ for the mutual impedance and that of a parameter):

$$a = 4\pi\sqrt{5} \times 10^{-4} \cdot D \cdot \sqrt{\frac{f}{\rho}}$$

- where $D = 2h_i$ in (m) for the impedance, $D = D_{ik}$ in (m) for the mutual impedance, and ρ ground resistivity in (Ω/m).

Note that $\Delta R'$ and $\Delta X'$ are cancelled out by a that extends toward infinity and where there is very low ground resistivity and, therefore, it acts as an ideal conductor. The Carson equations provide an infinite integral to $\Delta R'$ and $\Delta X'$, developed in the sum of four infinite series for $a \le 5$. For the power circuits at the network frequency, only a few terms of these infinite series are necessary.

To complete the analysis, a few comments can be made about the internal impedance $R'_{\text{int}} + jX'_{\text{int}}$ of the conductors.

With regard to the internal reactance, this is frequently combined in a single expression with the external reactance $\omega(\mu_0/2\pi)\ln(2h/r)$, substituting the radius r with the smallest "geometric mean radius" (GMR), included in the tables of conductors, in order to take the internal magnetic field into account:

$$\omega \frac{\mu_0}{2\pi} \ln \frac{2h}{r} + X'_{\text{int}} = \omega \frac{\mu_0}{2\pi} \ln \frac{2h}{\text{GMR}} \tag{4.7}$$

The concept of the geometric mean radius was originally developed for nonmagnetic conductors at the network frequency, in which the skin effect can be disregarded. In this case, its meaning is purely geometrical and the GMR is equal to the geometric mean distance between all of the elements in the cross section of the conductor, if this area is divided into an infinite number of equal and infinitesimal elements. For a nonmagnetic conductor that is circular, at low frequencies we have

$$\mathrm{GMR}/r = e^{-1/4}$$

a value that becomes equal to

$$\mathrm{GMR}/r = e^{-\mu_r/4}$$

if the conductor is made from a magnetic material with a relative permeability μ_r. In this case, it loses its geometric value, even when taking the skin effect into consideration.

Calculation (4.7) provides the conversion formula between the GMR and the internal reactance, which can be calculated for certain types of conductors, as part of the internal impedance $R'_{\mathrm{int}} + jX'_{\mathrm{int}}$. When nonmagnetic conductors turn out to be a small part of the total reactance, its exact calculation is not interesting. Instead, it is more important to calculate the R'_{int} value because there can be a considerable resistance increase with the frequency due to the skin effect. The internal impedance of a circular conductor at low frequencies can be calculated with the skin effect formulas, where R'_{int} is therefore more substantial than X'_{int}.

Aluminum cable reinforced in steel can generally be compared to tubular conductors when the influence of the steel core is negligible and there are more layers of aluminum, since the magnetization of the aluminum heart, caused by one layer being wrapped in a spiral in one direction, is balanced out by the next layer being wrapped in the opposite direction.

It can, however, be said that the skin effect assumes significant values at high frequencies, in which $R'_{\mathrm{int}} = X'_{\mathrm{int}}$, or even from 10 kHz. The difference between the two values is equal to about 2.2%, a value that drops to 0.2% at 1 MHz.

To come back to the calculation for conductor voltages toward the ground, the latter depend on the line load:

$$[v] = [P'][\mathrm{q}] \tag{4.8}$$

where [q] is the matrix loads per unit length of the conductors and [p] is the symmetric matrix of the Maxwell potential coefficients. Its elements can easily be deduced from the geometry of the masts and from the conductors' radius, under the following assumptions:

- The air is considered to be free of leakage and the earth is even at zero potential

- The radii are at least one order of magnitude lower than the distance between the conductors

Both of the assumptions are well founded in the case of overhead power lines. Consequently, the main diagonal elements are equal to

$$P'_{ii} = \frac{1}{2\pi\varepsilon_0}\ln\frac{2h_i}{r_i}$$

and the off-diagonal elements

$$P'_{ik} = P'_{ki} = \frac{1}{2\pi\varepsilon_0}\ln\frac{D_{ik}}{d_{ik}}$$

with ε_0 vacuum permittivity. The factor $1/2\pi\varepsilon_0$ in these equations is equal to $c^2(\mu_0/2\pi)$, where c is the speed of light.

The inverse relationship of (4.8) leads to the capacitance matrix $[C']$:

$$[q] = [C'][v], \quad \text{with} \quad [C'] = [P']^{-1} \tag{4.9}$$

The matrix $[C']$ is in the nodal form. This means that it can be inspected by placing the element on the diagonal C'_{ii} as the total capacity per unit length between the ith conductor and the other conductors, while the off-diagonal element $C'_{ik} = C'_{ki}$ is equal to the negative capacitance value per unit length between the ith and the kth conductors.

In alternating current conditions, the load vector, in phasor terms, is connected to the leakage current vector [-*di/dx*] according to

$$[Q] = -\frac{1}{j\omega}\left[\frac{dI}{dx}\right]$$

Consequently, the second differential equation system is given by

$$-\left[\frac{dI}{dx}\right] = j\omega[C'][V] \tag{4.10}$$

An equation that, together with (4.4), completely describes the steady-state behavior of a multiconductor line. Conductance G' has been disregarded in (4.10) because its influence can be considered negligible in overhead power lines, except at very low frequencies, when the behavior of the line is determined by R' and G', $\omega L'$ and $\omega C'$ become instead negligible. Therefore, when the effect of conductance G' is also considered, the whole equation becomes

$$-\left[\frac{dI}{dx}\right] = [Y'][V] \tag{4.11}$$

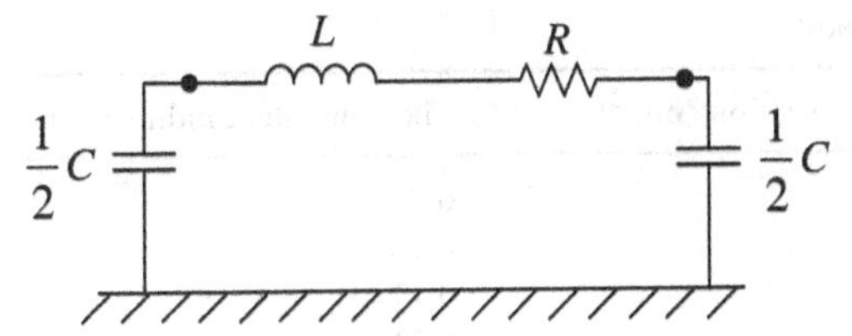

Figure 4.20 A circuit equivalent to the π of a line section.

where

$$[Y'] = [G'] + j\omega[C']$$

However, corrective terms are not considered when they take the effects of earth conduction on capacity into account, as frequencies below 100 kHz are negligible. In this case the capacities are constant and not dependent on the frequency, like they are with resistor and inductance series.

Once the parameters of the line per unit length have been defined, these are used to develop the line model determined by a specific length.

Given the complexity of the autotransformer system and the many conductors involved, the railway line can be discretized by identical circuits with the lumped element model (Figure 4.20), while maintaining accuracy for a steady-state study. In fact, the resistance and the inductance can be considered as approximately constant up to 1 kHz, much higher than the frequencies used in the alternating current traction systems.

The impedance and susceptance values can be obtained using the following Hasibar formulas:

$$Z_{\text{serie}} = R\cos^2\omega\tau - \left(0.5 + 0.03125\frac{R^2}{Z^2}\right)R\sin^2\omega\tau + j\sin\omega\tau\cos\omega\tau\left(0.375\frac{R^2}{Z} + 2Z\right)$$

$$\frac{1}{2}Y_{\text{shunt}} = \left[\left(-2 - 0.125\frac{R^2}{Z^2}\right)R\sin^2\omega\tau + j\frac{R}{Z}\sin\omega\tau\cos\omega\tau\right]/Z_{\text{serie}}$$

where

$$\tau = L_l \cdot \sqrt{L'C'}, Z = \sqrt{\frac{L'}{C'}}, \text{ and } R = L_l \cdot R'$$

where L_l is the length of the line.

Since the effective section where there is current flowing cannot be geometric, the equivalent radius needs to be calculated. The example in Table 4.1 shows the equivalent radii of the different conductors that make up the autotransformer system.

Table 4.1 Conductor Equivalent Radii

Conductor	The effective section (mm^2)	The equivalent radius (mm)
Contact wire	150	6.91
Messenger wire	120	6.18
Feeder	264.7	9.17
Grounding wire	150	6.31
Earthing rod	95	5.5
Track	7679	49.44

In most cases, conductors are made from a wire braid and not from a single cylindrical body, so when calculating the resistivity ρ $((\Omega \cdot \mathrm{mm}^2)/\mathrm{m})$, the coefficient of stranding K needs to be introduced, which may vary between 1.03 and 1.05, by increasing the value of the equivalent resistivity in proportion to the gross geometric section of the conductor.

4.4.4 Feeder

The conductor used for the feeder in Figure 4.21 is generally made up of a section of 264.6 mm^2 of aluminum and a section of 43.1 mm^2 of steel, for a total cross section of 307.7 mm^2.

Since the steel's resistivity is much greater than that of aluminum, almost all of the current will only flow in the circular aluminum rim, also because of the effect of the alternating currents. Consequently, the feeder can be modeled as an

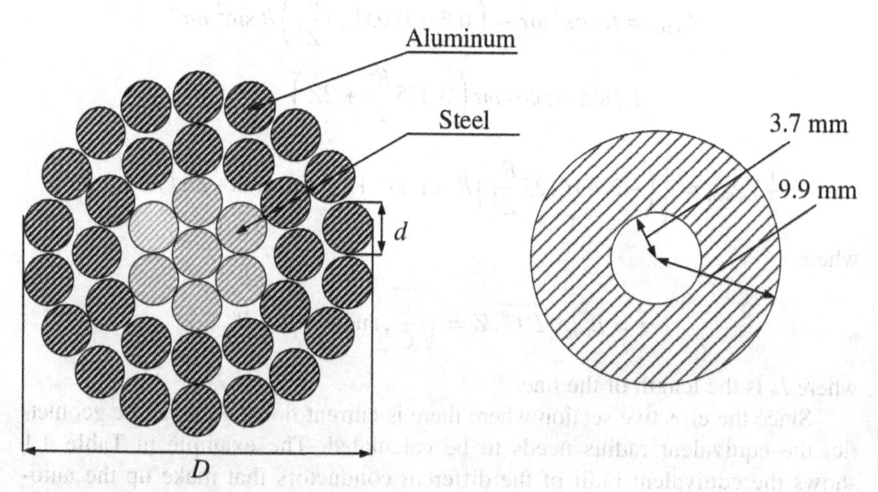

Figure 4.21 Cable used for the feeder conductor.

internal radius cable conductor of 3.7 mm and an outer radius equal to 9.9 mm, as shown in Figure 4.21.

4.4.5 Track

The track, compared to the other conductors seen previously, presents different problems. From an electrical point of view, the track is amongst the most important elements of the rail system, as is the electrical interface with the traction circuit, the signaling, and the ground. The signaling circuit uses the track as part of the transmitter–receiver circuit and the inductance of the track is included in the calculation of the resonant frequency of the tuned circuits in order to calculate the frequency separation without insulated joints of rail sections. For this reason, it is important to know the inductance value and also its behavior as a function of frequency and current. Also, the amount of return current flowing in the track or in the ground depends both on the longitudinal and transversal parameters. Considering that the track is formed of a magnetic material, such as steel, it is therefore necessary to take the skin effect into account even at 50 Hz and the variation of the relative permeability as a function of the current.

Generally, the rail used is made of UIC 60 steel, with a linear mass of about 60.34 kg/m; its cross-section is shown in Figure 4.22.

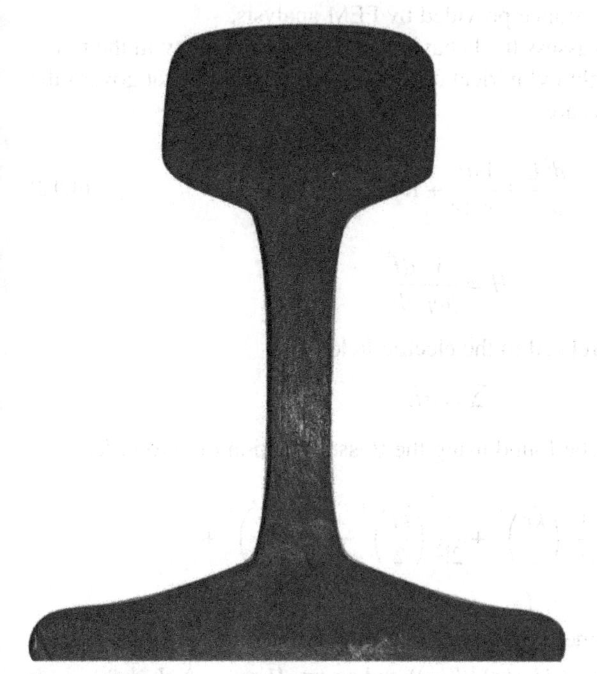

Figure 4.22 Cross section of UIC 60 steel rail.

The variation of the magnetic field produced by an alternating current produces an irregular current distribution on the cross section of the conductor. The currents tend to be concentrated near the surface of the conductor, causing an increase in the resistance value and a decrease in the inductance.

The skin effect is a very complicated phenomenon that can be analyzed in detail in certain cases: the shape of the conductor, linked to the concept of current penetration depth, is a major obstacle in the mathematical representation and definition of the impedance of the conductor function of frequency. The depth of penetration is defined as the radial distance from the surface of the conductor, toward the inside, in which the current density is reduced to a factor of $1/e$, where e is the base of the natural logarithm. A numerical method used to solve the skin effect problem in conductors with irregular section may be the finite element method (FEM), used to solve the differential equations that govern the magnetic field and electric current in a conductor. The finite element method is able to treat irregularly shaped rail, discretizing the conductor in numerous meshes.

Another method that can be used, which is less expensive computationally, can be the method of inductance coupling, which treats irregular sections of the conductor as the sum of n subconductors that are identical in square or rectangular shapes.

Since most calculation software will only represent circular conductors, we can consider the cylindrical conductor to be of an equal radius that can be calculated from the effective resistance provided by FEM analysis.

It is then necessary to assess the behavior of the current density in the rail.

With regard to a straight cylindrical conductor, the equations that govern the electric and magnetic fields are

$$\frac{d^2\overline{E}}{dr^2} + \frac{1}{r}\frac{d\overline{E}}{dr} + K^2\overline{E} = 0 \tag{4.12}$$

$$\overline{H} = \frac{1}{j\omega\mu}\frac{d\overline{E}}{dr}$$

and the current density is related to the electric field

$$\overline{\Delta} = \sigma\overline{E}$$

The solution of (4.12) can be found using the Bessel function of zero order:

$$J_0(\overline{k}r) = 1 - \frac{1}{1!^2}\left(\frac{\overline{k}r}{2}\right)^2 + \frac{1}{2!^2}\left(\frac{\overline{k}r}{2}\right)^4 - \frac{1}{3!^2}\left(\frac{\overline{k}r}{2}\right)^6 + \cdots$$

obtaining

$\overline{E} = cJ_0(\overline{k}r)$. The magnetic field can then be expressed using a Bessel function of first order $J_1(\overline{k}r) = -(dJ_0(\overline{k}r)/d(\overline{k}r))$ and so on, $\overline{H} = -c\frac{\overline{K}}{j\omega\mu}J_1(\overline{k}r)$.

The integration constant c is derived from circuit law relating to the perimeter of the conductor.

The values that are obtained for the electric field and, therefore, the current density are the following:

$$\begin{cases} \overline{E} = -\dfrac{K}{2\pi r_0 \sigma}\overline{I}\dfrac{J_0(\overline{k}r)}{J_1(\overline{k}r_0)} \\ \overline{\Delta} = \dfrac{K}{2\pi r_0}\overline{I}\dfrac{J_0(\overline{k}r)}{J_1(\overline{k}r_0)} \end{cases}$$

in which $K = \frac{j\omega\mu\sigma}{k}$.

As kr_0 has a very small value, that is, for frequencies, permeability, conductivity, and limited value radii, we have

$$\begin{cases} J_0(\overline{k}r) \simeq 1 \\ J_1(\overline{k}r_0) \simeq \dfrac{1}{2}kr_0 \end{cases}$$

and, therefore, $\overline{\Delta} = \overline{I}/\pi r_0^2$, that is, a constant value in the whole of the current density section.

The value of the electric field on the surface of the conductor multiplied by a section of length l, provides the voltage drop on that stretch, which is

$$E(r_0)l = \frac{\overline{K}l}{2\pi r_0 \sigma}\overline{I}\frac{J_0(\overline{k}r_0)}{J_1(\overline{k}r_0)} = \overline{I}(R + j\omega L_i)$$

This value can be modeled as an equivalent impedance in which parameter R is the equivalent resistance in alternating current (which produces the same losses by the Joule effect) and the parameter L_i is the internal inductance of the conductor. In practical applications, this makes the DC resistance equal to

$$R_0 = \frac{l}{\pi r^2 \sigma} = \frac{\rho l}{\pi r^2}$$

$$m = \frac{r}{2}\sqrt{\pi f \sigma \mu}$$

$$m_0 = \frac{r_0}{2}\sqrt{\pi f \sigma \mu}$$

where m and m_0 are dimensional quantities that represent, respectively, the value of the real part and the imaginary part of argument $\overline{k}r/2$ of the Bessel function J_0.

At high frequencies, R and ωL_i both tend to coincide with the value $m_0 R_0$:

$$R \simeq \omega L_i \simeq m_0 R_0 = R_0 \cdot \frac{r_0}{2}\sqrt{\pi f \sigma \mu} = \frac{1}{\sigma \pi r_0^2}\frac{r_0}{2}\sqrt{\pi f \sigma \mu} = \frac{1}{2r_0}\sqrt{\frac{f\mu}{\pi\sigma}} = \frac{1}{2\pi r_0 \sigma \delta}$$

where $\delta = \frac{1}{\sqrt{\pi f \sigma \mu}}$

is the skin depth. Therefore, at high frequencies, the effective resistance in AC is the one that would have a DC conductor composed of the surface layer of a depth equal to that of penetration. Furthermore, at high frequencies, the R increases with the square root of the frequency, while L_i decreases with the square root of f. In these conditions, the effective value of the current density is equal to

$$|\overline{\Delta}| = \frac{I}{2\pi r_0}\sqrt{\omega \sigma \mu}\sqrt{\frac{r_0}{r}}e^{-\frac{r_0 - r}{\delta}}$$

From which we can see, with regard to peripheral areas, a generally exponential decrease with increasing distance from the surface.

4.4.6 The Ideal Functioning of the Autotransformer System

This section presents the theoretical functioning principle of the 2×25 kV system, found in today's scientific literature, which will then be examined in more detail later in this chapter, removing the simplified assumptions introduced in order to go further than the simplified procedure and propose a more physically responsive model.

Figure 4.23 shows information on the basic diagram with an ideal breakdown of the currents. A step-down transformer in the substation with a central tap connected to earth feeds two conductors, the overhead contact line (OCL) at 25 kV and the feeder at −25 kV (the rated voltage is always 25 kV, but in

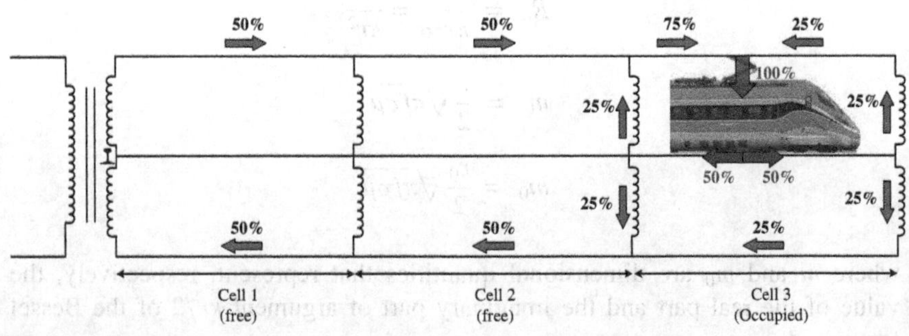

Figure 4.23 Power supply system functioning principle at 2×25 kV–50 Hz.

opposite phase compared to that of the contact line). Along the track 50/25 kV autotransformers, also equipped with center taps connected to earth, share the current absorbed by the train between the feeder and the contact line.

The line segments between one processing unit and the next are called cells. Assuming zero voltage drops on the transformers and on all conductors, it can be considered that the ideal distribution of the traction current is shown in Figure 4.23, when the line consists of three cells with the train in the middle of the third cell. In this situation the load is supplied bilaterally by two adjacent autotransformers. In order to obtain the distribution of the currents, let us start by assuming that the current absorbed by the train is equal to 100 A. This current returns through the track, dividing into two equal parts each of 50 A, since the impedance observed is the same on both sides, supposing that the train is halfway along. Of course, the distribution of current within the cell occupied by the train depends on its position. These 50 A are in turn broken down into two equal parts by the autotransformers so that the train receives bilateral power supply along the contact line. In the cell not occupied by the train, the current in the contact line and in the feeder is equal to half of that absorbed by the train, while in the track it is zero. The return of the current in this cell does not, therefore, affect the track, which eliminates the problem of interference to track circuits and reduces noise induced in the parallel conductors.

Finally, the 2×25 kV system, using a system with three voltage levels (the contact line, track, and power supply), makes it possible to transmit power at 50 kV and makes it available to the trains at the voltage of 25 kV. In this way it is possible to have a greater spacing of the TPSSs.

4.5 MATHEMATICAL–PHYSICAL STUDY OF THE FUNCTIONING

This section presents the study of power flow in the system, which involves verifying the actual distribution of the current between the various branches of the 2×25 kV system, both in the free sections and in those occupied by trains, starting from the ideal functioning of the 2×25 kV system and from the simple 25 kV system. This verification requires further study than what was described in the previous section, removing a lot of simplifying assumptions, in order to obtain a physical and mathematical representation of the actual functioning of the system.

4.5.1 Circuit Equations of the 2×25 kV–50 Hz System

With reference to the case of the 2×25 kV system, in order to determine the distribution of the currents, we will only consider the action of the train, modeled by an ideal current source. From the constraints arising from balancing magnetomotive forces of autotransformers and transformers, and from

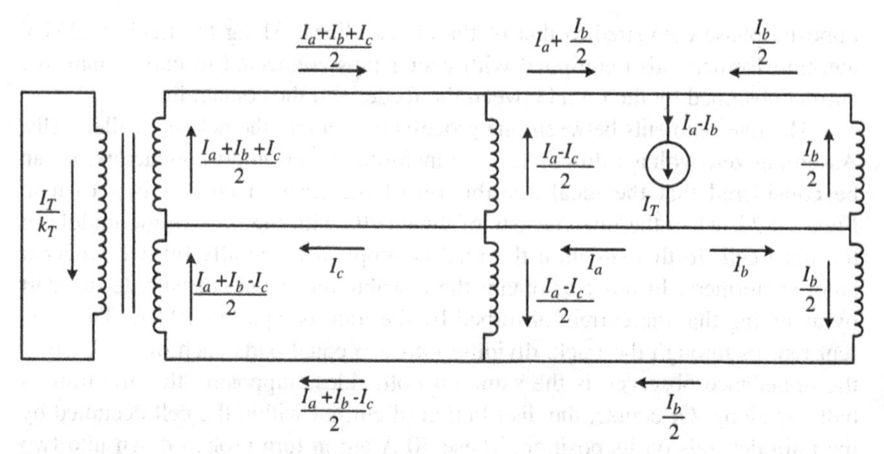
Figure 4.24 Distribution of currents in passive network systems.

Kirchhoff's laws on nodes, the currents are distributed among the various links, as in Figure 4.24, in which

$$I_T = I_a + I_b$$

is the current absorbed by the train and k_T is the transformation ratio of the TPSS transformer, and equals, for example, 132/27.5. It can therefore be noted that the current flows through the rails of the train toward the respective potential differences, redistributing themselves into the I_a and I_b elements as marked. In turn, the current I_a distributes partly into the autotransformer and partly into the I_c element, which passes through the track in the cell not occupied by the train. The components I_a, I_b, and I_c will be produced in compliance with the laws of Kirchhoff at voltages considering various couplings between the meshes.

The possible operating conditions imposed on the balance of magnetomotive forces are an operation at 2 × 25 kV, in which autotransformers work, and an operation at 25 kV, which instead leads to the exclusion of the autotransformers and the feeder, supplying power to the train from the OCL with return from the track in all sections through a half-winding of the TPSS transformer. Consequently, in the operation at 25 kV, the transmissible power is half that of the 2 × 25 kV operation.

To better understand this phenomenon also from a functional perspective, the distribution described in Figure 4.24 can, therefore, be seen as the sum of a component attributable to the operation of the ideal 2 × 25 kV system and one to the operation of a simple 25 kV system, as shown in Figure 4.25.

It can be noted that this distribution satisfies all the constraints previously introduced in calculating current distribution in the various branches. The current consumed by the I_T train is, therefore, produced by the sum of two currents, one, $I2 \times 25$, provided by an ideal 2 × 25 kV system and other, $I25$, from a

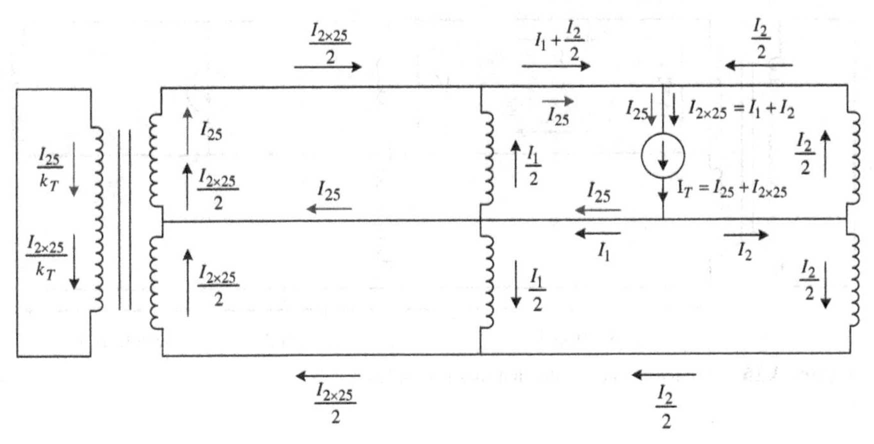

Figure 4.25 Redistribution of currents in the ideal 2 × 25 kV system and in the simple 25 kV system.

simple 25 kV system.

$$I_T = I_{2\times25} + I_{25}$$

Consequently, the $I_{2\times25}$ current will be a portion γ of the I_T train's total current:

$$I_{2\times25} = \gamma I_T$$

while the I_{25} current will make up the remaining part and, therefore, it will be produced by

$$I_{25} = (1 - \gamma)I_T$$

In turn, the $I_{2\times25}$ element is divided into two autotransformers in the elements I_1 and I_2, which can in turn be seen as fractions of the current $I_{2\times25}$ according to the factor α:

$$\begin{aligned} I_{2\times25} &= I_1 + I_2 \\ I_1 &= \alpha I_{2\times25} = \alpha\gamma I_T \\ I_2 &= (1-\alpha)I_{2\times25} = (1-\alpha)\gamma I_T \end{aligned}$$

The coefficient γ represents the operating portion that is the 2 × 25 kV system in comparison to the train's total power supply. It will therefore be important for this factor to be close to the unit so as to get as close as possible to the ideal functioning. The coefficient α, however, represents the distribution toward the two potential differences and, therefore, is proportional to the share of the 2 × 25 kV system's bilateral power supply. Unlike the coefficient γ, which tends to be constant, α is strongly influenced by the position of the train. Without removing generalities in the following discussion, for convenience we will consider the train at the middle of the second cell.

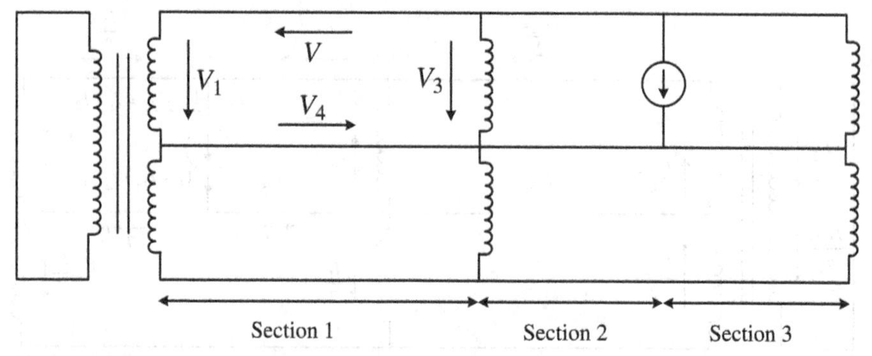

Figure 4.26 Voltages related to the first mesh considered.

The problem of studying the distribution of currents, therefore, leads to the calculation of the coefficients α and γ. Since there are two unknowns to be determined, it will be necessary to define a system of two equations that, considering the constraints on the previously introduced currents, must be derived from the laws of Kirchhoff on meshes.

The first law on meshes to be chosen is the one shown in Figure 4.26, for which it must be

$$V_1 + V_2 + V_3 + V_4 = 0$$

Since the system has been made passive, and thus the transformer has a short circuit on the primary side, the voltage of V_1 is substantially due to the voltage drop caused by reactance x_{TR} leakage between the primary winding and the secondary half-winding:

$$V_1 = x_{\text{TR}} \cdot \left(\frac{I_{2\times25}}{2} + I_{25}\right) = x_{\text{TR}} \cdot \left(\frac{\gamma I_T}{2} + (1-\gamma)I_T\right) = x_{\text{TR}} \cdot \left(1 - \frac{\gamma}{2}\right)I_T$$

However, with regard to the voltages V_2 and V_4, disregarding for the moment the resistances linked to reactors, these may be calculated as voltage drops on the reactances x_{LC1} and x_{B1} of the contact line and rail section 1 when taking the respective line inductances defined below into consideration:

$$V_2 = x_{\text{LC1}} \cdot \left(\frac{I_{2\times25}}{2} + I_{25}\right) = x_{\text{LC1}} \cdot \left(\frac{\gamma I_T}{2} + (1-\gamma)I_T\right) = x_{\text{LC1}} \cdot \left(1 - \frac{\gamma}{2}\right)I_T$$

$$V_4 = x_{B1} \cdot I_{25} = x_{B1} \cdot (1-\gamma)I_T$$

The voltage V_3 cannot simply be calculated as the voltage drop across the leakage reactance of the autotransformer in that the other portion of the winding is

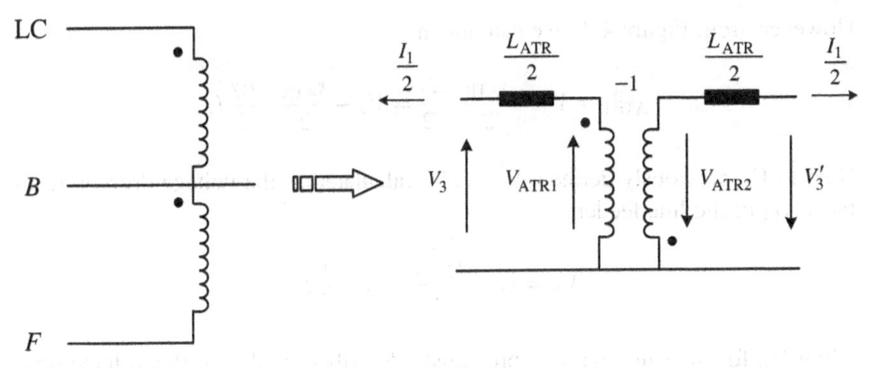

Figure 4.27 Circuit representation of the autotransformer.

not to be short-circuited. It is therefore necessary to determine the voltages at the ends of the autotransformer's windings. To this end, with reference to Figure 4.27, it can be noted how the autotransformer is comparable to a transformer that has a transformation ratio equal to −1.

It follows, therefore, that V_3 can be calculated as the algebraic sum between V_{ATR1} and the voltage drop across the leakage reactance of the autotransformer:

$$V_3 = V_{ATR1} - \frac{x_{ATR}}{2} \cdot \frac{I_1}{2} = V_{ATR1} - \frac{x_{ATR}}{2} \cdot \frac{\alpha\gamma}{2} I_T$$

In order to determine V_{ATR1} it is necessary to obtain V_3' by writing out the law of Kirchhoff on the mesh between the feeder and track section 1, as shown in Figure 4.28:

$$V_3' = V_4 - V_6 - V_5$$

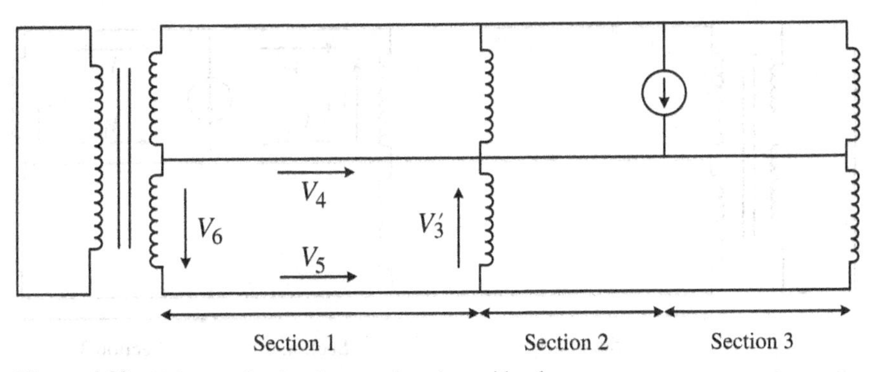

Figure 4.28 Voltages related to the second mesh considered.

However, from Figure 4.27 we can obtain

$$V_{\mathrm{ATR1}} = V_3' - \frac{x_{\mathrm{ATR}}}{2} \cdot \frac{I_1}{2} = V_3' - \frac{x_{\mathrm{ATR}}}{2} \cdot \frac{\alpha\gamma}{2} I_T$$

Next to V_4, previously defined, V_5 can be calculated as the voltage drop on reactance x_{F1} of the line feeder:

$$V_5 = x_{F1} \cdot \frac{I_{2\times 25}}{2} = x_{F1} \cdot \frac{\gamma}{2} I_T$$

while V_6, for the same reasons previously described, is due to the voltage drop caused by reactance x_{TR} of the leakage between the primary winding and the secondary half-winding:

$$V_6 = x_{\mathrm{TR}} \cdot \frac{I_{2\times 25}}{2} = x_{\mathrm{TR}} \cdot \frac{\gamma}{2} I_T$$

By combining the mentioned relationships, we can obtain the first equation of the system to calculate parameters α and γ:

$$x_{\mathrm{TR}}(1-\gamma) + x_{\mathrm{LC1}} \cdot \left(1 - \frac{\gamma}{2}\right) + 2x_{B1}(1-\gamma) - x_{F1}\frac{\gamma}{2} - x_{\mathrm{ATR}}\frac{\alpha\gamma}{2} = 0 \tag{4.13}$$

The second equation of the system can instead be obtained by considering the mesh in Figure 4.29, which, as can be seen, involves sections 2 and 3 with different distributions of the currents. It is not possible to write the equation on the mesh of a single section, as the voltage at the ends of the current are not known.

The resulting equation is therefore

$$V_3 - V_7 - V_8 - V_9 + V_{10} - V_{11} = 0$$

The voltage V_3 is already known from previous calculations.

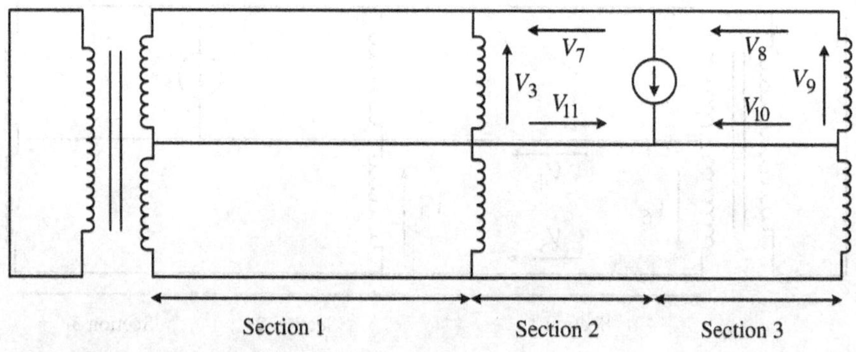

Figure 4.29 Voltages related to the third mesh considered.

Voltages V_7, V_8, V_{10}, and V_{11}, usually by disregarding the resistances related to reactances, can be calculated as the voltage drops on reactances x_{LC2} and x_{LC3} of the contact line and on the reactances x_{B2} and x_{B3} of track sections 2 and 3, whereas the respective line inductances

$$V_7 = x_{LC2} \cdot \left(I_1 + \frac{I_2}{2}\right) = x_{LC2} \cdot \left(\alpha\gamma I_T + \frac{(1-\alpha)\gamma I_T}{2}\right) = x_{LC2} \cdot \left(\frac{\gamma(1+\alpha)}{2}\right) I_T$$

$$V_8 = x_{LC3} \cdot \frac{I_2}{2} = x_{LC3} \cdot \frac{(1-\alpha)\gamma}{2} I_T$$

$$V_{10} = x_{B3} \cdot I_2 = x_{B3} \cdot (1-\alpha)\gamma I_T$$

$$V_{11} = x_{B2} \cdot (I_1 + I_{25}) = x_{B2} \cdot (\alpha\gamma I_T + (1-\gamma) I_T) = x_{B2} \cdot (1 + (\alpha - 1)\gamma) I_T$$

As for V_9, the previously expressed considerations for V_3 can be repeated:

$$V_9 = V_{ATR2} - \frac{x_{ATR}}{2} \cdot \frac{I_2}{2} = V_{ATR2} - \frac{x_{ATR}}{2} \cdot \frac{(1-\alpha)\gamma}{2} I_T$$

In order to determine V_{ATR2}, it is necessary to obtain V_9' by writing out Kirchhoff's law on the mesh between the feeder and track in sections 2 and 3, as shown in Figure 4.30:

$$V_9' = V_3' - V_{10} + V_{11} - V_{12}$$

The V_{ATR2} will instead be

$$V_{ATR2} = V_9' - \frac{x_{ATR}}{2} \cdot \frac{I_2}{2} = V_9' - \frac{x_{ATR}}{2} \cdot \frac{(1-\alpha)\gamma}{2} I_T$$

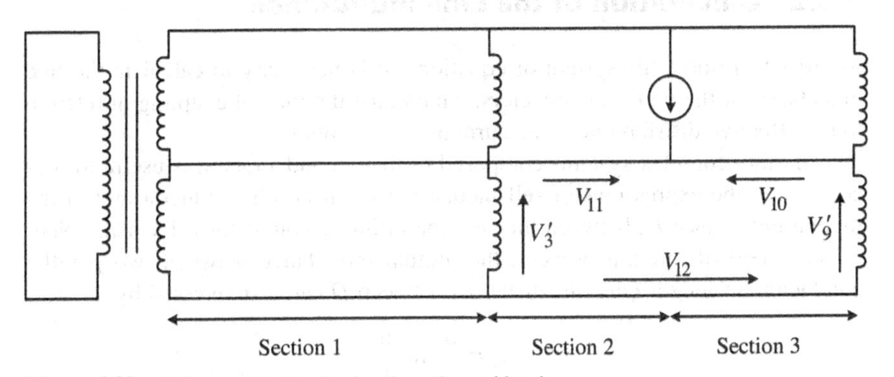

Figure 4.30 Voltages related to the fourth mesh considered.

In order to complete the mesh equation, we can calculate the V_{12}, which, as can be seen from Figure 4.30, involves two sections:

$$V_{12} = x_{F2} \cdot \frac{I_2}{2} + x_{F3} \cdot \frac{I_2}{2} = (x_{F2} + x_{F3}) \cdot \frac{(1-\alpha)\gamma}{2} I_T$$

By combining the above relationships, we can obtain the second equation of the system to calculate parameters α and γ:

$$\begin{gathered} x_{\text{ATR}} \frac{\gamma(1-2\alpha)}{2} - x_{\text{LC2}} \frac{\gamma(1+\alpha)}{2} - x_{\text{LC3}} \frac{\gamma(1-\alpha)}{2} - 2X_{B2}(1+(\alpha-1)\gamma) \\ +2X_{B3}(1-\alpha)\gamma + X_{F2} \frac{(1-\alpha)\gamma}{2} + X_{F3} \frac{(1-\alpha)\gamma}{2} = 0 \end{gathered} \tag{4.14}$$

The Equation 4.14, together with Equation 4.13, form the following system of equations, resolved by α and γ:

$$\begin{cases} x_{\text{TR}}(1-\gamma) + x_{\text{LC1}} \cdot \left(1 - \frac{\gamma}{2}\right) + 2x_{B1}(1-\gamma) - x_{F1}\frac{\gamma}{2} - x_{\text{ATR}}\frac{\alpha\gamma}{2} = 0 \\ x_{\text{ATR}} \frac{\gamma(1-2\alpha)}{2} - x_{\text{LC2}} \frac{\gamma(1+\alpha)}{2} - x_{\text{LC3}} \frac{\gamma(1-\alpha)}{2} - 2X_{B2}(1+(\alpha-1)\gamma) \\ +2X_{B3}(1-\alpha)\gamma + X_{F2} \frac{(1-\alpha)\gamma}{2} + X_{F3} \frac{(1-\alpha)\gamma}{2} = 0 \end{cases} \tag{4.15}$$

The equations of system (4.15) do not depend on the current of the train because the parameters of the line, as will be seen below, are a function of the coefficients α and γ but not of the current I_T.

4.5.2 Calculation of the Line Inductance

In order to resolve the system of equations, it is necessary to calculate the line inductance of the various conductors, which take the mutual coupling in relation to the effective distribution of the currents into account.

In more complex systems composed of many conductors, it is useful to proceed using the expressions of self-inductance L_{ii} of an ith conductor and of the mutual inductance L_{ij} between this and the ordinary conductor j. By using Neumann's field of integral process, the mutual inductance between two parallel conductors of length l and the distances between D can be expressed by

$$L_{ij} = \frac{\mu_0}{2\pi} \ln \frac{2l}{De}$$

while the self-inductance is

$$L_{ii} = \frac{\mu_0}{2\pi} \ln \frac{2l}{kr_0 e}$$

where e is the base of natural logarithms, r_0 is the radius of the conductor, and k is a factor that takes the current distribution inside the conductor into account. In particular, uniform distribution is valued at 0.7788 while cortical distribution tends to be 1.

If the system is solenoid, the forward currents are equal to the return current at any given time, therefore, in the calculation of the voltage drops between the L_{ij} and L_{ii} points, point $2l/e$ is eliminated for simplicity. In this case, we can simply use the expressions

$$L_{ij} = \frac{\mu_0}{2\pi} \ln \frac{1}{D}$$

$$L_{ii} = \frac{\mu_0}{2\pi} \ln \frac{1}{kr_0}$$

From the above formulas it is possible to calculate the inductance L_i of the "line," which multiplied by the pulsation ω and for the current in the conductor gives the total inductive voltage drop $\overline{V}_i$ for all the currents involved:

$$\overline{V}_i = j\omega \overline{I}_i L_{ii} + \sum j\omega \overline{I}_j L_{ij} = j\omega L_i \overline{I}_i$$

from which

$$L_i = L_{ii} + \sum L_{ij} \frac{\overline{I}_j}{\overline{I}_i} \ln \frac{1}{kr_0}, \text{ with } j = 1 \ldots n, j \neq i \tag{4.16}$$

Since the 2×25 kV system is connected to a single-phase network, the currents that circulate in it are always in phase or in counterphase. It follows that their relationship is always a real number that depends on the distribution determined by the above analysis.

In the following, line inductance in different sections of line will be calculated by applying (4.16) and by considering the distribution of currents indicated in Figure 4.25.

The actual configuration of the 2×25 kV system provides the coupling between six conductors to the direction of travel. However, in this discussion the following simplifying assumptions will be adopted:

- The contact line, composed of the messenger wire and the contact wire, will be represented as a single conductor. Nevertheless, the error committed is negligible because the distance between the two wires is much smaller than the distance between the other conductors.

- The two running rails are represented by a single equivalent conductor. The assumption is acceptable for the same reasons as the previous point.
- The earthing wires are disregarded since it is said that their contribution to the return of the traction currents is negligible.

The geometric parameters of the 2×25 kV system are listed as follows:

- r_{LC}: equivalent radius of the contact line conductor
- r_B: equivalent radius of the track conductor
- r_F: radius of the feeder conductor
- k_{LC}: current distribution factor in the equivalent contact line conductor
- k_B: current distribution factor in the equivalent track conductor
- k_F: current distribution factor in the feeder conductor
- D_{LCF}: distance between the equivalent contact line and feeder conductors
- D_{LCB}: distance between the equivalent contact line and track conductors
- D_{FB}: distance between the feeder and the equivalent track conductor

4.5.2.1 Section 1

Contact line

$$L_{LC1} = \frac{\mu_0}{2\pi}\left[\ln\frac{1}{k_{LC}r_{LC}} - \frac{I_{2x25}/2}{I_{25}+(I_{2x25}/2)}\ln\frac{1}{D_{LCF}} - \frac{I_{25}}{I_{25}+(I_{2x25}/2)}\ln\frac{1}{D_{LCB}}\right]$$

$$= \frac{\mu_0}{2\pi}\left[\ln\frac{1}{k_{LC}r_{LC}} - \frac{\gamma I_T/2}{(1-\gamma)I_T+(\gamma I_T/2)}\ln\frac{1}{D_{LCF}} - \frac{(1-\gamma)I_T}{(1-\gamma)I_T+(\gamma I_T/2)}\ln\frac{1}{D_{LCB}}\right]$$

By indicating the following quantities with δ

$$\delta = \frac{\gamma}{2-\gamma}$$

we can obtain the final expression of the contact line inductance in the first section:

$$L_{LC1} = \frac{\mu_0}{2\pi}\ln\frac{D_{LCF}{}^{\delta}\cdot D_{LCB}{}^{1-\delta}}{k_{LC}r_{LC}}$$

Track

$$L_{B1} = \frac{\mu_0}{2\pi}\left[\ln\frac{1}{k_B r_B} - \frac{I_{25}+(I_{2x25}/2)}{I_{25}}\ln\frac{1}{D_{LCB}} + \frac{I_{2x25}/2}{I_{25}}\ln\frac{1}{D_{FB}}\right]$$

$$= \frac{\mu_0}{2\pi}\left[\ln\frac{1}{k_B r_B} - \frac{(1-\gamma)I_T+\gamma I_T/2}{(1-\gamma)I_T}\ln\frac{1}{D_{LCB}} + \frac{\gamma I_T/2}{(1-\gamma)I_T}\ln\frac{1}{D_{FB}}\right]$$

from which we can obtain the final expression of the track inductance in the first section:

$$L_{B1} = \frac{\mu_0}{2\pi} \ln \frac{D_{\mathrm{LCB}}^{1/1-\delta}}{k_B r_B \cdot D_{\mathrm{FB}}^{\delta/1-\delta}}$$

Feeder

$$L_{F1} = \frac{\mu_0}{2\pi}\left[\ln\frac{1}{k_F r_F} - \frac{I_{25} + (I_{2\times25}/2)}{I_{2\times25}/2}\ln\frac{1}{D_{\mathrm{LCF}}} + \frac{I_{25}}{I_{2\times25}/2}\ln\frac{1}{D_{\mathrm{LCB}}}\right]$$

$$= \frac{\mu_0}{2\pi}\left[\ln\frac{1}{k_F r_F} - \frac{(1-\gamma)I_T + \gamma/2I_T}{\gamma I_T/2}\ln\frac{1}{D_{\mathrm{LCF}}} + \frac{(1-\gamma)I_T}{\gamma I_T/2}\ln\frac{1}{D_{\mathrm{LCB}}}\right]$$

from which we can obtain the final expression of the feeder inductance in the first section:

$$L_{F1} = \frac{\mu_0}{2\pi} \ln \frac{D_{\mathrm{LCF}}^{1/\delta}}{k_F r_F \cdot D_{\mathrm{FB}}^{1-\delta/\delta}}$$

4.5.2.2 Section 2

Contact line

$$L_{\mathrm{LC2}} = \frac{\mu_0}{2\pi}\left[\frac{1}{k_{\mathrm{LC}} r_{\mathrm{LC}}} - \frac{I_2/2}{I_{25} + I_1 + (I_2/2)}\ln\frac{1}{D_{\mathrm{LCF}}} - \frac{I_{25} + I_1}{I_{25} + I_1 + (I_2/2)}\ln\frac{1}{D_{\mathrm{LCB}}}\right]$$

$$= \frac{\mu_0}{2\pi}\left[\frac{1}{k_{\mathrm{LC}} r_{\mathrm{LC}}} - \frac{(1-\alpha)\gamma/2I_T}{(\alpha\gamma + [((1-\alpha)\gamma)/2])I_T + (1-\gamma)I_T}\ln\frac{1}{D_{\mathrm{LCF}}}\right.$$

$$\left. - \frac{\alpha\gamma I_T + (1-\gamma)I_T}{(\alpha\gamma + [((1-\alpha)\gamma)/2])I_T + (1-\gamma)I_T}\ln\frac{1}{D_{\mathrm{LCB}}}\right]$$

By indicating the following quantities with ε:

$$\varepsilon = (1-\alpha)\frac{\gamma}{2}$$

$$\frac{\alpha\gamma + (1-\gamma)}{(\alpha\gamma + (1-\alpha)\gamma/2) + (1-\gamma)} = \frac{1-\gamma(1-\alpha)}{1-(1-\alpha)\gamma/2} = \frac{\varepsilon}{1-\varepsilon}$$

we can obtain the final expression of the contact line inductance in the second section:

$$L_{LC2} = \frac{\mu_0}{2\pi} \ln \frac{D_{LCF}^{\varepsilon/1-\varepsilon} \cdot D_{LCB}^{1-2\varepsilon/1-\varepsilon}}{k_{LC} r_{LC}}$$

Track

$$L_{B2} = \frac{\mu_0}{2\pi}\left[\frac{1}{k_B r_B} - \frac{I_{25}+I_1+(I_2/2)}{I_{25}+I_1}\ln\frac{1}{D_{LCB}} - \frac{I_2/2}{I_{25}+I_1}\ln\frac{1}{D_{FB}}\right]$$

$$= \frac{\mu_0}{2\pi}\left[\frac{1}{k_B r_B} - \frac{[\alpha\gamma+(1-\alpha)\gamma/2+(1-\gamma)]I_T}{[\alpha\gamma+(1-\gamma)]I_T}\ln\frac{1}{D_{LCB}} - \frac{(1-\alpha)\gamma/2I_T}{[\alpha\gamma+(1-\gamma)]I_T}\ln\frac{1}{D_{FB}}\right]$$

from which we can obtain the final expression of the track inductance in the second section:

$$L_{B2} = \frac{\mu_0}{2\pi}\ln\frac{D_{LCB}^{1-\varepsilon/1-2\varepsilon}}{k_B r_B \cdot D_{FB}^{\varepsilon/1-2\varepsilon}}$$

Feeder

$$L_{F2} = \frac{\mu_0}{2\pi}\left[\ln\frac{1}{k_F r_F} - \frac{I_{25}+I_1+(I_2/2)}{I_2/2}\ln\frac{1}{D_{LCF}} + \frac{I_{25}+I_1}{I_2/2}\ln\frac{1}{D_{FB}}\right]$$

$$= \frac{\mu_0}{2\pi}\left[\ln\frac{1}{k_F r_F} - \frac{\alpha\gamma+(1-\alpha)\gamma/2+(1-\gamma)}{(1-\alpha)\gamma/2}\ln\frac{1}{D_{LCF}} + \frac{\alpha\gamma+(1-\gamma)}{(1-\alpha)\gamma/2}\ln\frac{1}{D_{FB}}\right]$$

from which we can obtain the final expression of the feeder inductance in the second section:

$$L_{F2} = \frac{\mu_0}{2\pi}\ln\frac{D_{LCF}^{1-\varepsilon/\varepsilon}}{k_F r_F \cdot D_{FB}^{1-2\varepsilon/\varepsilon}}$$

4.5.2.3 Section 3

Contact line

$$L_{LC3} = \frac{\mu_0}{2\pi}\left[\ln\frac{1}{k_{LC} r_{LC}} - \frac{I_2}{(I_2/2)}\ln\frac{1}{D_{LCB}} + \frac{I_2/2}{I_2/2}\ln\frac{1}{D_{LCF}}\right]$$

$$= \frac{\mu_0}{2\pi}\left[\ln\frac{1}{k_{LC} r_{LC}} - 2\ln\frac{1}{D_{LCB}} + \ln\frac{1}{D_{LCF}}\right]$$

from which we can obtain the final expression of the contact line inductance in the third section:

$$L_{LC3} = \frac{\mu_0}{2\pi} \ln \frac{D_{LCB}^2}{k_{LC} r_{LC} \cdot D_{LCF}}$$

Track

$$\begin{aligned} L_{B3} &= \frac{\mu_0}{2\pi}\left[\ln\frac{1}{k_B r_B} - \frac{(I_2/2)}{I_2}\ln\frac{1}{D_{LCB}} - \frac{(I_2/2)}{I_2}\ln\frac{1}{D_{FB}}\right] \\ &= \frac{\mu_0}{2\pi}\left[\ln\frac{1}{k_B r_B} - \frac{1}{2}\ln\frac{1}{D_{LCB}} - \frac{1}{2}\ln\frac{1}{D_{FB}}\right] \end{aligned}$$

from which we can obtain the final expression of the track inductance in the third section:

$$L_{B3} = \frac{\mu_0}{2\pi} \ln \frac{D_{LCB}^{1/2} \cdot D_{FB}^{1/2}}{k_B r_B}$$

Feeder

$$\begin{aligned} L_{F3} &= \frac{\mu_0}{2\pi}\left[\ln\frac{1}{k_F r_F} + \frac{(I_2/2)}{(I_2/2)}\ln\frac{1}{D_{LCF}} - \frac{I_2}{(I_2/2)}\ln\frac{1}{D_{FB}}\right] \\ &= \frac{\mu_0}{2\pi}\left[\ln\frac{1}{k_F r_F} + \ln\frac{1}{D_{LCF}} - 2\ln\frac{1}{D_{FB}}\right] \end{aligned}$$

from which we can obtain the final expression of the feeder inductance in the third section:

$$L_{F3} = \frac{\mu_0}{2\pi} \ln \frac{D_{FB}^2}{k_F r_F \cdot D_{LCF}}$$

Once all the values of the line inductance have been calculated and entered into the system (4.15), it is possible to calculate the coefficients α and γ, which determine the current distributions. Since the system equations are not linear, we proceed with its numerical resolution with the aid of calculation programs.

Considering the geometrical arrangement of the conductors in embankment and trench sections, can obtain the following distribution:

$$\gamma = 0.9538, \ \alpha = 0.5009$$

From the values obtained, it can be noted how the system does not deviate much from the ideal 2 × 25 kV operation, characterized by $\gamma = 1$ and $\alpha = 0.5$.

In Figure 4.31, the current trends in the contact line, the feeder and track are shown in the case of a real system.

However, it can be noted how the short-circuit voltage of the autotransformer considerably influences the current distribution. In fact, by increasing the short-circuit impedance (Z_{ATR}) the current tends to distribute itself more

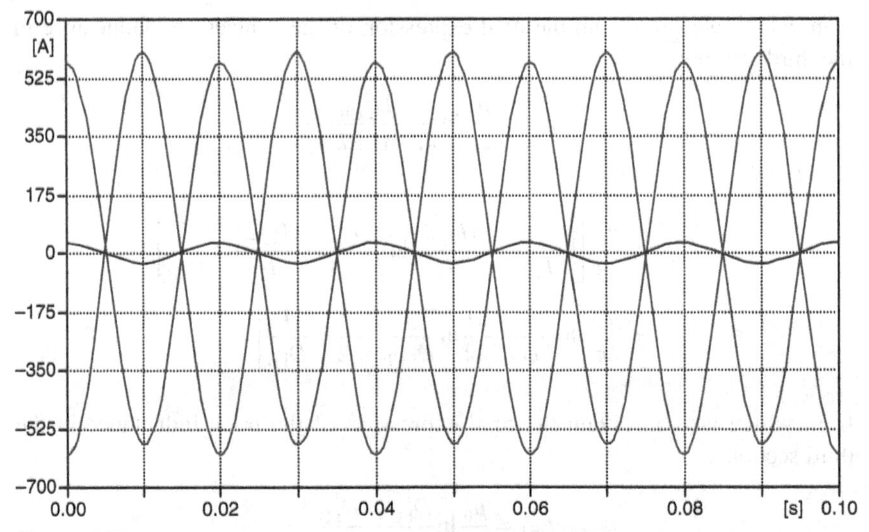

Figure 4.31 Current trends in the A_D (red), the feeder (green), and the track (blue) in a real system.

and more according to a traditional 25 kV system, going to affect the track even in cells containing no trains. For this reason, the short-circuit voltage of the autotransformers used in the object of study is rather low and exactly equal to 1%.

In support of the above, Figures 4.32 and 4.33 show the trends of gamma and alpha parameters to vary the Z_{ATR}, referring to the rated impedance Z_N of the autotransformer $Z_{ATRN} = Z_{ATR}/Z_N$. It can be easily noted how, with the increase in Z_{ATR} we increasingly end up further away from the ideal operation of 2×25 kV.

In particular, assuming, for example, that the value of the short-circuit impedance is a percentage of the autotransformer equal to 10%, we can obtain the following gamma and alpha values:

$$\gamma = 0.9003, \quad \alpha = 0.4759$$

To complete the analysis, we have also taken the dependence of gamma and alpha parameters from train movement into account (Figures 4.34 and 4.35) and from the position change of the feeder, which appears to be the sole conductor that can be moved (Figures 4.36 and 4.37). At first glance, it can be argued that these variables do not have a great influence on the two parameters α and γ.

From Figure 4.34 it can be noted that, with the approach of the train to the end of the cell, the gamma parameter tends to have a unit value, characteristic of the ideal operation.

From Figure 4.35 it can be noted that, as in theory, the alpha parameter varies between 1 and 0 with the train moving from the beginning to the end of the cell.

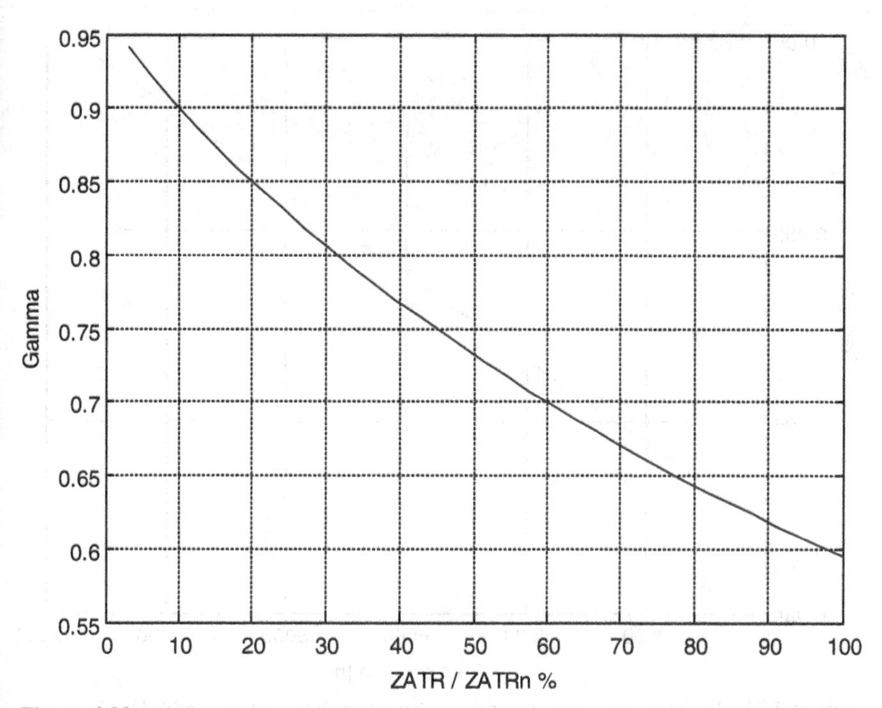

Figure 4.32 Evolution of the gamma parameter depending on Z_{ATR}.

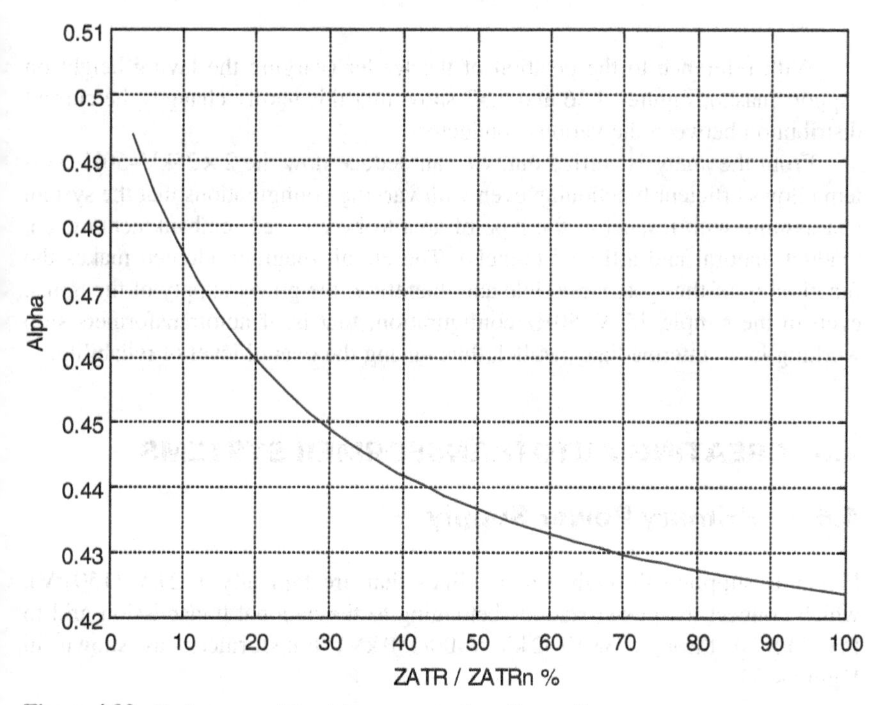

Figure 4.33 Performance of the alpha parameter depending on Z_{ATR}.

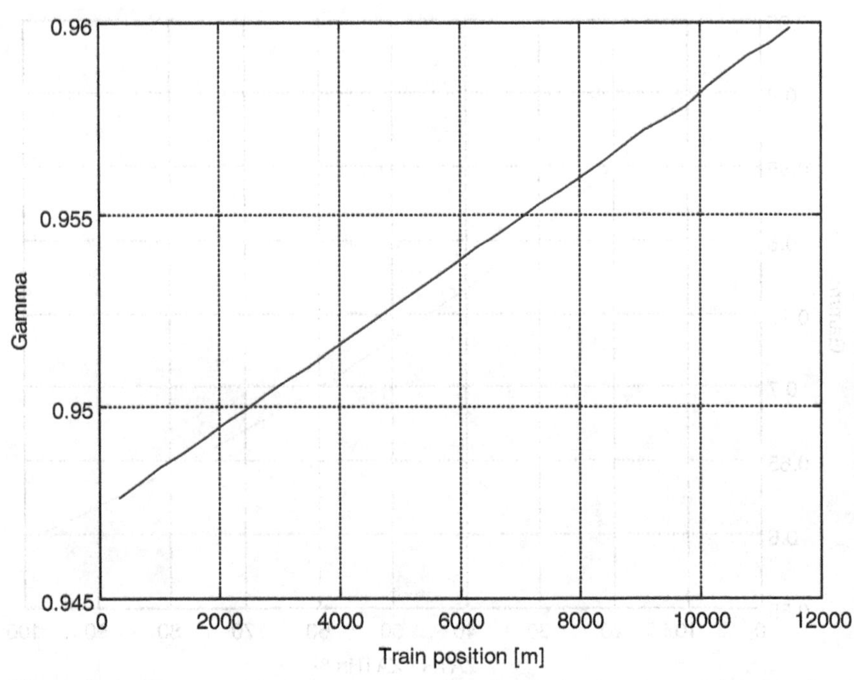

Figure 4.34 Evolution of the gamma parameter depending on the distance traveled by the train.

With reference to the position of the feeder (varying the laying height on support masts), Figures 4.36 and 4.37 show that this hardly changes the current distribution between the various conductors.

From the analysis carried out, we can deduce how the 2×25kV–50Hz system allows efficient functioning even with varying configurations that the system can assume with respect to the type of area to be covered (embankment/trench, viaduct, natural and artificial tunnels). The circuit diagram adopted makes the functioning of the system possible and, therefore, the power supply of the trains, even in the simple 25kV–50Hz configuration; that is, if autotransformers stop working in an intermediate parallel, thus raising the overall level of reliability.

4.6 CREATING AUTOTRANSFORMER SYSTEMS

4.6.1 Primary Power Supply

Power is supplied through primary lines that are typically 132 kV (150 kV), which connect the power stations belonging to the national transmission grid to the TPSSs) through 380/132 kV (400/150 kV) transformers, as shown in Figure 4.38.

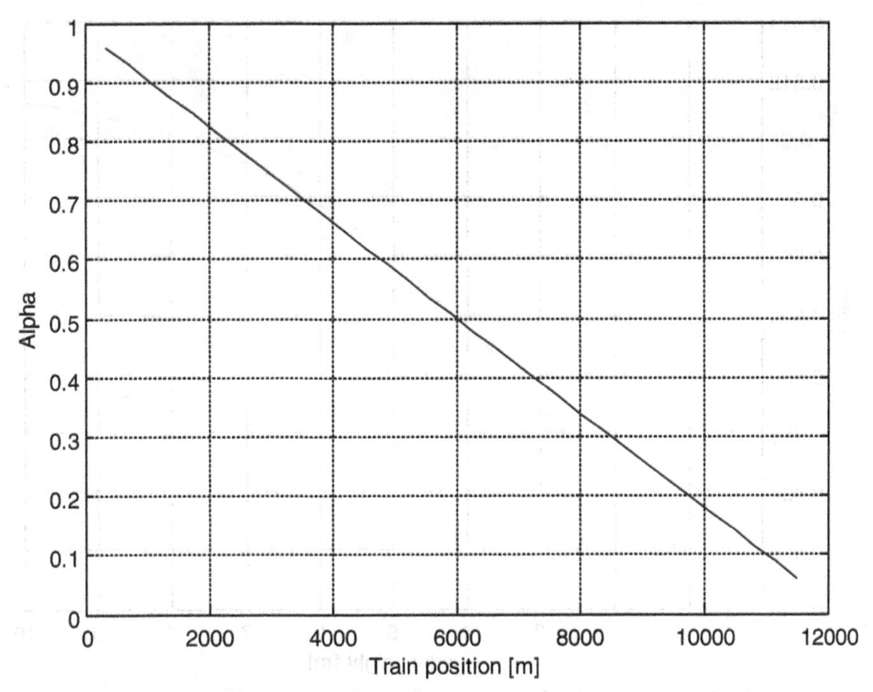

Figure 4.35 Performance of the alpha parameter depending on the distance traveled by the train.

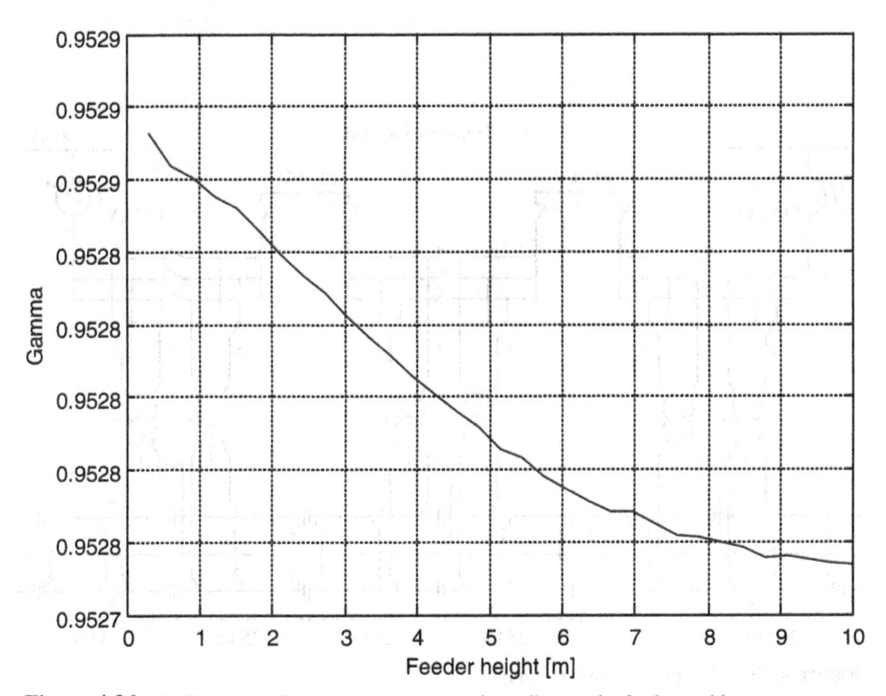

Figure 4.36 Performance of the gamma parameter depending on the feeder position.

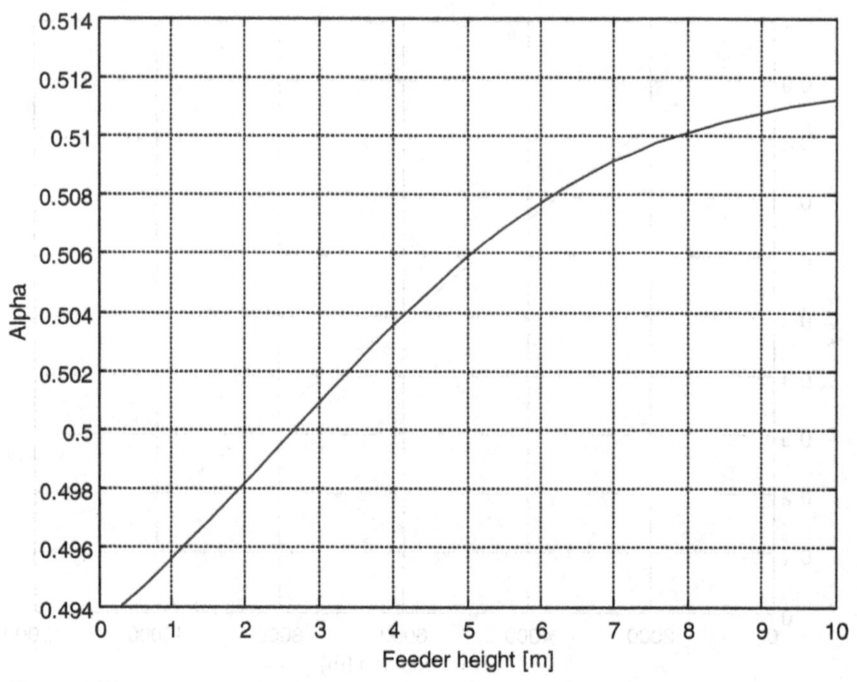

Figure 4.37 Performance of the alpha parameter depending on the feeder position.

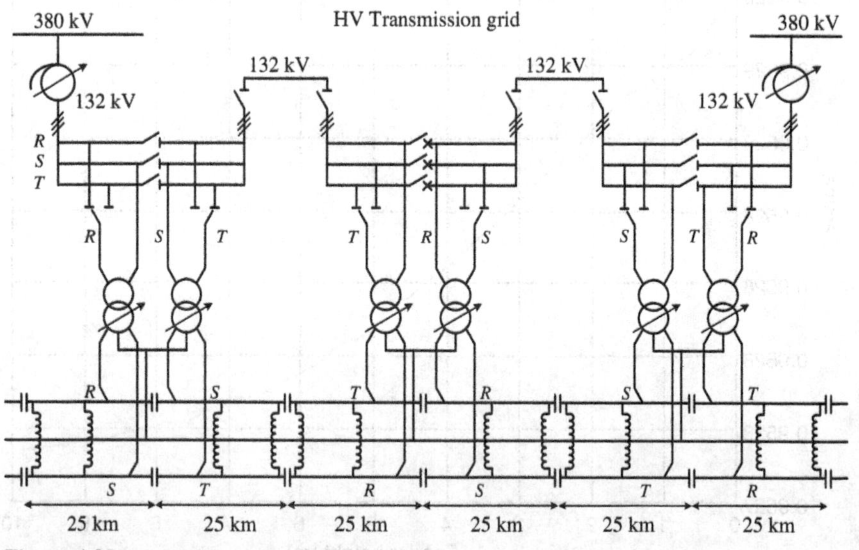

Figure 4.38 Ideal primary power supply.

In order to withstand the current unbalance caused by the connection of multiple single-phase loads, the AAT stations of the national transmission network, to which the line for supplying power to the train is connected, must present a sufficiently high short-circuit power.

The sections upstream and downstream of the substation must be divided with a neutral section, due to the different phases of supply voltage, and the potential difference at the terminals is equal to:

$$\Delta V = \sqrt{3}V = 42.5\,\text{kV}$$

The system as a whole is able to guarantee the continuity of power supply to the TPSSs even if a station in the national transmission network stops working or if a power line fails.

National transmission network stations are equipped with 380/132 kV autotransformers with a rated output of around 250MVA, which in turn feed power lines designed to carry a slightly lower power.

For these lines, masts with a reduced environmental impact are preferred, which are made of galvanized steel, and are tapered and polygonal. An example is shown in Figure 4.39, in comparison to a traditional lattice tower of Figure 4.39b.

In general, this type of mast is particularly appropriate in cases where it is necessary to mitigate the environmental or visual impact, to reduce or contain the amount of easement buffers, to move the conductors away from obstacles, to properly position the masts along the land belonging to the railway, and to reclassify existing power lines at nonrated voltages at a higher voltage.

This type of mast, which has the most similar phases compared to the traditional line, has a lower reactance resulting in greater transmission capacity and a significantly lower electromagnetic impact.

The use of masts with reduced environmental impact for the primary overhead power lines can therefore ensure environmental protection and high reliability, with the advantage of

Figure 4.39 View of (a) a traditional lattice tower and (b) a compact lattice with its picture.

Table 4.2 Electrical Characteristics of Power Lines for the Primary Line

Rated line voltage	132	kV
Maximum system voltage	145	kV
The maximum RMS of the current	800	A
Current density	1.36	A/mm^2

- low visual impact;
- fewer easement buffers;
- limited land occupation;
- small foundations;
- minor environmental damage and disruption during construction; and
- reduced electric and magnetic fields.

Lastly, in urban areas, as much as possible is done with cable sections in order to make the line less visible and intrusive. Typical electrical characteristics of the power lines are shown in Table 4.2.

4.6.2 Traction Power Substations (TPSS)

Substations located at the ends of lines are connected by a 132 kV (150 kV) power line to the nearest national transmission grid stations of 380 kV (400 kV), through a specific 380/132 kV (400/150 kV) autotransformer. All the TPSSs present along the railway line are connected to each other with a power line that is always at 132 kV (150 kV) when "entering and exiting." This configuration has the advantage of allowing a bilateral power supply to each TPSS and transfers the load unbalance to a relatively high short-circuit power node, the latter requiring a 380 kV voltage (400 kV). The double interconnection with the transmission system makes it possible to have a high level of continuity of operation. Different voltage values can be considered in relation to the availability of the network, provided they are of the same order of magnitude.

Since the TPSSs are placed at a greater distance (for example 50 km) compared to the case in DC, they will have a power that is able to obtain the same levels of specific power per kilometer of line. The typical value of installed power is 120 MVA, divided into two processing units of 60 MVA, as outlined in Figure 4.40.

As previously described, the two substation transformers are connected to the primary HV line according to the insertion at "V," with the phases rotated in successive substations. This linking strategy allows the power to be drawn on pairs of different phases every 25 km, in order to compensate the inevitable load unbalance on the primary network. The TPSS transformer (Figure 4.41) is single phase with external ONAN cooling and immersed in oil.

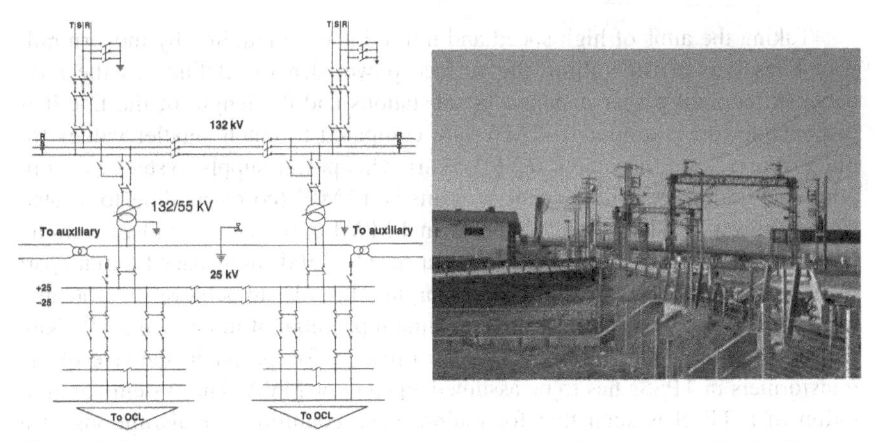

Figure 4.40 Simplified circuit diagram of a 2×25 kV TPSS and a photograph related to the connection of a TPSS.

Figure 4.41 Substation transformer.

Taking the aims of high speed and traffic intensity required by modern railway lines into consideration, the surface power density, defined as the ratio between the total power installed in substations and the length of the line it is responsible for, is around 1–2 MW/km, compared to much smaller values for lines with local traffic (about 0.5 MW/km). This power supply system can support, for example, the movement of trains at 12 MW (corresponding to a rated surface power density, totaling more than 1 MW/km on the two rails, assuming loads with a power factor of 0.95) set at a speed of 300 km/h spaced 5 min apart without any limits and with potential margins. In order to achieve the potential margin of 100%, thus achieving the maximum potential of more than 2 MW/km, with a single transformer functioning as a reserve for the other, the size of the transformers in TPSSs has to be assumed equal to 60 MVA. The system configuration of a TPSS is such that for maintenance conditions or disruptions of a transformation group, action can be taken without disconnecting the TPSS. In this case, the transformers can be overloaded at 50% for 15 min or 100% for 5 min. Each HV line stall is provided with disconnectors and breakers and the HV busbar is sectioned at the center with a motorized disconnector, which makes it possible to keep one of the two groups in service in the event of a busbar fault or equipment being rigidly connected to the busbar. In series with the disconnector, a second manually operated disconnector is provided that has the sole purpose of allowing trips on the motorized disconnector above, always keeping a transformer group in operation at 60 MVA. However, the two transformers are not designed to function in parallel, because they would have an excessively high short-circuit level, particularly with regard to the breaking capacity of onboard circuit breakers. A rough estimate indicates a short-circuit current of 20 kA and thus requires attention when choosing the substation circuit breakers, whose trip is particularly critical in the case of short circuits far away caused by high line impedance and unilateral power supply.

The transformation ratio is 132/55 kV with the low-voltage winding divided into two sections of 27.5 kV. The transformers are equipped with under load ratio variators with a voltage adjustment of $+4\times1.25\%$ and $-12\times1.25\%$ on the primary side to compensate for HV voltage variations.

For the 25 kV system, the pantograph limit values permitted for supplying power to the contact line according to regulation EN 50163 are as follows:

- The permanent maximum value is 27.5 kV (+10%), 29 kV for 5 min
- The permanent minimum value is 19 kV (−24%), 17.5 kV for 10 min

Under normal operating conditions, the mean value of the fundamental frequency measured over 10 sec shall be within a range of the HV supply network.

For systems with synchronous connection to an interconnected system:

- 50 Hz $\pm$ 1% (i.e., 49.5–50.5 Hz) for 99.5% of a year
- 50 Hz +4%/− 6 % (i.e., 47–52 Hz) for 100% of the time

For systems with no synchronous connection to an interconnected system (e.g., supply systems on certain islands)

- 50 Hz ± 2% (i.e., 49–51 Hz) for 95% of a week
- 50 Hz ± 15 % (i.e., 42.5–57.5 Hz) for 100% of the time

In order to allow the train to provide maximum performance, the power supply system is designed so that, with the exception of very limited sections, the pantograph has a voltage exceeding 22 kV.

The center tap of the secondary side of the transformer is connected to the track and to the earth link, the pole at 25 kV feeds the contact line while the pole at −25 kV is connected to the so-called negative power supplier (feeder) according to the diagram provided in the 2 × 25 kV system.

From the secondary substation busbar, in the case of a two-way line, four feeder posts are derived, each of which comprises a bipolar circuit breaker and two bipolar disconnectors, one of which is motorized and remote-controlled. The circuit breaker also protects the lines from short circuiting, while the circuit breaker ensures a lack of voltage when working on overhead conductors or in TPSSs. The protective equipment that acts on the feeder posts' circuit breakers has a reactance with rectangular characteristics and is capable of ensuring the operation even with parallel lines and extend through to the neighboring TPSS. Of the four feeders, two supply power to the uplines, while the other two supply power to the downlines, for a total of four contact lines and four feeders. The grounding circuit conductors are linked to the TPSS earth electrodes.

In short, all the TPSSs comprise the following:

- A 132 kV busbar
- Stall transformers at 132/2 × 25 kV (60 MVA)
- MV busbars ±25 kV
- Auxiliary service transformers
- Stall feeders at 25 kV
- Downline/upline parallel busbar feeders at 25 kV

4.6.3 Auxiliary Points

In normal operation, there is an expected change in the torque of power supply phases, both between the two groups of a TPSS and between the corresponding groups, each one supplying power to the line through the two TPSSs. Therefore, the contact line must be sectioned both in the TPSS and halfway along the roughly 50 km section. In order to be able to move the disconnections and better distribute the load in the event of abnormal conditions, such as a partial or total outage of a TPSS, disconnections are also made in the middle of each half-section of about 25 km and normally kept short-circuited. The overall power supply diagram is shown in Figure 4.42.

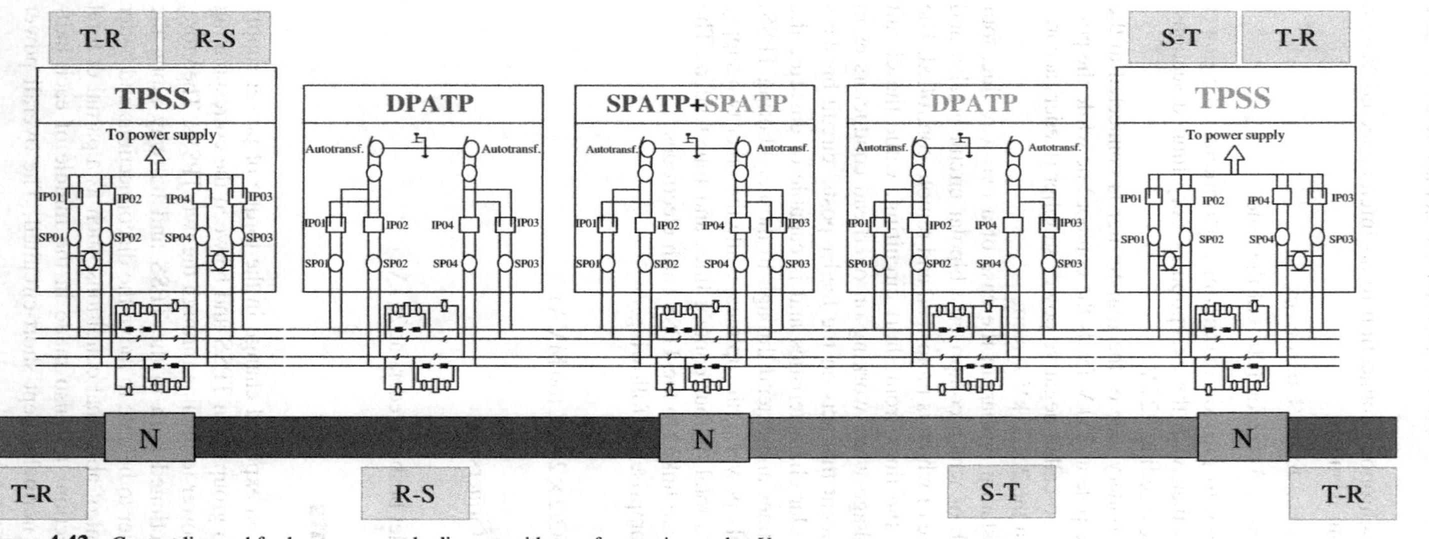

Figure 4.42 Contact line and feeder power supply diagram with transformers inserted at *V*.

There are "auxiliary points" along the line, classified as follows:

- A parallel and autotransformers point (PATP), which has the function of creating a parallel between the upline power supply and that of the downline and the power exchange between the contact line and feeder.
- An electric border point (EBP), which has the purpose of maintaining the mechanical and electrical separation between two conductors powered by different systems such as those in DC and AC, for example, the autotransformer system and that in direct current at 3 kV.
- A disconnection point (DP) and a disconnection and protection point (DPP), which carry out the electrical separation of two conductors from different electrical zones, powered, however, with the same phase voltage.

The parallel and autotransformers points, the electric border point, and the disconnection point therefore involve, in different ways, the electrical and mechanical separation of the conductors (contact line, feeder, rails, and earth) that flow into them. The neutral sections of PATPs or TPSSs and the EBP along the contact line are symmetrical, that is, they have the same structure both on the upline and on the downline, while the DPs are most often located on only one of the two tracks, as is the case for DPs delimiting the service points. The line sections between two auxiliary points or between a TPSS and an auxiliary point, that is, the sections that have the same electrical "zone," are called "full line sections."

4.6.3.1 Parallel and autotransformers point (DPATP)

Between two consecutive TPSSs there are three downline/upline PATPs, generally spaced 10–15 km apart (average wheelbase 12 km). Double and single parallel and autotransformers points (DPATP and SPATP) have the function of creating a parallel between the upline power supply and that of the downline, with regard to both the contact line and the negative power supply (feeder), and are equipped with two autotransformers at 15 MVA, 50/25 kV needed to transform the energy from the system at 50 kV, consisting of the contact line and the negative feeder, to the traction system at 25 kV (contact line and track).

The single PATP is made up of two identical modules, one for each of the disconnection, made up of the autotransformer connectable by means of bipolar shunting circuit breakers to the contact line and the feeder both on the even and odd tracks, achieving an upline/downline parallel (Figure 4.43). The feeder busbar of each module is derived from a single-phase transformer of 25/0.24 kV for auxiliary services. The SPATPs have a unique form identical to one of the two DPATP modules.

The "central" DPATP, namely, that positioned halfway between two TPSSs, is operated under normal conditions with the neutral section open, because the two line sections are fed from the two TPSSs in question and therefore have different phases (Figure 4.44). In this case, the system has the configuration of a

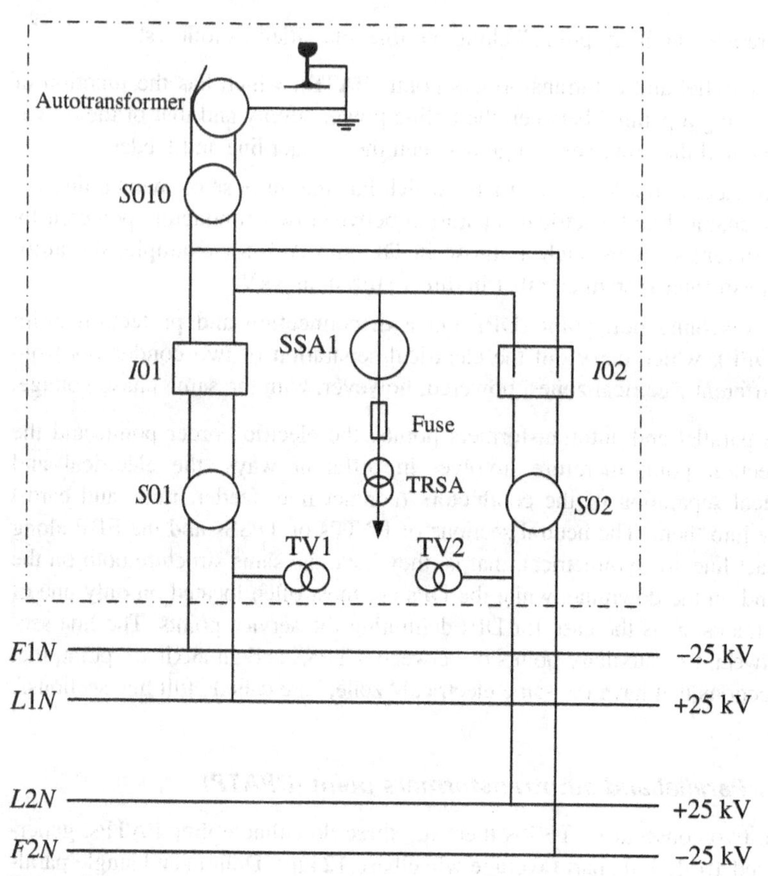

Figure 4.43 Upline/downline parallel and autotransformers point.

"DPATP with different line conductors supply" and work both in parallel "to the left" and "to the right" of the neutral section. In addition, in the case of trains passing through one of the two sections, the autotransformer dedicated to that portion of the line allows the reclosing in the feeder of the current coming from the TPSS of the section it is responsible for.

The DPATP closest to the TPSS is instead normally operated with the neutral section closed, since it must establish continuity; in other words, it must transmit the power supply from the section adjacent to and powered by the TPSS to the section that is not powered by any TPSS. In this case, the system has the configuration of an "only parallel autotransformers point with the possibility of disconnection if necessary." In general, only one of the two autotransformers is in operation, which, by means of the parallel, serves both the upline and downline, while the second autotransformer is only used as a reserve.

The sections (Figure 4.45) are created with double-space overhead and short nonpowered intermediate sections, called the neutral section. The train must

Figure 4.44 Autotransformers point halfway along the section.

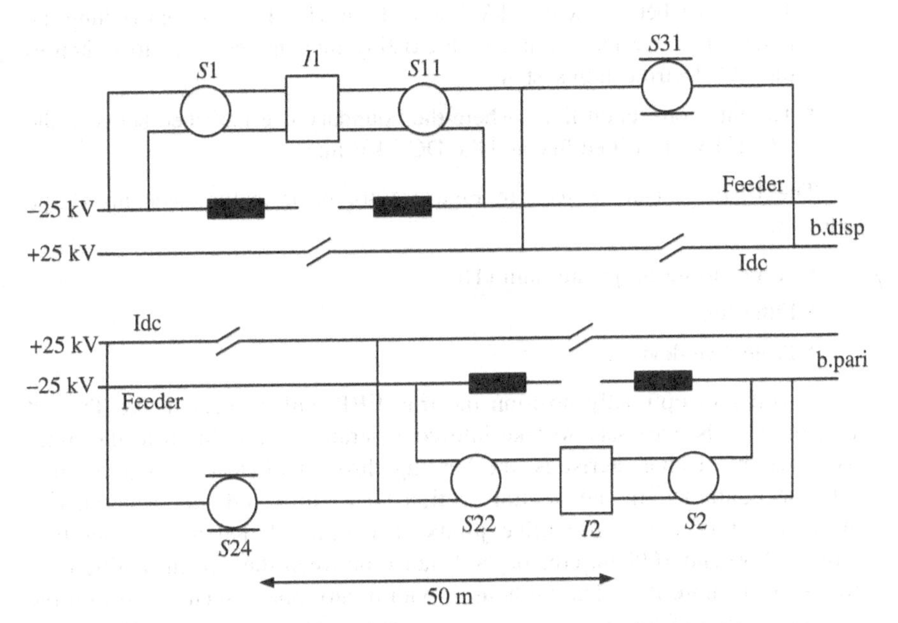

Figure 4.45 DPATP for a double-track line.

enter the neutral section without absorbing the current in order to avoid arcing between the powered mains and the neutral section. Switching off the drive and the main onboard circuit breaker will be controlled and/or automatically controlled via balises arranged in suitable positions on the track, so as to ensure maximum safety for carrying out the maneuver and at the same time minimizing the time taken to shut down the motors. The two sections are able to short circuit, the first by means of a double-pole circuit breaker (operable also under load), the second by means of a single-pole disconnector. The disconnection of the feeder, however, is unique, and is located at the first air gap encountered by the train in the legal sense (left), and carried out through mooring insulators without dead-ending a line.

A variant of the DPATP is the "single parallel autotransformers point" (SPATP), which aims to "close" the feeders at the end of the lines electrified at 2×25 kV, and is normally located in the immediate vicinity of the electric border points. In this case there is only one autotransformer and it is simultaneously put in upline/downline parallel with the feeder and the contact line.

4.6.3.2 Electric border point between AC and DC lines

EBPs aim to maintain the mechanical and electrical separation between two sections powered by two completely different and incompatible systems: the 2×25 kV–50 Hz AC system and the 3 kV DC system.

The EBP can therefore be found at the following:

- The border between the 25 kV line and the 3 kV line, or approaching the nodes or where interference with existing lines is necessary to switch to the DC electrification system.
- The interconnection lines, where the boundary is established between the AC 25 kV electrified line and the DC 3 kV line.

The EBP, as can be seen in Figure 4.46, is basically made up of the following:

- A transformer-separator unit (TS)
- Filter units
- Protective devices

In order to optimally position the true EBP with respect to the TS and filter units, it is necessary to take into consideration, in addition to the functional technical characteristics, the passing through of very long trains that must be contained in the neutral section so as to avoid the simultaneous short circuit of two consecutive joints. The optimal distance is therefore between 800 and 1000 m, creating a distance between the two insulating couplings of no more than 2 m, without providing any operational control on the isolated section of the track.

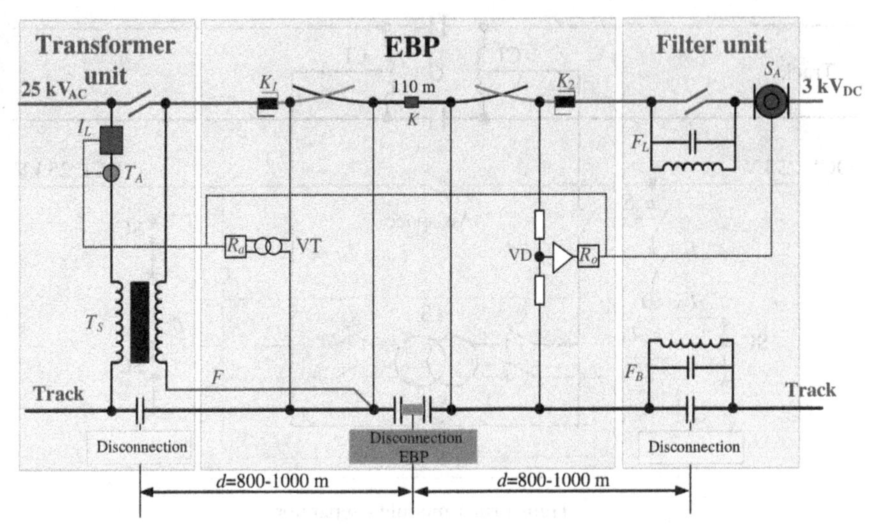

Figure 4.46 Operational diagram of the EBP on the interconnection track.

The border area must of course be traveled to low pantographs, necessary for when the pantographs must be swapped to adapt to the electrical and mechanical differences of the two lines. To ensure safety when the pantographs are lowered in the neutral section for phase change, automatic remote control are provided based on track balises.

Transformer-Separator Unit (TS) The transformer-separator unit represents the 25 kV side of the electric border point. The transformer unit has been designed for supplying power at 25 kV$_{AC}$ to a section of line, about a kilometer or so in length, electrically independent from the remaining system at 25 kV$_{AC}$, thanks to the presence of an insulated joint (Joint 1) that creates two electrically independent sections of the track. The unit is placed in the penultimate and ultimate automatic adjusters of the conductors prior to the EBP section, in connection to Joint 1 and the link is made through special parallel rail inductive boxes connecting the primary and secondary sides of the same transformer between the contact line and the track. The connection with the 25 kV contact line is carried out through the line equipment made up of the disconnector (L_D), the circuit breaker (L_{CB}), and the corresponding current transformer (C_T), as shown in Figure 4.47.

Such a solution firstly makes it possible to limit traction current leakages conducted in an alternating current to only a train's travel time through a (~1 km) section of interconnection track powered at 25 kV DC by TSs. It also makes it possible to limit the traction current in direct current toward the AC system.

The TS transformer unit of the interconnection tracks are single phase with separate windings, which naturally dry and cool in the air, and has a rated power

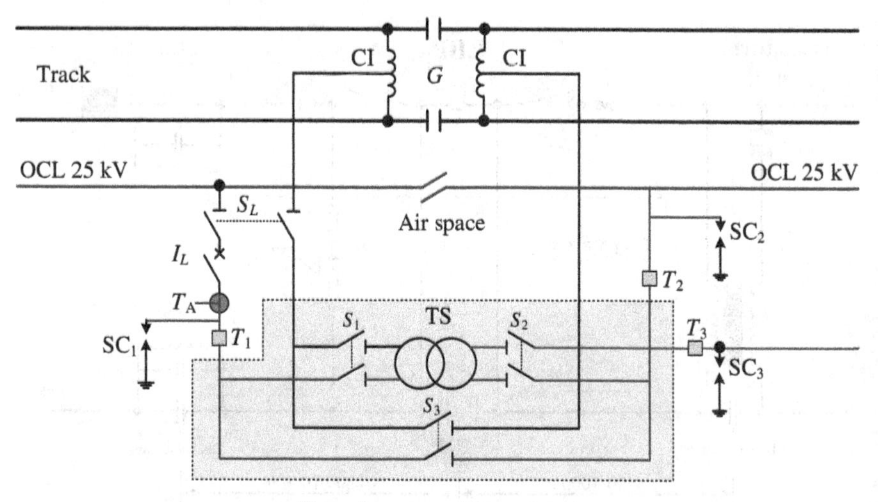

Figure 4.47 Transformer-separator unit, diagram of electric power connections.

of 2 MVA, able to be overloaded over 10 min, up to a power of 6 MVA for a time equal to at least 75 s and zero power for the remaining seconds to 600 s. The transformation ratio is 1:1 without intermediate taps; in fact for both windings, the rated voltage is equal to $V_n = 27.5\,\text{kV} \pm 10\%$. Finally, the short circuit voltage referring to power and the rated voltage is 5%.

The separator transformer, together with the disconnection, switching, and protection equipment, is contained in a specifically designed prefabricated module.

The TS transformer unit of the track opening to the full line is dry single phase with windings encapsulated in epoxy resin in a vacuum with a rated power of 3 MVA, which can be overloaded over the period of 5 min, up to a power of 9 MVA for a time equal at least to 40 s and zero power for the remaining seconds up to 300 s. The rated voltage for the primary winding is equal to $V_{nI} = (27.5 \pm 0.4)\text{kV}$, for the secondary winding it is equal to $V_{nII} = 27.5\,\text{kV}$. Lastly, the short-circuit voltage is 3.8% with a tolerance of $\pm 5\%$.

Filter Units The filter unit, on the 3 kV side, to be connected to both the contact line and the return circuit line (track) of the interconnected branch, separately carries out the electrical continuity of the contact line conductors, relating to the two automatic rail adjusters in correspondence of insulated joint 2, as shown in Figure 4.48, and at the same time prevents the passage of alternating currents, which are dangerous for the signaling system.

The filter units for each track are essentially made up of two filters that are functionally similar, which consist of an inductive resistor and a capacitor in parallel, tuned so as to ensure the reduction in interference at 50 Hz on the

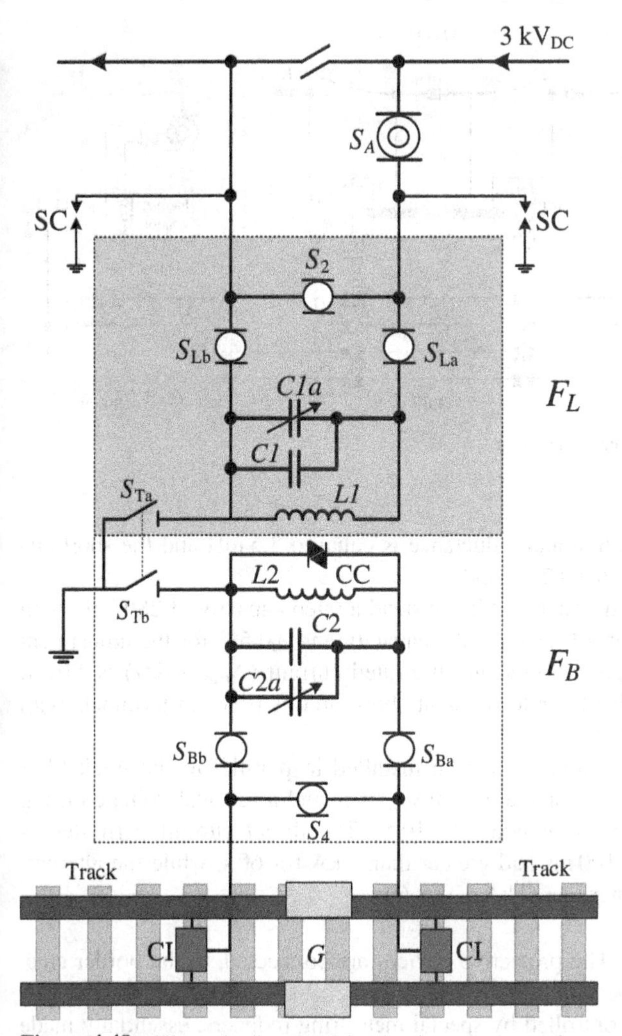

Figure 4.48 Filter unit, diagram of the electric power connections.

contact line (L_F) and the track (T_F). The main electrical diagram is shown in Figure 4.48.

The line filter (L_F) is made up of equipment and components connected to the 3 kV DC contact line. The track filter (T_F) is made up of equipment and components that carry out the electrical continuity of the track.

The filter unit resistor has been designed to be inserted on both the contact line and on the tracks of the railway line powered at $3\,kV_{DC}$, in order to reduce the intensity of the conducted currents that, coming from the system at $25\,kV_{AC}$, could propagate along the contact line and along the tracks of the $3\,kV_{DC}$ system, potentially interfering with machinery on board the rolling stock and fixed

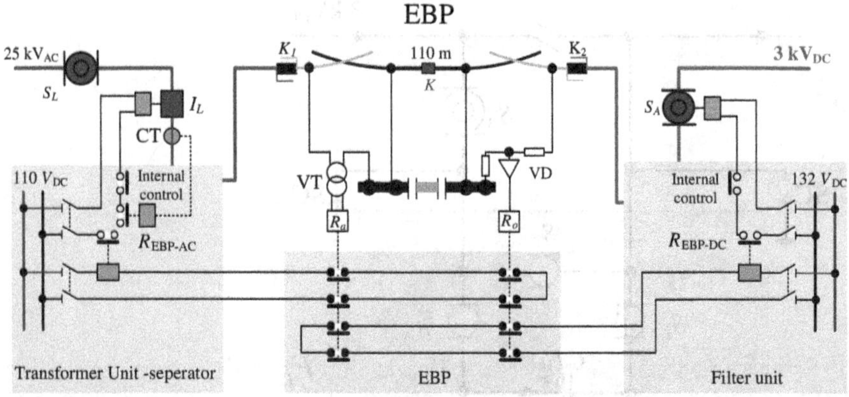

Figure 4.49 The EBP protection system.

signaling systems. The nominal inductance is equal to 3.5 mH and the short-circuit current is 30 kA_{rms} for 0.2 seconds.

The capacitor for the filter unit has instead a rated capacity of 2895 μF, with a range of variability of ±12.5% for the capacity and ±0.5% for the adjustment steps. The rated voltage is 900 V and the rated current (A_{rms}/50 Hz) is 110 A. The capacitors must also tolerate frequent short circuits between terminals with residual voltages of 200 V.

Finally, the bypass device must be installed in parallel to the track filter equipment and has a fuse voltage of 150 V ± 20% and a residual voltage during conductance that is lower or equal to 10 V. The direct current permitted is greater than 30 kA for 100 ms and greater than 1 kA for 60 s, while the alternating current load is greater than 2 kA_{rms} for 60 s.

Protective Devices The protective devices are connected, in the border area, between the contact lines and track.

Such devices are controlled by special measuring reducers, essentially made up of an external 25 kV voltage transformer (VT) and a 3 kV voltage divider (VD) protected in a suitable sealed container, connected according to the diagram in Figure 4.49.

Mode of Operation The transformer-separator unit can have three possible electrical configurations. First, there is the normal operating condition with the contact line's power supply guaranteed by the transformer-separator. Then there is the state of the equipment when the module's power supply is disconnected and when the power supply to the contact line in the EBP is interrupted. In this configuration, trains cannot interconnect. Lastly, there is the configuration in reserve that can be implemented to ensure, in any case, power supply to the contact line in the EBP if the transformer-separator unit is not working. The operation of the equipment can be electrically (local or remote) or manually controlled.

The filter unit also has three possible electrical configurations. First, there is the normal machinery operational condition with the continuity of the contact line and track guaranteed through the respective filters. Then there is the status of the equipment when the module's power supply is disconnected and grounded and when the contact line's power supply and the continuity of the track are interrupted. In this case, interconnection is switched off in the EBP and therefore the trains cannot proceed. Lastly, there is the configuration in reserve that can be implemented to ensure, in any case, the continuity of the contact line and the track, if the filter unit is not working. Disconnectors, with the exclusion of the earthing disconnector, can be electrically (local or remote) or manually controlled.

The central neutral section only guarantees the mechanical continuity of the contact line and the track and must be traveled over in inertia; it cannot be powered for any need. In the unfortunate event of a locomotive having to stop within this section, it must be towed away by a diesel locomotive.

4.6.3.3 Disconnection point (DP) and disconnection and protection point (DPP)

A disconnection point (DP) and a disconnection and protection point (DPP) carry out the electrical separation of two different electrical zones, powered, however, with the same phase voltage as shown in Figure 4.50. In this case the neutral section is missing and the post is reduced to being the overlap of the two conductors in question, as is the case for "disconnection sections" in 3 kV DC systems.

The DP is used whenever it is necessary to electrically separate a section of full line from a service point, without, however, having to avoid contact, even if it is only temporary, between the two conductors through the pantographs when the train passes through. Another application is the semidisconnection within a service point.

The DP is equipped with a remote-controlled bipolar disconnector that electrically connects the two sections of the contact line and the two feeder sections.

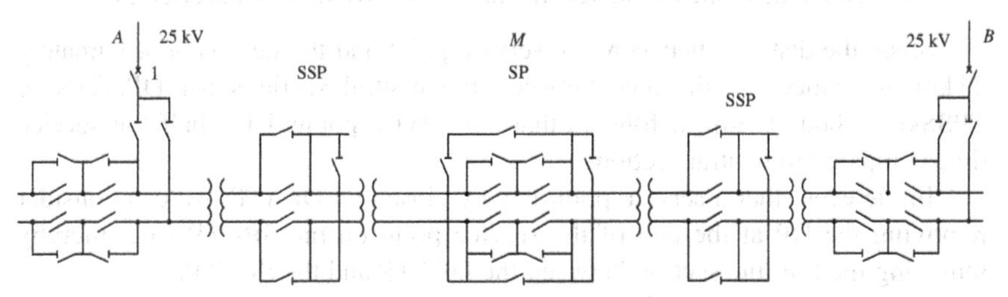

Figure 4.50 Diagram of the power supply for a section of double track between substations *A* and *B*, with a middle sectioning and parallel point and two parallel and autotransformers points.

This disconnector is normally kept closed to guarantee electrical continuity between the two conductors.

The DPP, which mechanically speaking is the same as the DP, is used as a line protector and is equipped with a bipolar or unipolar (in series) circuit breaker and disconnector, depending on the case (DPP protecting a EBP, located on the alternating current line or on the interconnecting line).

4.6.4 Service Point

Within each line "section" there are "service points," classified as follows:

- *Communication Points (CP)* spaced about 24 km apart, make it possible for trains to pass from one track to the other and constitute two upline/downline communications, one after the other and mounted in reverse order. In addition, the first communication is mounted in the same way as the last one at the previous service point and the longest stretch of track between two communication points is sectioned in half by the DP bipolar disconnector. This method minimizes the length of track in unlawful circulation if there is an interruption due to an outage in an electrical zone. Two other DPs provided with disconnectors are located at the ends of the CP, on the opposite track to the one holding the semidisconnector.
- *Movement Points (MP)* are used to connect plain line tracks with priority tracks and stabling tracks. Movement points are about 48 km apart and are placed to coincide with CPs. Their arrangement is similar to that of a CP but with the addition of two priority tracks, one inserted onto the plain line track on the upline and one on the downline, within communication points. Four 80 m-long track sections are also inserted onto priority tracks to make stabling points. The electrical structure is similar to that of the CP, with the addition of priority tracks.
- *Interconnection Points (IP)* allow the interconnection of alternating current lines with historical ones. A 25 kV interconnection line is started from the switches on the alternating current side, which is removed from the feeder and connected to the line powered at 3 kV through the EBP, downstream of the abovementioned switches. IPs naturally have no fixed service time.

Since the distance between one service point and the next is approximately 24 km and since the distance between the neutral sections (of DPATPs or TPSSs) is about 12 km, it follows that the service points have half the service time compared to neutral sections.

In the event that a service point is very close to a DPATP, we can consider removing the DP at the end of the service point on the SPATP side, thereby removing the full line section between the DPATP and the SPATP.

At service points, switches have been used to be able to sustain a train passing through up to a speed of 160 km/h, which naturally require a remarkable quality and reliability of the power supply for handling equipment (actuators) and dedicated heating, necessary for maintaining the system's operation in any environmental condition. Taking the length and number of actuators into consideration, it is easy to see how the power requirement that must be provided along the line is greatly increased compared to traditional systems.

4.6.5 Overhead Lines and Grounding Circuits

From an electrical perspective, the 2×25 kV contact line consists of various adjacent "sections," arranged to be powered by different phases separated by DPATPs or TPSSs. This separation is made, as previously described, through neutral sections at the DPATPs and TPSSs, carried out in such a way that the pantograph contact strip, when passing through, does not short circuit two different phases.

There are seven conductors for each track and precisely

- copper contact wire with a 120 m^2 section and a 25 kV power supply;
- copper messenger wire with a 150 mm^2 section and a 25 kV power supply;
- steel–aluminum feeder with a 307 mm^2 section and a −25 kV power supply (in phase opposition);
- aluminum alloy earthing cable with a 147.1 mm^2 section;
- copper or aluminum linear earth electrode with a 95 mm^2 section;
- external rail; and
- internal rail.

The conductors, with or without voltage, are supported by LS masts positioned typically at 50–60 m intervals in a straight line. On one piling there is the contact line, the messenger wire, the feeder, and the earthing cable (or wire). With regard to the geometry, the need to build a rail route across the land involves creating different types of routes that can be grouped into four classes known as "sections": embankment/trench, viaduct, natural, and artificial tunnels. Each of them corresponds to a particular geometrical arrangement of the conductors due to the different spaces available, especially for the feeder and the earthing wires. Figure 4.51 shows the arrangement of the conductors in an open-air section.

The contact line is usually a solid copper conductor and is shaped to be able to be supported lengthwise by the messenger wire through the droppers. The portion of the contact line, measured from the top of the rail, is rigorously stable for the entire line.

When a train passes through the line sections between one autotransformer and the other (or between an autotransformer and a TPSS), the feeder in this area contributes to the return of the current, while more generally it constitutes the return in sections not occupied by the train. A steel-cored aluminum conductor is used for the feeder in the ENEL standardization. The section of the line calls for

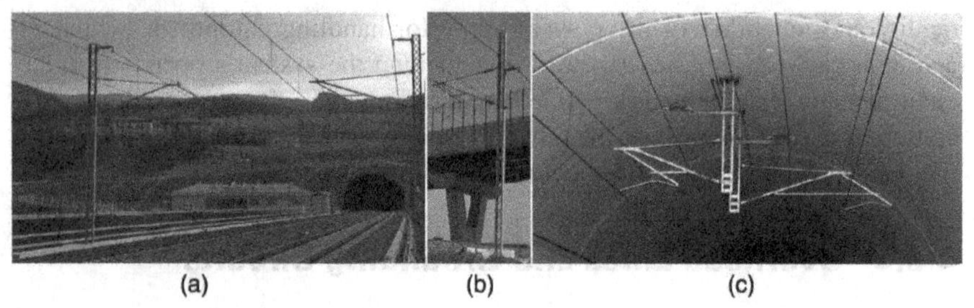

Figure 4.51 Arrangement of conductors in autotransformer systems. (a) Usual arrangement. (b) Bridge crossing. (c) Tunnel.

the feeder to be positioned on the top of the LS mast and supported by a bracket with a hanging insulator facing away from the pole, but in trenches and in the presence of counter walls, as is found on an open line, it is necessary to revise its position in order to respect the electrical dimensions. Therefore, it has been directed toward the inside of the pole and thus mounted directly above the hanger with a messenger insulator. Inside tunnels, it is always made up of an overhead conductor supported by the same type of insulators attached to the droppers. The insulators have the same characteristics as those of the contact line.

The values permitted for the line's power supply remain the same as those for the classic single-phase power supply of 25 kV at 50 Hz.

The grounding circuit is made up of two linear electrodes, two earthing wires, and the rails. The electrodes are made out of copper wire and are arranged in parallel at a specific depth in the ground, near and connected to the mast bases. Since the rails are not insulated by the ground, these linear electrodes, one for each of the two tracks, makes it possible to increase the natural leakage conductance of the track to the ground in a way which reduces the return/ground potentials and consequently avoids the risk of excessive contact voltages both in normal operation and in case of a fault. The linear electrodes are then connected to all the masts and the metal masses of equipment located along the line in order to protect workers from indirect contact. These linear electrodes also reduce the impedance of the masts to the ground, an important precaution for protection against lightning.

Earthing wires, however, are conductors made from aluminum alloy and are mounted on the same piling as the contact line and feeder. This conductor performs a triple function. First, it is used as the return conductor for the auxiliary services along the line, since, by virtue of its location next to the active line conductors, it attracts a good share of the return current, reducing the amount that is leaked into the soil and consequently decreasing interference. Second, it is a ground conductor for protection against the risk of voltage on the masts, since it directly connects the masts to the track; therefore, if one of the insulators breaks, it has a short circuit resulting in immediate protective action. Finally, it has a keraunic protection function, because it is an effective protection against lightning.

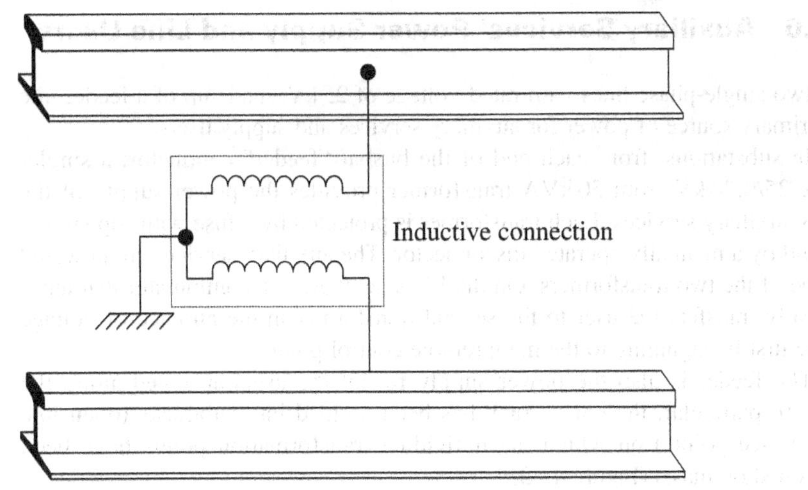

Figure 4.52 Connection of the rails to the ground network through inductive connection.

The rails are made of 60 UNI steel. All the masts of each track are connected to each other by the earthing wire, which is connected to the track through inductive connections approximately every 1500 m: this is necessary for signaling systems to work correctly. There are parallel connections between the upline and the downline conductor at the middle and end of each 1500 m section.

The inductive connections are equipped with two terminals that are each connected to one of the rails, plus a center tap connected to the earth, Figure 4.52.

They are designed so as to function as a low-pass filter. In fact, they offer low impedance to the passage of the traction current at 50 Hz and hinder the passage of signaling currents at audio frequency. There are the following two types of inductive connections:

- *Barrier*: Makes it possible to balance the traction return currents between the rails and connect the rails to the ground.
- *High Voltage Return* (for TPSS and DPATP): Makes it possible to connect the center taps of the transformer and autotransformer to the rails.

With regard to their construction, they are made up of two half-windings on a ferromagnetic core with an air gap to prevent saturation and to maintain the linear characteristic even with high currents.

Referring to the case of a double-track line, the contact line is electrically connected to the TPSS via eight conductors (four feeders, each made up of a bipolar line that supplies power to a catenary and feeder in the same section); the center taps of the two transformer windings are connected to the ground and both tracks: this link is made by a cable, and connects transformers of the contact lines of the down- and uplines (or autotransformers in the case of DPATP) to the tracks, with the interposition of return inductive cases.

4.6.6 Auxiliary Services' Power Supply and Line Users

The two single-phase lines with rated voltage of 25 kV, made up of a feeder, are the primary source of power for auxiliary services and supply users.

In substations, from each end of the busbar "feeder" conductor, a single-phase 25/0.24 kV from 50 kVA transformer provides the power supply of the TPSS auxiliary services. Each transformer is protected by a fuse with trip signaling and by a manually operated disconnector. The auxiliary services are powered by one of the two transformers. On the LV side, there is a commutator that automatically transfers the user to the second transformer in the absence of voltage on the first by signaling to the main remote control point.

The feeder is also the power supply for all the users arranged along the track. In particular, the entire track has been divided into subtracks (open line and service points) on which the individual transformation points have been positioned on masts (Figure 4.53).

The 2 × 25 kV–50 Hz system enables the provision of electricity simply coming from the feeder through mast transformers protected by fuses and disconnectors. Many systems are powered by this source: lighting systems, power systems, emergency systems, signaling systems, the heating of switches, GSM-r radio base stations, and data acquisition units for line diagnostics. The power supply from the feeder, therefore, makes energy available at any point along the line, but introduces potential problems regarding the quality of the energy supplied, since the feeder can easily be subjected to disturbances present in the line that are essentially caused by trains, as well as to frequent detachment maneuvers.

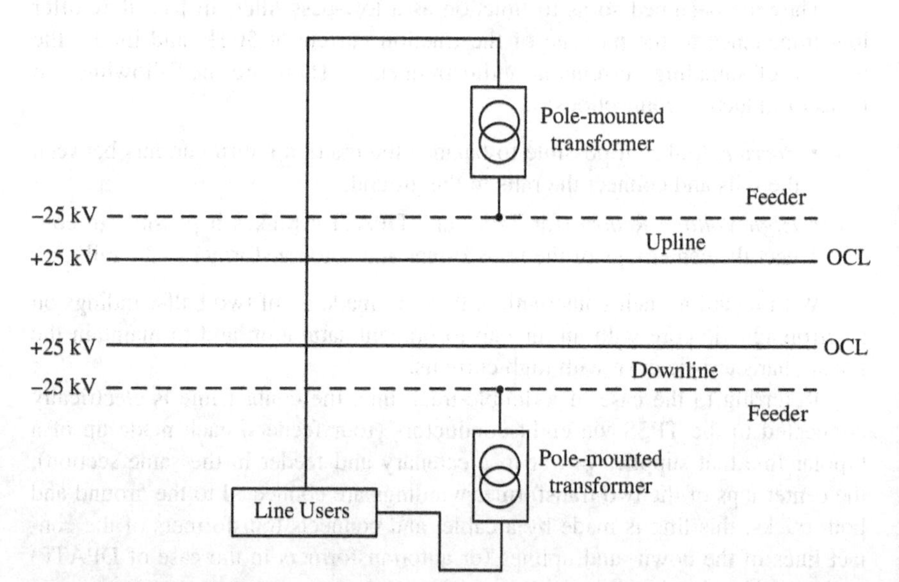

Figure 4.53 Transformation points along the line.

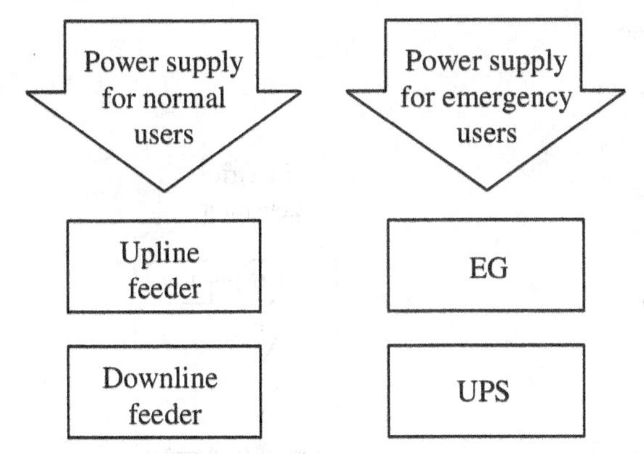

Figure 4.54 Power supplies for normal and emergency users.

However, since this power supply affects essential users, for safety reasons, it is clear that it is necessary to minimize the time that the supply is out of service if there is a fault on the main power line. In order to ensure supply continuity for essential users, it is necessary to provide a backup power source, as well as buffer batteries for the essential users, as shown in Figure 4.54.

Each affected user (FPP, auxiliary services buildings (ASB), and electric heating (EH)) has loads that can be divided into normal, priority, and essential users. Normal utilities are, for example, the nonemergency lighting and driving power, auxiliary services, monitoring the data acquisition unit (DAU), and the heating of switches. A priority utility is, for example, conditioning (CD), while essential utilities include signaling systems (SS), telecommunications (TLC), electrical traction (ET), safety lights, fire prevention systems, auxiliary services, automation, and diagnostics and maintenance (D&M).

As an example, Figure 4.55 shows a single-wire electrical diagram of the FPP power supply.

As can be seen from Figure 4.55, the continuity of the power supply is therefore guaranteed by two UPSs and an EG.

4.6.7 UPS

As shown by the single-line diagram in Figure 4.55, UPSs, used in FPP in an "online" configuration, guarantee the uninterrupted supply of power to the utilities connected to the "essential" busbar at 400 V AC, which is three phase with an accessible neutral. The main power of the FPP is drawn directly from the 2 × 25 kV system of electric traction through the return feeder at the rated voltage of 25 kV AC single phase, and then transferred to the busbar system after appropriate transformation in LV (rated 230 V AC single phase). The emergency power supply is instead provided by an engine generator whose output is 380 V AC and three phase. Consequently, the static control units can alternatively be

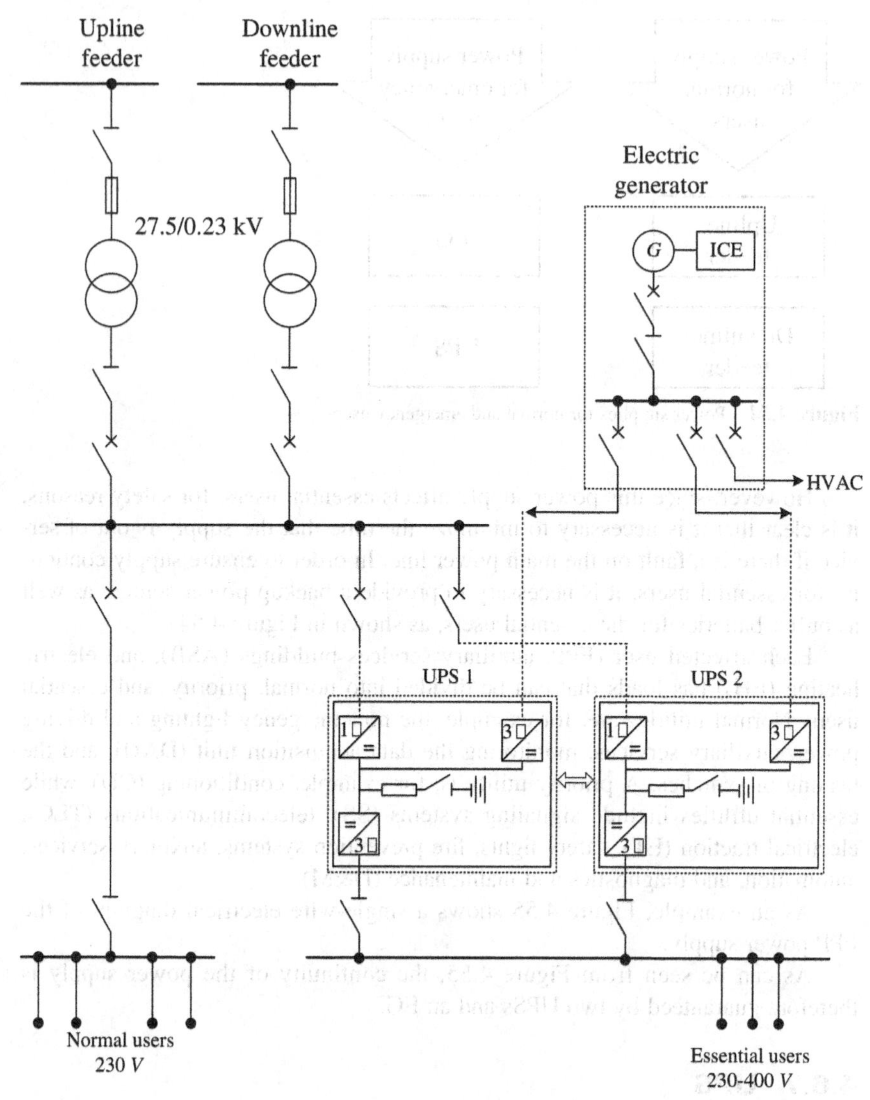

Figure 4.55 Single-wire electrical diagram of the FPP power supply.

powered by both systems, and are designed and constructed according to a dedicated solution. In particular, the control units are equipped with a double input stage rectifier. To ensure a higher reliability and availability of the system's "essential" power supply, each FPP has two static control units, each designed for 100% of the maximum load required. In case of failure or maintenance of the input stages of a UPS, the continuity of energy supply to the load is ensured by the second UPS.

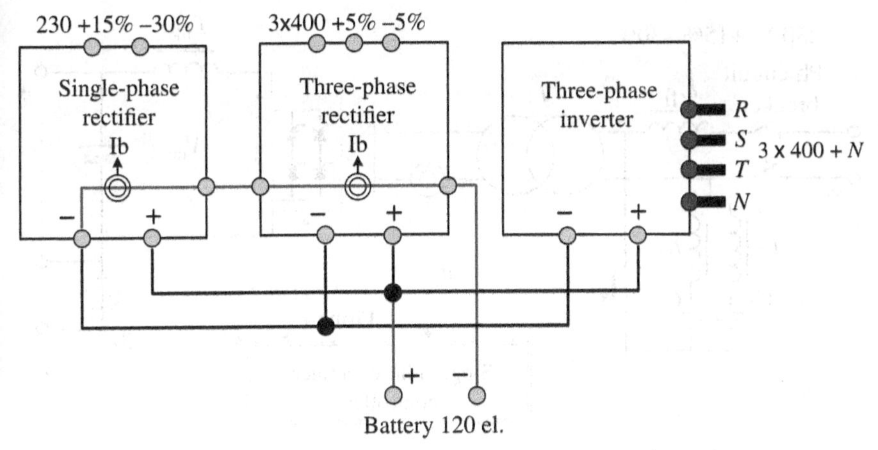

Figure 4.56 Interconnection of the three modules that constitute the continuous input stages.

Considering that the UPS may be easily subjected to loss of power from the feeder, it is necessary to pay particular attention to the batteries, the duration of which must be able to ensure the supply of essential utilities when the EG outlet takes over. For this reason, each UPS is equipped with a stationary sealed lead rechargeable battery with gas recombination; they are able to ensure autonomy of power load for 30 min. With regard to the quality of the energy, the output stages of the UPS are able to work with nonlinear loads that contain distortion in the form of an output voltage wave up to 8%.

The UPS is formed by three functional blocks, the interconnection of which is shown in Figure 4.56:

- Single-phase rectifier charger
- Three-phase rectifier charger
- Three-phase inverter with static circuit breaker

The two modules, single and three phase, work on an exclusive basis, which means that they can never be simultaneously powered. As can be seen, both the single-phase three-phase rectifier are able, if properly powered, to deliver power to the inverter and to the battery in order to charge it. If both single-phase and three-phase power supplies fail, the battery would continue to create a reserve of energy in order to keep the inverter working for the entire time it is operating autonomously.

4.6.7.1 Single-phase rectifier charger

The single-phase rectifier charger draws power through a line from the transformer MV/LV 27.5/0.24 kV. Since the rated voltage of 230 V_{AC} can undergo

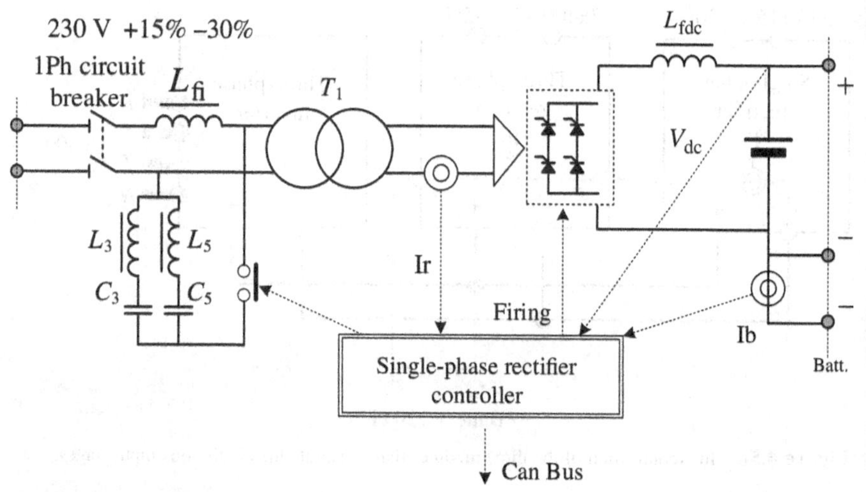

Figure 4.57 Single-phase rectifier.

very wide variations (contained between −30 and +15%), the input section of the single-phase rectifier, made up of a multisocket transformer T1 and the conversion bridge (Figure 4.57), functions as a multistep single-phase rectifier converter that can be completely controlled, or the most appropriate step will be activated depending on the level of the supply voltage.

The section consists of Lfi, L3-C3, and L5-C5 and performs the filter functions of the 3rd and 5th harmonic in order to absorb the disruption that, vice versa, would be injected conversely into the line by the rectifier stage. This filter, which is normally disconnected and has a rectifier that works in a vacuum, is activated from about one fifth of the maximum power that the machinery can supply. The control logic managed by a microprocessor will control the power in such a way as to make the rectifier capable of delivering the necessary energy to operate the inverter and to recharge the batteries, in addition to performing a diagnostic test on the efficiency of the battery.

4.6.7.2 Three-phase rectifier charger

A three-phase rectifier charger is shown in Figure 4.58.

In this case, the power is supplied by an engine generator capable of delivering enough power to a practically stabilized voltage (380 V ± 5%). The generated power is rectified by a completely controlled twelve-phase converter, with the two conversion bridges connected in parallel and controlled so as to implement a load distribution of 50% to each. The control logic performs the same functions on the battery as the previous single phase.

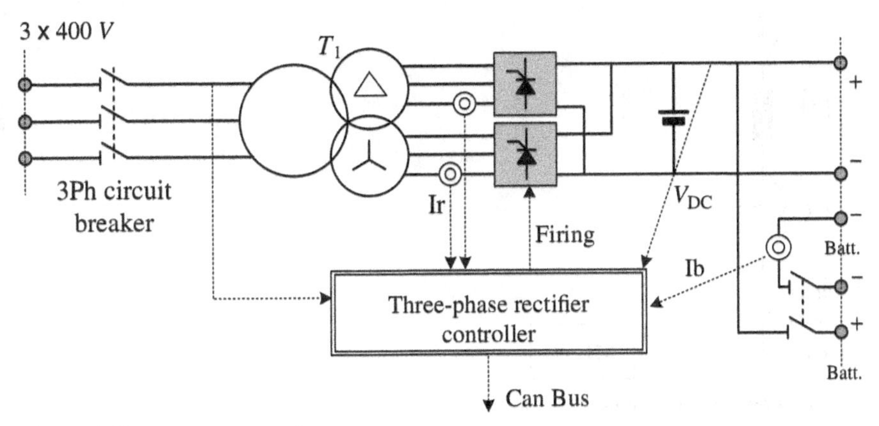

Figure 4.58 Three-phase rectifier.

4.6.7.3 Three-phase inverter with static circuit breaker

The three-phase inverter module with a static circuit breaker is powered by the voltage supplied by the rectifier, or by the single-phase or three-phase rectifier, or by the battery. It accepts a DC supply voltage that can vary between 280 and 210 V_{DC}, corresponding to the charging voltages at the end of a battery's life to 120 units.

The functional configuration of the inverter module is shown in Figure 4.59.

The control logic, as well as managing the switching bridge and the IGBT capable of generating a set of three symmetrical three-phase voltages from all of the transformers/output filters, manages the static circuit breaker trip in coordination with its counterpart in the inverter of the second UPS, so that only one of

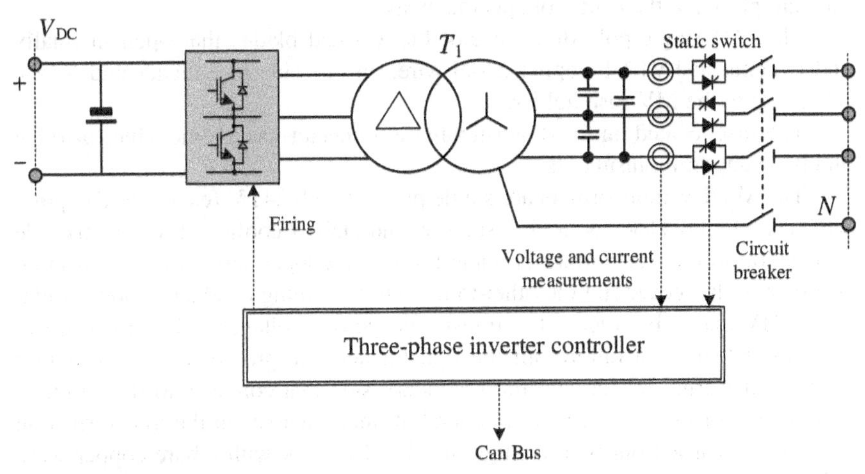

Figure 4.59 Three-phase inverter.

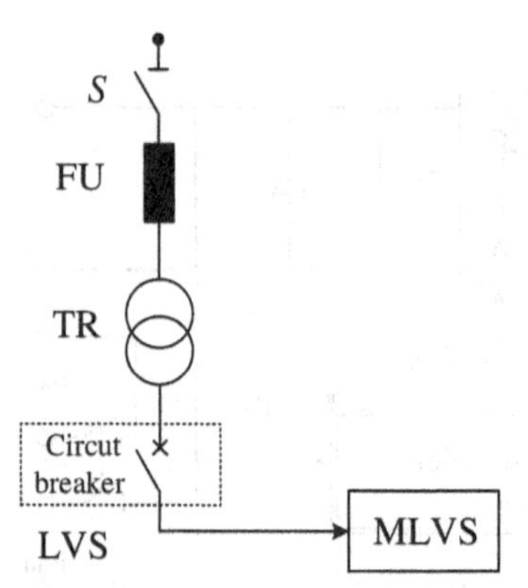

Figure 4.60 Pole transformation points.

the two transfers power to the utilities, while the other constitutes the first hot reserve.

4.6.8 Pole Transformation Points

Transformation points (TP), installed in the vicinity of the rail and whose diagram is shown in Figure 4.60, are made up of a single or double support pole depending on the power and the weight of the transformer, a single-pole disconnector with postponed maneuver, a fuse with a fuse holder, a single-phase transformer, and an LV framework at the base of the pole.

A single pole is normally used for transformers up to 200 kVA, while a double pole for transformers above 250 kVA. The masts are conical poles in galvanized steel, with a polygonal section, their above ground height is 8.70 m, and they are placed at the center of specific areas.

The MV single-pole disconnector has vertical blades that open manually and is connected with a copper aerial wire, on one side to a feeder and on the other side to the MV fuse holder.

The fuse located on the downstream disconnector has a removable cartridge and two supporting insulators.

The MV/LV transformers are single phase 27.5/0.24 kV (except in RD posts which are 27.5/0.4 kV), with oil insulation and ONAN cooling. The primary side connection to the transformer is made through a copper aerial wire, connecting one pole to the feeder and the other to the return earthing conductor, both available on HV masts. In order not to transfer dangerous voltages to LV downstream systems if there is a failure, all the transformers are provided with insulation between the primary side and the secondary side that connects to the earth, as well as that of its housing and in general all metal masses in the transformation post; this is carried out with a single insulated wire or with a bare copper wire, using the ground plate purposely located on each HV pole. The two poles of the secondary side of the transformer are connected to the framework placed at the base of the pole by means of a single-pole cable. They are self-extinguishing and fire retardant, varying in shape depending on the size of the transformer.

The LV framework for the exterior is intended for disconnecting the power supply from the MV/LV transformer on pole and leads to the main LV distribution board (MLVS) within the FPP and ASB buildings. Within the LV switchgear (LVS) at the base of the pole, there is a protective automatic circuit breaker with motorized control.

In the case of transformation points for supplying power to train-to-ground transmission systems along the line and in tunnels less than 5 km long, one of the LVSs at the base of the post, in order to retain the same construction and mode of installation, is larger in size and also internally houses all the protection circuit breakers of the various circuits and the UAD, in addition to a PLC unit to allow the control of the systems from a different location.

With regard to the principle of operation, disconnector S works by means of a rigid rod controlled from a box placed at the base of the pole. There is a terminal in the control box that is used by the command and control circuit to open and close the disconnector. The LV circuit breaker, located in the LVS at the base of the TP, is interlocked with disconnector S so that the operation of the latter can only take place after the circuit breaker is opened. The LVS auxiliary services at the base of the pole are powered by the switchgear near the building (FPP or ASB). In this way the power supply of the control circuit for the electromechanical interlock between the disconnector and the circuit breaker is always available, even when the disconnector is open and the power supply for the whole transformation point is switched off.

4.6.9 LV Section

The power supply management system makes it possible to remotely control the main functions of the equipment, that is to say, to intervene in disconnecting and protecting controls that have motorized actuators. The equipment that carries out the command and control logic of the LV section is contained in a special switchgear (PLCS). The control logic ensures the automatic reconfiguration of the system in a degraded mode. The local system of command and control is implemented in such a way that a generic out of service of the PLC initiates duplicate logic wired in relay, normally in standby mode. If the control logic malfunctions, all the operations can be performed either manually or by remote control.

The normal distribution of power is single phase. The main low voltage switchboard (MLVS) is composed of a single busbar system, on which two main circuit breakers are installed (CB1, CB2), which connect the two power lines to the busbar coming from TPs on the up and downlines (Figure 4.61).

The operation of the system logic requires that the flow of energy occurs on one of the two power supply lines with the possibility of automatic switching

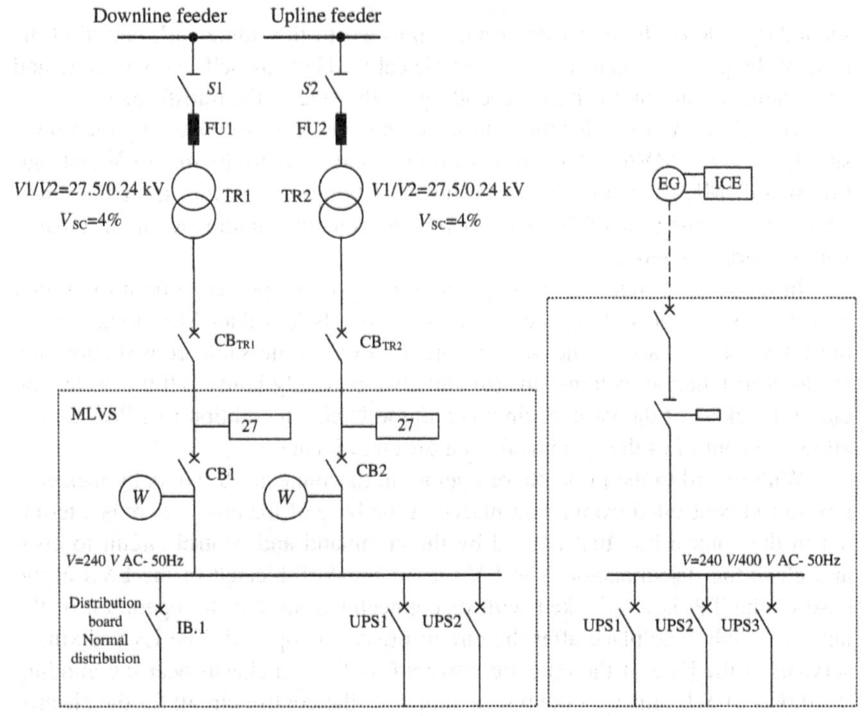

Figure 4.61 Single-wire electrical diagram of the MLVS and EG switchboard with their power supplies.

from one line to another if a feeder loses its voltage. The presence of voltage is detected by two minimum voltage relays on lines coming into MLVSs. The designation of the main power line can be controlled remotely or locally, via a switch on the front of the PLCS framework.

A three-phase engine generator set represents the emergency power, which can function for about 36 h without the need for fuel supplies.

In case of a power failure on both TPs, after 2 s and having checked that the switches CB1 and CB2 are open, the operating logic starts the EG.

If during the start-up phase of the EG, the voltage returns on one or on both feeders, the procedure would be continued until the load is taken by the EG itself. The power supply to essential utilities during the start-up transition or to the shutdown of the EG is guaranteed by two UPSs.

Finally, the normal and emergency power supply systems can be disconnected at the end of maintenance. Since in this case remote management is not envisaged, neither system contains motorized elements.

Chapter 5

Single-Phase Networks at Railway Frequency

Railway frequency systems were developed at the beginning of railway line electrification in order to raise contact line voltage, and thus transmissible power voltage, by means of electromechanical transformers, and simultaneously use commutator motors similar to the direct current ones, which used to allow for finer speed regulation and a sole contact wire.

These systems are now prevalent in Europe (Austria, Germany, Norway, Sweden, and Switzerland) with a frequency of $16^2/_3$ Hz and in the United States with a frequency of 25 Hz.

In principle, railway frequency systems are similar to alternating current ones in the case of centralized distribution or more similar to direct current ones in the case of distributed conversion.

In the case of centralized distribution, the generation, high-voltage transmission, and substations constitute a near-isolated system, entirely functional at the railway frequency. For which the substations achieve only the function of voltage conversion and contact line protection.

In split conversion, however, the substations are directly connected to the three-phase industrial network and must, therefore, achieve the conversion of phase and frequency.

5.1 CENTRALIZED DISTRIBUTION

The centralized distribution adopted in Europe (Austria, Germany, Switzerland, and subsequently in Sweden) and in the United States is characterized by the presence of a high-voltage, single-phase distribution network, functioning, respectively, at 110 kV, $16^2/_3$ Hz and 138 kV, 25 Hz, for the exclusive use of the railway system.

Electrical Railway Transportation Systems, First Edition. Morris Brenna, Federica Foiadelli, and Dario Zaninelli.

This HV line network feeds all the substations, equipped with single-phased transformers and distributed along the railway lines. The different operating frequency requires that the system be isolated with few points of interchange with the industrial network. At the beginning of electrification, this fact did not constitute a problem as the industrial network was not yet sufficiently developed and, therefore, capable of dealing with the railway load, both as values of power and availability in proximity to the traction lines. Therefore, the rail system was powered by dedicated hydroelectric or thermoelectric generation groups or stations. Subsequently, with the increase of railway load, which required greater power than was available, and the rapid development of industrial systems and the consequent spread of the electric transmission network, it was more convenient to power the railway lines by the industrial network through rotating frequency, and subsequently static, conversion groups. (Figure 5.1)

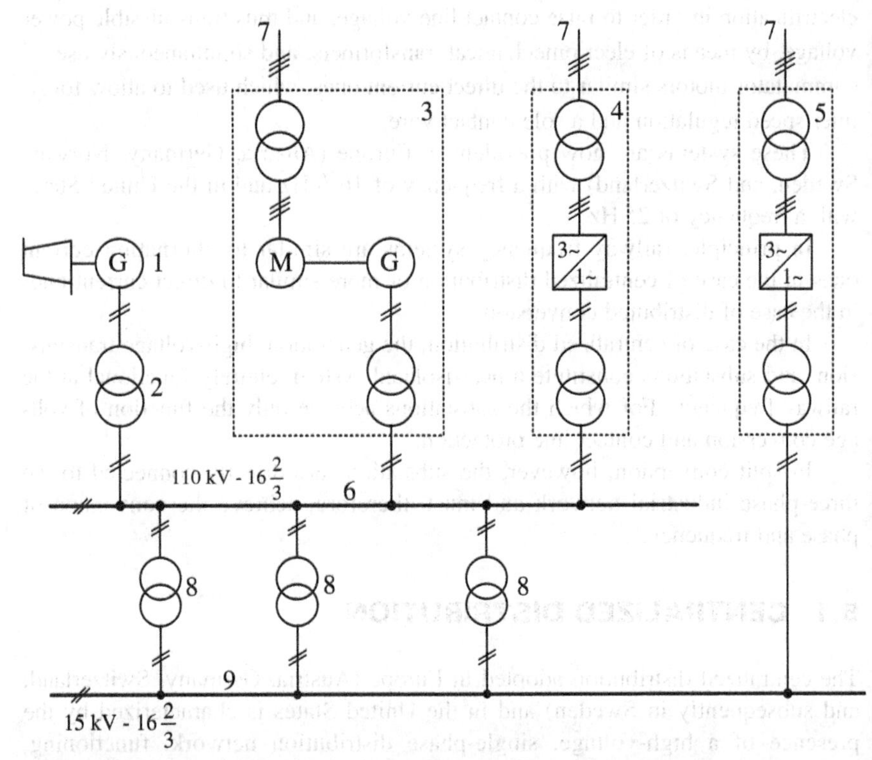

Figure 5.1 Single-phase, alternating current system at railway frequency with centralized power. 1: Single-phase traction generator sets, in thermoelectric or hydroelectric stations; 2: station step-up transformers; 3: centralized phase and frequency conversion stations with rotating asynchronous/synchronous groups; 4: centralized static-conversion group stations, connected to the high-voltage single-phase network; 5: centralized static-group conversion stations, connected to the contact line; 6: single-phase primary lines; 7: high-voltage industrial three-phase network; 8: single-phase substations; and 9: contact line.

A centralized network can absorb energy, under high load conditions, from the public three-phase networks through the conversion stations; if, however, there is an excess of power in the railway stations, this can be transferred to the three-phase networks.

Single-phase systems are electrically separate with respect to others and have their own frequency regulation f_1, with higher percentually admitted variations than those typical to industrial networks.

Converter stations must therefore ensure an "*elastic*" relationship between the instantaneous frequencies f and f_1; that is, variations are allowed of frequency

$$\Delta f = f - \frac{f}{f_{1n}} \cdot f_1$$

in which the frequency of the industrial network f is considered constant and f_{1n} is the nominal railway frequency.

In Germany, Switzerland, and Austria, the elasticity of rotating groups has been used to change the nominal frequency from $16^2/_3$ Hz to 16.7 Hz. The frequency $16^2/_3$ Hz is equal to 1/3 of 50 Hz, that is, the mains frequency. This frequency can be obtained in rotating groups with a twelve-pole motor working at 50 Hz, connected to a four-pole generator that therefore generates $16^2/_3$. The rotation speed of the group is thus 500 rpm. The corresponding speed of the four-pole generator, if it generated 50 Hz, would be equal to 1500 rpm. Therefore, the frequency ratio may be obtained from the ratio 500/1500. Under this condition, the rotation of the motor is always synchronous with the 50 Hz system and, therefore, any perturbations always involve the same phase. For this reason, in Europe, the frequency has been changed to 16.7 Hz, corresponding to a ratio of 501/1500, for which the motor's rotation velocity is no longer synchronous with the network and any perturbations are distributed evenly across the three phases.

In the United States, however, the system always functions synchronous to the network and the conversation is obtained with synchronous 24-pole motors and 10-pole generators rotating at a speed of 300 rpm.

Until the 1980s, for phase and frequency conversion, rotating groups were created, starting with the station's three-phase high-voltage busbars, constituted of protective systems, three-phase converter transformer, three-phase motor, synchronous single-phase generator coaxial with the motor, single-phase step-up transformer, and protective systems connected to the high-tension, railway frequency network through the output busbars.

The conversion group motor is asynchronous in systems in which the railway frequency is not synchronous with that of the network, as in the case of part of Europe, or a synchronous motor in case of synchronistic operation.

The rotor winding of the asynchronous motor is powered, through the slip ring, by a three-phase system of currents, the frequency of which is regulated as a function of flow. At one time this was obtained with a rotating machine, though currently is obtained through an electronic converter.

For the excitation of the alternator as well, a three-phase controlled bridge is used, powered through a transformer in parallel to the asynchronous motor.

Starting in the 1990s, electronic converter groups were put into service, derived from those that had already been functioning for several years in the substations of distributed conversion networks, as shown further, and, above all, from the single-phase converters developed for onboard operation.

An exclusively traction-dedicated network will inevitably have fairly limited power, both in absolute value and in relation to the peak power absorbed by train sets. A network of this kind requires that the load has a high power factor and low harmonic content.

5.1.1 Contact Line Power Supply

The centralized distribution system in itself has the prerogative of supplying all TPSSs in phase and with the same frequency; this enables, as in direct current, the bilateral supply of the contact line, having an advantage both in the lower voltage drops and in the distribution of the current absorbed by each traction vehicle between the two TPSSs.

By way of example, European TSI standards allow, with respect to 15 kV nominal voltage, for the following variations:

- Maximum permanent value 17.25 kV (+15%), for 5 min 18 kV.
- Minimum permanent value 12 kV (−20%), for 10 min 11 kV.

The frequency may vary by approximately ±2% for nonsynchronous systems and ±1% for synchronous systems.

Two single-phase transformers, each capable of powering the concerned sectors on its own, are usually installed in the substations. The average distance L between the substations is 60–80 km.

The line circuit is more complex than in direct current, among other reasons due to the considerable distance between substations, which leads to the division of each sector into multiple sections, thus resulting in the limitation of the de-energized section in the event of failure or maintenance work.

To limit voltage drops, connections are made in parallel between the two tracks in double-track systems, in correspondence with the sectioned-off sites.

5.2 THE DISTRIBUTED CONVERSION SYSTEM

In a distributed conversion system, the substations are connected directly to the industrial network and fulfill the dual function of transformation and conversion of phase and frequency (Figure 5.2).

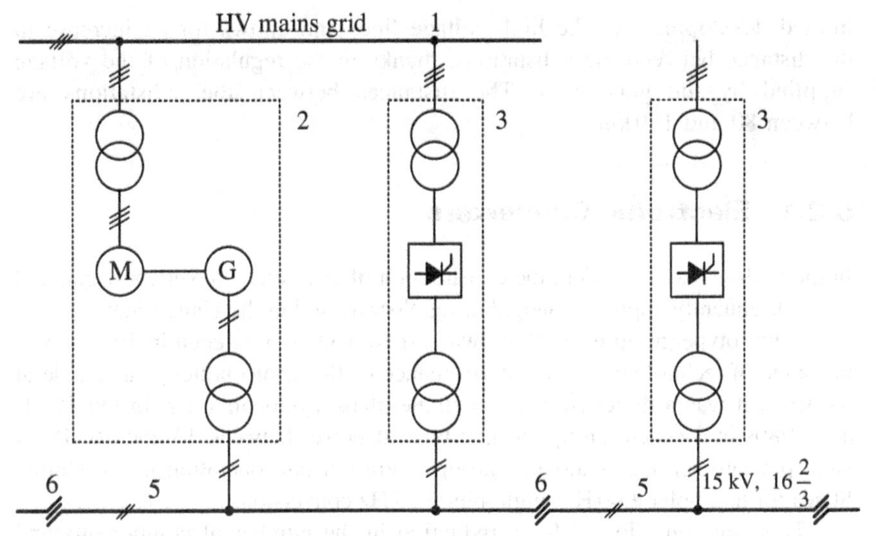

Figure 5.2 Single-phase, alternating current system with split conversion. 1: Three-phase high-voltage industrial network; 2: conversion substations equipped with rotating synchronous/asynchronous groups; 3,4: substations equipped with electronic converters; 5: contact line; and 6: neutral sections of the contact line.

Unlike what happens in centralized conversion, in this case the frequency ratio is rigid, that is, it is always the case that

$$\Delta f = f - \frac{f}{f_{1n}} \cdot f_1 = 0$$

for this reason, the rotating conversion groups are constituted by two synchronous machines.

The bilateral supply of one section included by two substations may be adopted, as in the centralized system, provided that the two substations be connected to the same three-phase network (example: positions 2 and 3 and of Figure 5.2); otherwise, the differences in phase and frequency require the sectioning off of the contact line (Figure 5.2, position 6) and the insertion of a neutral section as in industrial frequency systems.

Because the substations power independent sectors and the frequency f_1 is determined by the three-phase network, regulation is simpler than in the centralized system, as it is limited to control over output voltage, which is generally kept constant.

The system is applied in Sweden, Norway, part of Germany, and in the United States.

The distributed conversion solution obviously presents greater cost of the substations than centralized conversion; however, it requires the highly limited development of the high-voltage lines and allows for an increase in the distance between the substations, thanks to the regulation of the voltage supplied by the generators. The distances between the substations are between 80 and 100 km.

5.2.1 Electronic Converters

In the mid-1970s in Sweden, the construction of electronic converters began and was subsequently applied in neighboring Norway and in the United States.

A prototype group of 6 MVAs was experimented in Sweden in 1973; given the stage of technology and the performance of the semiconductors available at the time, it was a direct-conversion single-phase cycloconverter. In the 1970s and 1980s in Sweden, groups of this type of power between 13 and 16 MVAs were put into service; analogous groups were put into operation in the United States for three-phase 60 Hz/single-phase 25 Hz conversion.

This solution allowed for a reduction in the number of components and was compatible with the rigid frequency ratio between the two networks, but was not capable of absorbing the pulsations of the active single-phase power, which were thus discharged over the three-phase network. In fact, instantaneous single-phase power oscillates with a double frequency and, therefore, causes a second harmonic. For example, in 25 Hz systems, a cycloconverter produces a 50 Hz harmonic that, being composed of the 60 Hz industrial mains frequency, causes voltage oscillation at 10 Hz or 110 Hz, that is, interharmonics of the mains frequency. These interharmonics are difficult, if not impossible, to eliminate and cause disturbances to the loads connected to the industrial network.

Considerable improvements have been achieved with indirect conversion – that is, two stage with a direct current intermediate circuit used by voltage source converters (Figure 5.3). The DC-link is capable of decoupling the single-phase

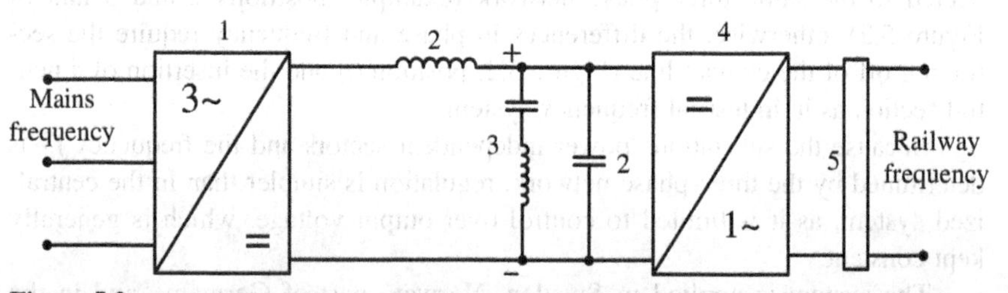

Figure 5.3 Basic diagram of a two-stage electronic converter inserted between the public three-phase network and the single-phase traction network at railway frequency. 1: Controlled rectifier; 2: inductive and capacitive elements of the direct current intermediate circuit; 3: second-harmonic LC filter; 4: single-phase inverter; and 5: filter.

network from the three phase and greatly limiting the second harmonic caused by the single-phase absorption, through special tuned LC filters inserted within. These filters, similar to those installed onboard vehicles, will be examined in Chapter 11.

Double-stage electronic converters can function both with a rigid and elastic ratio of frequencies f and f_1, given that the necessary regulation systems are accounted for.

This flexibility of use, which has allowed for the initiation of the aforementioned experiments in centralized systems, constitutes a great advantage not found with rotating groups, as the two types of operation entail a difference in machinery (synchronous/asynchronous or asynchronous/synchronous).

The constructive techniques are derived from those employed in the direct current, high-voltage energy transmission field, which carries out a double three-phase/direct and direct/three-phase conversion.

The three-phase stage, connected through the input transformer to the public network, can be made with thyristor-controlled rectifiers, operating in natural commutation with twelve-phase reaction.

Double-stage converters based on three-level voltage source converters have recently been adopted both on the three-phase and single-phase side, as shown in Figure 5.4.

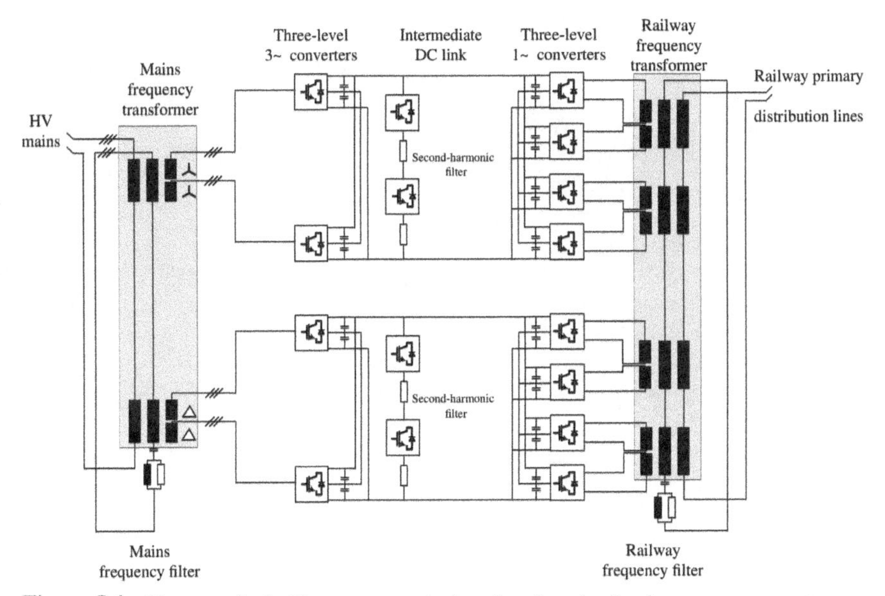

Figure 5.4 Diagram of a double-stage converter based on three-level voltage source converters.

Both input three-phase and output single-phase converters, of the four-quadrant type, are connected in parallel in the DC-link and each to its own dedicated three-phase or single-phase transformer winding. Thus, it is possible to reduce the harmonic content on the public network through the twelve-phase reaction and serially connect the single-phase transformer's secondary windings such that it can be directly connected to the primary lines. The power of this new converter generation can reach as much as 100 MW.

Chapter 6

Electromagnetic Compatibility

Today we are witnessing an ever-widening diffusion of static devices, which have given rise to increasingly better performing and easily managed electrical systems that were unimaginable with traditional electromechanical systems. However, this has caused some significant side issues, such as intolerable harmonic interference for certain utilities.

The world of traction is characterized by particular difficulties: operating issues, the presence of loads that vary in time and space, and the atmospheric and mechanical stresses the dedicated systems are subjected to.

In view of these considerations, greater operating limits have always been tolerated in traction systems compared with traditional industrial systems in order to permit an acceptable level of service continuity without compromising safety.

However, the need to standardize the components used, particularly in the auxiliary services along the line, has introduced problems relating to their difficult coexistence with systems specifically designed for use in the field of electric traction. To solve these problems, various regulations such as the European Technical Specification for Interoperability (TSI) set limits on both radiated and conducted noise and the various components' capacity to be immune to such interference.

In order to comply with regulations and avoid creating problems in the electric system components' operation or generate malfunctions in the signaling system managing traffic control, the harmonic interference in the power supply system must be predicted. One effective tool for doing so is modeling and simulating the entire system. The importance of the choice and complexity of the model lie first in understanding the phenomenon, and second in the possibility to highlight relevant properties that are not immediately identifiable. Considering the propagation and uptake of interference, a "rail domain" is composed of electrical and electronic power installations as well as communication and control

Electrical Railway Transportation Systems, First Edition. Morris Brenna, Federica Foiadelli, and Dario Zaninelli.

systems; it interacts electromagnetically with its operating environment as an agent causing interference and as the object of interference from other signals existing in the area and from other sources. It is thus easy to understand how the single subsystem can be both the source of interference and the victim, creating a complex cycle of cause and effect.

To better understand the phenomenon and assess all the causes that concur in the determination of the system's harmonic behavior, the different issues must first be dealt with separately.

Electric traction presents peculiarities that make it far less predictable compared with that which occurs in other application areas, such as in the industrial sector. First of all, the extent of the systems being discussed should be taken into account and, in particular, the movement of the rolling stock present in space and time, in addition to the engines' many operating modes. Then both the coexistence of power and signal currents and their contiguity should be considered in relation to the small amount of available space within the rolling stock that enhances the effects of radiated noise.

In particular, compatibility issues have been detected in the new systems both regarding the coexistence of the many electrical subsystems constituting the system, which have different tasks but must operate simultaneously, and the proximity with the older lines that are powered differently.

Moreover, the high number of conductors in the systems (contact wires, load-bearing wires, feeders, guard wires, and linear dispersers), increasingly prevalent worldwide, certainly complicates the study. In fact, to calculate the magnetic field the currents flowing in all conductors when the load varies must be identified, that is, to vary the number, speed, and position of the trains. The magnetic field current sources vary considerably from point to point and from instant to instant, also making noise containment difficult in relation to the various harmonic components present in the current and the length and position of the interference areas' parallels.

The advent of power electronics has led to a notable increase in harmonic interference in supply systems. Each type of converter emits a particular harmonic spectrum of the voltage and current that not only depends on the type of converter, but also on the impedance value of the equivalent system that powers it. These technologies are used both onboard rolling stock and in fixed installations, therefore verifying the components' correct simultaneous behavior is necessary for excluding mutual interference. Modern rolling stock for railway services is equipped with pulse-width modulation (PWM), static, and multistage converters. Since, by their nature, the converters modify the shape of currents and voltages, harmonic current components are produced that interfere with the utilization circuit and power supply circuit with a number of resonance frequencies due to the distributed line parameters (capacity and inductors) and concentrated elements such as transformers, and so on.

Another electromagnetic compatibility problem is interference on the track circuits (CdB), which are a part of the signaling system, caused by the train drive circuit. The power supply system and the CdB partly share the same conductors, which are the rails. Tests have confirmed that the CdB's correct operation is influenced by the occurrence of a current unbalance between the two rails that alter the energy level in the digital signal band used.

The broad variety of issues related to the study of electromagnetic compatibility in railway systems have only been briefly presented here. It is therefore necessary to deal with individual issues separately to better understand their origins and effects, to finally be able to the study it as a whole.

6.1 INTERFERENCE PHENOMENA

Electromagnetic coupling between a source of interference and a generic victim system is manifested whenever an interaction occurs between the electromagnetic field generated by the source and the victim system, thus creating a transfer of energy between the two that adversely alters the physical characteristics and/or performance of the disturbed system. In classifying the different electromagnetic coupling mechanisms, a distinction is commonly made between those phenomena generated by electromagnetic fields that change slowly or rapidly over time.

In the former, since the wavelengths of the magnitudes involved are much greater than the linear dimensions of the systems concerned, the Maxwell differential equations that describe the behavior of the electric and magnetic field can be regarded as essentially decoupled, which means that the basic coupling typology can be dealt with separately. Therein follows a description of the interference phenomena that are part of this category: conducted interference phenomena, where the source and victim share one or more conductors, and induced phenomena, where the victim is concatenated with the magnetic induction flux associated with the source current, and capacitive phenomena caused by the varying electric fields that induce current.

The overall result of these three phenomena, always present simultaneously in real cases, is given by the superposition of the individual contributions' effects. All phenomena associated with industrial frequency electromagnetic fields, where the wavelengths involved are on the order of 6×10^6 m, fall into this class.

For fields changing rapidly over time, on the other hand, the electric and magnetic fields are strongly coupled and give rise to essentially radiative phenomena that require resorting to traditional electromagnetism to treat. Even if irradiated-type interference phenomena do not constitute part of this text, they will be briefly described for completeness in this chapter.

6.1.1 Conducted Interference Phenomena

Conducted interference phenomena are caused by the sharing of one or more conductors by both a source and victim. Conducted interference can be generated by the injection of distorted current into the power lines from devices with nonlinear characteristics, or by transient overvoltage caused by maneuvers, atmospheric phenomena, and so on.

Conduction mainly concerns the problem of currents that can propagate from one system to another on electric boundaries between rail systems and different types of power supply. Both power supply systems would be damaged by mutual interference, and thus by the presence of a current that is not their own and that changes the operating conditions. For example, in the case of switching between an AC and a DC system, the presence of harmonics from the AC side in the feed line section in DC must be avoided, as this would cause disruption in the operation of the signaling system and repetition onboard. To this end, analyzing the currents absorbed by the train is useful not only as the absolute amplitude value but also as the Fourier analysis spectrum in order to compare the amplitudes of the harmonics with a special mask that indicates the alternate current limit that convoys can bear when powered with direct current.

6.1.1.1 Protection of interconnection areas

In order to avoid the disturbances described above, the interconnection areas must be equipped with electric border points that insulate the two lines from penetration by either a direct or an indirect current. The detailed description of the measures to be taken for the electric border points has already been given in Chapter 4.

Following is a first study that evaluates the characteristics of the electric border point filter unit and calculates the transfer function. As previously mentioned, the presence of harmonic interference in the line might cause problems in the proper operation of the various subsystems that make up the system, for example, the signaling circuit, the converters onboard trains, or the auxiliary loads supplied directly from the feeder.

Impedance of the transmission line connecting the generator with the bus bar causes power losses that are due to the fundamental current produced by the generator and the harmonic load currents without useful effects. It is therefore obvious that dissipation on transmission lines will be greater when they are fed nonlinear loads, such as with traction vehicles. The harmonic components that follow the line downstream to upstream in part cancel each other, as such components are generally not all generated with the same phase: they belong to loads of a different nature, and consequently harmonic components with rather modest currents may also reach the generator. There are, however, critical cases that require the correct design and sizing of protective filters, for example, in resonance conditions.

Looking at the network from the load side, it is possible to derive the equivalent circuit in Figure 6.1, in which the presence of resistive loads is overlooked.

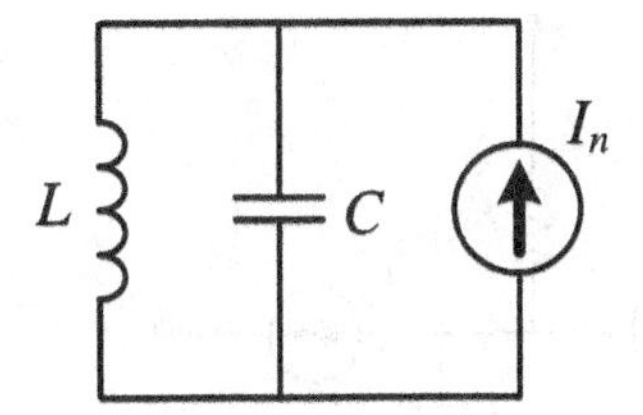

Figure 6.1 Equivalent load of the system-side circuit.

The most severe condition is obtained with parallel resonance, that is, in the case that

$$X_L = X_C \Rightarrow 2 \cdot \pi \cdot f_r \cdot L = \frac{1}{2 \cdot \pi \cdot f_r \cdot C}$$

The resonance frequency is equal to

$$f_r = \frac{1}{2 \cdot \pi \cdot \sqrt{L \cdot C}}$$

In resonance conditions, the impedance of the network viewed from the load tends to infinity, thus it has a high level of voltage distortion with a small current. This resonance could cause signal alteration in the communication systems of older lines working at 50 Hz as carrier frequency.

The following are taken into consideration for calculating the resonance frequency:

- X_{L50} and X_{C50} (reactance at 50 Hz);
- $h = f_r/50$ (p.u. value of the resonance frequency with respect to the mains frequency).

From these data, the following is obtained:

$$h \cdot X_{L50} = \frac{X_{C50}}{h} \Rightarrow \frac{1}{h} \cdot \frac{V^2}{X_{L50}} = h \cdot \frac{V^2}{X_{C50}}$$

In summary, we have

$$\frac{S_k}{h} = h \cdot Q_C \Rightarrow h = \sqrt{\frac{S_k}{Q_C}} \Rightarrow f_r = 50 \cdot \sqrt{\frac{S_k}{Q_C}}$$

These relations give the value of the resonance frequency as a function of the short-circuit power (S_k) and the reactive power supplied by capacitors at the basic frequency (Q_c). It is clear that high-power connection points should be

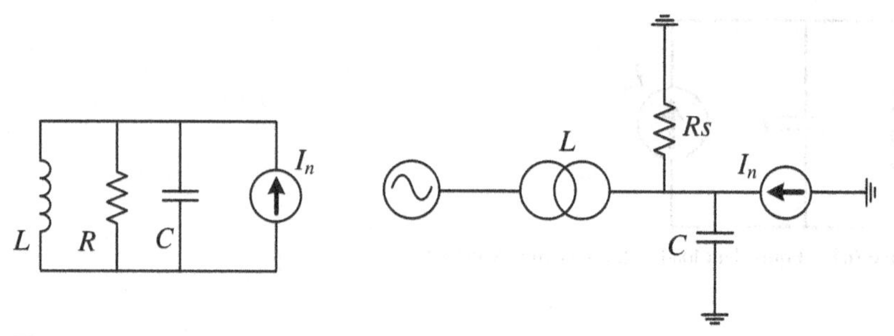

Figure 6.2 Insertion of resistive loads in the system.

sought in order to increase the harmonic order in which resonance occurs. Naturally, this condition is not always easy to determine because the trains move along the line and at the most distant points from the TPSS, the short-circuit level may not be so high. Furthermore, each rolling stock should be tested under different operating conditions and network configurations in order to exclude the risk of resonance in the spectrum of the signaling frequencies.

If resistive loads (Figure 6.2) are taken into consideration, the possible operating conditions are as follows:

- With zero load, R is infinite (open circuit).
- With small loads, R is high.
- With large loads, R is low.

Of course, more serious conditions with modest loads bring Z up to high values, while loads of greater power contribute to dampening these resonance phenomena.

At nth frequency, $X_{Ln} = X_{Cn}$ (series resonance) is possible, and the branch thus assumes zero impedance and consequently the noise circulates in the preferential branch toward capacity C.

Such resonance, which therefore manifests itself in the case of basic multiple harmonic components, could strongly enhance the voltage, but involves harmonics with a higher order that can be attenuated by the skin effect. It is clear how these phenomena can create problems, especially for telecommunications systems.

Finally, consider the filter unit consisting of two tuned filters, one installed on the contact line and one on the track, as shown in Figure 6.3.

Considering the equivalent circuit of Figure 6.4, the value of the equivalent resistance R_{eq} can be calculated if the electrical characteristics of the various components are known:

- Rated inductance: 3.5 mH.
- Winding resistance: DC at 75 °C equal to ~9.8 mΩ, increase in AC (50 Hz) at 75 °C: ~ +0.78 mΩ.

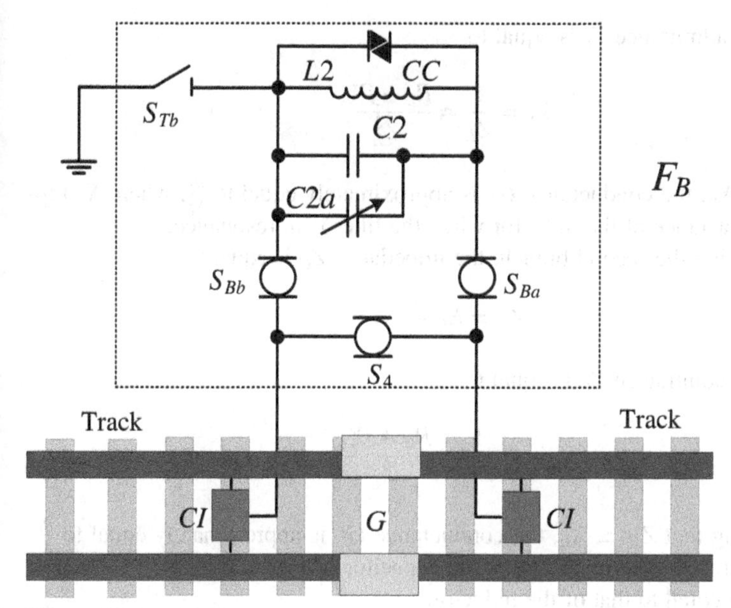

Figure 6.3 Track-side filter.

- Nominal capacity C: 2895 μF.
- Equivalent series resistance of capacitor: 3 mΩ.

Considering the first branch, the impedance Z_L is equal to

$$Z_L = R_L + jX_L$$

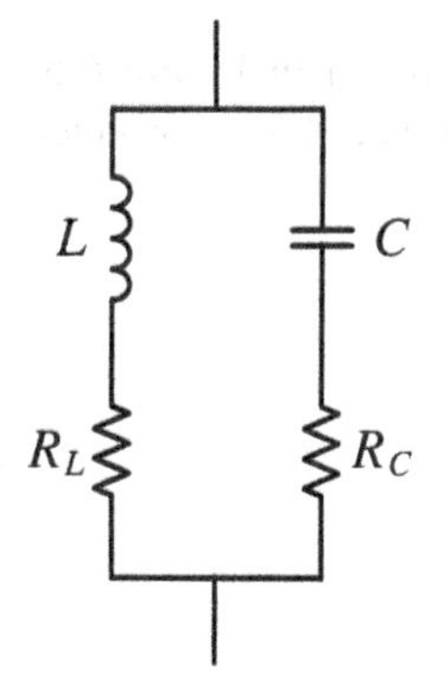

Figure 6.4 Full equivalent circuit of the filter.

and thus, the admittance Y_L is equal to

$$Y_L = \frac{1}{Z_L} = \frac{R_L - jX_L}{Z_L^2}$$

Since $Z_L \simeq X_0$, the conductance G_L is approximately equal to $\frac{R_L}{X_0^2}$, where X_0 represents the reactance of the inductor when the filter is in resonance.

Considering the second branch, the impedance Z_C is equal to

$$Z_C = R_C - jX_C$$

and thus, the admittance Y_C is equal to

$$Y_C = \frac{1}{Z_C} = \frac{R_C + jX_C}{Z_C^2}$$

Observing that $Z_C \simeq X_0$, the conductance G_C is approximately equal to $\frac{R_C}{X_0^2}$, where X_0 represents the reactance of the capacitor when the filter is in resonance and therefore equal to that of the inductor.

Consequently,

$$Z_L \simeq Z_C \simeq X_0 = 1,1\,\Omega$$

The equivalent conductance G_{eq} is

$$G_{eq} \simeq \frac{R_L + R_C}{X_0^2} = \frac{13.58 \cdot 10^{-3}}{1.21} = 11.22 \times 10^{-3}\ \text{S}$$

from which the equivalent resistance R_{eq} of the two elements is obtained:

$$R_{eq} = \frac{1}{G_{eq}} = 89.13\,\Omega$$

The equivalent circuit can then be reduced to that shown in Figure 6.5, where only the first-calculated resistance R_{eq} is considered. The shunt resistance

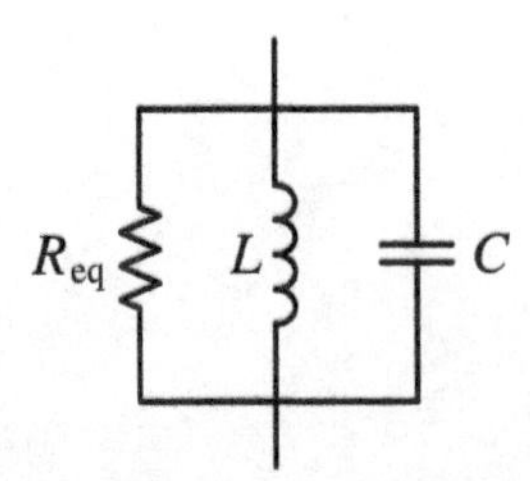

Figure 6.5 Reduced equivalent circuit of the filter.

Table 6.1 Minimum Values Required for the Ohmic Resistance of the Joint at Different Humidity Levels

Humidity (%)	Joint resistance (MΩ)
60	600
65	70
70	7
75	3
80	1.5

of the joint gives a negligible effect, as can be seen from its values reported in Table 6.1 for humidity.

Introducing the frequency deviation δ in p.u. of a generic angular frequency ω of a generic disturbance with respect to the resonance one $\omega_r = \frac{1}{\sqrt{LC}}$:

$$\delta = \frac{\omega - \omega_r}{\omega_r}$$

and the quality factor Q of the filter, in this case given by the relationship between the susceptance B_0 and the conductance G_{eq}:

$$Q = \frac{B_0}{G_{eq}} = 81.019$$

where

$$B_0 = 2 \cdot \pi \cdot f_0 \cdot C = \frac{1}{2 \cdot \pi \cdot f_0 \cdot L}$$

the admittance of the filter is equal to

$$Y(f) = G + j\left(\frac{-1}{2 \cdot \pi \cdot f_0(\delta + 1)L} + 2 \cdot \pi \cdot f_0(\delta + 1)C\right) = G + j\left((\delta + 1)B_0 - \frac{1}{(\delta + 1)}B_0\right)$$
$$= G\left(1 + j\left((\delta + 1)Q - \frac{1}{(\delta + 1)}Q\right)\right)$$

In order to calculate the bandwidth, defined as the range of frequencies for which $|Z(f)| \leq \sqrt{2} \cdot R$ applies, we get

$$1 + j\left((\delta + 1) - \frac{1}{(\delta + 1)}\right)^2 Q^2 = 2$$

$$\left[\frac{\delta + 2}{\delta + 1}\right]^2 \delta^2 Q^2 = 1$$

Since in general $\delta \ll 1$, that is, the disturbance is near the resonant frequency of the filter, we get

$$\delta = \pm\frac{1}{2Q} = \pm 0.00617$$

which are the extremes of the bandwidth. Higher values of Q involve very selective filters with respect to disturbances and, in this case, it can be said that the filter adopted is rather selective, as will be further verified below when the transfer function is analyzed.

It is important to assess the influence of the values L, C, and f on δ:

$$\delta = \frac{f - f_r}{f_r} = \frac{f}{f_r} - 1 = 2 \cdot \pi \cdot f \cdot \sqrt{L \cdot C} - 1 \Rightarrow \delta = \delta(f, L, C)$$

Thus, δ is a function of the three independent variables L, C, and f. In particular, f is the frequency of the interference to be filtered, which should be equal to

$$f_{rn} = \frac{1}{2 \cdot \pi \cdot \sqrt{L_n C_n}}$$

where L_n and C_n are nominal values, while f_r is the actual resonant frequency of the filter equal to

$$f_r = \frac{1}{2 \cdot \pi \cdot \sqrt{LC}}$$

where L and C are actual values.

The nominal operation point of the filter will then be in the condition f_{rn}, L_n, and C_n, a condition in which the current absorption I_n and $\delta = 0$. However, considering that the actual operating point cannot be such, it is possible to calculate the total differential δ calculated in the nominal point (P_n) to see which offset exists between the noise to be filtered and the actual frequency the filter is tuned to. The total differential is equal to

$$d\delta(f, L, C)_{P_n} = \frac{\partial}{\partial f}\delta(f, L, C)_{P_n} df + \frac{\partial}{\partial L}\delta(f, L, C)_{P_n} dL + \frac{\partial}{\partial C}\delta(f, L, C)_{P_n} dC$$

which is equal to

$$d\delta(f, L, C)_{P_n} = \frac{df}{f_{rn}} + \frac{1}{2}\left(\frac{dL}{L_n} + \frac{dC}{C_n}\right)$$

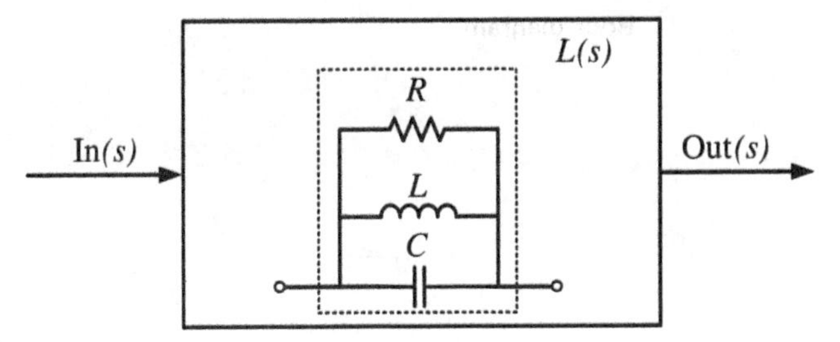

Figure 6.6 The filter model to calculate its transfer function.

Going from differentials to finite variations, we get

$$\Delta\delta = \frac{\Delta f}{f_{rn}} + \frac{1}{2}\left(\frac{\Delta L}{L_n} + \frac{\Delta C}{C_n}\right)$$

from which it can be seen that the variation of the values of L and C has a lower weight with respect to the frequency change f.

In order to assess the behavior of the filter, the transfer function is deduced below with a filter as a closed box where there is an input and an output signal (Figure 6.6). The input signal In(s) corresponds to the supply current, while the output signal Out(s) corresponds to the voltage.

The transfer function is

$$L(s) = \frac{\text{Out}(s)}{\text{In}(s)} = \frac{1}{C} \cdot \frac{s}{s^2 + \frac{1}{RC}s + \frac{1}{LC}} \tag{6.1}$$

The poles of the transfer function seem to be complex and conjugated and correspond to

$$p_{1,2} = \frac{-\frac{1}{RC} \pm \sqrt{\frac{1}{R^2C^2} - \frac{4}{LC}}}{2}$$

which when calculated by entering the nominal values of the RLC parameters gives

$$p_1 = -8.6 - 314.15i$$

$$p_2 = -8.6 + 314.15i$$

The function only has one zero at the origin. The Bode diagram of the transfer function given in (6.1) is shown in Figure 6.7.

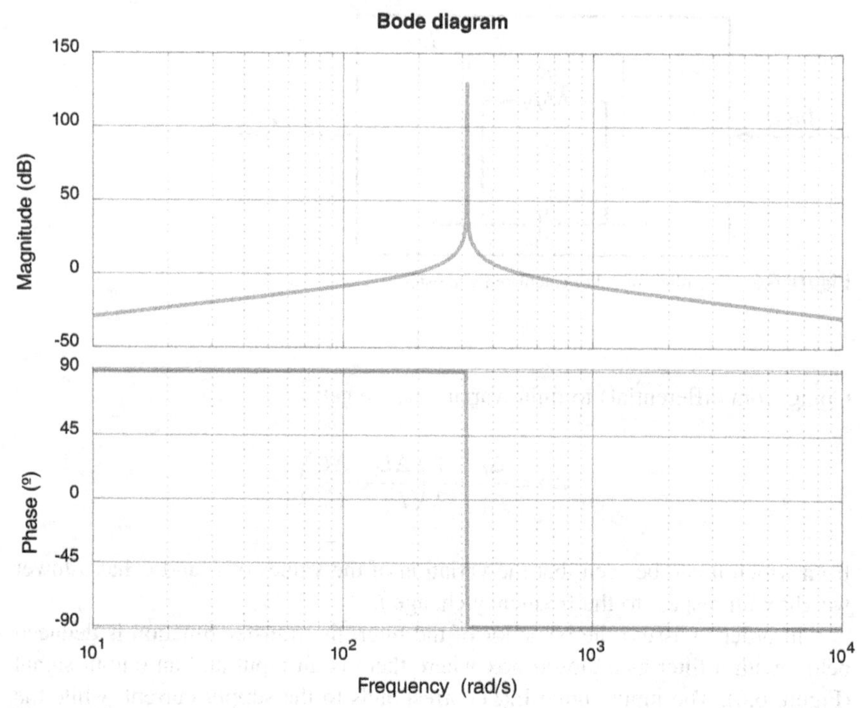

Figure 6.7 Bode diagram of the filter transfer function.

From the Bode diagram shown in Figure 6.7, it can be seen that very selective filters have been chosen. There is in fact a strong increase in the impedance at the frequency of 50 Hz. It is essential to achieve a perfect tuning at the cutoff frequency in these filters.

6.1.2 Induced Type Interference Phenomena

Inductive coupling occurs when the magnetic flux associated with the source current is linked with a second system. An electric traction line is a source of electromagnetic interference, as it can induce an electromotive force (emf) on the conductors located nearby, and to a greater extent in those parallel to the line itself (Figure 6.8). In fact, it is crossed by the traction current at the rated frequency, to which a certain harmonic content is superimposed that is generated by the traction converters, or from the same primary network. Other interference can occur even for short-circuit current primers or for switch operations.

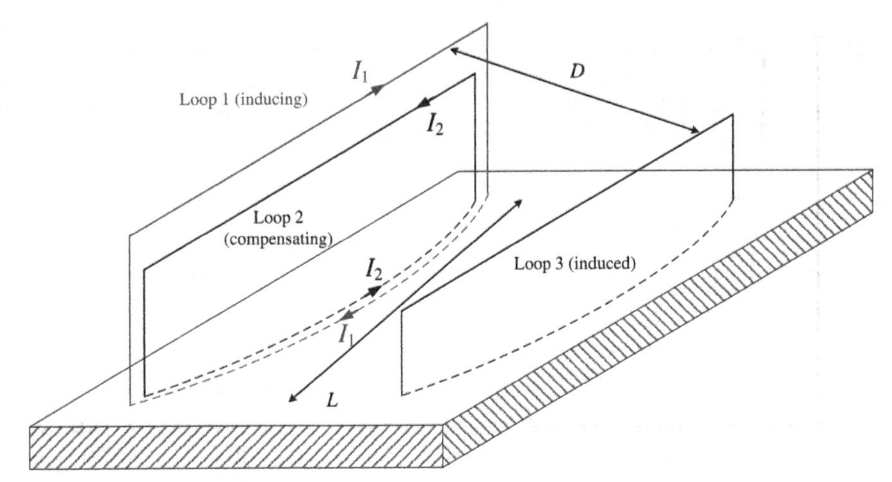

Figure 6.8 Induced interference.

The general expression of the effective value of the induced emf takes the following form:

$$E = K \cdot \omega \cdot M \cdot \ell \cdot I$$

where

- I: RMS value of the current at a certain frequency f (A);
- M: mutual inductance per unit length (H/km);
- ℓ: length of the parallel portion (km);
- K: reduction factor that translates compensation effects caused by circulating currents in other conductors.

The induction is produced by the AC system, which is continuously generated by closed loops crossed by traction current and delimited by the TPSS supply, the contact line, the loads consisting of the trains in transit and by the track, or the return circuit. Figure 6.9 shows an example of the qualitative variation of the induced voltage on a conductor, parallel to an electric traction line at $25\,kV_{AC}$, crossed by current, as a function of the distance from the line itself, resulting in associated values at distances of less than 15 m, that is, comparable with the size of the inducing loop.

In real cases, of course, the inducing current, the amount of parallelism, the average distance, and other side parameters such as soil resistivity should be taken into account.

In order to calculate the value of the induced electromotive force, it is necessary to calculate the values of the conductor inductance and, for this purpose, we will make use of the Carson–Clem theory.

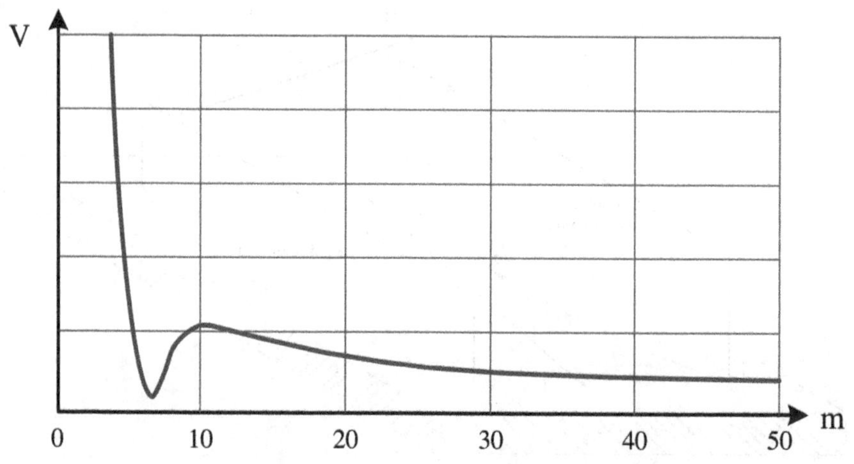

Figure 6.9 Results of the voltage induced on a conductor parallel to a 25 kV line.

The basic formulas for the calculation of the impedance and mutual impedance are given by Carson expressions that take the following form:

- Impedance associated with the external flux:

$$Z_{Ei} = j\omega \frac{\mu_0}{2\pi} \ln \frac{2h_i}{r_i} + 2(\Delta R_{ii} + j\Delta X_{ii}) \tag{6.2}$$

- Mutual impedance:

$$Z_{ij} = j\omega \frac{\mu_0}{2\pi} \ln \frac{D_{ij}}{d_{ij}} + 2(\Delta R_{ij} + j\Delta X_{ij}) \tag{6.3}$$

where

- $\omega = 2\pi f$ with f frequency in Hz;
- μ_0 is the permeability of the vacuum;
- h_i is the average height of the conductor i from the ground;
- r_i is the radius of conductor i;
- ΔR and ΔX are the Carson correction factors that take the effects of the soil into account;
- d_{ij} is the distance between the conductor i and the conductor j;
- D_{ij} is the distance between the conductor i and the conductor image j.

The impedance associated with the outer flux and the mutual impedance can also be expressed as a function of x, as shown in the following expression:

$$Z = 10^{-4}\left(\pi^2 f + j\omega 2\ln\frac{1.852}{x}\right)\Omega/\text{km} = 10^{-3}\left(0.157\omega + j\omega 0.2\ln\frac{1.852}{x}\right)\Omega/\text{km}$$

where $x = \alpha d$ for the mutual impedance calculation, while $x = \alpha r$ if the impedance is associated with the external flux (Z_E), or $x = x' = \alpha r'$ when the impedance associated with the external flux that takes the conductor's internal reactance into account needs to be calculated.

Substituting the values of x for the different cases, and introducing the value of the equivalent depth of the return current's hypothetical path, we get

- Impedance associated with the external flux:

$$Z_{Ei} = \left(10^{-3} \cdot 0.987 \cdot f + j\omega 2 \cdot 10^{-4} \ln \frac{D_e}{r_i}\right) \Omega/\text{km}$$

- Impedance associated with the external flux together with the conductor's internal reactance:

$$Z' = \left(10^{-3} \cdot 0.99 \cdot f + j\omega 2 \cdot 10^{-4} \ln \frac{D_e}{r'}\right) \Omega/\text{km}$$

- Mutual associated impedance:

$$Z_{ij} = \left(10^{-3} \cdot 0.99 \cdot f + j\omega 2 \cdot 10^{-4} \ln \frac{D_e}{d_{ij}}\right) \Omega/\text{km}$$

Let us consider, for example, the case of the lines supplied with the autotransformer system: the Carson–Clem formulas can be applied with a good deal of accuracy, considering that a distance of less than a few hundred meters between the conductors and industrial frequency are involved.

In fact, considering an inducing frequency of 50 Hz, we get

- Mutual impedance between two conductor–earth loops (Ω/km):

$$Z_m = 0.05 + j0.0628 \left[\ln\left(\frac{A}{D}\right) + 0.5\right]$$

where A is the equivalent notional depth of the return line through the ground and D is the distance between the two conductors (m).

- Impedance characteristic of a conductor–earth loop (Ω/km):

$$Z_p = R + 0.05 + j0.0628\left[\ln(A/r) + \mu/4 + 0.5\right] \tag{6.4}$$

where R is the resistance of the conductor (Ω/km), μ is the magnetic permeability of the conductor, and r is the radius of the conductor, which, in the case of noncircular conductors can be assumed to be equal to the perimeter/2π.

The formula (6.4) does not take the skin effect into account, which in the case of a magnetic material, such as rail, also has an influence at 50 Hz, or a change in permeability as a function of the current. In this first phase, referring to Chapter 6 for further details and references to actual studies in the international literature, the formula (6.4) can also be considered valid with good approximation for rail, assuming the conductors have a magnetic permeability value between 6 and 10 times that of the vacuum and a conductor resistance R between 0.1 and 0.2 Ω.

In the specific case under examination, assuming the following values:

- $\rho = 150\,\text{m}\Omega$ (average value of resistivity);
- $A = 400\sqrt{\frac{\rho}{f}} = 56.57\sqrt{\rho} = 693$;
- $D = 1.5\,\text{m}$ (distance between the two rail joints of a track);
- $R = 0.15\,\Omega$;
- $r = 0.11\,\text{m}$;
- $\mu_r = 7$;

we get

- impedance of the rail–earth loop:

$$Z_{p(r-t)} = 0.15 + 0.05 + j0.0628\left[\ln(693/0.11) + 7/4 + 0.5\right] = 0.2 + j0.69$$

- mutual impedance between the two rail joints:

$$Z_m = 0.05 + j0.0628\left[\ln(693/1.5) + 0.5\right] = 0.05 + j0.417$$

In order to roughly calculate the voltages induced by the power lines on parallel conductors, first we will consider the case of the classic single-phase system consisting of a contact line cantilever supplied with return through the track, as shown in Figure 6.10.

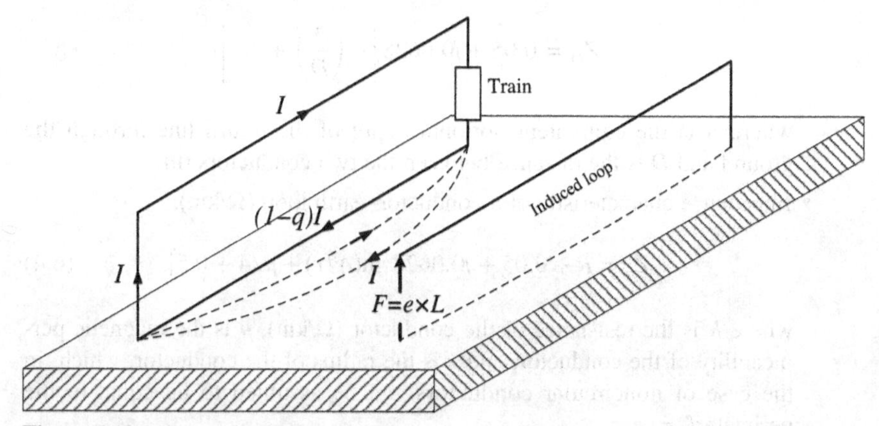

Figure 6.10 Diagram of single-phase line.

The inducing line consisting of track and contact line can be seen as two inducing single-phase lines with return to ground crossed by currents in the opposite direction. The current value I in the contact line will not be the same as that in the track because the track is not perfectly insulated from the ground and thus a qI part of the return current does not pass through the rail but through the ground. Consequently, the return current through the track is the remaining part equal to

$$I_{track} = (1 - q)I$$

The term $(1 - q)$ is called the "ground loop compensation factor" and leads to a reduction in the compensator effect of the return-ground loop. In fact even at great distances, the compensating effect cannot be total because the total current does not cross the "compensating" loop.

If for a conductor we consider a parallel one at a distance D_1 from the contact line and a distance D_2 from the track with an end connected to the ground, an induced voltage V_{ind} will be created at the other end that is proportional to the length of the conductor and to the longitudinal induced electromotive force equal to $E = Z_m I$ (V/km), in which Z_m is the mutual impedance between the inducing loop and the induced loop. Considering the inducing effect of loop 1 (contact line–ground) crossed by the current I and the compensating effect of loop 2 (track–ground) crossed by the current $(1 - q)I$ in the opposite direction and applying the principle of superposition, the total voltage induced by the two loops is given by

$$E = [Z_{m1} - Z_{m2} \cdot (1 - q)]I$$

where Z_{m1} corresponds to the mutual inductance between the contact wire–ground loop (crossed by the current I) and the induced loop; however, Z_{m2} corresponds to the mutual inductance between the track–ground loop (crossed by a current equal to $(1 - q)I$) and the induced loop (Figure 6.10).

Substituting the values of Z_{m1} and Z_{m2}, we get

$$\begin{aligned} E = &\left[\left[0.05 + j0.0628\left(\ln\left(\frac{A}{D_1} + 0.5\right)\right)\right]\right. \\ &\left.-\left[0.05 + j0.0628\left(\ln\left(\frac{A}{D_2} + 0.5\right)\right)\right](1 - q)\right]I \\ = &\, j0.0628 \ln\left(\frac{D_2}{D_1}\right)I + qI\left[0.05 + j0.0628\left(\ln\left(\frac{A}{D_2} + 0.5\right)\right)\right] \end{aligned} \quad (6.5)$$

It can be noted how there is a slow decrease in E with distance, according to a logarithmic law. Two terms can be discerned from the formula (6.5). The first represents the case of two single-phase inducing lines with return to ground crossed by the same current I but in the opposite direction. Its value does not depend on A, and thus earth resistivity, and attenuates rapidly with increasing parallel distance, in that the ratio D_2/D_1 rapidly moves toward 1. The second term is, however, proportional to the fraction q of current leaked into the ground.

Making a comparison between the two terms, it is clear that even at relatively small distances, the first term can be considered negligible. For a distance of 10 m between the two conductors, for example, the value of the second term is 17 times that of the first.

Thus, the second term only needs to be considered:

$$E = qI\left[0.05 + j0.0628\left(\ln\left(\frac{A}{D_2} + 0.5\right)\right)\right]$$

Consequently, the longitudinal induced electromotive force, in the case of single-phase traction, greatly depends on the amount of current leaked into the ground.

To reduce the interference, the portion of the current qI that passes into the ground should be reduced, and in a classic single-phase system this is equal to about 0.4.

In fact, assuming that track dispersion is very high, q is expressed by the following formula:

$$q = \frac{(Z_b - Z_{cb})}{Z_b}$$

where Z_{cb} is the mutual impedance between the contact line–ground loop and the track–ground loop, while the impedance of the track–ground loop Z_b is equal to

$$Z_b = \frac{(Z_p - Z_m)}{2}$$

in which Z_p is the impedance of each rail–ground loop of a track and Z_m is the mutual impedance between the two rail–ground loops of a track.

Using a relative magnetic permeability of the conductors equal to 7, ground resistivity of 150 Ωm, and a frequency of 50 Hz, we get

$$Z_p = 0.69\,\Omega/\text{km} \quad Z_{cb} = 0.34\,\Omega/\text{km}$$

$$Z_m = 0.42\,\Omega/\text{km} \quad Z_b = 0.56\,\Omega/\text{km}$$

from which we get a q value of 0.4. In this case, considering an approximate current value as that which is absorbed by a high-speed train ($I = 250$ A) and a distance between the inducing line and the induced line of 100 m, we get an

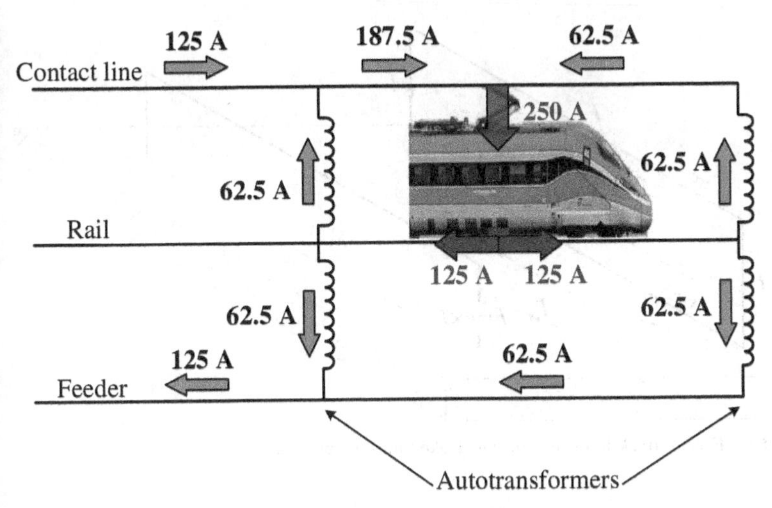

Figure 6.11 Ideal current distribution in an autotransformer system.

electromotive force equal to

$$e(I = 250\text{ A};\quad D = 100\text{ m};\quad q = 0.4) = 15\text{ V/km}$$

One solution adopted in order to reduce this amount is to use return strands with an override operation provided by a suitable transformers, which allow the amount of return current passing through the ground to be reduced by 80%, obtaining a q equal to 0.1 and an electromotive force of about

$$e(I = 250\text{ A};\quad D = 100\text{ m};\quad q = 0.1) = 3.8\text{ V/km}$$

Adopting an autotransformer system definitely allows the situation to be improved, although the study of the phenomenon is complicated by the presence of autotransformers and feeders. With this type of system and the presence of the return strand, a q value of around 0.1 is the objective sought.

In order to determine the interference caused by the line, reference is made to the theoretical distribution of the currents, focusing only on the affected cell occupied by the train, as shown in Figure 6.11.

Three key loops can be identified in the linear system:

- Feeder–track loop crossed by a current of 62.5 A
- Contact line–track loop crossed by a current of 187.5 A
- Contact line–track loop crossed by a current of 62.5 A

For each loop, the return current leaked into the ground and the term q should be taken into account, as shown in Figure 6.12.

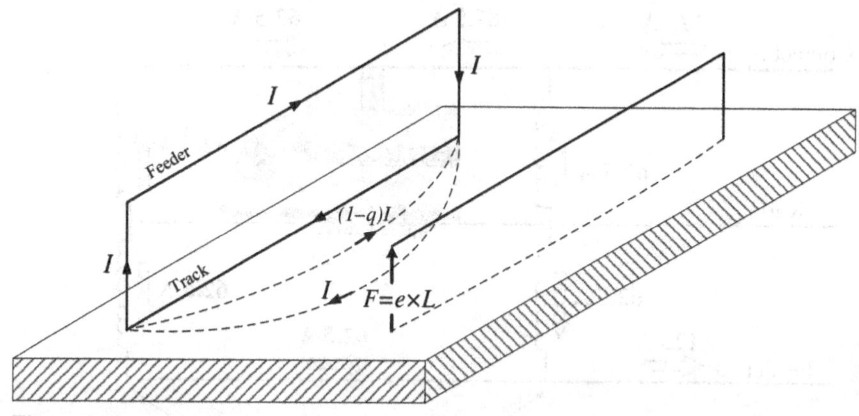

Figure 6.12 Feeder–track loop and currents leaked into the ground.

Thus, assuming a q value of 0.1, it is possible to calculate the induced electromotive forces:

$$e_{\text{feed-track}}(I = 62.5\text{ A};\quad D = 100\text{ m};\quad q = 0.1) = 0.96\text{ V/km}$$

$$e_{\text{ocl-track}}(I = 187.5\text{ A};\quad D = 100\text{ m};\quad q = 0.1) = 2.87\text{ V/km}$$

$$e_{\text{ocl-track}}(I = 62.5\text{ A};\quad D = 100\text{ m};\quad q = 0.1) = 0.96\text{ V/km}$$

Since the loops are not all overlapping, the total electromotive force induced in the first section of the cell is equal to

$$e_{\text{tot 1}} = e_{\text{ocl-track 2}} - e_{\text{feed-track 1}} = 2.87 - 0.96 = 1.91\text{ V/km}$$

The two single terms do not add up because the interference they have induced tends to offset the sum (feeder compensator effect).

In the second part of the cell, the total electromotive force is equal to

$$e_{\text{tot 2}} = e_{\text{feed-track 1}} + e_{\text{ocl-track 3}} = 0.96 + 0.96 = 1.92\text{ V/km}$$

The interference induced by an autotransformer system is therefore equal to 50% of the interference induced by a classic single-phase power supply system:

$$\frac{e_{2\times25}}{e_{\text{single phase}}} = \frac{1.92\text{ V/km}}{3.8\text{ V/km}} \cong 50\%$$

This feature was of course expected, as autotransformer systems with the same power requirement are still transmitted at twice the voltages and the current value is reduced by half. In addition, the mutual inductance due to the catenary-feeder loop is reduced, as autotransformer systems with autotransformers have these two conductors close together.

Naturally, the induced interference increases when the area of the loop, the frequency, and the value of the current increase, and thus depends on the train's operational situations. Ideally, autotransformers restrict the current flow only to the cell occupied by the train that is supplied on a bilateral basis, creating a partial compensation effect. Interference would only increase if autotransformers belonging to a single PATP were out of service, as the area of the inducing loop would be increased.

6.1.2.1 Protection of parallel lines

An example of induced interference in the domain considered is the situation that arises when alternating current electrified lines are parallel and close to traditional lines. In this case, the track circuit of the traditional line is subject to induced interference by the traction current at 50 Hz.

In this case the signaling system is protected from induced interference by changing the frequency of the encoded current: the system is supplied at a frequency of 83.3 Hz (instead of 50 Hz) so that the ground and onboard system operation is not subject to interference from the induced electromotive forces.

With regard to the communication lines or any other conductor (e.g., pipelines) that runs parallel to the railway line for a stretch S that is long enough, electrical insulation and earthing (of the induced conductor/line) must be seen to in order to avoid injury to people or damage to systems.

Induced voltage limit values are established for the different situations and therefore the parallel section between two electric sections must be no greater than

$$S = \frac{F_{\text{lim}}}{e_{\text{ind}}}$$

where

- S = the parallel stretch between two electric sections (km);
- F_{lim} = induced voltage limit value (V);
- e_{ind} = induced electromotive force (V/km).

Moreover, in a telephone line, the induced electromotive forces give rise to so-called noise voltages that worsen the intelligibility of the communication. Taking the human ear's sensitivity to different sound frequencies into account, the International Advisory Committee for Telegraphy and Telephony has set noise coefficients that can be used to calculate the psophometrically weighted noise voltage or equivalent:

$$U_{\text{ps}} = \frac{1}{C_{800}} \cdot \sqrt{\sum U_f \cdot C_f}$$

where

- U_{ps} = psophometrically weighted noise voltage;
- U_f = noise voltage due to the induced electromotive force;
- C_f = coefficient of human hearing frequency f.

The psophometrically weighted noise voltage or equivalent is, in essence, a noise voltage at 800 Hz that produces the same noise effects as all U_f voltages.

Some of the measures for the reduction of noise caused by the traction line that are used when designing telecommunication systems are

- use of shielded cables for the construction of telecommunication systems with earthing sheaths. The smaller the longitudinal resistance of the shield, the better the shielding action;
- use of optical fiber cables, which are made with dielectric materials and are therefore not subject to induced emf;
- used for telecommunications connections in radio links using frequencies between 30 MHz and 14 GHz that are not subject to interference from the electromagnetic fields produced by the traction line (e.g., GSM-R 900 Hz);
- use of "pilot wire" laid in the vicinity of telecommunications cables and earthed at the ends, so that the emf induced on such telecommunications cables has an opposite effect than that of the traction line;
- possible installation of overvoltage limiters (generally the same ones that protect the cables from atmospheric discharges) against voltage spikes that can occur when there is a fault.

6.1.3 Capacitive Interference Phenomena

Capacitive coupling is caused by the generation of electrostatic electric fields and occurs when an electric field source has a non-negligible parasitic capacitance toward the victim circuit, causing unwanted interference voltages on the latter. The parameter that characterizes this type of coupling is naturally the coefficient of mutual capacitance between the inducing and the induced circuit.

This coupling mode is implemented if the railway lines are adjacent to the interfering catenary line (inducing electric field source) and the conductors in the

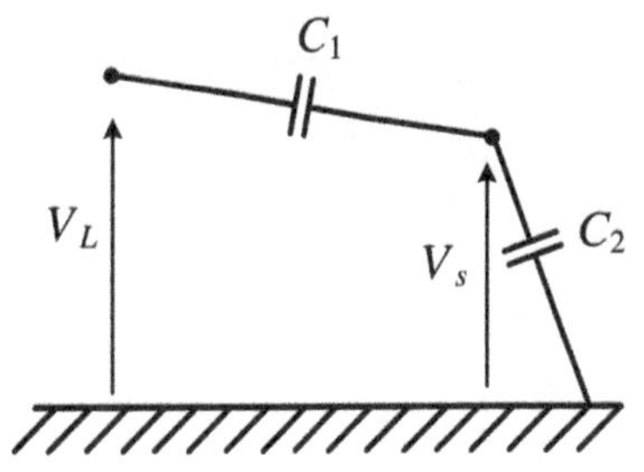

Figure 6.13 Capacitive couplings of two conductors.

line receiving interference (rail, signaling and telecommunication cables, metal masses).

It is known that a multiconductor overhead power line has distributed capacitive parameters. Thus, we have the conductors' capacity toward the ground, the capacity between conductors, and the capacity between conductors and external metal structures. The noise transmitted by capacitive coupling becomes significant for high voltages. In fact, the noise is given by the voltage distribution on the capacity according to the capacitive divider rule. With reference to XXFigure 6.13, in which C_1 is the capacity between the two conductors and C_2 that between conductor and earth, assuming that $C_1 \ll C_2$ we can write

$$\frac{V_s}{V_L} \approx \frac{C_1}{C_2}$$

The value of the capacity, and therefore the division of the currents, is inversely proportional to the distance between the metal parts (conductors and/or metallic structures).

The capacitive coupling becomes important as the frequency increases, as it increases the displacement currents. One possible solution is to increase the distance between the source and the victim or install shielding with at least one end connected to the ground. In Figure 6.14, for example, some of the capacity related to a route of a railway line section is given.

6.1.4 Radiated Interference Phenomena

In the presence of rapidly varying electromagnetic fields, the electrical size of the victim system and its distance from the source can be comparable or greater than the wavelength of the disturbing electromagnetic field, therefore the coupling between the source and victim must be analyzed while considering the radiated EM field and the propagation mechanisms that characterize it.

The pantograph of traction vehicles powered by an electric traction line is a cause of rather significant radiated noise. The electromagnetic emission in radio

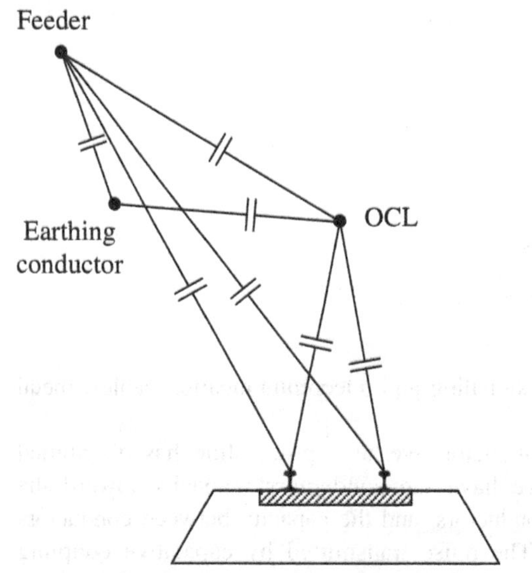

Figure 6.14 Possible capacitive couplings for a single line.

frequency is caused by the arc that occurs following the detachment of the pantograph, when this is absorbing current. The detachment of the pantograph frequently occurs, particularly at high speed, because of contact line's nonuniform compliance, vibrations induced in it by other means, or due to the accidental lowering of the pantograph. Although these are arcs of very short duration, the highly nonlinear nature of the arc recalls distorted currents with radio frequency spectra that cause radiated emissions.

This category also includes the phenomena related to the switching transients in electronic converters, which can interfere with the signaling and telecommunications of the noisy line.

6.1.5 Electromagnetic Fields Inside the Train

There are magnetic fields inside the train: In the case of a DC power supply, they are only generated by the auxiliary and drive systems; however, if there is an AC power supply with the autotransformer system at mains frequency, the magnetic field onboard the train is produced mainly by the currents of the traction line covering all areas of the train.

Trains' bodies are generally made with aluminum alloys that have very modest shielding properties for the low-frequency magnetic field, and the field produced by the traction line may penetrate inside the train without suffering severe attenuation and thereby affecting passengers; hence, we must also consider this aspect when designing trains.

Since there are not many remedies against magnetic fields, preventing their formation is essential. In this regard, both traction and auxiliary service converters take on a great deal of importance in the design of trains, and must be arranged so as to mitigate the electromagnetic fields in the areas where passengers and operating staff are present as much as possible.

6.2 STRAY CURRENTS

Stray currents are an unavoidable problem for every rail system where the running rails also fulfill the return conductor function. Especially when the system is supplied by direct current, these currents can cause or accelerate the corrosion of the systems themselves and any metallic structures nearby.

The upgrading of rail electrification and the high density of buried metallic structures, especially in urban areas, therefore creates the necessity to be increasingly more committed to limiting the impact of rail infrastructure on surrounding areas.

Metal reinforcements in concrete structures require careful monitoring in order to deal with the phenomenon of corrosion, as well as the use of appropriate tools to predict its occurrence in the design stage of railway systems, thus also limiting economic consequences.

The importance of the basic parameters for assessing how the traction current will return to the substations must be emphasized: the study of the track, the earthing system of the substation, and the structures affected by stray currents from the track. At the design stage, feedback about this "return" system should come alongside any feedback relating to substations, catenary, and traction vehicles.

Stray currents from DC traction systems are also referred to as stray because their magnitude and origin are difficult to predict and will vary over time.

The two running rails, in addition to performing their primary function to mechanically support the rolling stock, are also used as return conductors for the traction current.

This design choice results in an undoubted economic advantage, since the system does not need a dedicated return conductor, which is actually in the road drive systems. The disadvantages involved must, however, be taken into consideration.

First, the increase/decrease of rail voltage can reach dangerous thresholds for the safety of people present near the ballast, caused by the possibility of them stepping on and/or making contact with the track. The voltage of the latter essentially depends, in fact, on the current passing through it and its electrical resistance. Although it is normal to connect the running rails in an attempt to minimize the electrical resistance of the return circuit, a considerable voltage continues to exist between the track and the ground. The increase in rail voltage not only causes safety problems but also causes stray currents, as this voltage is manifested through the imperfect rail–earth insulation modeled through the dispersion conductance (shunt conductance) that is evenly distributed between track and ground. We must also consider how the track–earth voltage is heavily dependent on the earthing system adopted. With regard to the track–ground resistance, there is a conflict between the requirements for safety and those needed to limit stray currents: to limit them as much as possible, the highest possible degree of rail insulation toward the ground is necessary, yet such a configuration would lead to dangerous rail voltage values.

In this respect, the legislation thus lays down the minimum levels recommended for the track–ground resistance. Since the rails have the same voltage in the same cross sections, they can be regarded as a single conductor for the study of stray currents.

Furthermore, stray currents cause changes in the electrical system's subsurface with possible corrosion, which is usually indicated with the term electrical interference (UNI 9783-90) or electrolytic corrosion on the buried metal structures. Interference phenomena are also found in reinforced and prestressed concrete structures.

6.2.1 Origin of Stray Currents

A simple diagram of the phenomenon is illustrated in Figure 6.15, which refers to a traction system in which the positive circuit is constituted by an overhead contact line.

The current coming from the positive rail of the substation supplies the vehicle via the overhead contact line; as mentioned, the return to the substation takes place through the rails. Since these are never effectively insulated with respect to the ground, a certain amount of current will leave the track and enter the soil, especially in the area around the train. The current,

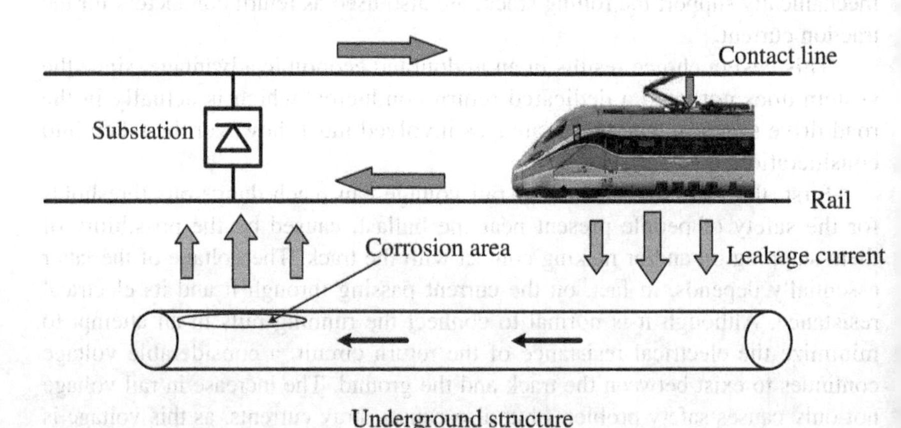

Figure 6.15 Stray currents' origin.

in fact, will follow the path of least resistance, in particular along underground metal tubes and pipes arranged along the line and in the reinforcing bars if there are any reinforced concrete structures. Near a TPSS (Traction Power Substation) the "stray" current will leave the latter and return to the rails through the ground, through which it will finally reach the negative bus bar of the substation.

This sketch, although convenient for understanding the origin of stray currents, does not properly illustrate the complexity of the situations that occur in reality. In fact, especially in urban areas this is characterized by a considerable density of both traction systems and metal structures. In the situations described and in the presence of multiple traction vehicles on each line, it is difficult to predict the path of the current flowing through the nearby metal structures after it leaves the rails.

DC stray currents dispersed in the ground can cause serious damage: In areas where they leak, the metal structures show signs of corrosion caused by an electrolytic effect. If the voltage caused by stray currents exceeds the activation potential, then the abovementioned metal areas act as anodes in relation to the ground, which, in turn, acts as electrolytes, resulting in a dissolution process. As can be deduced from Figure 6.15, the corrosion hazards are related to

- rails in areas far from the substation, in the presence of a train in absorption (which then inputs current to the track);
- steel structures in areas near the substations; in such areas, in fact, the current leaves the system to return to the negative TPSS power supply (feeder).

In recent times this issue, which is already particularly significant for the electrolytic corrosion it causes on metal pipes and cables' metal coatings when near railway and tram lines, has also become an issue for the reinforced concrete structures at train stations, also taking the by now generalized use of tunnels and viaducts in reinforced concrete into account. The risk of the reinforcements' corrosion is relative to both stray currents from the line under consideration and the stray currents from the rails of other neighboring railway lines.

In general, the state of interference of tunnels and concrete viaducts can thus be derived from two possible sources of stray currents:

- An "internal" current caused by the continuous stray currents from the tracks that run inside the tunnel/viaduct.
- An "external" current that could result from traction systems with DC power that is external to the tunnel/viaduct, trams and the trains, or other systems that use direct current.

This second case also concerns tunnels and viaducts of electrified lines with alternating current that are located in the vicinity of electrified lines with direct current.

The problem is not therefore only in railways and tramways in urban areas, where the risk of corrosion concerns a large number of buried structures, often densely arranged underground (subway tunnels, bridges, concrete structures used for the foundations of buildings, water and gas pipes), but the entire rail transport system with direct current, in which the major risks relate to reinforced concrete bridges and viaducts as well as underground metallic structures, and with alternating current, affecting features in the vicinity of the DC electrified lines.

The effects of the stray currents should not therefore be considered only in terms of influences and possible protections relating to corrosion. There are numerous examples in which stray currents from a DC traction system continue to generate considerable compatibility problems with the signaling system of another nearby railway system, in which case costly measures need to be taken in order to ensure safety.

6.2.2 Implications for the Transport System Infrastructure

This category includes all the remedies available to the infrastructure manager to minimize stray currents acting on railway system elements. They can be chosen in the design phase or be altered or improved in the case of already existing structures. The following are the main parameters of the transport system that affect stray currents.

6.2.2.1 Insulation of the return to ground circuit

The first element in the strategy for limiting stray currents at their source is the provision of high levels of insulation between the rails and the ground while maintaining the step and contact voltage under specific levels.

The standard method to achieve this is to ensure high resistance between rail and sleeper; this depends on the characteristics of the sleepers, the shutter devices of the rails, and the ballast.

Insulation is usually provided between the track and fixing system and between the fixing system and the sleepers; the insulation between fixers and sleepers must not compromise the fixers' ability to withstand the forces they are subjected to when a train passes (Figure 6.16).

The electrical resistance given by the sleepers must not be neglected: Concrete sleepers insulate a ballast better than wooden sleepers, which tend to become more conductive first on the surface, then internally due to their deterioration over their useful life. In some light transit systems, the rails are encapsulated in polymers not only to reduce noise and vibration but also to increase insulation (Figure 6.17).

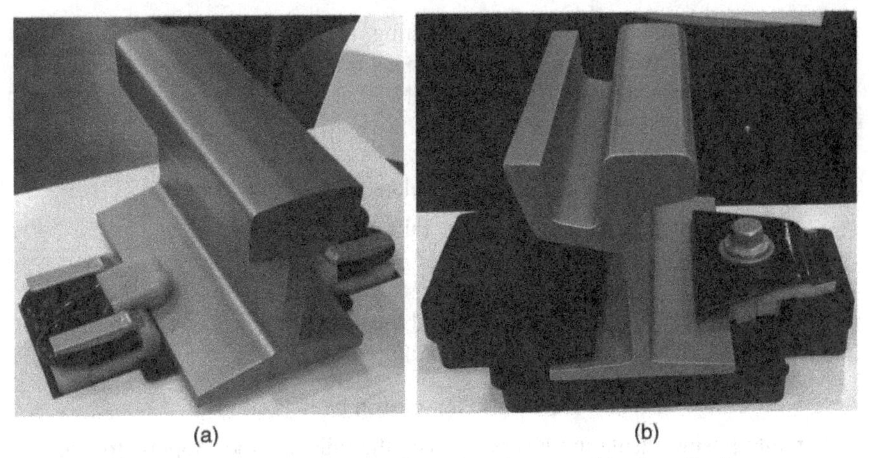

Figure 6.16 Insulated fixing systems for railway (a) and tramway (b) rails to the sleepers.

Figure 6.17 Rails encapsulated in polymers.

The appropriate measures for reaching the conductance values recommended for railways and road and rail mass transit systems recommended by industry standards such as the European EN 50122-2 are

- a clean ballast;
- wood or concrete sleepers with insulated fixing;
- an adequate distance between the rails and ballast;
- effective water drainage system.

The measures to achieve the best (and thus lower) values of conductivity for the road-type track systems are, for example,

- covering the rails with an insulating resin bed;
- interposing insulating layers between the rails and the support structures.

6.2.2.2 Return circuit conductivity

A second element to limit stray currents at their source is to choose rails that allow a low resistance return path for the traction current, usually using the largest section compatible with economic considerations.

Note that rails designed for a longer useful life normally have a higher electrical resistivity and wear increases their resistance, decreasing the useful section for the design calculations.

The importance of achieving the lowest possible resistance for the return circuit should be emphasized. For this purpose, the rails should be welded or electrically connected by means of longitudinal low-resistance connections, so that the total longitudinal resistance of the rails is not increased by more than 5%.

6.2.2.3 Line operating voltage

This parameter is closely related to the current absorbed by the traction vehicle. Like the absorbed voltage, the current required for the train will be inversely proportional to the supply voltage. This suggests that increasing the operating voltage helps to reduce the stray current phenomenon.

The discharge current from the rails, which when substations are very far from each other may constitute up to half of the return current, could thus be drastically reduced. At the same time, given the standardization that has now been reached in operating voltages of DC powered systems, the real possibility of acting on this lever is considerably reduced.

6.2.2.4 Distance between substations

The ever-increasing demand for the power that is required to operate the lines implies that more recent rail systems are designed so that their substations have lower power and are less spaced out for a given load. In this way stray currents are also reduced as a result of the reduction in rail voltage with respect to the ground at each point of the return current's primary route, resulting in improved safety. The extra costs due to the greater number of substations are partly reduced by the need for a smaller number of straightening units for each TPSS and the possibility of installing less expensive, more standardized modular equipment. With less distance between TPSSs, it may also be possible to provide a lower voltage rating for each TPSS, since the system can continue operating, albeit with reduced performance, if a single TPSS breaks down.

Studies in the field have shown that, given the ampere-hour (Ah) for stray currents when trains pass, the two parameters that most influence the value of the stray current are the rail-to-ground resistance and the distance between TPSSs.

6.2.2.5 Current absorbed by traction vehicles

As the operating voltage is virtually a fixed parameter, the current absorbed by traction vehicles substantially depends on the power they require. A substantial increase in power, due primarily to the ever greater power required for the auxiliary services of individual traction units, and the increase in the number of vehicles in circulation determine a considerable intensity in the current that passes through (and thus which may leave) the tracks.

6.2.2.6 Track drainage

As already seen, high levels of insulation can be obtained if the ballast below the rail level is kept clean. In these conditions, an electrical resistance of several hundred ohms-km can be reached. However, resistance can be reduced to very low values in wet conditions; to avoid this situation, an effective tool is needed to drain water from the track.

6.2.2.7 Stray current protection systems

If the rail–earth insulation is good, part of the return current will follow reclosing paths beyond the running rails.

To protect critical structures or underground utility services, the creation of "preferential" paths for stray currents is being used more and more by placing meshes under the rails that convey the currents or use the steel frames in the concrete slabs for the same purpose.

Rather than making the meshes electrically continuous and connecting them directly to the substation's negative rail, it is common practice to separate them at distances of about 300 m and connect each of them to the collector cables parallel to the rails and connected to the TPSS's negative rail through a diode.

Division into sections has some advantages:

- It facilitates the installation of each loop.
- It allows periodic electrical inspections during the life of the system to verify and monitor the performance obtained.
- It makes it easy to replace parts in case of deterioration or breakdown.
- It minimizes inductive coupling effects with signaling systems.

6.2.2.8 Substation earthing system

The design of an earthing system must reach a compromise between two conflicting requirements: minimizing stray currents and ensuring the maximum safety of staff and equipment.

To achieve this objective, the earthing system must be designed to meet the following criteria:

1. Under normal operating conditions, the earthing system must minimize stray currents. This can be obtained by ensuring the system is insulated. There is thus no intentional connection between the rail and the ground.
2. In abnormal operating conditions, characterized by dangerous values of the rail–ground voltage, the system must be directly earthed, short-circuiting the negative polarity to the ground to eliminate the dangerous voltage. This situation should be achieved automatically, in the shortest possible time, through protection relays and short-circuiting devices. Once the danger has passed, the system will automatically return to its initial state, that is, insulated.

If using fuse or short-circuit systems (Soulè fuse), these will have to be replaced after a short time.

6.2.3 Implications on Underground Structures Located Near the Transport System

This category includes all measures that, although not directly concerning the railway system, allow the intensity of the stray current to be reduced.

Such measures are adopted in an effort to try to prevent the stray current by adjusting the resistance it would encounter if it were to leave the running rails.

Increasing the electrical resistance of the current's alternative return circuits ensures that the return current reaches the TPSS's negative bus bar mainly through the rails.

In practice, this also includes, for example, underground structures in the vicinity of the railway line or even parallel to it, such as fire-fighting pipes in tunnels or tunnel reinforcements.

The resistance between the ground and buried metal structures or the longitudinal electrical resistance of the metal structure to be protected may also be increased.

6.2.3.1 Increase in electrical resistance between the ground and buried metal structures

In order to obtain adequate electrical resistance, it is necessary to choose a good coating that ensures effective electrical insulation of metal constructions, effective protection against moisture, and a good ability to withstand the possible stresses from the system's environment. This choice, which must always be made, is more effective if associated with the correct use of insulating joints.

Coatings, in addition to being very expensive, cannot be completely trusted because local defects in the insulating protective coating can always occur accidentally. At these defective points, the current emerging from metal structures would favor an extremely hazardous concentration of corrosive effects; the current collection area is indeed always large enough and makes the amount of current discharged in the vicinity of the substation significant.

It is thus necessary to use cathodic protection at the same time. The loss of the protective features of the coatings is rarely due to the chemical decay of the material forming the coating. It is usually produced by the following mechanical or physical causes:

- Mechanical damage of the coatings during transportation and installation of structures.
- Action of stones or rocks during installation and service.
- Action of forces exerted by the ground.
- Fragility of the coating at low temperatures, or its softening at high temperatures.
- Poor adherence of the coating due to inadequate surface preparation.

In some cases, the damage is caused by the cathode operation of the insulated structure in overprotective conditions (cathodic disbonding).

6.2.3.2 Insulation resistance

The protective capacity of a coating can be expressed by two complementary quantities: efficiency ξ and insulation resistance R_0:

- Efficiency ξ represents the fraction of the surface actually covered by the coating.
- Insulation resistance R_0 is the "equivalent" ohmic resistance of 1 m^2 of surface area.

Using ΔV to show the ohmic voltage drop of the coating caused by the passage of a current density, we get

$$R_0 = \frac{\Delta V}{i}$$

where

$$i = i_0 \cdot (1 - \xi)$$

with the current density for bare metal. Thus,

$$R_0 \cdot (1 - \xi) = \frac{\Delta V}{i_0}$$

In the following we will refer to the apparent insulation R_0' resistance where the coating is perfectly intact. When defects are present, the practical insulation resistance is determined by the following formula, valid for N defects per square meter of radius r_m:

$$R_0 \cdot = \frac{\rho \cdot (4t + r_m)}{4N\pi r_m^2}$$

in which t represents the thickness of the coating and ρ the ground resistivity.

6.2.3.3 *Increase in longitudinal resistance of steel structures*

An increase in the electrical resistance of the interfered structure reduces the collected current and thus that which is discharged in the area around a substation. An increase in resistance is obtained by resorting to the electrical separation of contiguous portions of structures with the interposition of insulating joints, such as those used to make metal pipes.

It is appropriate to choose the nature, distribution, and number of joints, taking care to insulate areas with metallic structures that are adjacent to the joints themselves, or inserting ohmic resistances in parallel with the joints that have a sufficiently high value to increase the longitudinal resistance of the structure overall.

Chapter 7

Elements of Transport Technology

7.1 INTRODUCTION

Electric traction has become increasingly important for the collective transport of people and goods, as it effectively contributes to mitigating congestion and pollution caused by road traffic. Throughout its history it has significantly evolved, leveraging the progress made in mechanical, electrical, electronic, and computer engineering fields. Developments in technology have made it possible to improve the railway sector in terms of energy efficiency, safety, comfort, and services.

The following is a description of the principal mechanical and electrical characteristics of a traction vehicle, and the equations that govern its motion in order to determine its distinctive features, depending on the required performance.

7.2 THE MECHANICAL ASPECTS OF ELECTRIC TRACTION VEHICLES

Vehicular motion exploits tractive force on kinematic pairs, which is transmitted through various interfaces such as the wheel/road interface in road transport and the steel wheel/rail interface in the case of train locomotives. As mentioned in Section 7.3, a rail vehicle has a number of wheelsets (Figure 7.1), each of which consists of an axle and two wheels, which constitutes a single rigid body so that the angular velocity Ω_1 of the left wheel is always identical to that of the right wheel Ω_2, namely, the wheelset angular velocity Ω (rad/s). Given D (m) as the driving wheel rolling diameter, the travel speed in m/s, is given by Equation 7.1:

$$v = \Omega \cdot {}^{D}\!/_{2} \tag{7.1}$$

Electrical Railway Transportation Systems, First Edition. Morris Brenna, Federica Foiadelli, and Dario Zaninelli.

Figure 7.1 Bogie of an EMU. (a) Motor bogie. (b) Trailer bogie.

Because of wear due to rolling and mechanical braking, when it is implemented by means of friction shoes (cast iron or synthetic) pressed against the rims, D decreases with operation over time compared with the initial value. Wear also degrades the profile of the guiding flange, requiring reprofiling of the rims by lathe machining. After a number of lathe machining operations, D will have degraded beyond the permissible minimum value and the wheel must be replaced, thus causing D to return to its nominal "as new" value.

There are also vehicles in which the two wheels are not rigidly linked to each other (so-called independent wheel systems) and in a generic instant the following may apply: $\Omega_1 \neq \Omega_2$. The wheeset as such ceases to exist.

While the automatic guidance of rail vehicles is assured by wheel flanges, road transport depends on solid rubber or pneumatic tires and paved surfaces. In these systems, the two wheels of each axle are mutually independent and if the wheels are powered, they are connected by a differential gear, or driven individually by a drive motor. In railway vehicles, each wheelset is connected to the frame by means pin/bearing kinematic pairs; the pin is part of the wheelset and is called the spindle; the bearing (the outer ring of which is fixed) is enclosed in the bushing, which is, in turn, connected to the frame by means of an appropriate suspension and guidance system, and transmits to it the various vertical and horizontal forces.

The complex of wheelsets is known as the wheel arrangement. There are vehicles with rigid axles, in which the wheelsets are part of a single frame, such as freight cars and carriages with two wheelsets and some types of shunting diesel locomotives. In this case, the distance between the front and rear wheelsets in each bogie is called the wheelbase; when there are more than two rigidly connected wheelsets, to facilitate the negotiation of bends a provision is made for some lateral displacement, or the flanges may be eliminated on one or more of the intermediate wheelsets.

The rigid wheelset solution is the simplest, but it is acceptable only for modest speeds and only for vehicles not designed to carry passengers. To obtain a satisfactory dynamic performance also at high speeds, it is necessary to resort to the use of bogies.

7.3 RAIL VEHICLES WITH BOGIE STRUCTURES

In most railway rolling stock (traction vehicles and towed vehicles), the wheel arrangement is divided into two or three bogies, each of which may have two or three wheelsets (Figure 7.1), which the car-body rests on (Figure 7.2); the bogie wheelbase varies from approximately 1.80 m (tramway vehicles) to 2.5–3.0 m (railcars and locomotives).

There are articulated vehicles in which the ends of two adjacent car-bodies are supported on the same bogie (Figure 7.3).

In the case of locomotives, the separation of the car-body in two sections connected by an articulation complicates several structural and engineering aspects. For this reason, a number of single car-body locomotives have been developed with bases on a three-bogie structure, with the central bogie having lateral freedom of movement when cornering.

The car-body–bogie linkage ensures the transmission of the various forces, leaving the two elements freedom to rotate around a vertical axis,

Figure 7.2 Locomotive with two bogies and four wheelsets.

which allows for cornering; the distance between the vertical axes of rotation of two adjacent bogies corresponds to the distance between bogie pivots (Figures 7.2 and 7.3).

A suspension system known as the central or secondary suspension is arranged at the interface point where the car-body rests on the bogie. The

Figure 7.3 Articulated light rail electric vehicle.

springs may be of the leaf type, steel helix, rubber/steel, rubber filled with air at variable pressure depending on the load (air suspension), or a combination of the above. Another suspension stage is interposed between the bogie frame and the bushings; the suspension on the bushings is also called the primary.

The parts of the vehicle that are borne directly or indirectly by at least one suspension system are known as the sprung mass (such as the car-body and the bogie frames). Conversely, the parts that rigidly bear on the rail (such as the bushings and the wheelsets) constitute the unsprung mass. The distinction is important, for example, when assessing the impacts that take place during motion at the inevitable track irregularities.

7.4 ROLLING STOCK WHEEL ARRANGEMENTS

The wheelset can be driven when torque is applied, or load-bearing otherwise. Normally rolling stock wheel arrangement layout is identified by an alphanumeric code indicating the distribution of the wheelsets among the various bogies or partial frames, distinguishing between driven and load-bearing wheelsets.

Each group of wheelsets belonging to the same bogie or frame is marked by a letter of the alphabet if the wheelsets are driven, or by a simple number if they are load-bearing: the number or the order of the letter corresponds to the quantity of adjacent wheelsets (e.g., *B* corresponds to two wheelsets; *C* corresponds to three wheelsets). Moreover, the letters without subscripts refer to groups of driven wheelsets mechanically coupled to one another by means of connecting rods or gears, while the zero subscript denotes an individually or independently controlled wheelset, each of which is driven by a motor. Some examples are listed here, classified according to traction vehicle.

- $B_0 + B_0$ or simply B_0B_0 = locomotive or railcar multiple unit with two bogies, each comprising two driven wheelsets with individual controls; in total we thus have four traction motors. This is the most frequently adopted solution for trams, metros, railcar multiple units, and railway locomotives.
- BB = locomotive or railcar multiple unit with two bogies, each comprising two driven coupled wheelsets. This is the preferred configuration adopted for railcar multiple units and locomotives with single-motor bogies.
- $B_0B_0B_0$ = locomotive or railcar multiple unit equipped with three bogies, each having two wheelsets with individual wheelset control. There are articulated dual car-bodies whose articulation pivot coincides with the central bogie or single car-body locomotives.
- BBB = locomotive or railcar multiple unit fitted with three single-motor bogies, each with two wheelsets.

- C_0C_0 = locomotive with two bogies, each with three wheelsets, having six traction motors.
- CC = locomotives with two single-motor bogies, each with three axes.

The mass borne by the individual wheelsets, known as mass per wheelset, varies depending on the nature of the vehicle and cannot exceed certain limits imposed by the constructional characteristics of the lines (normally 20–22.5 t/wheelset in standard lines). Other limit values apply for high-speed trains (15–18 t/wheelset), for metro rolling stock (10–12 t/wheelset), and for trams (6–9 t/wheelset).

The wheel diameters vary as a function of wheelset load and performance specifications: from 0.68 m for trams and similar vehicles, to 0.74–0.82 m for metro lightweight locomotives, 0.86–1.04 m for heavy electric locomotives, electric multiple units (EMUs), and lightweight locomotives, and 1.25 m for high power locomotives.

7.5 CLASSIFICATION OF ROLLING STOCK

Railway vehicles can first be divided into two major categories: standard rolling stock and electric multiple units.

Standard rolling stock consists of a locomotive, or vehicle housing all the traction equipment, and several carriages or wagons, depending on whether the train in question is used for transporting passengers or freight. The advantage of this solution is that it allows flexibility when forming the train; however, it penalizes performance in terms of acceleration since it limits the number of available drive wheelsets. Consequently, such trains are currently used for goods and some long-distance passenger services, although, in the latter case, the use of EMUs is becoming more widespread.

An EMU consists of several self-propelled units in a fixed formation, where the traction is distributed along the length of the train and the motors are housed by bogies, generally of different carriages. In contrast to ordinary convoys, where the traction power is exclusively concentrated in the locomotive, EMUs do not require a locomotive, which results in a drastic reduction of the axial loads on the rails and their supporting structures, such as the bridges. This, in turn, results in lower mechanical stress being applied to the infrastructure, even at high speeds. The distributed traction also helps to increase the vehicle's adhesion mass, thus improving acceleration performance.

One drawback of such systems is that, since each unit of an EMU has its own specific function, all the carriages must remain connected at all times. Therefore, once the composition has been defined, it can no longer be varied.

EMUs used on regional and metropolitan lines feature specifications designed to optimize characteristics such as acceleration, braking, load capacity, passenger boarding, and alighting times during stops.

(a)

(b)

Figure 7.4 Articulated EMUs.

EMUs can be divided into two broad categories, depending on the manufacturing solution: *Articulated EMUs*.

In this solution two adjacent car-bodies share the same carriage (Figure 7.4). This solution is increasingly used for regional and metropolitan electric services and in some cases also for high-speed trains: a well-known example is the electric trains used on the French TGV high-speed service.

The first advantage of this solution is the reduction in the number of bogies, and hence the cost. Also, articulation reduces the lateral oscillations of the bogies at high speeds and results in less rolling and pitching, thus increasing

(a)

(b)

Figure 7.5 EMU with independent carriages.

passenger comfort. On the other hand, it is necessary to use shorter car-bodies in order to respect the curve limits, which results in less available space for installing access doors, thus making it more difficult for passengers to board and alight. It also compromises versatility when maintenance work must be carried out, since it is not possible to divide the train, which requires a specially equipped plant dedicated to the specific type of train.

EMUS with Independent Carriages In this solution each car-body rests on two bogies, so that from a mechanical point of view, the car-bodies are independent of one another. In any event, the train will only function in a fixed formation, as the various traction and auxiliary equipment is distributed throughout the convoy (Figure 7.5).

Thus, an EMU with independent carriages can be equipped with longer, more capacious car-bodies, with multiple passenger boarding points, thereby reducing the dwelling time at stations. They are also simpler to service, even at nondedicated maintenance plants, since the train may be easily divided.

A drawback of this system is the need for a greater number of bogies and antioverlap systems between the car-bodies in the event of more complex accidents, resulting in increased costs.

Light rail vehicles (LRVs) represent a particular type of EMU that is suitable for use on tramway routes, where the need for fully lowered floors for ease of passenger boarding in urban areas dictates the use of bogies having independent wheels without axles (Figure 7.6).

Figure 7.6 Light railway vehicles (LRV).

In this case, a widely used solution is to mount short car-bodies on the bogies and join them together by means of longer car-bodies and appropriate couplings.

The fact that equipment may not be installed in the central part of the bogies means that the traction motors are positioned on the outer sides.

7.6 THE WHEEL–GROUND KINEMATIC PAIR

Due to the elastic deformations of the rim and the rail, contact between the wheel and the rail occurs in a finite area A, called the footprint, where weight G is distributed with a pressure $p = dG/dAp$ variable from point to point throughout A (Figure 7.7); the average pressure $p_{av} = G/A$ can reach values on the order of 500–1000 N/mm^2.

The footprint shape is similar to a circle or an ellipse when the wheel is fully shaped, and it approaches a rectangle with increasing wear of the rim; on this area the pressures are distributed with an approximately parabolic relationship. When the wheel is in motion, the pressure distribution is altered by the elastic hysteresis effect of the wheel and the rail materials.

The resultant force of the elementary ground reactions then shifts in the direction of movement by a small distance δ and produces an opposite torque $G{\cdot}\delta$.

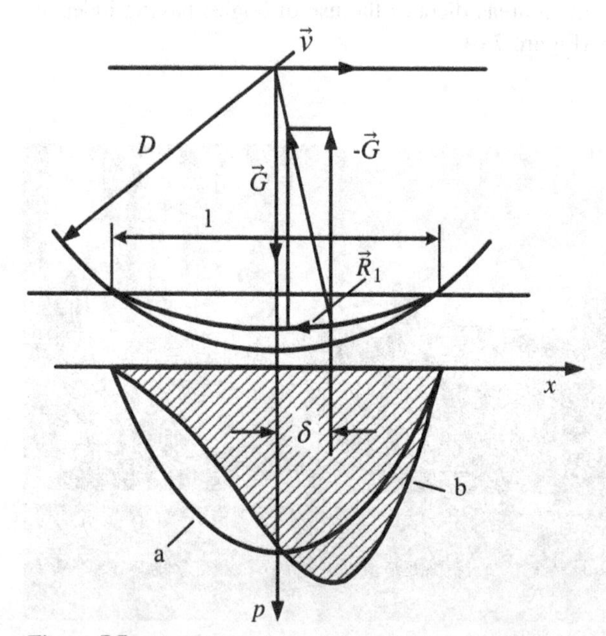

Figure 7.7 Distribution of pressure p in the footprint. *a*: Diagram $p = f(x)$, for $v = 0$; *b*: $p = f(x)$ diagram, for $v \neq 0$.

For pneumatic tires, the contact area is much larger than for railway wheels; the average pressures P_{av} estimated as indicated above depend on the inflation pressure and are on the order of a few tens of N/cm^2.

7.7 VEHICULAR MOTION

The study of the characteristics of vehicular motion involves the following forces (Figure 7.8):

- Active forces F, whose direction is that of the velocity vector $\vec{v}$.
- Passive forces or resistances R, whose direction is opposite to that of the vector $\vec{v}$.
- Inertia forces.

The active forces, in turn, may have the direction of vector $\vec{v}$ if they are driving (traction forces), or the opposite direction if they are resistant (braking forces).

For the sake of simplicity, the vehicle or combination of vehicles shall be considered as a material point P, having vehicle mass m. F then represents the resultant of all the active traction forces and R is the resultant of all passive forces. If the braking forces are zero, the rectilinear equation of motion is (Figure 7.8a):

$$F - R = m_e \cdot a \tag{7.2}$$

where a is the acceleration, considered positive if directed as $\vec{v}\left(\frac{dv}{dt} > 0\right)$, and m_e is the equivalent mass of the vehicle, which takes not only the actual mass into account but also the effect of the rotating parts (Section 7.21).

When the vehicle brakes, it is subject to both R and the braking force F_b – considering again for simplicity (7.2) as valid the set F_b as the negative active force and substitute $-F_b$ for F (Figure 7.8b). The braking acceleration or deceleration a_b is also negative and is then given by

$$a_b = \frac{F_b + R}{m_e} \tag{7.3}$$

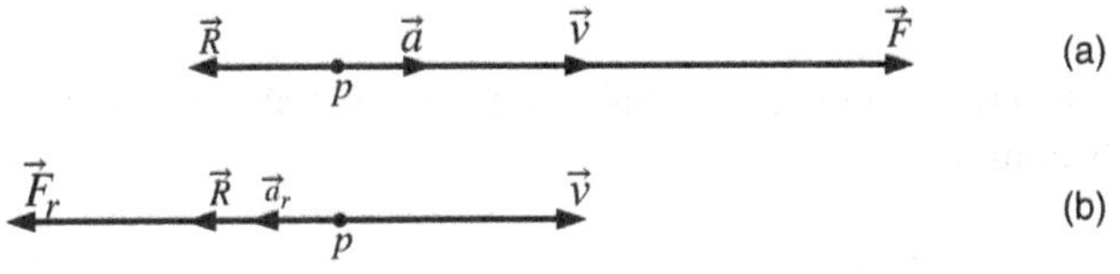

$\vec{R}$ $\vec{a}$ $\vec{v}$ $\vec{F}$ (a)

p

$\vec{F}_r$ $\vec{R}$ $\vec{a}_r$ $\vec{v}$ (b)

p

Figure 7.8 Diagram of active and passive forces applied to the material point P. (a) Starting and forward motion. (b) Braking.

7.8 THE ADHESION FACTOR

The active tangential forces F and the vertical load G interact between wheel and ground through the contact area A. The phenomenon by which this occurs is called adhesion. The force G acts on A with pressure p such that $G = \int_A p \cdot dA$. Likewise, given t as the unitary tangential force: $t = dF/dA$, then $F = \int_A t \cdot dA$.

Given a wheel subject to weight G and tangential force F, adhesion applies if it rolls. As the value of F increases, rolling continues until the adhesion limit F_{ad} is reached, beyond which grip is lost and the wheel begins to slip.

The adhesion coefficient is defined as

$$f = {}^{F_{ad}}/_{G} \tag{7.4}$$

For useful adhesion to exist, the following condition must be satisfied:

$$F \le F_{ad}; \quad \text{therefore, } F \le f \cdot G \tag{7.5}$$

Once this adhesion limit has been exceeded, slippage occurs and the tangential force is reduced to the value:

$$F' = f' \cdot G < F_{ad'}$$

given that the kinetic friction coefficient f' is less than f.

The conditions for adhesion must also be satisfied during braking; when the braking force F_b is transmitted through the contact footprint, if it exceeds the value $F_{ad'}$, the wheel will tend to lock. In the extreme case, if the railcar brakes are implemented by friction, for example, by means of shoes acting on the wheel rims or on disks extending from the wheelbase axles, the angular velocity goes to zero and the wheels skid. Skidding during braking may damage the surface of the rim, as destructive friction takes over in the contact footprint.

For a given wheel or wheelset, condition (7.5) must always be respected: if the force F varies over time between a minimum F_{min} and a maximum F_{max}, the most restrictive condition is

$$F_{max} \le f \cdot G$$

Usually, it is convenient to refer to the tangential force average value F_m, so that the above expression becomes

$$F_m \le \frac{F_m}{F_{max}} \cdot f \cdot G; \quad F_m \le f_0 \cdot G$$

where

$$f_0 = \frac{F_m}{F_{\max}} \cdot f$$

is the equivalent average value of the adhesion coefficient.

Under the conditions considered above, natural adhesion is said to apply. When the forces F are transmitted, in whole or in part, by means of various elements excluding the vehicle weight-bearing wheels, then the adhesion is considered as artificial. For example, in rack railways, F is exerted by dedicated toothed cogs and in funiculars, or cable railways, by the hauling cable.

Natural Adhesion Consider a driving wheel in motion, of which D is the effective rolling diameter, to be introduced in (7.1) to calculate the forward velocity v. Two areas can be distinguished in the contact footprint:

- An area of adhesion A_1, which is formed toward the front of the footprint with respect to the direction of travel, in which the relative speed is zero. Throughout A_1: $t < f_1 \cdot p$, where f_1 is the slippage onset coefficient of friction.
- A rear zone (area of slip) A_2, where minimal irreversible slippages occur (creep).

By effect of the combined (reversible) elastic deformations and microslippages, the wheel's circumferential velocity v_r, under traction, is slightly greater than the forward velocity v. The so-called creep velocity may be expressed in absolute terms as the difference:

$$v_s = v_r - v \quad (7.6)$$

and in relative terms, by means of the expression:

$$\sigma = \frac{v_r - v}{v_r} = 1 - \frac{v}{v_r} \quad (7.7)$$

The longitudinal force F transmitted within the contact footprint depends on σ: for $\sigma = 0$, then $F = 0$.

As F increases, so does σ, the creep is initially rather limited, the slippage area A_2 grows, at the expense of A_1 until it fills the entire footprint A, for a given value of F. Beyond this limit, a further increase of F is observed, in a growing field of σ (macroslippages) until the upper limit F_{ad} is reached, beyond which, real slippage takes over, with values of F decreasing with increasing σ (Figure 7.9). In braking, the phenomenon is similar, except that σ is negative, since $v_r < v$.

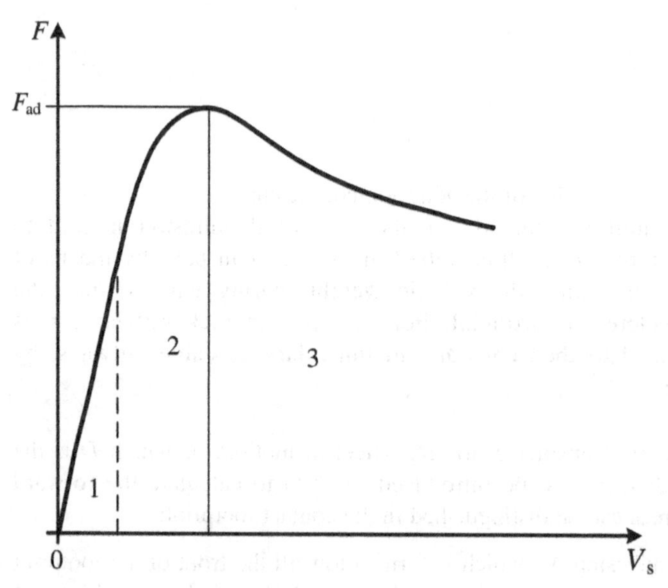

Figure 7.9 Creepage and slippage. 1: Microslippages; 2: macroslippages; 3: slippage.

7.9 THE ADHESION CONDITIONS OF INDIVIDUAL RAILCARS AND TRAINS

In Chapter 1, a single wheel or wheelset was considered, subject to traction or braking forces.

For a traction vehicle consisting of N driving wheelsets, for the purposes of verification of the traction adhesion conditions, it is necessary to distinguish between the total mass m_e and that which bears on the driving wheelsets, known as the adhesion mass m_{ad}.

Driving wheelsets are always of identical construction, in respect of both the active forces acting on their rims and the mass per wheelset, given that the mass m_{ad} is evenly spread across them. We note that today's electric locomotives always operate in full adhesion, that is, all their wheelsets are driven.

In general, given a train with an overall mass m, the adhesion mass m_{ad} is that which is borne by the N driving wheelsets of the locomotive(s); then the ratio

$$\alpha = {}^{m_{ad}}/_{m} \quad (\alpha \leq 1) \tag{7.8}$$

is known as the adhesion ratio. It can range from 1, for full adhesion trains composed of only locomotives, to values on the order of 10% and even less for heavy trains hauled by locomotives. The foregoing applies to motion under traction; in braking, normally all of the train wheelsets are braked, so $\alpha = 1$.

When considering forward motion under traction, in the general case the adhesion conditions indicated in 7.5 must be satisfied for each wheelset, that is, for the ith wheelset: $F_i < f \cdot G_i$. Since, for calculation purposes, it is convenient to consider the total traction force, then

$$F = \sum_{1}^{N} F_i$$

and the total adhesion weight is

$$G_{ad} = g \cdot m_{ad} = \sum_{1}^{N} G_i$$

then (7.5) can be expressed in the generic form as

$$F \leq f_0 \cdot G_{ad} \tag{7.9}$$

where the coefficient f_0 takes account of

- the diversity in instantaneous values of f for the individual wheelsets;
- the uneven distribution of the adhesion weight G_{ad} among the various wheelsets caused by, among other factors, pitching (to be examined in Section 7.19);
- the partitioning of the total traction force F, which for various reasons could result in the ith wheelset having a value greater than F/N.

In this regard, it may be advantageous to introduce coupling between wheelsets; as a limited example, consider a single-motor two-wheelset bogie (e.g., a locomotive with *BB* or equivalent wheel arrangement), the condition to be satisfied is

$$F_{bogie} = F_1 + F_2 \leq f \cdot (G_1 + G_2)$$

where F_1, G_1 and F_2, G_2 are the forces and loads, respectively, of the first and second wheelset. The rigid coupling between two wheelsets allows for the application of the overall adhesion weight $G_1 + G_2$.

Naturally, it must be considered that, due to the pitching phenomenon mentioned previously, the adhesion weight of the locomotive is not evenly distributed between the front and rear bogies.

In conclusion, the average values f_0 of the adhesion coefficient that are used in calculations must be reduced with respect to those applicable for single wheelsets – this is due to the previously mentioned lack of uniformity and, as seen in Section 7.8, the time-varying averages of the forces that must be considered rather than their instantaneous values.

The optimal exploitation of adhesion, that is, the possibility of generating considerable traction forces, is a significant issue in modern locomotive design, with powers that extend to 1–1.5 MW/wheelset, considering that in Europe the applicable limits are 21–22.5 t/wheelset. Suffice it to say here that, in establishing the limits for the exploitation of adherence, it is important to keep the numerous and variable factors that impact the phenomenon in mind as well as the intrinsic actuation and regulation characteristics; in fact, these determine the behavior of the system, should the adhesion limit be exceeded with the occurrence of slippage.

Modern electronic actuators and the introduction of efficient traction control and antiskid systems have sustained enormous progress in this field; optimal results may be achieved by three-phase drive systems, especially in the area of individual traction motor control, as it enables instantaneous optimization of the tractive effort on a single wheelset as a function of the available adhesion.

7.10 THE ADHESION COEFFICIENT

The value of the adhesion coefficient f depends on the type of wheel/ground interface; for a railway wheel it normally takes on lower values compared with the tire–road interface. Hence, the problems associated with adhesion are of fundamental importance in rail, electric, or diesel traction, and to a large extent they determine the design of traction vehicles.

For a given kinematic pair, the coefficient f essentially depends on the forward motion velocity v and decreases as v increases; for this reason, the formulas or experimental curves give the values of the f as a function of v.

Several other factors affect the coefficient of adhesion; a number of these are listed as follows:

1. The terrain surface conditions.

   f increases with dry surfaces and decreases as surfaces become wet. In the presence of sludge or grease f degrades significantly. It must be noted that in the case of road surfaces, f may drop to exceptionally low levels in the presence of ice, whereas ice does not have any significant effect on rail transport given the high pressures that occur in the contact footprint.

2. In railways, an appropriate dusting of dry silica sand on the track improves the coefficient f.

   Consequently, it is standard practice to apply sand to rail tracks by appropriate means (sandbox) when adhesion conditions are difficult.

3. The variations of the instantaneous vertical load G on the wheels, due to the vehicle shock absorbers and shocks. The relationship between the traction force applied to a wheelset and the load it bears can overcome the adhesion limit, causing slippage. The same phenomenon can occur

for localized poor track conditions. This entails reducing, as seen in the previous section, the average adhesion coefficient for the purposes of the calculations.

4. When cornering, lateral sliding on the rails can occur due to mechanical linkage between the wheels of the same wheelset, which causes a reduction of the average adhesion coefficient referenced to each wheelset.
5. As always in the case of railways, the motion transmission system to the driving wheelsets. As noted in Section 7.9, the coupling between wheelsets obtained by means of linkages, chains, and gears allows for higher average adhesion coefficients compared with those achievable by means of individual wheelset control.

7.11 PRACTICAL VALUES FOR THE ADHESION COEFFICIENT

In order to determine the adhesion coefficient as a function of forward velocity, it is normal practice to use formulas or experimental curves. It should be noted that experimental tests can provide a considerable spread of results, as shown in Figure 7.10.

Rail Transport Important experiments were performed by Metzkow, Müller, Curtius, and Kniffler (1943) and Nouvion and Bernard (1960).

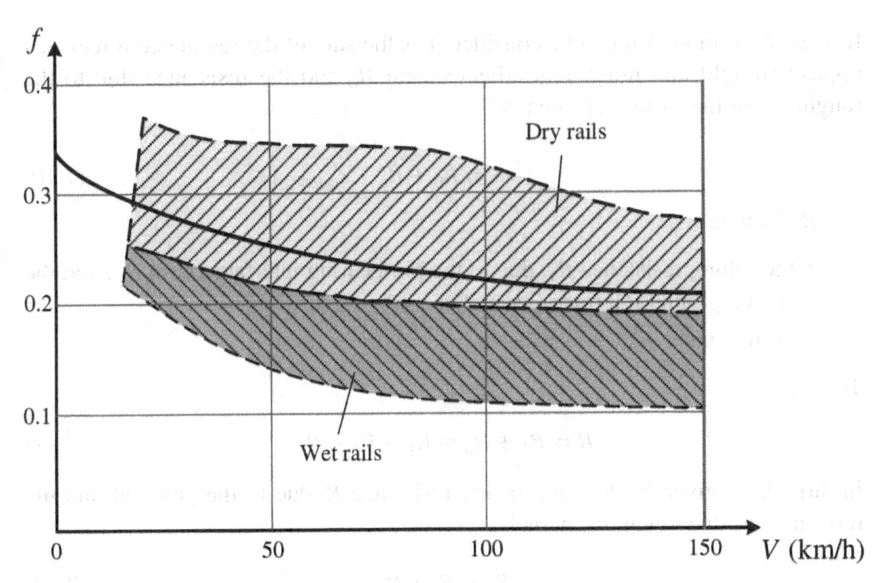

Figure 7.10 Adhesion coefficient f.

Müller's formula provides the result:

$$f = \frac{f_0}{1 + 0.011 \cdot v} (v \text{ in km/h})$$

where it can be assumed that $f_0 = 0.25$ for wet rails, $f_0 = 0.33$ for dry rails.

The values of f obtained from the work of Curtius and Kniffler are indicated in Figure 7.10. For locomotives with traction motors connected in parallel, such as those driven by single-phase AC, reliable adhesion coefficient data can be extrapolated from the continuous line curve that corresponds to the empirical expression:

$$f = \frac{7.5}{v + 44} + 0.161 \quad (v \text{ in km/h})$$

Roads The adhesion coefficient of tires is strongly influenced not only by speed but also by the type and conditions of the paved surface, the type of tread, and other factors. With very low speeds and for different degrees of roughness (the higher values are obtained with rough surfaces), f can vary from 0.60 to 0.85 for dry surfaces, from 0.40 to 0.65 for wet and (clean) surfaces, from 0.30 to 0.40 for muddy or soiled surfaces, and from 0.15 to 0.25 for oil-stained or icy surfaces.

7.12 RESISTANCE TO MOTION

Resistance to motion R can be considered as the sum of the resistance forces that oppose straight and horizontal advancement R_0 and the resistance due to the roughness of the surface R_e, that is,

$$R = R_0 + R_e \tag{7.10}$$

R_0 includes

- the rolling resistance R_1, due to the pin-bearing interface (term R_1') and the wheel–ground interface (term R_1'');
- air resistance R_2.

Therefore,

$$R = R_1 + R_2 = R_1' + R_1'' + R_2 \tag{7.11}$$

In turn, R_e is given by the sum of the resistance R_i due to the gradient and the resistance R_c due to curves, namely,

$$R_e = R_i + R_c \tag{7.12}$$

In addition to the resistant forces R, frequently the relative resistances are also considered (Section 7.3):

$$r = {}^{R}\!/_{G} \tag{7.13}$$

in the cases for which the R terms are proportional to the weight G of the vehicle.

Resistance due to the Pin-Bearing Interface The resistant torque due to pivot-bearing friction is $F_T \cdot d/2$, where d is the diameter of the pin and F_T is the tangential friction force. The latter is given, in turn, by $F_T = f' \cdot G'$, where f' is the coefficient of friction and G' is the vertical load on the bearing.

This resistant torque is matched by the resistance to motion R'_1; if D is the diameter of the running wheels, the following equivalence applies

$$f' \cdot G' \cdot {}^{d}\!/_{2} = R'_1 \cdot {}^{D}\!/_{2}$$

which yields the resistance

$$R'_1 = \frac{d}{D} \cdot f' \cdot G'$$

If, for simplicity, we neglect the difference between the load on the bearing G' and that on the ground G, the expression becomes

$$R'_1 = \frac{d}{D} \cdot f' \cdot G \tag{7.14}$$

Since the R'_1 is proportional to the weight G, the relative resistance can be considered

$$r'_1 = \frac{R'_1}{G} = \frac{d}{D} \cdot f' \tag{7.15}$$

In railway vehicles, normally $d/D = 1/8–1/10$.

The coefficient f' for rolling bearings, in normal use today, is on the order of 0.01; hence, the relative resistances R'_1 are typically in the range $1–1.5 \times 10^{-3}$, slightly increasing as speed increases.

Resistance due to the Wheel–Ground Interface It depends on the following causes:

1. *Rolling friction.* During motion, the pressure distribution in the contact footprint is modified, as shown in Figure 7.7; the vertical ground reaction $-\vec{G}$ will move forward in the direction of motion, giving rise to

a resistance torque $G \cdot \delta$, equal to R_1'', $D/2$. The resistance to motion can then be derived as

$$R_1'' = \frac{2 \cdot \delta}{D} \cdot G \tag{7.16}$$

and the relative resistance is

$$r_1'' = \frac{R_1''}{G} = \frac{2 \cdot \delta}{D} \tag{7.17}$$

2. *Swaying motion.* In rail vehicles, since there is some measure of unavoidable clearance between the wheel flanges and the rails, this may give rise to oscillatory lateral movement that, besides causing a disturbance and degradation of ride quality, also gives rise to an additional resistance to motion due to the resulting slippage between wheels and rails.
3. *Shock.* The velocity vector $\vec{v}$ may be subject to sharp variations caused by surface irregularities such as the rail joints. The corresponding impact kinetic energy, equal to half the product of the relevant impacting mass and Δv^2, is dissipated, giving rise to an additional resistance that equals the ratio between said kinetic energy and the average path between two successive shocks. Only the unsprung mass is considered, that is, that which bears directly on the ground (Section 7.3).

   In the case of railways, R_1'' takes the additional resistances caused by the swaying and shock effects into account and it increases with velocity, but is the least significant component of R_1. The relative resistance r_1'' is less than 0.001.

   For road vehicles, however, r_1'' dominates with respect to r_1' and can take on values greater than 0.010.

Rolling Resistance When driving or braking toques are not acting on a wheel, the distribution of forces is as shown in Figure 7.11. In addition to the vertical load $\vec{G}$, there is also the ground reaction which, as said in Section 7.6, moves in the direction of motion with respect to the force $\vec{G}$ (Figure 7.7). It can be assumed that this reaction is given by the vector $-\vec{G}$ and the resistance torque R_1'', $D/2$. The wheel is subject to the resistance torque caused by friction in the pivot-bearing interface, corresponding to resistance R_1' and is proportional to the load G' borne by the pin. For simplicity, consider a single resistive torque T_1 as the sum of the previous two, proportional to the vertical load G and corresponding to the overall rolling resistance, then

$$R_1 = \frac{T_1}{D/2} = R_1' + R_1'' \tag{7.18}$$

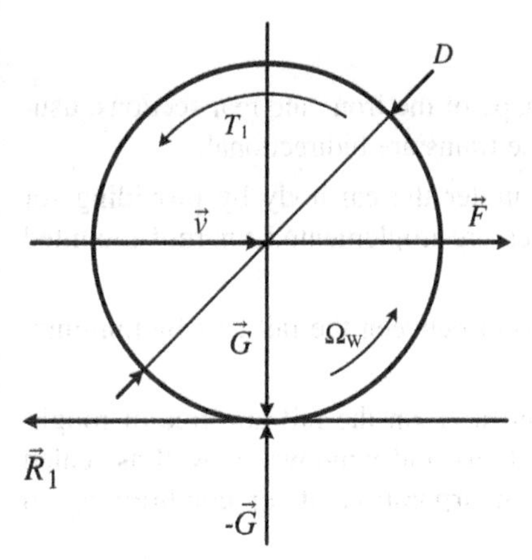

Figure 7.11 Load-bearing wheel forces.

The relative resistance is

$$r_1 = {}^{R_1}\!/_G = r_1' + r_1'' \tag{7.19}$$

To counteract $\vec{R}_1$, a horizontal driving force $\vec{F}$ with modulus: $F = R_1$ must be applied to the spindle. This represents the hauling action exerted by the bogie or the vehicle; it is evident that the torque formed by the forces $\vec{R}_1$ and $\vec{F}$ are equal and opposite to T_1.

What has been illustrated for a single wheel also applies to a wheelset.

With a good approximation, R_1 can be considered to have negligible variation in relation to velocity.

7.13 AIR RESISTANCE

A moving vehicle will encounter frontal resistance in its motion that is caused by the medium it moves in; this resistance is proportional to the area of its cross section, a shape coefficient and, with good approximation, the square of its speed.

This frontal resistance combines with others due to the roughness of the sides and suction intakes at the rear of the vehicle, as well as air turbulence throughout the bogie structure and in the space between the floor and the ground. It is important to also consider the vortexes that tend to form between the many railcars of a train and the effects of wind drag.

Therefore, the total resistance R_2, proportional to v^2, does not depend on the weight of the train, but only on its shape, its cross section, and its overall length.

As the forward velocity increases, R_2 becomes an increasingly greater component of the total combined drag force R_0; its effect becomes dominant at high speeds, mandating a series of measures for its reduction.

Typical such measures include

- precise study of the aerodynamic shape of the front and rear sections, usually identical in rolling stock because trains are bidirectional;
- reducing turbulence in the spaces under the car-body by providing for deflective fairings. Similar remedies are implemented for roof-mounted equipment;
- reduction of vortexes that tend to form between the rail cars by minimizing external wall discontinuities;
- implementation of smooth lateral surfaces on the railcars, free of roughness as far as possible, with flush doors and windows as well as sealed windows. The latter requires the incorporation of air conditioning in coach design.

Remarkable results have been achieved in implementing such measures for high-speed rolling stock, as will be highlighted in the next section.

It is worth mentioning that aerodynamic phenomena in high-speed trains not only give rise to a considerable increase of the resistance R_2 but they also cause inside and outside pressure variations and turbulence when trains encroach and pass through tunnels. In order to reduce these effects, high-speed rail coaches are internally pressurized and tunnels are built with wider wall clearances compared with traditional lines.

7.14 RESISTANCE TO FORWARD MOTION

For the estimation of straight-line resistance to forward motion R_0 on a horizontal plane as defined by (7.11), it is common practice to employ empirical formulas that give values of R_0 as a function of velocity, for different types of vehicles, with expressions such as:

$$R_0 = a + b \cdot v + c \cdot v^2, \quad \text{or} \quad R_0 = a + c \cdot v^2$$

where the magnitudes of a, b, and c are appropriately selected according to the nature of the rolling stock and type of train.

Sometimes to achieve greater approximation the resistances to forward motion related to the locomotive R_{0L} and the hauled rolling stock R_{0R} are considered separately. R_{0L} is significantly influenced by the term R_2; the following binomial formulas may be used, which are valid for modern locomotives with reasonable provision for aerodynamic criteria (V in km/h; R in N; G_L = weight of the locomotive, in kN):

- $R_{0L} = 4.5 \cdot G_L + 0.46 \cdot V^2$, for four-wheelset locomotives weighing between 800 and 880 kN.

- $R_{0L} = 4.2 \cdot G_L + 0.72 \cdot V^2$, for six-wheelset locomotives weighing between 1100 and 1200 kN.

For passenger trains with a hauled load G (kN), the applicable formulas are of the form:

$$R_{OR} = (a + b \cdot V) \cdot G + c \cdot V^2$$

The size of c depends on the type of carriage and their number, which affects the resistance R_2. In some cases, the influence of the term V^2 is accentuated, which also takes the effect of wind into account, setting $(V + V_0)^2$ in place of V^2, with $V_0 = \text{const}$. A typical formula that can be recalled is

$$R_{OR}(\text{N}) = (1.9 + 0.0025 \cdot V) \cdot G(\text{kN}) + 4.7 \cdot (n + 2.7) \cdot A \cdot \left(\frac{V + 15}{10}\right)^2$$

where n is the number of the carriages and A is the equivalent cross section.

Frequently, it is simpler to consider the relative resistance

$$r_0 = {}^{R_0}\!/_G = r_1 + {}^{r_2}\!/_G \tag{7.20}$$

which is independent of weight G. In theory, this would not be permitted because, as is known, R_2 is not proportional to G; however, acceptable results are obtained for homogeneous compositions by appropriately choosing the numerical value of the various factors. For example, for modern passenger trains (with V in km/h) it can be assumed that

$$r_0(\text{N/kN}) = (1.25\ -2.0) + (0.016\ -0.025) \cdot \left(\frac{V}{10}\right)^2$$

for normal freight trains, with railcars bearing average loads:

$$r_0(\text{N/kN}) = (1.5\ -2.5) + (0.05\ -0.06) \cdot \left(\frac{V}{10}\right)^2$$

for freight trains composed of special wagons:

$$r_0(\text{N/kN}) = (1.5\ -2.0) + (0.024 - 0.040) \cdot \left(\frac{V}{10}\right)^2$$

In the latter case, the unitary resistance is reduced; if the load is considerable, then

$$r_0(\text{N/kN}) = 1.2 + (0.020\ -0.025) \cdot \left(\frac{V}{10}\right)^2$$

At the onset of motion (under inrush driving power conditions), freight trains have additional resistances to motion that are not negligible compared with the values indicated above. These can be on the order of 5 N/kN at speeds ranging between 0 and 10 km/h.

The relative resistance to forward motion is taken into account, for simplicity, also for complete trains, provided they are of homogeneous composition, for example, electric locomotives and railcars, or also high-speed trains consisting of aerodynamic head and end locomotives and a number of intermediate railcars. Indeed, with the variation of the composition of these special categories of rolling stock, there is a corresponding variation, although not proportionally, of both the mass and the portion of the resistance R_2 attributable to lateral forces.

For the lightweight rail contexts such as tramways, metros, and regional commuter rail lines without particular aerodynamic features, given the relatively limited speeds involved the following expression can be used (with V in km/h):

$$r_0(\text{N/kN}) = (2.5 \ - 3.0) + 0.04 \cdot \left(\frac{V}{10}\right)^2$$

The relative resistances to forward motion of HS aerodynamic trains are calculated with the formula

$$r_0(\text{N/kN}) = (0.6 \ - 1.5) + (0.007 \ - 0.022) \cdot \frac{V}{10} + (0.0125 \ - 0.017) \cdot \left(\frac{V}{10}\right)^2$$

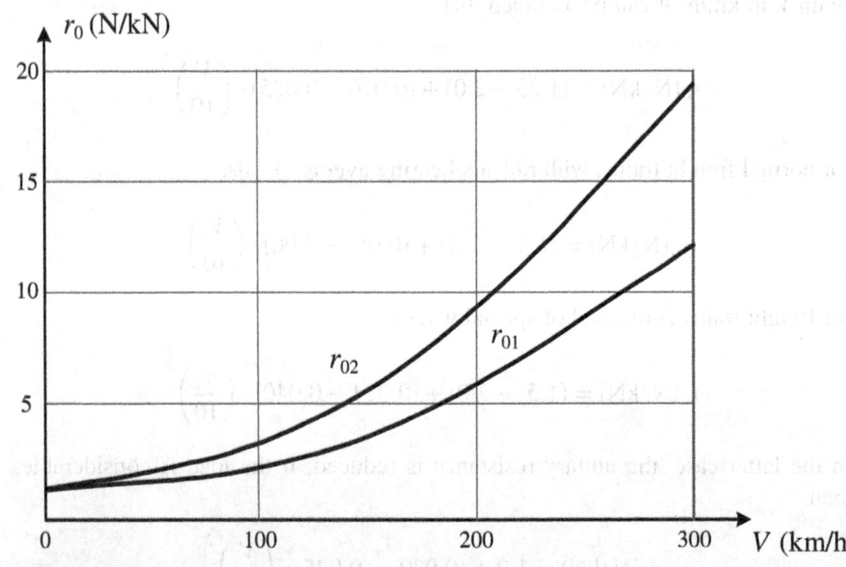

Figure 7.12 Values for relative resistance to horizontal motion in a straight line, in the open (r_{0a}) and in tunnels (r_{0g}), acquired by FS for ETR 500 HS electric trains.

During transit through tunnels, air resistance R_2 increases considerably; in fact, with the tunnel widths in use today, it approximately doubles.

The expressions above confirm the importance of studying the shape profiles of high-speed rolling stock as mentioned in Section 7.13; in fact, at 300 km/h the resistance R_2 actually constitutes 90–96% of the total R_0. The improvements applied to HS rolling stock have reduced the values of r_0 compared with those given by the previous expression; Figure 7.12 illustrates the curves obtained by the FS analysts, which can be interpreted by means of the binomial formulas:

- in the open: $r_{0a}(\mathrm{N/kN}) = 1 + 0.0125 \cdot \left(\frac{V}{10}\right)^2$
- in tunnel: $r_{0g}(\mathrm{N/kN}) = 1 + 0.0207 \cdot \left(\frac{V}{10}\right)^2$

The ratio r_{0g}/r_{0a} varies from 1.5 to 1.6 as speed varies from 120 to 300 km/h.

7.15 INCIDENTAL RESISTANCES

Gradient Resistance The component of the weight $\vec{G}$ that is parallel to the road is (Figure 7.13)

$$R_i = G \cdot \sin\alpha = G \cdot \frac{\mathrm{tg}\,\alpha}{\sqrt{1 + \mathrm{tg}^2\alpha}} \tag{7.21}$$

Considering that the gradient is given by: $i = \mathrm{tg}\,\alpha$, we have

$$R_i = \frac{i \cdot G}{\sqrt{1 + i^2}}$$

which yields the relative resistance

$$r_i = \frac{R_i}{G} = \frac{i}{\sqrt{1 + i^2}}$$

For railways with naturally adhesion, the maximum gradient is on the order of 25–35‰. For secondary lines, the gradient may exceptionally extend to

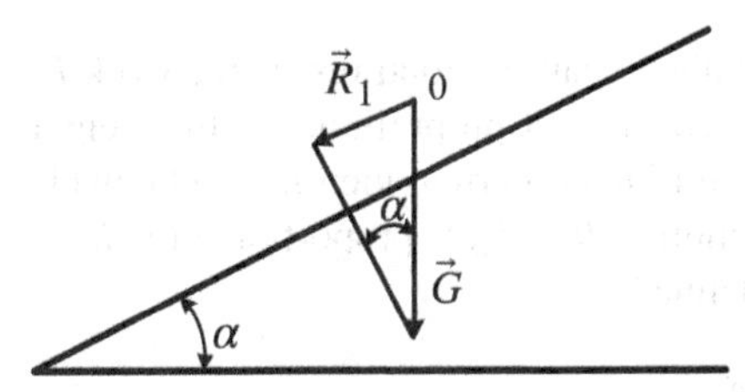

Figure 7.13 Components of *R*.

50–60‰. In this case, the α angle is so small, that by approximation

$$R_i \approx G \cdot \text{tg}\ \alpha$$

hence

$$R_i = i \cdot G; \quad r_i = i \tag{7.22}$$

Steeper gradients are achievable in road traction and in artificial adhesion railways; in such cases, for r_i the full expression (7.21) must be used.

Resistance due to Cornering When rolling stock negotiates a curved section of track, the cornering resistance R_c depends on several factors; the first of these is the parallel nature of the wheelsets that does not allow the wheels to have a pure rolling motion. The relative sliding that occurs between wheel and rail gives rise to lost work due to friction and, therefore, to a resistant force.

Moreover, since the two wheels on the same wheelset are integral with each other, while the vehicle travels through a curved bend of mean radius ρ or sliding will occur due to the difference between the paths traveled by the outer and inner wheels, this path difference is given as

$$\Delta = \phi(\rho + s) - \phi(\rho - s) = 2 \cdot \phi \cdot s$$

where $2s$ is the track gauge and ϕ is the partial sector circumference of the curve, measured in radians. The corresponding lost work to friction gives rise, as above, to a resistant force. It should be noted that the semiconical taper with which the rims are constructed should compensate, within certain limits, the path difference Δ, thanks to the play allowed between flange and rail; in fact, the wheelset will naturally "hunt" for an equilibrium position in which the outer wheel has a greater rolling diameter than the inside one. Ideally, for standard gauge rail ($2s = 1435$ mm) with curvature $\rho > 300$–350 m, the bends should be banked. In reality, some degree of sliding occurs in any case due to the rigid locking of the wheels on the wheelset axle.

Finally, some consideration must be given to the transverse forces that the lateral surface of the rail head exerts on the wheel flange, by means of which the vehicle is driven when cornering; sliding occurs in the flange/rail contact interface, which gives rise to friction work.

The various contributions considered above cause a total resistance work L_c that is essentially friction and that can be considered proportional to the weight G of the convoy. If S is the space traveled during cornering, then L_c can be made to correspond to an average resistance to motion $R_c = L_c/S$, proportional to G.

Thus, the relative resistance can be defined

$$r_c = {}^{R_c}\!/_{G}$$

Table 7.1 Cornering Resistance

	$10^3 \cdot r_c$		
ρ (m)	Railway company	Von Röckl	$0.8/\rho$
250	3.4	3.5	3.2
300	2.8	2.8	2.65
350	2.4	2.2–2.3	2.3
400	2.0	1.9	2.0
500	1.5	1.5	1.6
800	0.8	0.9	1.0
1000	0.5	0.7	0.8

Analytic evaluation of r_c is considerably difficult; hence, it is common practice to consult experimental formulas, which include those of Von Röckl

$$r_c = {}^{a}\!/_{(\rho - b)} \tag{7.23}$$

in which the terms a and b take on the following values:

- for standard gauge ($2s = 1435$ mm):

  $a = 0.65$ m, $b = 55$ m for $\rho \geq 350$ m; $b = 65$ m for $250 < \rho < 350$ m;

- for narrow gauge ($2s = 1000$ mm):

  $a = 0.50$ m; $b = 30$ m.

Others use simple formulas, such as

$$r_c = {}^{c}\!/_{\rho} \tag{7.24}$$

setting $c = 0.8$ m for standard gauge.

Some railway companies adopt the values reported in Table 7.1, alongside that are indicated, by way of comparison, those given by formulas (7.23) and (7.24).

In road vehicles, the driving axles always incorporate differentials; hence the additional cornering resistance is negligible in the context of the overall resistance to forward motion.

Incidental Resistance Incidental relative resistance due to the gradient (i) and planimetric variations is given by

$$r_e = {}^{R_e}\!/_{G} = {}^{(R_i + R_c)}\!/_{G} = i + r_c$$

When traveling downhill, the resistance R_i becomes a driving force; by convention, it is considered as a negative resistance and the generic expression becomes

$$r_e = \pm i + r_c \tag{7.25}$$

where i takes on the positive or negative sign according to whether the gradient is respectively uphill or downhill. A fictitious gradient is frequently referenced

$$i' = r_e = \pm i + r_c$$

which simultaneously takes the gradient and the additional cornering resistance into account.

7.16 OVERALL RESISTANCES

The total resistance to motion can be expressed as the sum of all terms proportional to the weight G, except for R_2, that is,

$$R = r_1 \cdot G + R_2 + (\pm i + r_c) \cdot G$$

Within the approximation constraints noted in Section 7.14, it is possible to establish an overall relative resistance that varies with velocity v

$$r(v) = {}^{R}\!/_{G} = r_0(v) \pm i + r_c \tag{7.26}$$

assuming proportionality between the total resistance and G, then

$$R = r \cdot G = (r_0 \pm i + r_c) \cdot G \tag{7.27}$$

7.17 TRACTIVE EFFORT DIAGRAM OF TRACTION VEHICLES

As already mentioned, the active force F that a driving wheelset develops due to its driving torque is transmitted tangentially to the ground through the contact footprint, called wheel tractive force.

The curve which characterizes the total wheel tractive force as a function of the forward velocity constitutes the mechanical characteristic of the traction vehicle (Figure 7.14).

The developed mechanical power, that is, the power to the wheels, is given by

$$P_c = F \cdot v \tag{7.28}$$

Expressing F in kN and V in km/h, P_c (kW) $= F \cdot V/3.6$.

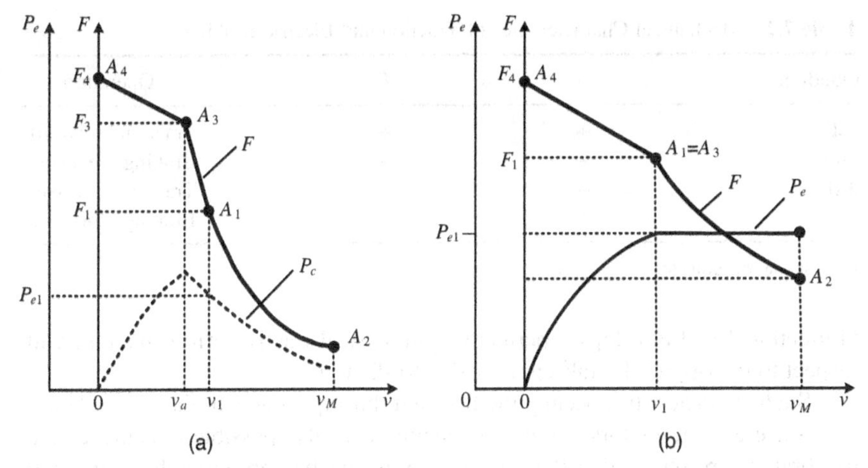

Figure 7.14 Mechanical characteristic of a traction vehicle. (a) Traditional drives. (b) Three-phase electronic drives.

In traditional driving vehicles equipped with commutator motors with series excitation, the mechanical characteristic has the typical curve in Figure 7.14a, where the following can be noted:

- A_1 velocity v_1 force F_1 – operation of the electric motors at rated levels, corresponding to power delivered to the wheels as $P_{c1} = F_1 \cdot v_1$.
- A_2 maximum velocity v_M – operation of the motors at the maximum permitted velocity.
- A_3 velocity v_a force F_3 – limit point for use of the characteristic $F(v)$.

For $v < v_a$ the value F_a of the tractive force varies from F_3 to the maximum F_4 (point A_4; $v = 0$); as a limiting condition $F_a = F_3 = const$ within the start-up interval $0 - v_a$. The section of the characteristic $A_4 - A_3$ is obtained by a suitable regulation of the traction electric drive and must be compatible with

- the adhesion limit, that is, $F_a \leq f \cdot G_{\text{ad}}$;
- the maximum values of torque that the traction motors can develop.

In the section $A_4 - A_3 - A_1$, the motors work in overload, that is, for a limited service interval; the power, equal to the nominal value at the point A_1, decreases as the velocity increases from v_1 to v_M. Traction vehicles with electronic drive and three-phase motors have a mechanical characteristic of the type shown in Figure 7.14b: points A_3 and A_1 effectively coincide and throughout the entire velocity range $v_1 - v_M$ the power $P_{e1} = \text{cost.}$ is developed.

Reversibility In vehicles driven by electric motors, it is easy, as will be shown later, to reverse the direction of motion; that is, the drive is inherently

Table 7.2 Mechanical Characteristics of Traction and Electric Braking

Quadrant	v	F	Operation
1st	+	+	Traction "forward"
2nd	−	+	Braking "reverse"
3rd	−	−	Traction "reverse"
4th	+	−	Braking "forward"

v: Abscissa; F: ordinate.

bidirectional and develops a perfectly symmetrical tractive effort diagram with respect to the origin of v (abscissa) and F (ordinate).

Such characteristics occupy the first and third quadrants in the $v-F$ plane.

Since electric motors are also reversible, it is also possible to achieve electric braking, provided that the power circuits can be appropriately configured. For a given direction of motion, the reversal from electrical traction to regenerative braking corresponds to the inversion of the sign of the force F; therefore, in this case the tractive effort diagram occupies all four quadrants of the $v-F$ plane, as shown in Table 7.2.

The Driving Wheel As indicated in Figure 7.15, in addition to the vertical forces $\vec{G}$ and $-\vec{G}$ applied to a driving wheel, the following are also applied:

- The drive torque T, corresponding to the tractive force on the wheel rim:

$$F = \frac{T}{D/2}$$

- The adhesion force, transmitted between the wheel and the ground, given by $\vec{F}_a = \vec{F} - \vec{R}_1$ (representing the force that the ground exerts on the wheel).

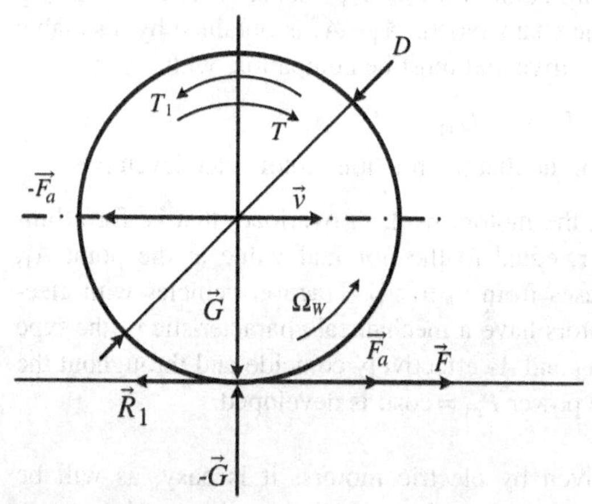

Figure 7.15 Forces applied to a driving wheel.

- The resisting torque T_1, corresponding to the resistance $\vec{R}_1$.
- The horizontal force $-\vec{F}_a$, applied to the spindle, which is the reaction due to the resistance to motion and the forces of inertia relative to the vehicle and any hauled rolling stock.

The condition for adhesion is thus: $F_a \leq f \cdot G$, for simplicity R_1 is usually considered as included in $-\vec{F}_a$, hence the adhesion condition becomes $F_a \leq f \cdot G$.

7.18 DETERMINING THE MECHANICAL CHARACTERISTIC

Consider a vehicle provided with N identical traction motors (of the commutator type with series excitation), having the mechanical torque/rotational speed characteristic of Figure 7.16, where the points A_1 $A_{2'}$ $A_{3'}$ A_4 correspond to those considered in Figure 7.14a. The motors normally transmit motion to the axes through gears; ρ represents the gear transmission ratio

$$\rho = \Omega / \Omega_W \tag{7.29}$$

where Ω and Ω_W are the angular velocities of the motor and the wheels of diameter D. Usually, the mechanical transmission is such that $\Omega > \Omega_W$, that is,

$$\rho > 1$$

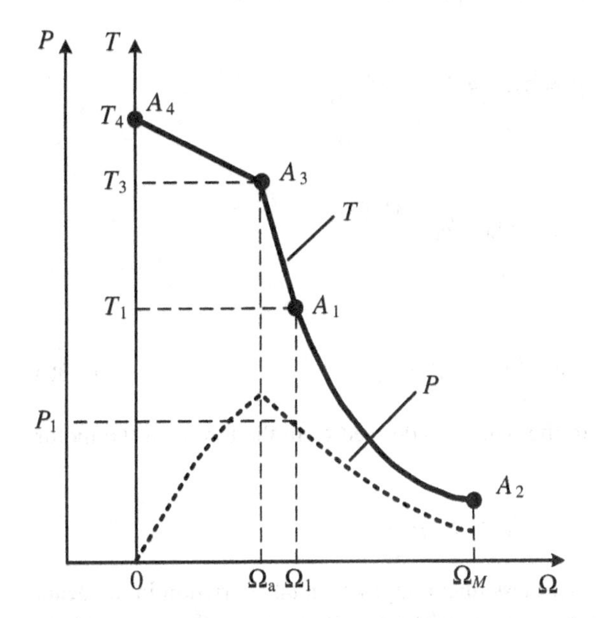

Figure 7.16 Mechanical characteristic of a traction commutator motor with series excitation.

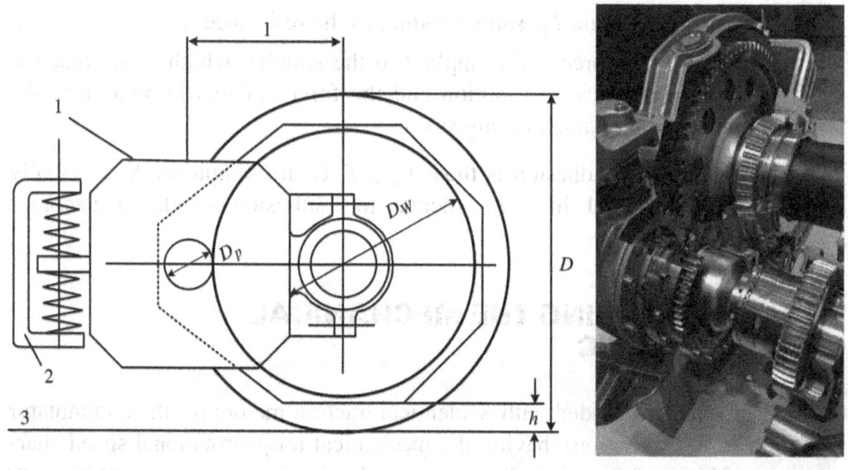

Figure 7.17 Diagram of a transmission unit (traction motor in semirigid suspension). 1: Traction motor; 2: bogie frame; 3: top of rail; *D:* wheel rolling diameter; D_W: pitch diameter of the gear wheel; D_p: pitch diameter of the pinion/wheelbase; *h*: clearance above the top of rail.

When the motor is directly connected to the wheelset, then $\rho = 1$; it is referred to as "direct drive" transmission.

The simplest transmission structure consists of a pair of cylindrical gears, of modulus m' with the pinion having Z_p teeth and pitch diameter $D_p = m' \cdot Z_p$ and driven gear wheel having Z_W teeth and pitch diameter $D_W = m' \cdot Z_W$ (Figure 7.17). In this case

$$\rho = {}^{\Omega}/_{\Omega_W} = {}^{D_W}/_{D_p} = {}^{Z_W}/_{Z_p}$$

The travel speed is

$$v = \Omega_W \cdot {}^{D}/_{2} = \frac{\Omega \cdot D}{2\rho}$$

Letting $k_\Omega = {}^{2\rho}/_{D}$, then

$$v = {}^{\Omega}/_{k_\Omega}; \quad \Omega = k_\Omega \cdot v \tag{7.30}$$

The driving torque applied to the wheels is derived from the torque at the motor shaft with the relationship

$$T_W = \eta_g \cdot \rho \cdot T$$

where η_g represents a coefficient less than unity, which takes friction in the gears into account and corresponds to the mechanical efficiency of the gear unit; in fact, for traction it is expressed as

$$\eta_g = \frac{\Omega_W \cdot T_W}{\Omega \cdot T} = \frac{T_W}{\rho \cdot T}$$

The tractive force to the wheel rims for N driving wheelset is then given by

$$F = N \cdot \frac{T_W}{D/2} = \eta_g \cdot 2 \cdot N \cdot \frac{\rho \cdot T}{D}$$

Let

$$k_W = 2N \cdot \frac{\rho}{D}$$

then

$$F = \eta_g \cdot k_W \cdot T \tag{7.31}$$

In determining the gear ratio ρ, the following must be considered:

- the maximum velocity of the vehicle v_M is linked to the maximum motor speed by the relationship

$$\Omega_M = k_\Omega \cdot v_M \tag{7.32}$$

- the maximum tractive effort on the wheel rims F_M is related to the maximum torque T_M that the motor can develop by the relationship

$$F_M = \eta_g \cdot k_W \cdot T_M$$

The values k_W and k_Ω cannot be chosen arbitrarily. In fact, for a given wheel diameter D, the following dimensional constraints apply (Figure 7.17):

1. The distance h between the gear unit housing and the top of the rail cannot be less than the specific minimum clearance allowed for the given track.
2. The spacing between the gears

$$I = \frac{D_p + D_W}{2} = m' \cdot \frac{Z_p + Z_W}{2}$$

must be compatible with the bulk of the motor and the diameter of the wheelset.

3. The diameter of the pinion D_p cannot be less than certain limits related to the size of the motor shaft and the torque to be transmitted.

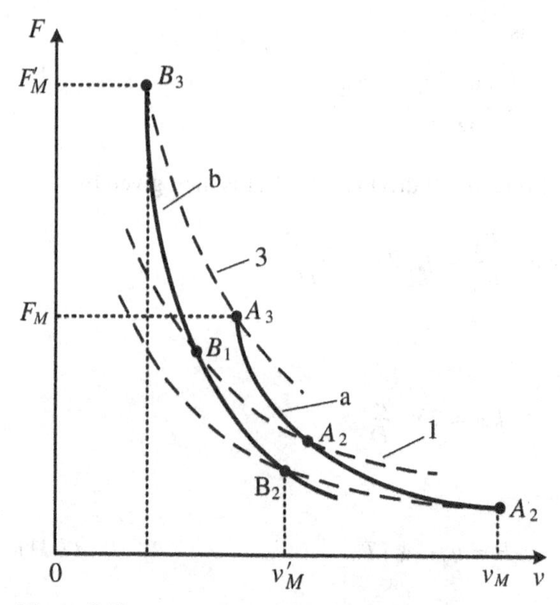

Figure 7.18 Tractive effort diagram of the same type of locomotive, with different gear ratios. *a*: For high-speed services; *b*: for heavy freight services; 1: hyperbola $P_{c1} = \text{const.}$; 3: hyperbola $P_{cm} = \text{const.}$

In practice, there is an upper limit for the ratio ρ/D, and, hence, for the coefficients k_Ω and k_W. In special cases, it is necessary to have high levels of k_Ω and k_W, in this case a double gear reduction is implemented with gear ratios ρ_1 and ρ_2 in series, for which $\rho = \rho_1 \cdot \rho_2$.

To adapt locomotive performance (hauling capacity and speed) to various service requirements (passenger trains: high v; freight trains and shunting: high F), different gear ratios can be established, so as to have machines that, while belonging to the same type, are specialized for different services (Figure 7.18).

There have also been cases where onboard mechanisms were installed for changing the transmission ratio, of course this would occur with the vehicle halted.

These devices are no longer needed, as the mechanical characteristic of today's locomotives with electronically controlled three-phase motors cover, in the field $v_1 - v_M$ at constant power (Figure 7.14b), the full range of services. This operational "flexibility" is important for a traction vehicle from the point of view of its use, and, among other things, allows for appreciable type unification.

7.19 VARIATIONS IN WHEELSET LOAD

A locomotive (or EMU) capable of developing a total traction at the wheel rims $\vec{F}$ exerts a force $\vec{F}_g$ at the coupler on the hauled railcars that is equal to the

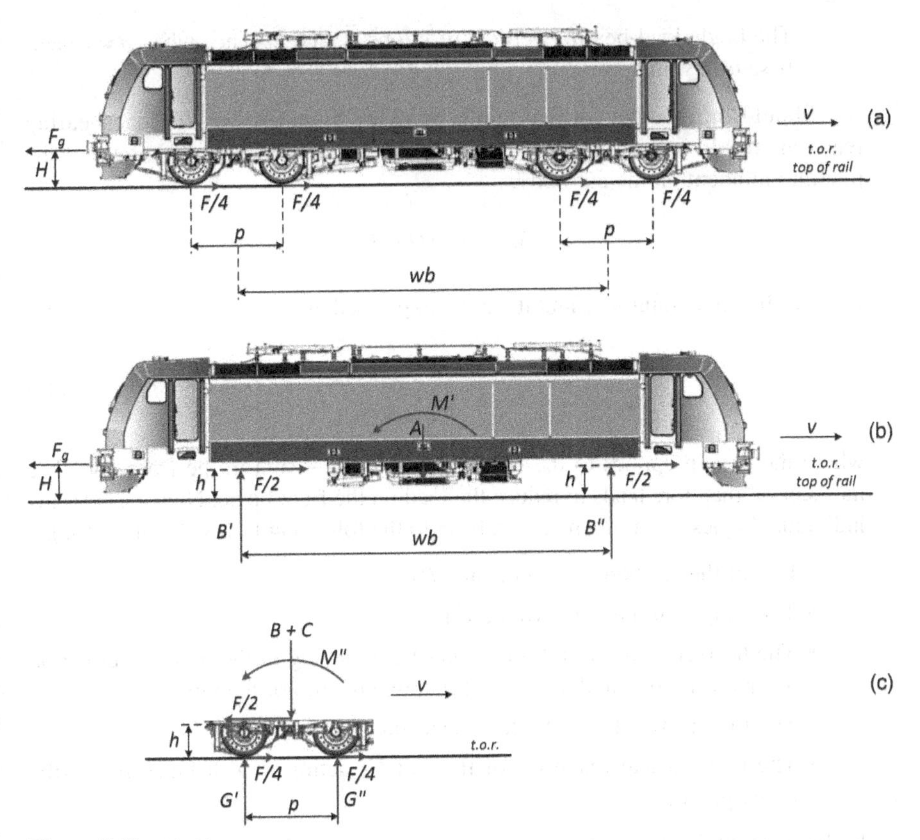

Figure 7.19 Pitching moments.

difference between $\vec{F}$ and the resistances to motion $\vec{R}_L$ of the locomotive alone, that is, $\vec{F}_g = \vec{F} - \vec{R}_L$. Naturally, the hauled railcars exert a force $-\vec{F}_g$ on the locomotive coupler in the opposite direction to that of motion. Suppose now that the traction unit is structured on bogies with wheel arrangement B_0B_0 (Figure 7.19).

On the car-body, if we neglect the resistances to motion of the locomotive for simplicity, allowing $F_g \approx F$, the following forces are applied (Figure 7.19b):

- Its own weight $\vec{A}$.
- The reaction at the coupler of value $F_g = F$ applied at height H above the top of the rails. H is normally equal to 1–1.1 m and has a constant value for a given railway network, or even at an international level, to ensure the exchange of traction vehicles and rolling stock.
- The two tractive forces $\frac{\vec{F}}{2}$ that the bogies transmit to the car-body, applied at the point of car-body/bogie interface, at a height h above the top of the rail.

- The bogie load-bearing vertical reactions $\vec{B}$, at a distance wb (wb = wheel-base of a vehicle).

Should the tractive force be null, applying symmetry to the load-bearing reactions would indicate that $B_0 = A/2$. As a result of the moment, known as the pitching moment on the car-body,

$$M' = F \cdot (H - h)$$

It varies by an amount $\pm q_1$ and it can be expressed as

$$\begin{aligned} M' &= q_1 \cdot wb = F \cdot (H - h); \quad q_1 = F \cdot \frac{H - h}{wb} \\ B &= B_0 \pm q_1 = {}^{A}\!/_{2} \pm q_1 \end{aligned} \tag{7.33}$$

where the + sign applies to the rear bogie and the − sign to the front one. The moment M' therefore tends to reduce the load on the front bogie. Now consider the individual bogies, each of which is subject to the following forces (Figure 7.19c):

- Part of the car-body weight (force $\vec{B}$).
- The weight of the bogie (force $\vec{C}$).
- The horizontal reaction $\frac{\vec{F}}{2}$ of the car-body, acting in the opposite direction to that of motion and at a height h above the top of the rail.
- The two tractive forces to the wheel rims $\frac{\vec{F}}{4}$.
- The two vertical reactions of the rail $\vec{G}$, acting at a distance p (p = the bogie pitch).

In the event of $F = 0$, by symmetry, the load on each wheelset would correspond to a quarter of the weight G_1 of the vehicle; hence,

$$G_0 = {}^{B_0}\!/_{2} + {}^{C}\!/_{2} = {}^{A}\!/_{4} + {}^{C}\!/_{2} \tag{7.34}$$

Due to the effect of the pitching moment

$$M'' = \frac{F}{2} \cdot h$$

the rear wheelset would be loaded more than the front one; setting $\pm q_2$ as the corresponding variations of the rail vertical reactions, then

$$\begin{aligned} M'' &= q_2 \cdot p = \frac{F}{2} \cdot h; \quad q_2 = \frac{h}{p} \cdot \frac{F}{2} \\ G &= {}^{B}\!/_{2} + {}^{C}\!/_{2} \pm q_2 \end{aligned}$$

(+ for the rear wheelset, − for the front).

Based on (7.33) and taking into account (7.34),

$$G = {}^{A}\!/_{4} \pm {}^{q_1}\!/_{2} + {}^{C}\!/_{2} \pm q_2 = G_0 \pm {}^{q_1}\!/_{2} \pm q_2 \tag{7.35}$$

Then the load variations on the wheelsets, depending on the couples M' and M'', in general terms can be expressed as

$$G = G - G_0 = \pm^{q_1}/_2 \pm q_2 = \frac{F}{2}\left(\pm\frac{H-h}{wb} \pm \frac{h}{p}\right) \tag{7.36}$$

where the signs are as follows: −, − for the first wheelset; −, + for the second; +, − for the third; +, + for the fourth.

Concerning adhesion, the worst situation is for the first wheelset to be the less loaded one; the corresponding absolute value of its change in load results as

$$\overline{q} = \frac{F}{2}\left(\frac{H-h}{wb} \pm \frac{h}{p}\right) \tag{7.37}$$

The influence of the bogie (couple M'') is normally greater than that of the car-body (couple M'), given that the wheelbase wb is much longer than the bogie pitch p. When designing the mechanical parts of traction vehicles, a number of specific precautions must be taken to reduce the value of $\overline{q}$. The design of modern locomotives has seen the introduction of the so-called low-traction solution; according to this approach, the direction of application of car-body/bogie hauling force is very low compared with the rail level; in other words, height h is kept small. In this way, the influence of the bogie (which is dominant) is greatly reduced; despite the slight increase of the influence of the car-body, the load transfer away from the first wheelset is reduced.

The same considerations can be applied for braking, with the caveat that the braking force applied to the wheels is directed in the opposite direction to that of motion, so that there is less load transferred away from the rear bogie; this applies particularly to the last wheelset of the vehicle.

If the resistance to motion of the traction vehicle is not negligible compared with the traction at the coupler, or if significant inertia forces should intervene during the various motion phases, then these forces must also be taken into consideration.

Ultimately, for an isolated locomotive, the tractive force at the coupler would be null, hence the only applicable forces would be the inertia forces and the resistance to motion.

7.20 THE TRACTION DIAGRAM

Consider a train of weight G in motion on a line with

$$i' = \pm i + r_c = \text{const}$$

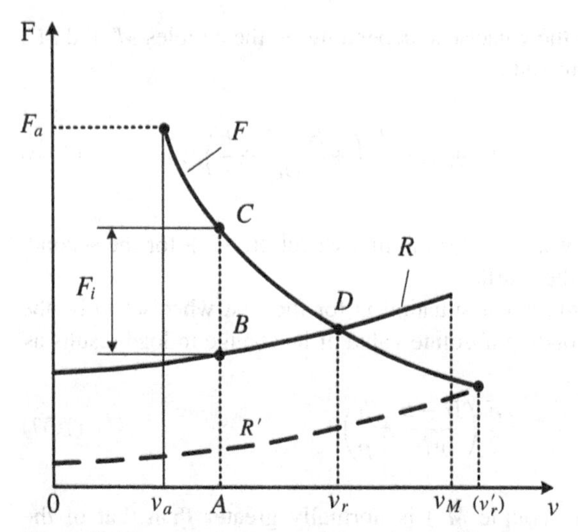

Figure 7.20 Mechanical characteristic $F(v)$ and resistance to motion $R(v)$.

The resistance to motion is given by (Figure 7.20)

$$R(v) = (r_0 \pm i + r_c) \cdot G$$

Since $F - R = m_e \cdot a$, the segment BC of Figure 7.20 represents the accelerating force $F_i = m_e \cdot a$, which in this case is positive, given that $F > R$.

The motion is therefore accelerated, with an acceleration given by

$$a(v) = \frac{F(v) - R(v)}{m_e} \tag{7.38}$$

which varies from a maximum for $v = 0$ to zero as the velocity reaches its steady state value v_r, for which

$$F = R; \quad a = 0; \quad v = v_r = \text{const}$$

The steady-state velocity depends on the value of the resistances to motion $R(v)$. For example, if there should be a reduction in gradient i (curve R' of Figure 7.20), then $v'_r > v_m$, in which case the traction should be reduced in order not to exceed the maximum allowed velocity v_m.

Given the definition $a(v) = dv/dt$, by integration

$$t = \int \frac{dv}{a(v)} + \text{const}$$

From which the velocity can be expressed as a function of time $v(t)$.

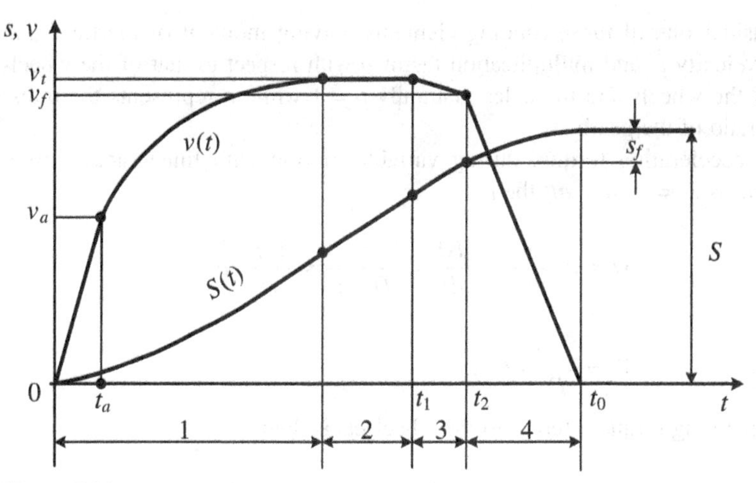

Figure 7.21 Traction diagram. 1: Start-up; 2: steady state; 3: coasting; 4: braking ($t_f = t_0 - t_2$).

Moreover, since $v(t) = ds/dt$, further integration yields the distance traveled as

$$s(t) = \int v(t) \cdot dt + \text{const} \tag{7.39}$$

The traction diagram (Figure 7.21) includes the curves $v(t)$ and $s(t)$

7.21 START-UP

Start-up Phases Start-up refers to those motion phases in which the velocity changes from zero to the steady-state value v_r. Essentially two phases can be distinguished (Figures 7.20 and 7.21)

- from 0 to v_a in which the tractive force is, on average, equal to F_a. Because the velocity is low, R does not vary significantly (negligible air resistance); the accelerating force $F_i = m_e \cdot a = F_a - R$ and hence the acceleration a remain essentially constant. As a characteristic value of the vehicle's start-up acceleration, the value a is assumed for this phase, since the condition of naturally accelerated motion can be considered valid;
- in the interval from v_a to v_t which is governed by the natural mechanical characteristic of the motors, the acceleration decreases gradually to zero.

Equivalent Mass The equivalent mass m_e takes the inertia of the rotating parts of the vehicle into account (motors, gearboxes, wheelsets), which require an additional start-up accelerating torque.

Consider one of these rotating elements, having moment of inertia J and angular velocity ω and multiplication factor ρ with respect to that of the wheelbase: for the wheels and the axles, naturally $\rho = 1$ while ρ represents the transmission ratio of the gearbox.

The accelerating torque, during variable motion with linear acceleration $a = dv/dt$, is $T' = J \cdot d\omega/dt$; then

$$\Omega = \rho \cdot \frac{v}{D/2}; \qquad \frac{d\Omega}{dt} = \frac{2 \cdot \rho \cdot dv}{D \cdot dt} = \frac{2 \cdot \rho}{D} \cdot a$$

$$T' = \frac{2 \cdot \rho}{D} \cdot J \cdot a$$

The accelerating torque referred to the wheelsets is then

$$T'' = \frac{2 \cdot \rho}{D} \cdot J \cdot a$$

And the force applied to the wheels is

$$F'' = \frac{T''}{D/2} = \frac{4 \cdot \rho^2}{D^2} \cdot J \cdot a$$

To the overall inertia forces $m \cdot a$ must be added all the terms F'', for which the total accelerating tractive force is given as an absolute value by

$$F_i = m \cdot a + \sum F'' = \left(m + \sum \frac{4 \cdot \rho^2}{D^2} \cdot J \right) \cdot a$$

For simplicity of calculation, the equivalent mass is defined as

$$m_e = \frac{F_i}{a} = m + \sum \frac{4 \cdot \rho^2}{D^2} \cdot J = m \cdot \left(1 + \frac{1}{m} \cdot \sum \frac{4 \cdot \rho^2}{D^2} \cdot J \right)$$

Therefore,

$$m_e = m \cdot (1 + \varepsilon), \quad \text{with} \quad \varepsilon = \frac{1}{m} \cdot \sum \frac{4 \cdot \rho^2}{D^2} \cdot J \tag{7.40}$$

For example, simplifying the wheel as a homogeneous disk of uniform thickness and with $\rho = 1$,

$$J = \frac{m \cdot (D/2)^2}{2} = \frac{m \cdot D^2}{8}; \quad \varepsilon = \frac{1}{m} \cdot \frac{4}{D^2} \cdot \frac{m \cdot D^2}{8} = \frac{1}{2}; \quad m_e = 1.5 \cdot m$$

In fact, it can take on the following values, referenced to the vehicle or the entire train:

- Locomotives

$$\varepsilon = 0.15 - 0.20$$

- EMUs, trolleybuses

$$\varepsilon = 0.09 - 0.11$$

- Coaches and railcars

$$\varepsilon = 0.03 - 0.06$$

- Complete trains

$$\varepsilon = 0.05 - 0.09$$

The coefficient ε is considerably higher for the traction vehicle, given the determining influence of the moments of inertia and the traction moments which rotate with angular velocities that are multiples of the wheelset angular velocities according to ρ.

Inertial Forces and Limit Acceleration According to (7.40), the accelerating force is given by

$$F_i = (1 + \varepsilon) \cdot m \cdot a \tag{7.41}$$

comparing its value with the rectilinear resistances to motion on a horizontal plane $R_0 = r_0 \cdot G = r_0 \cdot m \cdot g$, for a given mass m, it is evident that F_i is the dominant term. For example, let: $r_0 = 0.005$; $\varepsilon = 0.10$ and $a = 1 - 1.2\,\text{m/s}^2$, then $F_i/R_0 = 22.4 - 26.9$.

If the resistances to motion are neglected during the start-up phase, that is, $F \approx F_i$, then the adhesion conditions can be expressed as follows:

$$F_i \leq f \cdot \alpha \cdot G; \quad (1 + \varepsilon) \cdot m \cdot a \leq f \cdot \alpha \cdot m \cdot g$$

which provides the important expression

$$a \leq \frac{f \cdot \alpha}{1 + \varepsilon} \cdot g \tag{7.42}$$

which indicates the start-up acceleration as a function of both the average coefficient of adhesion and the adhesion ratio α.

7.22 THE DECELERATION AND BRAKING PHASE

Inertial Coasting When traction ceases and in the absence of braking, a vehicle is subject only to the overall resistance to motion. If there is no appreciable downhill gradient, that is, while the resistance $R = (r_0 + i + r_c) \cdot G$ is positive (gradient $i \geq 0$ or, if negative, having an absolute value of $r_0 + r_c$), then the vehicle's motion is said to be in the coasting mode and decelerating, therefore

$$a_c = -\frac{R}{m_c} = -\frac{(r_0 + i + r_c) \cdot G}{(1 + \varepsilon) \cdot m} = -\frac{r_0 + i + r_c}{(1 + \varepsilon)} \cdot g \tag{7.43}$$

For flat and straight railways ($i = r_c = 0$; $r_0 \approx 0.005$), deceleration a_c is on the order of $4 - 5\ \mathrm{cm/s^2}$, hence very small.

Braking Deceleration only by coasting a_c would result in excessive stopping distances. For the vehicle to stop within a reasonable distance, it is necessary to apply a braking force F_b, whose direction is opposite to that of the velocity vector $\vec{v}$; as mentioned in Section 7.7, we consider F_b as a negative acting force. The braking deceleration a_b is given by (7.3). During a braking sequence with initial velocity v_b (Figure 7.21) until the vehicle stops, a_b is normally considered constant and motion as uniformly delayed. The braking time and the braking distance are then given by

$$t_b = {}^{v_b}\!/_{a_b}; \quad s_b = \frac{1}{2}\,{}^{v_b^2}\!/_{a_b} \tag{7.44}$$

Braked Wheel In this case the distribution of the forces is indicated in Figure 7.22. The wheel is subject to the following forces:

- The vertical forces $\vec{G}$ and $-\vec{G}$, already considered in the cases of load-bearing and traction wheels.
- The resisting torque T_1, corresponding to the resistance $\vec{R_1}$.
- The braking torque $T_b = F_b \cdot {}^{D}\!/_{2}$. The overall reaction of the ground on the wheel is thus given by the resistance force $\vec{F_r} = \vec{F_b} + \vec{R_1}$.
- The horizontal force $-\vec{F_r}$ applied to the spindle in the direction of motion, which is equal in magnitude to the ground reaction $\vec{F_r}$.

Limiting Deceleration Given that in rail vehicles all driving and load-bearing wheelsets are always braked ($\alpha = 1$), for adhesion conditions to apply, it is necessary that

$$F_b \leq f \cdot G$$

from which, neglecting the resistance to motion,

$$m_e \cdot a_b \leq f \cdot G; \quad a_b \leq \frac{f}{1 + \varepsilon} \cdot g \tag{7.45}$$

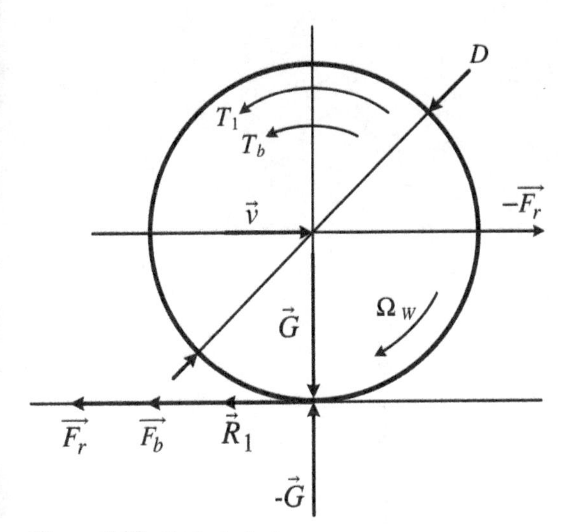

Figure 7.22 Braked wheel.

In rail vehicles where the limiting deceleration imposed by the distance between two stops (7.45) must be exceeded, an electromagnetic shoe system must be fitted (Section 7.24).

7.23 AVERAGE AND COMMERCIAL SPEEDS

If S is the length of a single line section between two stops (Figure 7.21) and t_0 is the time taken to travel it, then the average speed is

$$v_m = {}^{S}\!/_{t_0}$$

For an arbitrary path, including many sections, the commercial speed is defined as the ratio

$$v_c = \frac{\sum S}{\sum t_0 + \sum t_s} \tag{7.46}$$

where the terms t_s represent the durations of the intermediate stops.

In urban transport, it is normal practice to reference the average values of S and t_s; hence (Figure 7.23),

$$v_c = \frac{S}{t_0 + t_s} \tag{7.47}$$

These parameters are of considerable importance due to the relatively short average length S of these sections (250–400 m in trolleybus and tram services,

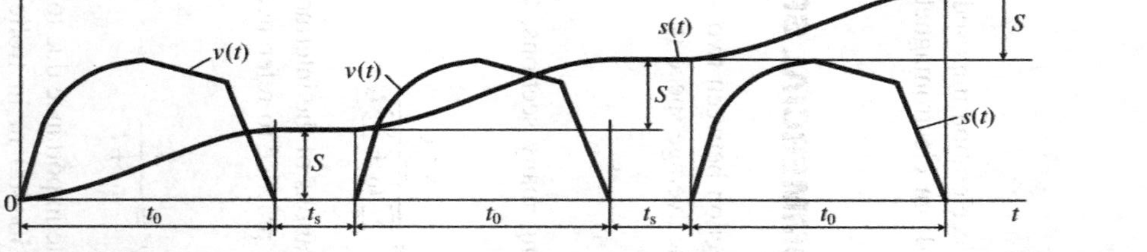

Figure 7.23 Traction diagrams for an urban line with equal sections of length S. t_0: Section transit time; t_S: station stop time

500–1000 m in urban metros). Consider a vehicle equipped with N identical traction motors:

1. High acceleration and deceleration performances must be achieved in order to ensure good average speeds. With regard to the start-up phase, recalling (7.42), rail vehicles should be associated with high adhesion ratios α, subject to other rolling stock economic and service-related constraints.

   For a given total installed power, in order to reduce the initial material investment and maintenance costs, power capacity should be spread over the smallest possible number of drives and driving wheelsets; as a limiting case, the most cost effective solution would be for trains to be composed of one locomotive plus a certain number of carriages. In any case, the preferred choice for urban transport requirements consists of individual vehicles, usually articulated (typical examples are trams and light metro vehicles), or trains composed of EMUs and coaches.

   For $\alpha = 1$, achievable start-up accelerations on the order of $1.5\ \text{m/s}^2$ are possible. This value coincides with the typical physical comfort limit for onboard passengers. In this regard, it should be noted that for this type of transport the fact that the majority of passengers will be standing during peak times is accepted (seated passengers normally do not exceed 20–30% of the total capacity). The density of standing passengers is estimated at the rate of 4–6 persons per square meter of free surface.

   Acceptable accelerations are considered to be on the order of $1.2 - 1.3\ \text{m/s}^2$ with $\alpha = 0.7 - 0.8$ in urban transport systems and $0.8 - 1.0\ \text{m/s}^2$ with $\alpha = 0.5 - 0.6$ in suburban lines.

   In trolleybuses, thanks to the high tire/road adhesion coefficient, it is usually sufficient to have only one driving wheelset;
2. The reduction of stop times t_s is achievable by facilitating access to the coach vehicles (number and width of doors, reducing the height differences between station platforms and the coach floor levels, where possible making them flush and with careful design of step profiles), and optimizing the door opening and closing mechanisms in order to make them faster and safer, subject to remote control and monitoring. The relevant onboard system always provides for devices that individually control the closing position of the doors, enabling departure only with the doors fully closed.

7.24 BRAKING SYSTEMS

Mechanical and Electric Braking All vehicles are equipped with mechanical friction braking systems. In the case of rolling stocks, these consist of cast iron or synthetic shoes that act on the wheel rims, or through gaskets on disks extending from the wheelset axles. Such brakes are typically activated by compressed air. Onboard pneumatic systems, powered by compressors installed in the traction vehicle, normally operate at pressures of 700–900 kPa (7–9 bar). In urban vehicles not equipped with pneumatic systems (referred to as "all electric" systems), the mechanical brake is controlled and operated by auxiliary circuits and electrical or electrohydraulic servos.

In certain countries, there are also "vacuum" braking systems in which the braking action is achieved by creating a "depression," that is, a momentary pressure level that is lower than the atmospheric one.

As a detailed description of such mechanical braking systems is beyond the scope of this book, some brief notes will suffice to provide some background information. Compressed air braking solutions for rail vehicles must be as follows:

1. Continuous, that is, actionable from a single command post. For this purpose, a pneumatic duct runs throughout all the train vehicles, the sections of which are interconnected by means of flexible couplers. This duct serves both to feed the individual vehicle tanks by means of the traction vehicle compressors and to transmit the braking and brake-release controls. Since the pressure variations are transmitted through the duct at maximum speeds of 250–270 m/s, for very long trains (over 500 m), a problem arises due to the braking action not being simultaneous for all the vehicles.

   In freight trains where the transmission speeds are lower and the train lengths are considerable, appropriate constraints are applied to the rate of change of the acting cylinder pressure during braking and brake release. The conditions are more favorable for passenger trains, since they allow for a greater rate of change of braking pressure.

   Electropneumatic systems are employed in high-speed trains as well as in urban and suburban transport where braking promptness is essential. Here the braking action is again achieved with compressed air fed through a centralized duct, but the activation commands are transmitted electrically;
2. Automatic, in the sense that the brake must activate immediately if the duct pressure flow should be interrupted. For this reason, in a pneumatic braking system, activation causes a reduction of pressure in the duct;
3. Capable of gradual variation, that is, it must be possible to moderate the braking effect both during braking and brake release;
4. Inexhaustible, in the sense that braking efficiency must not degrade below 10–15% of its nominal level, even after several short interval repeated braking and brake release events.

For the case of electric traction, given that the electric motors are reversible, the wheelsets can be equipped with regenerative electric braking solutions.

However, it must be emphasized that ultimate safety trip is always entrusted to mechanical brakes acting on the driving wheelsets:

- During normal operation, to integrate the action of the electric brake, whose effectiveness is lost at low speeds.
- In the event of failure of the electric brake, automatically taking over at any speed.

In some rare cases, brake systems based on eddy current principles (Foucault current) have been implemented; these are generated in cohesive rotating disks extending from the wheelset axles by electromagnets powered from onboard auxiliary circuits. It should be noted that the abovementioned braking systems, mechanical, electrical, and eddy current mechanisms all depend on wheel/ground adhesion and must therefore comply with the limits examined in Section 7.22.

Electromagnetic Friction Braking In addition to the braking systems based on mechanical or electrical actions on the wheelsets of tram vehicles, and in some cases on metro and railway carriages, there is also an electromagnetic braking system that forces shoes into direct contact with the rails, causing a frictional resistance that is independent of adhesion factors (Section 10.6.6). Each shoe has two steel plates that are attracted to the rail by magnetic induction with a force F_0, dragging on the rail's upper surface creates friction and a consequent braking force $F' = f' \cdot F_0$ that is independent of adhesion.

By fitting two shoes on each bogie, the system greatly reduces the braking distance, since its action supplements those of the other braking systems that exploit adhesion; in urban vehicles this can lead to deceleration performances greater than 2 m/s^2.

A negative aspect of this system is the significant variability of the coefficient of friction f between the shoe and the rail as a function of speed. The effectiveness of the brake is greatly reduced at high speeds (e.g., at 70 km/h $f' = 0, 10$, compared with values greater than 0.20 for $v < 20$ km/h); on the other hand, the deceleration is rapidly reinforced as the speed decreases, hence this solution becomes quite "brutal" for standing passengers during the stopping phase. In urban transport, electromagnetic friction braking is therefore used as an emergency brake to be able to cope with sudden obstacles or unusually unfavorable adhesion conditions.

7.25 OPERATIONAL SPEED LIMITS

A moving train must not only observe the condition

$$v \leq v_M$$

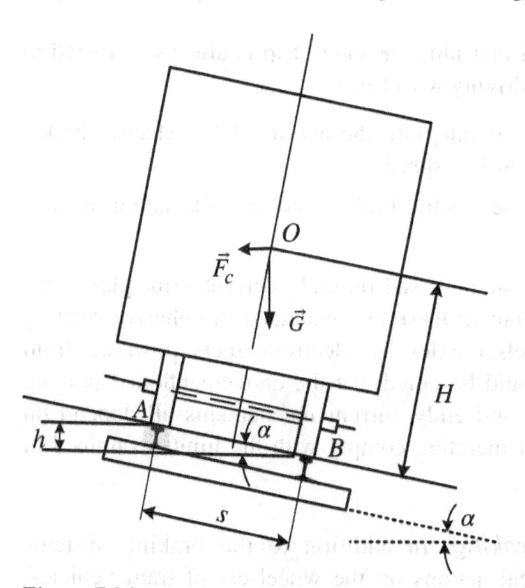

Figure 7.24 A railway vehicle during cornering. *O*: Center of gravity; *h*: superelevation; *s*: distance between the wheel/rail contact points *A* and *B*.

in which v_M represents the maximum speed of the traction vehicle, but must also respect the speed limits imposed by the elevation profile of the line and the curves. Detailed examination of this topic is beyond the scope of this book; however, some brief notes will be presented.

Consider a vehicle of mass m, traveling at speed v on a curved section of line with radius of curvature ρ.

The vehicle is subject to the centrifugal force (Figure 7.24) given by

$$F_c = m \cdot a_c = m \cdot {}^{v^2}\!/_{\rho} \tag{7.48}$$

acting on the center of gravity O, located at height H above the top of the rails; the centrifugal acceleration is $a_c = {}^{v^2}\!/_{\rho}$. Normally when cornering, the external rail is elevated with respect to the inside one; the track is inclined relative to the horizontal by an angle

$$\alpha = \arcsin\left({}^{h}\!/_{s}\right)$$

where h is the superelevation and s is the distance between the A and B wheel contact areas and the rail. For standard gauge rail, s is assumed to be 1.5 m. In rail networks the maximum elevation is $h(\max) = 0.16$, hence sin α (max) $= 0.1067$; α (max) $= 6°$.

Assume that the floor of the vehicle is kept parallel to the rail level, therefore at the same angle a to the horizontal; in this direction, the vehicle's weight

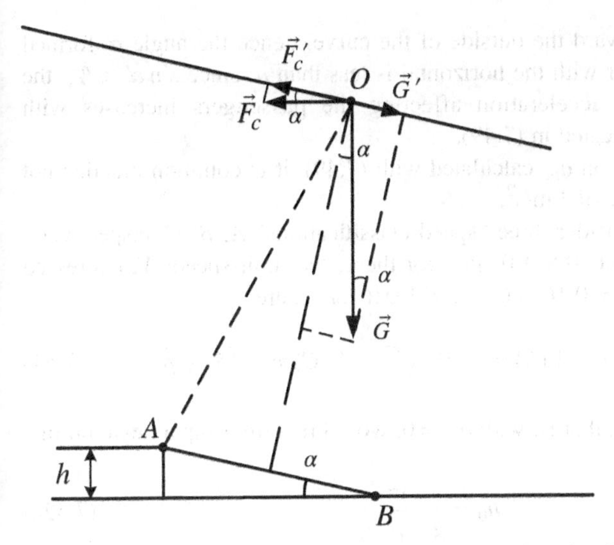

Figure 7.25 Forces acting during cornering. O: Center of gravity of the vehicle; A and B: ideal wheel/rail contact points; h: superelevation.

G acting through the center of gravity O and the centrifugal force F_c have the components (Figure 7.25):

$$G' = G \cdot \sin\alpha = m \cdot g \cdot \sin\alpha, \quad F_c' = \frac{m \cdot v^2}{\rho} \cdot \cos\alpha - m \cdot g \cdot \sin\alpha$$

Letting $\cos\alpha \approx 1$; $F_c' \approx F_c$, given the small values of α and defining the term for uncompensated lateral acceleration as

$$a_{nc} = {}^{F_{nc}}\!/_{m}$$

Therefore,

$$a_{nc} = {}^{v^2}\!/_{\rho} - g \cdot \sin\alpha = {}^{v^2}\!/_{\rho} - {}^{g \cdot h}\!/_{s} \tag{7.49}$$

Depending on the characteristics of the line (armature, ballast, foundation) and the vehicles, certain values of a_{nc} are allowed, taking into account that passengers do not tolerate uncompensated lateral accelerations exceeding 1–1.2 m/s^2. Having established permissible a_{nc}, the speed limit is given by

$$v = \sqrt{a_{nc} + g \cdot {}^{h}\!/_{s}} \cdot \sqrt{\rho} \tag{7.50}$$

It should be noted that during cornering, the so-called "flexibility," due to the elasticity of the suspensions, causes the vehicle car-body to rotate with respect to the bogie toward the outside of the curve, hence the angle α' formed by the plane of the floor with the horizontal is less than a; since $\sin\alpha' < {}^{h}/_{s}$, the uncompensated lateral acceleration affecting the passengers increases with respect to the value indicated in (7.49).

For this reason, for an a_{nc} calculated with (7.49), it is common practice not to exceed the limit value of 1 m/s^2.

It is possible to consider three "speed classifications" A, B, C, respectively, characterized by $a_{nc} = 0.6–0.8–1.0$ m/s^2; for these, the limit speeds V, expressed in km/h, and valid for $h = 0.16$ m ($g \cdot {}^{h}/_{s} = 1.046$ m/s^2), are:

$$V(A) = 4.62 \cdot \sqrt{\rho};\quad V(B) = 4.89 \cdot \sqrt{\rho};\quad V(C) = 5.15 \cdot \sqrt{\rho} \tag{7.51}$$

A perfect compensation, that is, with $a_{nc} = 0$, would require a superelevation of

$$h_0 = \frac{s}{g} \cdot \frac{v^2}{\rho} \tag{7.52a}$$

which, by expressing h_0 in mm, V in km/h, leads to

$$h_0 = 11.8 \cdot {}^{V^2}/_{\rho} \tag{7.52b}$$

Introducing the superelevation deficiency as

$$h_i = \frac{s}{g} \cdot a_{nc} \tag{7.53}$$

then Equation 7.49 yields

$$h_i = h_0 - h \tag{7.54}$$

Given a section of line with vehicle speed $v = \text{const.}$, determined according to the allowed a_{nc} (max) and h (max) for bends with minimum curvature radius ρ (min), in the other wider curves with greater radius $\rho = K\,\rho$ (min), then by setting the superelevation as $h = h$ (max)/K, the acceleration a_{nc} is reduced by the same proportion (see Equation 7.49).

For trains running at speeds v' below the limit (7.50), acceleration a_{nc} decreases; in a bend with curvature radius ρ (min) and superelevation h (max), it will tend to null at the speed

$$v_0 = \sqrt{\frac{g \cdot h}{s}} \cdot \sqrt{\rho},\; V_0(\text{km/h}) = \sqrt{\frac{h(\text{mm}) \cdot \rho(\text{m})}{11.8}}$$

If $v' < v_0$, the superelevation h gives rise to an overcompensation, particularly for freight trains, which normally do not exceed the speed of 120 km/h; these tend to cause considerable wear on the internal rail. The determination of the superelevation values is thus the result of a careful examination of the requirements: in HS lines, if the line must carry mixed passenger-freight traffic with highly differentiated speeds, it is necessary to limit the maximum h to no more than 80–100 mm; diversely, 160–180 mm can be specified for specialized lines carrying high-speed passenger traffic.

The rail is able to withstand values of a_{nc} greater than those allowed for passengers; for safe cornering, the following is required:

- The total lateral force Y transmitted from each wheelset to the rails does not exceed the limit:

$$Y_{\lim} = 0.85 \cdot \left({}^{G}\!/_{3} + 10\right)$$

where both Y and G are expressed in kN and G is the vertical load to ground on a given wheelset.

- The critical speed for overturning must not be exceeded, as this is the limiting condition for which the wheel load tends to null on the inner curve rail (in this case, the resultant of the forces $\vec{F}_c$ and $\vec{G}$ shown in Figure 7.25 act through the line joining the center of gravity O with the point of contact A).

The accelerations a_{nc} corresponding to these limits are considerably higher than those permitted for the comfort of passengers; therefore, it is possible to increase cornering speeds with respect to (7.50) provided that the vehicle carbody is inclined at an angle γ toward the inside of the curve, so that the floor forms an angle of $\alpha+\gamma$ with the horizontal. In these trains, passengers are subjected to an uncompensated acceleration of

$$a_{nc}^{*} = {}^{v^2}\!/_{\rho} - g \cdot \sin(\alpha + \gamma)$$

Setting the approximation of 1 for the cosines of α and γ, we have

$$a_{nc}^{*} = {}^{v^2}\!/_{\rho} - g \cdot \sin\alpha - g \cdot \sin\gamma = a_{nc} - g \cdot \sin\gamma$$

given that a_{nc} is the acceleration sustained by the wheelsets and the rails. In the case of the ETR 460 electric multiple unit, $\gamma = 8°$, for which, even accepting an a_{nc} on the order of 1.8 m/s^2, the a_{nc}^{*} is limited to less than 0.5 m/s^2. The increase in a_{nc}, by virtue of (7.50), allows for a speed increase of 20% compared with the speed classification C.

In conclusion, it should be mentioned that the transition from straight-line travel to cornering and vice versa requires a transition section leading into/out of the curve of appropriate length L_r, in which both the superelevation h_x and the curvature ${}^{1}\!/_{\rho_x}$ increase linearly from zero (in the straight section) to the values h

and $^1/_\rho$. In this manner, the "backlash" or "shock"

$$j = {}^{a_{nc}}/_{t_r}$$

does not exceed the acceptable limit values on the order of 0.25–0.40 m/s^3, where t_r is the time required to travel the transition section at the speed v, such that $t_r = {}^{L_r}/_v$. Indicating $i = {}^h/_{L_r}$ as the gradient of the transition section, then

$$j = \frac{a_{nc} \cdot v}{L_r} = \frac{a_{nc} \cdot v \cdot i}{h}$$

given the previous discussion concerning the constancy of the ratio ${}^{a_{nc}}/_h$ in a given section of line traveled at a constant speed, if i = const, then j = const. The linearity of the function $^1/_{\rho_x} = f(x)$ leads to the establishment of a curve that corresponds to a cubic parabola.

7.26 MOTION TRANSMISSION

The transmission of motion between the drive motors and the wheelsets normally occurs, as seen in Section 7.18, by means of gears. Construction criteria may differ; in some systems the motors are completely suspended, in others they are borne directly by the wheelset (nose suspension). The distinction is important because the unsprung mass differs considerably for the two cases; this affects both the driving characteristics of the vehicle and the operation of the motors. Herein follows a brief discussion of the characteristics of the more common systems.

Motors with Nose Suspension The arrangement is shown in Figure 7.17. On one side the motor is elastically supported by means of springs fixed to the bogie frame; on the other, it bears directly on the wheelset by means of two arms equipped with plain or rolling bearings. The part of the motor's weight that is borne by the wheelset thus increases the rigid mass of the vehicle.

This very simple suspension system is generally used for speeds not exceeding 60–80 km/h.

The transmission system normally consists of a pair of cylindrical gears located at one side of the motor and enclosed in a housing. When a double gear reduction is required (for the reasons highlighted in Section 7.18, that is, very high values of the ratio ρ/D), the gear wheel of the first pair and the pinion of the second are keyed on a countershaft that is fixed by the motor support arms.

Motors That Are Fully Suspended on the Bogie If the motor is fixed to the bogie frame, it is completely suspended. It can be arranged transversely with respect to the bogie, that is, with its axis parallel to the wheelset axle.

In so-called hollow axle transmissions, the motor, through a pair of cylindrical gears, drives a hollow axle shaft that surrounds the wheelset and is supported by the motor housing itself (Figure 7.1); in this case also the gear unit is completely suspended.

The distance between the wheelset axles and the hollow shaft can vary due to the oscillations between the frame and wheelset allowed by the primary suspension; of course, sufficient distance must be assured to prevent the hollow shaft and the wheelset axles coming into contact. The drive torque is transferred from the hollow shaft to the wheels through a deformable system that allows for adequate freedom of movement; typical solutions include metal springs and rubber buffers as well as articulated systems such as universal joints and discs.

An intermediate system between nose and integral suspension is achieved by supporting the hollow shaft directly on the wheels by means of rubber elements. In this case, the motor + gear unit + hollow shaft complex is partially suspended.

There are also solutions in which the motor is again fixed to the bogie frame, and therefore completely suspended; a deformable transmission element, capable of absorbing the relative displacements, is located between the motor shaft and the gear unit; the gear wheel is keyed on the wheelset. In turn, the gear unit housing rests on the wheelset by means of roller bearings and is supported on the bogie frame by means of articulated arms; also in this case the gear unit is partially suspended. Consider a vehicle equipped with *N* identical traction motors.

The helical gear units consist of two gears (pinion + driven gear) or, when the distance is very large, by three gears (pinion + intermediate gear + driven gear). Also in this case, the transmission ratio preserves the value $p = Z_r/Z_p$ indicated in Section 7.18.

In so-called longitudinal layouts, the motor is fixed to the bogie frame with its axis parallel to the track; the motion is transmitted to the wheelset by means of a telescopic cardan shaft, a torsional elastic coupling, and a conical reducer that is partially or fully suspended.

Motors Fully Suspended on the Car-Body There are also solutions in which the motor is mounted on the car-body frame rather than on the bogie, that is, a double suspension solution. Transmission requires a mechanical system capable of absorbing displacements related to the suspension and the rotation of the bogie with respect to the car-body.

Single-Motor Bogies In single-motor bogies, the motor may be mounted

1. with its axis parallel to that of the wheelset. This solution is usually adopted for heavy locomotives or heavy EMUs; the bogie can have two or three wheelsets. The motor is completely suspended. Transmission to the wheelset is through a helical gear unit and a system of joints capable of absorbing the relative displacements between the motor and the wheelset, which is located above or below the gear unit;

2. with its axis aligned longitudinally. This layout is adopted in two-wheelset bogies used in tram and metro vehicle drives. Also in these cases, the motor is completely suspended.

7.27 PERFORMANCE REQUIRED FROM A TRACTION DRIVE

To select an electric drive for use as a traction device, it is important to know what tractive effort diagram limits it must have in order to meet the required performance. To do this, it is necessary to plot the qualitative resistant tractive effort diagram and then use this to select the drive characteristics.

In general, as we have seen, vehicle motion may be broken down into the following stages:

- *Start-up* – during which the speed varies continuously from standstill to the cruising speed.
- *Cruising* – during which the speed is maintained constant at the cruising value.
- *Coasting* – during which speed is reduced by resistance to motion only.
- *Braking* – during which the speed is reduced from cruising to standstill.
- *Stop* – that is, when the speed is zero.

It is possible to determine the required acceleration values from the value of the cruising speed and the duration of the various motion phases, as defined in the assigned specifications. Observing the law of mechanical equilibrium at the drive axis,

$$T - T_r = J_{\text{tot}} \cdot \frac{d\Omega}{dt}$$

where

- T corresponds to the torque produced by the drive;
- T_r corresponds to the resulting resistant torque applied to the machine axis;
- J_{tot} corresponds to the total moment of inertia applied to the machine axis;
- Ω corresponds to the machine rotation speed.

It can be noted that, at a constant resistant torque, each increase in the vehicle's acceleration or deceleration must be matched by a corresponding increase in the driving or braking torque T delivered by the drive. Therefore, a drive must be capable of delivering a higher torque than the resistant torque in order to enable it to meet the acceleration/deceleration requirements. It follows that the

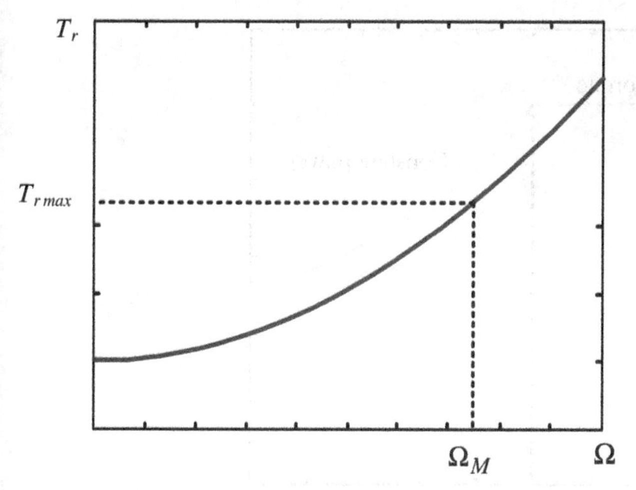

Figure 7.26 Typical diagram of motion resistance forces.

greater the torques and speeds, the greater the power that the drive must be designed to deliver.

Resistant Effort Diagram and Limit of a Drive Figure 7.26 illustrates a typical resistant tractive effort diagram that, as can be seen, increases with speed.

Electrical drives have design values for quantities such as power, voltage, current, and so on, which should not be exceeded in order to avoid oversizing the drive components or damaging them. From a mechanical point of view, these limits translate to a nominal electromagnetic torque T_{en} and a nominal speed Ω_b, known as the base speed. For as long as the speed Ω remains below the nominal value, the drive can develop the nominal torque; in this condition it is said to be operating in *normal* or *constant torque mode*. When the drive develops the nominal torque at the rated speed, it delivers the nominal power P_n and absorbs the nominal current I_n. By using the appropriate control techniques, which are described in the next chapters, it is possible to increase the speed above the base value, decreasing the torque at the same time so that the developed power is equivalent to the nominal value, without exceeding the voltage and current limits. In this way it is possible to extend the operating range of the drive, stating that it is operating in *extended or constant power mode*. The typical resulting tractive effort diagram is illustrated in Figure 7.27.

Adaptation of a Drive's Limit Tractive Effort Diagram In Figure 7.28 the base speed Ω_b and nominal torque T_n of the limit tractive effort diagram coincide with the maximum speed Ω_M and maximum resistant torque $T_{r\,\max}$ of the resistant tractive effort diagram; this means that the drive has been designed

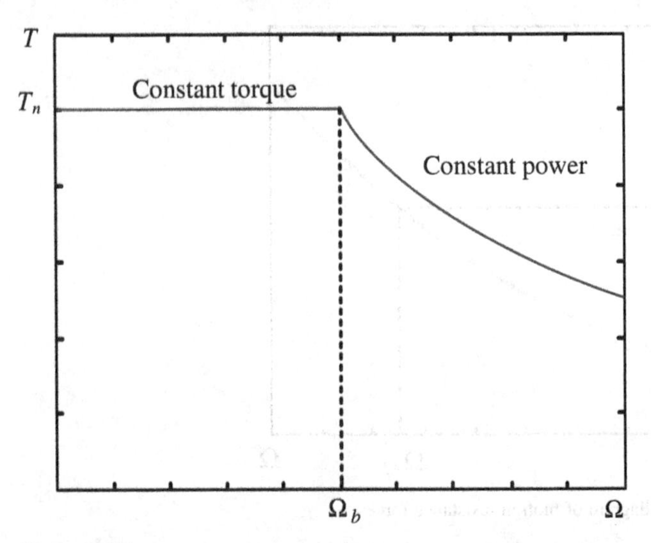

Figure 7.27 Tractive effort diagram limit.

with a nominal power rating P_n that is equivalent to the maximum load power at the cruising speed $P_{\max}$. In this way it is possible to be certain that the driving torque is always greater than the resistance by a margin that is sufficient to overcome the force of inertia and allow the vehicle to accelerate throughout the speed range. The performance in terms of acceleration will be limited by the maximum torque that the drive is capable of delivering, that is, T_n.

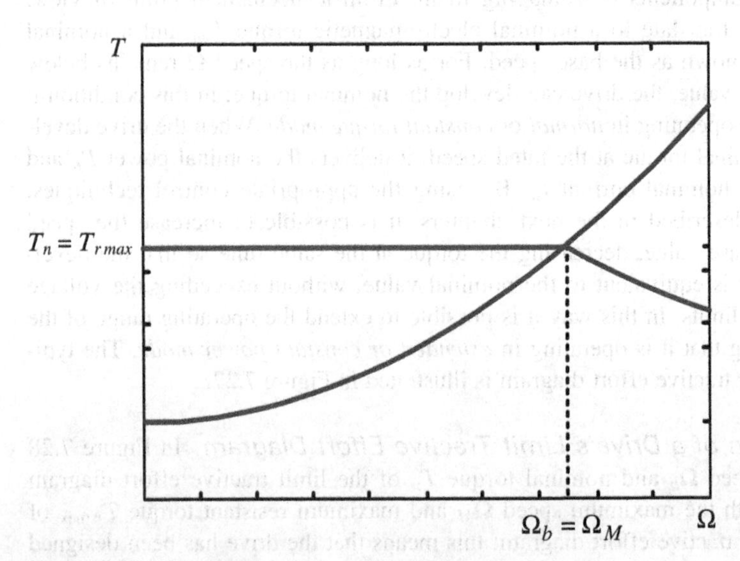

Figure 7.28 Tractive effort diagram when $T_n = T_{r\,\max}$ and $\Omega_b = \Omega_M$.

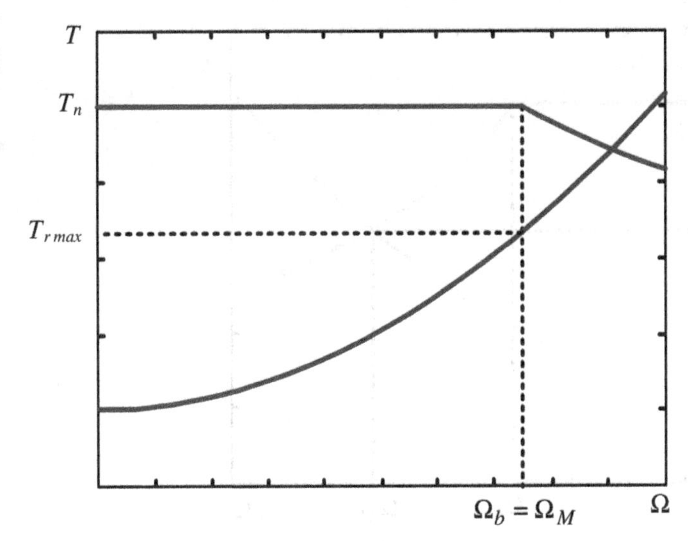

Figure 7.29 Tractive effort diagram when $T_n > T_{r\ \max}$ and $\Omega_b = \Omega_M$.

In order to improve acceleration performance, consider the situation illustrated in Figure 7.29, where the nominal torque T_n developed by the machine has been increased, resulting in an increase in the nominal power P_n for which the drive should be designed.

Doing so will improve the performance in terms of acceleration during start-braking phases, at the cost, however, of compromising operation at higher speeds. Therefore, this situation is well suited to traction applications where the start-up and braking phases are preponderant with respect to the cruising condition, such as a short-haul vehicles for urban or metropolitan transportation.

By taking advantage of constant power mode, the drive can be designed for a nominal power rating P_n equivalent to the maximum load $P_{\max}$, selecting the nominal torque T_n and speed Ω_b values, so that their product is always equal to the nominal power P_n (Figure 7.30). This provides good performance during all motion phases, without the need to oversize the drive. Therefore, in addition to the applications that have already been mentioned, this situation also adapts well to long-haul services and high-speed applications, where the cruising state is prevalent over the starting/braking phases.

A fundamental characteristic of the electric drive is its elasticity β, that is, the capacity to generate constant mechanical power by varying the torque and speed parameters over a wide operating range while maintaining the electrical design of the machine itself substantially unaltered in terms of rated voltage and current. Elasticity is defined as the ratio between the maximum vehicle speed and the base speed:

$$\beta = \frac{v_M}{v_b} = \frac{\Omega_M}{\Omega_b}$$

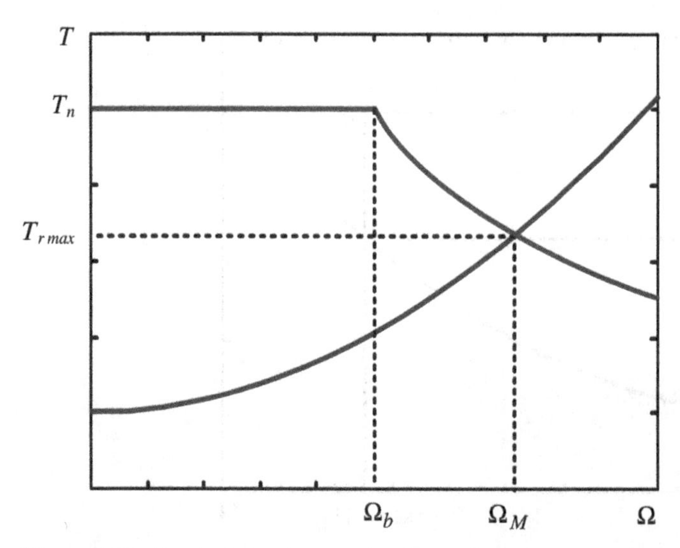

Figure 7.30 Tractive effort diagram when $T_n > T_{r\,\max}$ and $\Omega_b < \Omega_M$.

and is equivalent to the ratio between the maximum rotation speed and the basic rotation speed, as electric vehicles have fixed transmission ratios.

Therefore, vehicles used on lines with frequent stops, such as regional and underground trains, are characterized by a low value of β due to the need to prioritize torque, and hence acceleration, over speed. Conversely, high-speed vehicles should operate at constant power over a wide range of speeds, which means that β may reach values close to 3, or higher in some cases.

7.28 INTRODUCTION TO TRACTION DRIVES

The drive of a train consists of the traction motors, the static traction converters (inverters), the auxiliary static converters (choppers and inverters), the braking rheostat, and the electronic traction control unit (eTCU) that is used to control them.

For a block diagram of the traction unit, see Figure 7.31.

Various types of electric motors are used in electric railway traction applications; these motors, which are described in detail in the following chapters, are listed below:

- DC motors
- three-phase induction motors
- permanent magnet synchronous motors

In the early 1900s, electric traction technologies made use of series excited DC motors for series excitation or single phase, commutator AC

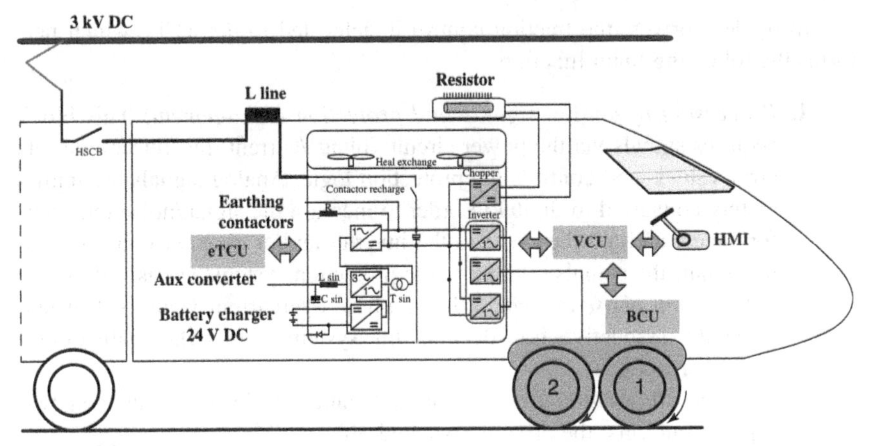

Figure 7.31 *TCU.*

motors. The former were powered at relatively low voltages, with rheostatic regulation, and were largely free from any critical issues, while the latter had significant limits relating to switching in the commutator, which tended to dramatically reduce their performance. On the other hand, the use of alternating current, which facilitates current interruption operations, made it possible to operate the power supply system at higher voltage levels, resulting in lower losses on the contact line and increased power transportability. The solution adopted at the time was to create a single-phase power supply system at reduced railway frequency with respect to the industrial frequency. This eliminated the problems in the commutator, but introduced the disadvantage of having to set up a special frequency system. As a result, it was necessary to construct an infrastructure dedicated to the production and transportation of energy solely in support of the rail system, or an interconnection system with the industrial mains network using phase and frequency converters consist of rotating groups.

At the same time, experiments were being conducted into the use of onboard three-phase induction motors with dual contact lines. The induction machine had obvious advantages of robustness, reliability, and flexibility. However, the system presented difficulties when collecting and handling vehicles at fixed speeds, and was thus abandoned.

It was therefore decided to re-electrify these lines with the other national systems in use, such as the 3 kV DC system, or the single-phase AC systems. Only toward the 1970s, with the advent of the inverters, was it possible to reevaluate the use of three-phase induction motors, exploiting the performance of a variable voltage and frequency drive.

A recent innovation has seen the introduction of permanent magnet synchronous drives.

In modern drives, the traction control is delegated to the TCU, which performs the following main functions:

1. *Processing of analog signals and protection management:* This block acquires signals via the power circuit voltage/current transducers, detects error values, and controls the protection logic. Analog signals are immediately converted to digital in order to increase the signal/noise ratio performance, and are then sent to the inverter and/or chopper controller. In particular, the signals relating to the protection systems are as follows:
   - *Maximum phase current*: When a phase overcurrent value is detected, via the respective transducers, this system causes the main circuit breaker to open.
   - *Motor phase unbalance*: If an unbalance is detected between two phase currents, the drive is switched off.
   - *Filter overvoltage*: If the voltage exceeds the maximum system voltage limit, the braking chopper is activated for a given period. If the voltage continues to rise, first the converters are blocked and then the main circuit breaker is opened.
   - *Drive cards diagnostics*: In the event of a failure of any of the drive cards, the pulses are blocked and the main circuit breaker is opened.

   Whenever the main circuit breaker is opened, the braking chopper intervenes in order to discharge the filter capacitors. The capacitors are also fitted with resistors, which are connected with them permanently in parallel in order to ensure that they are discharged more rapidly.
2. *Inverter control*: Motor control can be achieved by techniques such as FOC (field-oriented), Volt/Hertz (scalar flux control), and DTC (direct torque control), which will be described hereafter.
3. *Braking chopper control*: This function controls the electrodynamic braking function (ED brake) and protects against input voltage overloads. This unit detects the voltage of the DC link inverter input and controls the power dissipated on the brake chopper in order to limit overvoltages.
4. *Pulse generation control logic*: This unit generates the command sequence sent to the individual inverter and chopper power semiconductors, starting from the inverter control PWM modulator waveforms and the chopper control pulse train. These signals are sent to the drivers that drive the power modules.
5. *Antijerk ramps*: In response to a transition/coasting/braking or a variation in the requested performane value, the TCU increases/decreases the force level setting according to a linear ramp that limits the acceleration gradient (jerk) to the set point so as not to compromise passenger comfort.
6. *Antislipping/skidding function*: The TCU controls slipping/skidding on the train wheel in order to maximize vehicle performance in response to

the adhesion conditions of each individual wheelset. In particular, slipping occurs when the TCU is in traction mode and the torque delivered by the motor to the wheelset is too high for the current adhesion conditions. Skidding occurs when the TCU is in braking mode and the braking torque delivered by the motor to the wheelset is too high for the current adhesion conditions. In fact, each bogie wheel may be in a slipping or skidding state at any given instant. The control unit detects these conditions by comparing the speed and acceleration measurements from the individual vehicle wheels with the predetermined reference limits.

When a slip (or skid) is detected, the drive (or braking) torque is reduced in order restore correct wheel/rail adhesion. Once the correct value has been restored, the torque in question is increased again until it reaches the required value.

The "antiskid" system algorithm is based on the following functions:

- Wheel diameter measurements (by means of sensors)
- Vehicle speed and acceleration calculation
- Slip/skid detection
- Slip/skid correction

In particular, the slip/skid state is detected when the difference between the speeds of the two wheelsets on the bogie exceeds a certain threshold ΔV. In order to discriminate between slip or skid, the sign of the acceleration is taken into account; in fact, this value is calculated over an appropriate time interval prior to exceeding a certain predetermined threshold, decreased in order to allow for the slope, and to a predetermined limit which approximates the resistance to motion.

Chapter 8

DC Motor Drives

DC motor drives were the first to be developed, thanks to both the availability of a direct current source, either via the contact line or the batteries, and the capacity to regulate vehicle motion in terms of acceleration and speed by means of purely electromechanical systems. Thus, these drives played a leading role in all areas of urban transport, and a significant portion of the rail sector, until the advent of power electronics. Even today their use is widespread, particularly on older tram networks and underground railway systems.

8.1 CONSTRUCTION FEATURES

Considering that drive motors are mounted on vehicles, they must meet strict manufacturing and operating requirements arising from the limited space available, the considerable mechanical stresses they are subjected to due to the effects of motion (vibration, shock), and the continuous and broad variations of the load and power supply conditions.

The stator casing and front shields are of cast or forged steel, the rotor has roller bearings or ball bearings. Main parts of a stator and of a rotor for a DC machine are represented in Figures 8.1 and 8.2, respectively.

The number of poles p is usually equal to 4; for greater powers $p=6$, and rarely $p=8$. The pole pitch $\tau = \pi \cdot {}^{D}/_{p}$ (D = diameter of the armature) is therefore considerably wider. The four-pole motor often has a square section and the poles are always laminated. The armature is laminated and made of 0.5 mm thick sheet metal with open slots; in addition to the diameter D, the overall width L of the sheet roll pack is considered. The peripheral speed v of the armature can reach 50–60 m/s. Wave-type windings (series) and lap (parallel)-type windings are normally used and splitted in a internal ways.

The commutator, of a particularly robust construction, usually has a diameter D_k, which is less than that of the armature ${}^{D_k}/_{D} = 0.75-0.9$. Its peripheral velocity v_k can reach 40–50 m/s. The carbon-type brushes consistently work

Electrical Railway Transportation Systems, First Edition. Morris Brenna, Federica Foiadelli, and Dario Zaninelli.

Figure 8.1 Cross section of a compensated four-pole traction motor stator. 1: Yoke and frame; 2: main field pole; and 3: interpole.

with considerable pressure due to the vibrations and shocks that the machine is subjected to.

The drive motors are provided with an equal number of interpoles and main poles (the latter are also referred to as field poles). For greater powers, or when there are special operating requirements, the compensating winding is adopted, placed in slots obtained in the field pole shoes.

The construction is normally of a protected type, since free access to the internal parts is impeded, without however hindering the passage of cooling air.

8.2 NOMINAL DATA

In traction vehicles powered by the contact line, the motors operate at variable voltage and constant current, naturally within the limits of variation of the V_L line voltage; it is important to keep in mind that the voltage at the terminals can be varied for speed control.

In the first case, among the specifications on the data plate, the nominal voltage V at the terminals A–C is found; instead, if V_L is the line voltage, $V = V_L$ or $V = {}^{V_L}/_2$ can be used (half voltage motors). The second case is normal when $V_L = 3000$ V; this results in $V = {}^{3000}/_2$ V. The latter notation shows that the motor, while functioning at 1500 V, must be isolated for the line voltage of

Figure 8.2 Four-pole traction motor rotor. 1: Active conductors; 2: front connections; and 3: commutator.

3000 V, with consequences relating to size. As regard the power P (kW) output shaft, the corresponding current absorption I (A) and the velocity of rotation n (s^{-1}) generally refers to continuous service; limited duration operation of 60 min (hourly service) may also be considered.

The continuously variable voltage and current operation is precisely that of diesel–electric vehicles or vehicles powered by auxiliary power converters.

8.3 MOTOR SCHEMATICS

The series field motor includes the following (Figure 8.3):

- An armature AB circuit formed by the armature winding (armature core or stack), winding of the interpoles and the possible compensator winding. A motor provided with compensator winding is referred to as compensated. The resistance of this circuit will be denoted as r_i.
- Field coils or BC field, of resistance r_f.

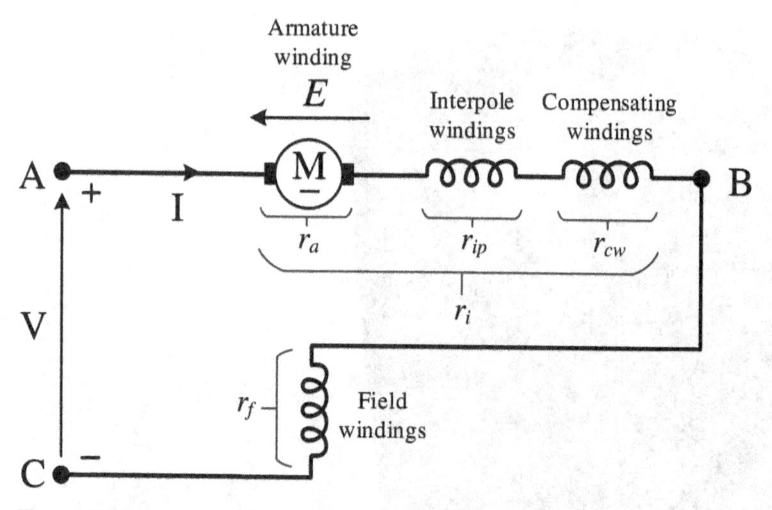

Figure 8.3 Schematics of a direct current traction motor.

Because the two circuits are in a series, the total internal resistance of the motor is equal to

$$r = r_i + r_f \tag{8.1}$$

The field winding, consisting of N_1 and the middle tap/pole, is crossed by the field current i, for which the mmf field is $M = N_1 \cdot i$. In the series field motors, for their operation, the i is equal to current I absorbed by the motor; therefore, under these conditions, the following applies:

$$M = N_1 \cdot I = f(I) \tag{8.2}$$

It follows that the field coils consist of a relatively limited number of coils, of an elevated section (ribbon), and are therefore very strong. This is a particularly advantageous feature for machines such as drive motors, which are subject to high dynamic stresses. In locomotives with electronic equipment, series equivalent motors with separate fields may be used, for which

$$M = K \cdot I \tag{8.3}$$

since K is constant. The operation does not differ from that of series motors field.

8.4 MAGNETIC CIRCUIT

If the magnetic circuit of the motor has a field with only mmf of inductor M (no-load operation), the field curve, which gives the peripheral distribution $B(x)$ to the air gap according to the x-axis, will have the pattern shown in Figure 8.4; within it $\tau = \pi \cdot {}^{D}/_{p}$ is the pole pitch, b the pole arc, B_M the maximum magnetic

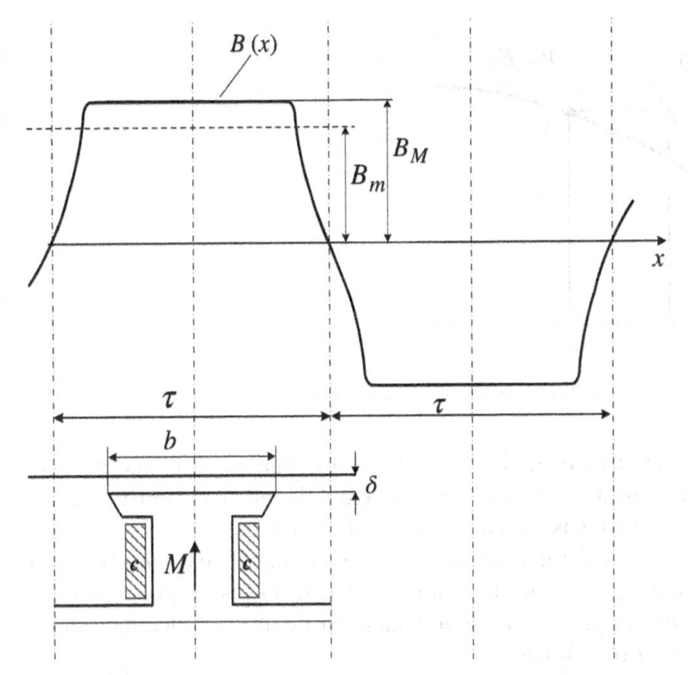

Figure 8.4 Field curve.

field induction at the center of the pole, where the air gap δ is usually constant. In the adjacent pole pitch, where the mmf $-M$ acts, the trend of $B(x)$ is equal in absolute value and of an opposite sign.

Along the periphery of the rotor there are $p/2$ complete cycles, with the electrical angle 2π corresponding to each; since the amplitude of a double pole pitch is geometrically

$$\frac{2\tau}{D/2} = \frac{2\pi}{p/2}$$

and we have the following correspondence between electric angles α_e and geometric or "mechanical" angles α_m:

$$\alpha_m = \frac{\alpha_e}{p/2} \tag{8.4}$$

The average induction value B_m can be considered approximately equal to 70% of B_m.

Below, it will be assumed that the flux of each pole ϕ (Wb) is given by the simplified equation

$$\Phi = \tau \cdot L \cdot B_m = \frac{\pi \cdot D \cdot L}{p} \cdot B_m \tag{8.5}$$

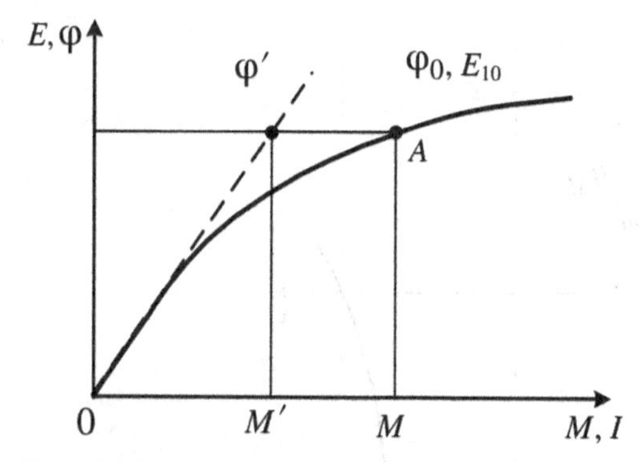

Figure 8.5 Characteristic of no-load magnetization, for Ω_1 const.

acceptable for essentially qualitative considerations. The curve of the no-load flux Φ_0[1] based on mmf M, or of the field current I, is the characteristic of no-load magnetization, which has the known trend of Figure 8.5.

If point A represents the normal operation of the motor, the ratio between the mmf corresponding to M and the mmf M' that would be required, for the same flux, if only the air gap was present (characteristic Φ'), is called the saturation ratio, on the order of 1.3–1.5.

8.5 NO-LOAD OPERATION

In no-load operation, that is, with no induced current flux, it is necessary to separate the armature circuit of the motor field winding and power the latter with current i (Figure 8.6).

In these conditions, if the motor is driven at the velocity Ω (rad/s) or n (s^{-1}), the following emf alternate frequency is induced in the armature core or stack:

$$f = p \cdot {}^{\Omega}/_{4\pi} = p \cdot {}^{n}/_{2} (\mathrm{Hz}) \tag{8.6}$$

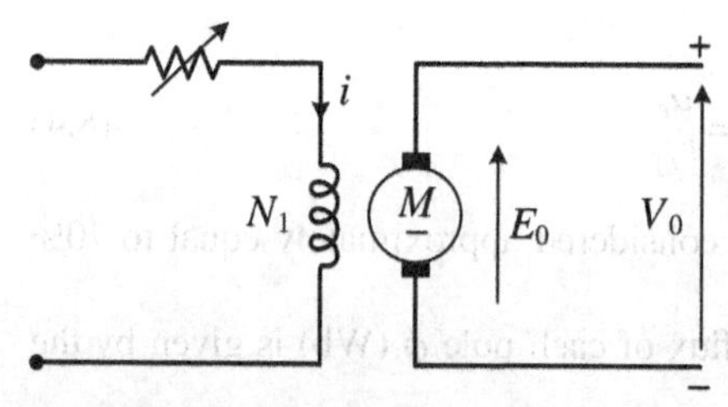

Figure 8.6 Schematics for the no-load test.

[1] The flux Φ_0 is called no-load as it is due to only the inductor mmf, without the armature being crossed by current.

The induced emf in each active conductor, as is well known, has a value of

$$e = B \cdot L \cdot v \tag{8.7}$$

Since the peripheral speed

$$v = \Omega \cdot \frac{D}{2} = \pi \cdot D \cdot n \tag{8.8}$$

is constant, since Ω is constant, the emf e is in proportion to B and it varies in time according to a curve $e(t)$ which, apart from the scale, reproduces the curve of the field; Figure 8.4 shows a complete cycle.

The correspondence between $B(x)$ and $e(t)$ is such that the semiperiod $^1/_2 \cdot f$ corresponds to the pole pitch and the maximum field induction B_M to the maximum emf in a conductor. Thus, the average value of e will be

$$e_m = B_m \cdot L \cdot v \tag{8.9}$$

Substituting the values of v and B_m, given by (8.8) and (8.5):

$$e_m = \frac{\Phi \cdot p}{\pi \cdot D \cdot L} \cdot L \cdot \pi \cdot D \cdot {}^{\Omega}/_{2\pi} = \frac{p}{2\pi} \cdot \Omega \cdot \Phi$$

The brushes collect a total unidirectional emf E that is in any case equal to the sum of the instant emf induced in each of the N/a conductors internally. It may also be stated that the E is equal to N/a times the mean value e_m, meaning

$$E = \frac{N}{a} \cdot e_m = \frac{p}{2\pi \cdot a} \cdot N \cdot \Omega \cdot \Phi = k \cdot \Omega \cdot \Phi \tag{8.10}$$

where $k = p \cdot {}^{N}/_{2\pi \cdot a}$ is a constant depending on the motor winding data.

During no-load operation, if an mmf $M = N_1 \cdot i$ machine field is created, rotating at revolutions $\Omega_1 = \text{const}$, an fem is induced

$$E_{10} = k \cdot \Omega_1 \cdot \Phi_0$$

proportional to the no-load flux Φ_0. The no-load characteristic $E_{10} = f(i)$ can then be represented, changing the scale, by the magnetization characteristic $\Phi_0(M)$ of Figure 8.5.

8.6 NO-LOAD LOSSES

8.6.1 Mechanical Losses

In no-load operation, at speed Ω with emf E_0, initially there are mechanical losses P_M due to friction in the bearings and air resistance (ventilation). The first term can be considered approximately proportional to the speed, according to the cubic speed. The curve $P_M(\Omega)$ represents the trend shown in Figure 8.7.

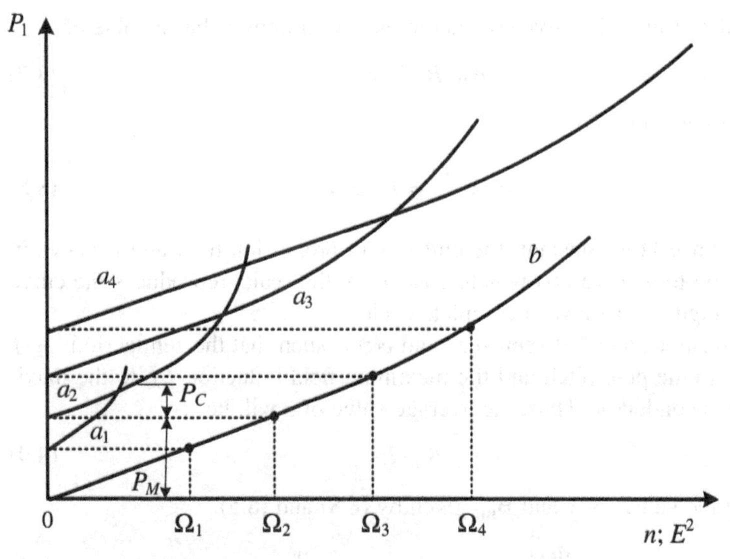

Figure 8.7 No-load losses. a_1–a_4 Curves in losses: $P_M + P_C = f(E^2)$, for Ω_1–Ω_4 = const; b) Mechanical losses $P_M = f(n)$.

8.6.2 Rotor Core Losses

In no-load operation, there are also rotor core losses P_C caused by the cyclic magnetization of the active parts of the rotor and the pole shoes in the vicinity of the air gap:

$$P_C = P_h + P_{ec}$$

where

P_h = hysteresis losses
P_{ec} = eddy current losses

The former are proportional to the area of the hysteresis loop and the internal frequency f, thus to the angular speed of the rotor Ω.

The eddy current losses are proportional to the square of the maximum field induction, to the square of the frequency (thus of Ω), and the square of the thickness of the metal sheets.

Usually, the specific rotor core losses are calculated P'_C in W/kg with the one-part formula:

$$P'_C = C \cdot \left(\frac{f}{50}\right)^{\alpha} \cdot B_M^{\beta}$$

where C is the loss figure of the metal sheets in W/kg, at a frequency of 50 Hz and the maximum induction of 1 T.

Normally, the exponents have the following variables: $\alpha \approx 1.3$; $\beta \approx 2$. Keeping in mind that $E \sim f \cdot B_M$, the following is also true:

$$P'_C \sim f^{\alpha-\beta} \cdot E^{\beta} \sim f^{-0.7} \cdot E^2 \tag{8.11}$$

The metal sheets of the armature have a thickness of about 0.5 mm and the following loss figures:

- $C = 3.6$ W/kg, if normal metal sheets (approximately 2.4 W/kg are dissipated by hysteresis; 1.2 W/kg for eddy currents).
- $C = 2$–3 W/kg, if these are partial alloy sheets.

No-load tests show the loss curves P_C for $\Omega = \text{const}$ as a function of E^2 (Figure 8.7).

For low inductions, these losses increase proportionally to the square of E, therefore, in agreement with (8.11) when saturation occurs, the exponent β gradually assumes increasing values up to 3–5–7.

8.6.3 No-load Test

According to the norms, a no-load test of the motor may be performed with the regime of Figure 8.6, powering it to voltage V_0 and allowing it to rotate with net shaft torque equal to zero. In fact, it has been noted that the power absorbed should be zero, but in said case it would be necessary to drag the rotor with an auxiliary motor.

Under the test conditions, the motor draws a small current I_0 and a power $P_{a0} = V_0 \cdot I_0$ that is the sum of the mechanical losses, the rotor core losses, and the per unit p.u. copper losses due to I_0. Since the latter is very small, the corresponding per unit p.u. copper losses equal to $r_i \cdot I_0^2$,[2] are negligible, where

$$P_{a0} \approx P_M + P_C \tag{8.12}$$

The emf developed by the motor differs from the power supply voltage V_0 due to the internal voltage drop, which is also negligible, thus the following may be stated:

$$V_0 = E_0 + r_i \cdot I_0 \approx E_0$$

The no-load test is performed by varying V_0 and i, in order to have $\Omega_1 = \text{const.}$; the curve $V_0 = f(i)$ represents, with sufficient approximation to the above, the characteristic of no-load magnetization $E_{10} = k \cdot \Omega_1 \cdot \Phi_0$ (Figure 8.5).

Bringing back, as a function of V_0^2 (therefore, E^2), the values $P_{a0} = V \cdot I_0 \approx P_M + P_C$, the no-load losses' curves at constant revolutions are

[2] r_i is the equivalent resistance of the armature circuit.

obtained (Figure 8.7), from which it is possible to obtain the diagram $P_M = f(\Omega)$.

8.7 LOAD OPERATION

8.7.1 Armature Core or Stack Reaction

When the current I flows in the armature, a mmf (armature core or stack reaction) is generated within it, which distorts the field produced by the mmf of the inductor. The performance of the armature mmf M_i, whose axis is shifted by $\tau/2$ with respect to the axis of the field poles, is indicated in Figure 8.8.

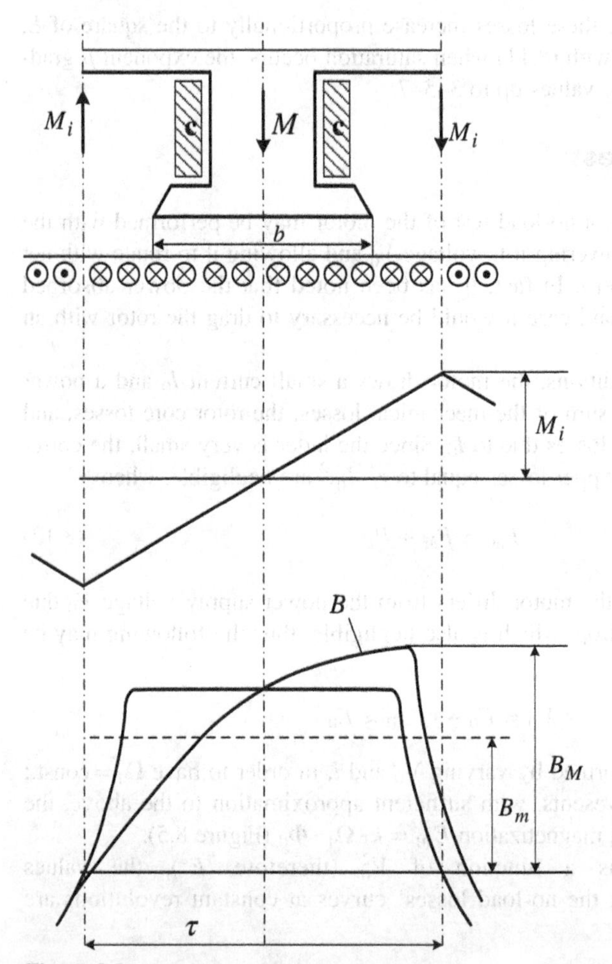

Figure 8.8 The field curve in loaded condition.

Called $I_c = {}^{I}/_{a}$, the current flowing through the individual conductors of the armature, the linear current density in A wires/m is

$$A = \frac{N \cdot I_c}{\pi \cdot D} = \frac{N \cdot I}{\pi \cdot a \cdot D} = \frac{N \cdot I}{a \cdot p \cdot \tau} \tag{8.13}$$

The armature mmf assumes the maximum value

$$M_i = \frac{\tau \cdot A}{2} = \frac{N \cdot I}{2 \cdot a \cdot p} \tag{8.14}$$

at the centerline interpolation (neutral area, inversion of I_c).

Due to the effect of M_i, which is a twisting mmf, the dependent field curve has a trend like that in Figure 8.8.

Supposing that the flux is concentrated within the pole arc b, with constant air gap so that it may be considered proportional to the average induction in b, the magnetization characteristic can then represent, as a function of the mmf M, both the flux Φ and induction B (Figure 8.9).

In a no-load situation, with mmf $M = OA$ there is a constant induction equal to AA' along the polar arc. If the armature is crossed by the current, the armature must be added algebraically to the M and mmf that assumes, at the ends of the pole arc, the values $\pm\frac{b \cdot A}{2}$; the resulting fmm, therefore, varies from OB to OC, the armature field from BB' to CC'.

The following two are the effects of the armature core or stack reaction:

1. The ratio B_M/B_m increases with respect to the no-load operation, as can be seen from Figure 8.8. In practice said ratio goes from 1.5 to 2–2.5. At

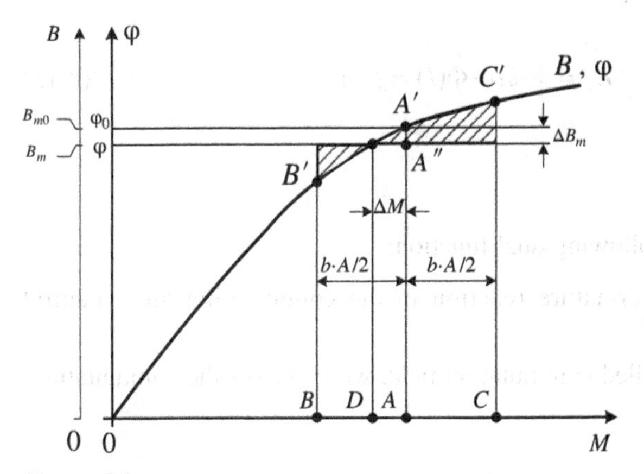

Figure 8.9 Torque effect and demagnetization of the armature core or stack reaction.

constant average induction, therefore, the maximum induction increases under one end of the pole extremities; this increase is more felt, in relative value, in the straight portion of the magnetization characteristic, that is, for low values of B_m.

2. While in the no-load operation, with the assumptions above, the average induction B_{m0}, and thus the flux Φ_0, are represented by the ordinate AA', in the load operation the B_m is given by the average ordinate AA'' of the tract $B'A'C'$: an ordinate that, due to the curvature of the magnetization characteristic, is lower than AA'. It follows that the dependent flux Φ, proportional to AA'', is lower than Φ_0. The armature reaction, while being due to a twisting mmf, thanks to the saturation of the magnetic circuit, thus carries out a demagnetizing action, which involves a decrease in the flux proportional to $A'A''$. It can also be said that the armature mmf has longitudinal component demagnetizing $\Delta M = \text{AD}$. The demagnetizing effect is most felt in the knee of the characteristic and is small in the straight section, where instead the relationship B_M/B_m is higher.

8.7.2 Load Magnetization Characteristic

As previously stated, the flux load Φ depends on both the points of the magnetization curve, which is considered to be the value of the armature mmf, thus of the current I. Since in series motors the armature and field currents coincide,[3] the flux is uniquely dependent on the single function I, namely, $\Phi = f(I)$, unless the field conditions of the machine vary, as is the case with field regulation (Section. 8.14). Consequently, it can be considered a single magnetization curve load at speed Ω_1 = const, which gives the emf E_1 based on the current I (Figure 8.10):

$$E_1 = k \cdot \Omega_1 \cdot \Phi(I) = f(I) \tag{8.15}$$

8.7.3 Interpoles

The interpoles have the following dual function:

- To neutralize the armature reaction in the commutating area (neutral area)
- To produce a so-called commutation field, which favors the commutation

[3] Or they are directly proportional to each other, in series motors with separate field.

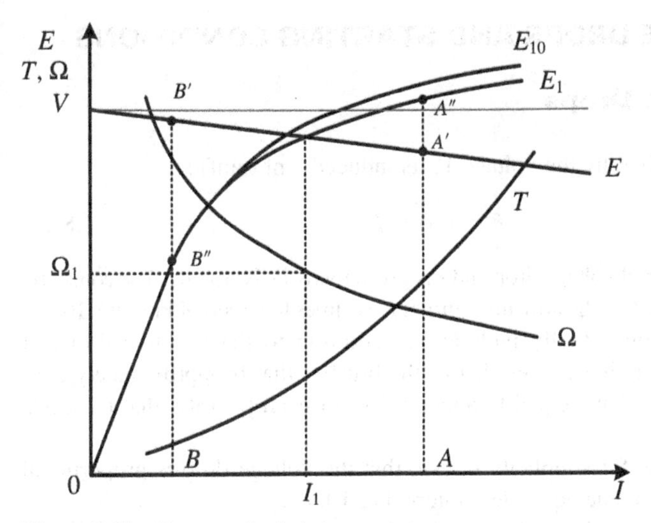

Figure 8.10 Electromechanical characteristics.

Their axis coincides with the centerline between two poles, and then with the axis of M_i, being, as stated, their mmf

$$M_{ip} = N_{ip} \cdot I$$

opposite to M_i, generally, in traction motors without compensator winding we have

$$M_{ip}/M_i = 1.2 - 1.3 \tag{8.16}$$

the remaining mmf $M_{ip} - M_i$ opposite to M_i, produces the commutation field.

8.7.4 Compensator Winding Effect

The mmf M_c compensator winding traversed by the armature current I serves to neutralize the armature core or stack reaction along the polar arc, meaning

$$M_c = \frac{b}{\tau} \cdot M_i \tag{8.17}$$

In compensated motors, the value (8.16) of the mmf of the interpoles poles is reduced by the M_c, by which

$$M_{pa} = (1.2 - 1.3) \cdot M_i - M_c \tag{8.18}$$

Thanks to the interpoles and compensator winding, the twisting effects and demagnetizing armature reaction are therefore practically eliminated; the on-load flux is almost the same as the no-load one.

8.8 VOLTAGE DROPS AND STARTING CONDITIONS

8.8.1 Voltage Drops

If the motor is supplied by the voltage V, its induced emf (emf) is

$$E = V - r \cdot I \tag{8.19}$$

where $r \cdot I$ is the internal voltage drop and r is given by (8.1). Normally, $V = \text{const.}$, by which the emf E, based on the armature current I, assumes the trend of Figure 8.10.

The voltage drop actually includes, in addition to the resistive drops of armature and field windings, even that of the brushes that is approximately constant on variations of I and equal to about 1.5 V for each polarity, that is, at 3 V total.

We will assume, for simplicity's sake, that the voltage drop is proportional to I, considering a suitable equivalent internal resistance.

The relative voltage drop, given by

$$\varepsilon = \frac{r \cdot I}{V} \tag{8.20}$$

since $V = \text{const}$, is proportional to the current I; at nominal current I_n, the following normally appears:

$$\varepsilon_n = r \cdot {}^{I_n}/_V = 0.04\text{–}0.7$$

8.8.2 Starting Conditions

Powered by nominal voltage V and with a locked rotor, that is, with $\Omega = 0$ and $E = 0$, the motor would absorb the short-circuit current,

$$I_{sc} = \frac{V}{r} = \frac{I_n}{\varepsilon_n} \tag{8.21}$$

which is too high for the motor[4] and corresponding to an incompatible torque with the adhesion limits. The inrush current must not exceed a predetermined value

$$I_a = K \cdot I_n \tag{8.22}$$

with the overload coefficient K normally allowed in the starting conditions of between 1.4 and 1.8. These values are the maximum performance required; to ensure the necessary gentle starting conditions, it is however necessary to be

[4] For $\varepsilon_n = 0.05$, $I_{sc} = 20 \cdot I_n$ would result.

able to start-up even with currents significantly lower than the nominal value I_n. In so-called approaching phases (Section 8.17), the following must be done:

$$I_{A\min} = K_{\min} \cdot I_n \tag{8.23}$$

with $K_{\min} < 1$

The limitation of the inrush current is obtained

1. in traditional drives (Section 8.12), by inserting a resistor R in the series with the motor, so that:

$$I_A = \frac{V}{r + R} \tag{8.24}$$

2. in electronic drives, by applying to the motor, through an electronic converter, a minimum voltage of

$$V_{\min} = r \cdot I_A \tag{8.25}$$

Limiting the inrush current in the approaching values (8.23) is not a simple problem, and is linked in conventional drives to the size of the rheostat and, in electronic ones, to the choice of the switching frequency and the dynamic characteristics of the semiconductors used.

8.9 SPEED CHARACTERISTIC

In correspondence with the current $I = OA$ (Figure 8.10), we have

$$E = k \cdot \Omega \cdot \Phi(I) = AA' \quad \text{with } V = \text{const.}$$
$$E_1 = k \cdot \Omega_1 \cdot \Phi(I) = AA' \quad \text{with } \Omega_1 = \text{const.}$$

Dividing member to member, the speed of rotation Ω on the voltage V can be derived:

$$\frac{E}{E_1} = \frac{\Omega}{\Omega_1}; \quad \Omega = \Omega_1 \cdot \frac{E}{E_1} = \Omega_1 \cdot \frac{V - r \cdot I}{E_1} \tag{8.26}$$

That is

$$\Omega = \Omega_1 \cdot \frac{AA'}{AA''}$$

The rotation speed Ω_1 is obtained at the I_1 axis of the intersection point of the curve $E_1 = f(I)$ with the straight line $E = f(I)$.

For currents greater than I_1, for example $I = OA$, we have

$$\Omega = \Omega_1 \cdot \frac{AA'}{AA''} < \Omega_1$$

The speed then decreases as the current increases. The slope of the rotation speed curve is modest; however, in this field, said curve intersects the horizontal axis of the short-circuit current $I_{sc} = V/r$. For currents lower than I_1, for example $I = OB$, we have

$$\Omega = \Omega_1 \cdot \frac{BB'}{BB''} > \Omega_1$$

As the current decreases, the speed increases and asymptotically tends to infinity for $I \rightarrow 0$.

8.9.1 Air Gap Torque

To each active conductor armature crossed by the current $I_c = {}^{I}/_{a}$, the average tangential force is applied,

$$F_1 = L \cdot B_m \cdot I_c = L \cdot B_m \cdot {}^{I}/_{a}$$

which corresponds to the drive torque $T_1 = F_1 \cdot {}^{D}/_{2}$; for (5.5), we have

$$T_1 = \frac{D \cdot L}{2 \cdot a} \cdot \frac{\Phi \cdot p}{\pi \cdot D \cdot L} \cdot I = \frac{p}{2\pi \cdot a} \cdot \Phi \cdot I$$

Drive torque $T = T_1 \cdot N$ developed by active N conductors, is then

$$T = T_1 \cdot N = \frac{p \cdot N}{2\pi \cdot a} \cdot \Phi \cdot I = k \cdot \Phi \cdot I \tag{8.27}$$

having indicated a coefficient dependent on the construction characteristics of the motor with $k = \frac{p \cdot N}{2\pi \cdot a}$.

Equation (8.27) shows the torque at the air gap which, multiplied by the angular speed Ω (rad/s), yields the air gap power:

$$P_t = T \cdot \Omega = \frac{p \cdot N}{2\pi \cdot a} \cdot \Phi \cdot I \cdot \Omega = \frac{p \cdot N}{2\pi \cdot a} \cdot \Omega \cdot \Phi \cdot I = E \cdot I \tag{8.28}$$

Since the flux load is dependent on the armature current, in the series motor the torque depends only on the value of the absorbed current, that is $T = f(I)$. The performance of T is shown in Figure 8.10.

8.10 POWER LOSSES AND EFFICIENCY

The power $P_a = V \cdot I$ absorbed by the motor fed to the voltage $V = \text{const.}$ varies linearly based on the current.

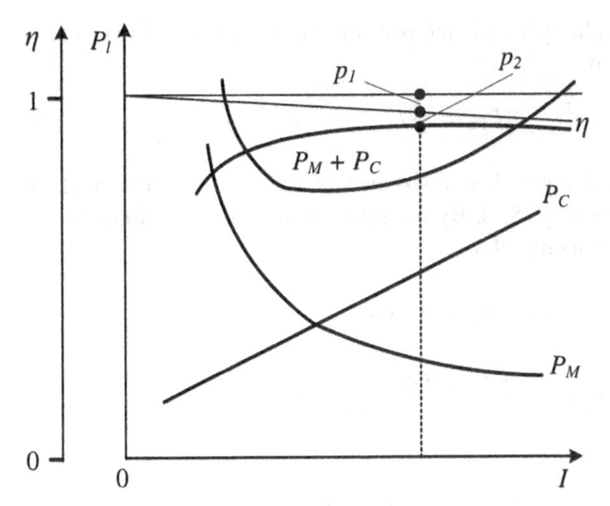

Figure 8.11 Power losses and efficiency (V = const).

The mechanical power to the net shaft torque is given by

$$P = \eta \cdot P_a = (1 - p_l) \cdot P_a \tag{8.29}$$

where

$\eta = 1 - p_l$ is the motor efficiency

p_l is the per unit p.u. losses, given by the ratio between the power lost P_l and the absorbed one P_a, namely,

$$p_l = {}^{P_l}/_{P_a}.$$

Constant load and power supply voltage yields the following:

1. Mechanical losses P_M, which depend on the speed of the rotor and are detected in the no-load test. The trend of these losses, as a function of the current I, is indicated in Figure 8.11.
2. Rotor core losses P_C, which depend on the speed and on the $E = V - r \cdot I = k \cdot \Omega \cdot \Phi$ emf. They are found, like the previous ones, in the no-load test.
3. Per unit p.u. copper losses in the armature windings, the inductor, the interpoles, and any compensator.
4. Per unit p.u. copper losses due to the contact of the brushes with the commutator which, as stated in Section 8.8, are approximately equal to 3 W/A.
5. The additional losses, which norms recommend considering conventionally equal to 0.5% of the net power.

Considering, for simplicity's sake, the per unit p.u. copper losses (3) and (4) expressed by a single term

$$P_{\mathrm{Cu}} = r \cdot I^2$$

in which r is the internal equivalent resistance of the motor, which will be assumed equal to that given by (8.1). By the same manner, the additional losses (5) will also not appear explicitly, thus

$$P_l = P_M + P_C + P_{\mathrm{Cu}}$$

$$p_l = \frac{P_M + P_C + P_{\mathrm{Cu}}}{P_a} = p_1 + p_2$$

having placed:

$$p_1 = \frac{P_{\mathrm{Cu}}}{P_a}; \quad p_2 = \frac{P_M + P_C}{P_a}$$

Because

$$p_1 = \frac{r \cdot I^2}{V \cdot I} = \frac{r \cdot I}{V}$$

Per unit p.u. copper losses are equal in number to the voltage drop ε (8.20); as being V = const., they are proportional to I (Figure 8.11).

The efficiency curve

$$\eta = 1 - p_l = 1 - p_1 - p_2$$

based on I has the trend indicated in Figures 8.11 and 8.12. Knowing the efficiency, the power may be calculated with (8.29), obtaining the curve shown in Figure 8.12.

8.11 TRACTIVE EFFORT DIAGRAM

The value of the torque to the motor shaft can be derived from the net power P with the ratio:

$$T = \frac{P}{\Omega} \tag{8.30}$$

The difference between the net torque and air gap torque (8.27) is linked to the difference between internal air gap power P_t, and net power P, a difference equivalent to mechanical losses and rotor core losses. In fact,

$$P_t = E \cdot I = (V - r \cdot I) \cdot I = P_a - P_{\mathrm{Cu}}; \quad P = P_a - P_{\mathrm{Cu}} - P_M - P_C;$$
$$P_t - P = P_M + P_C$$

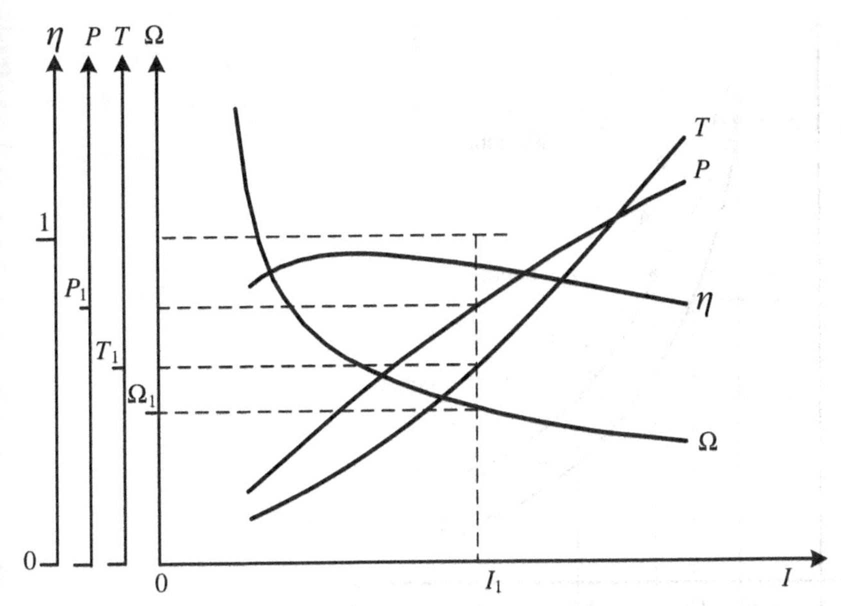

Figure 8.12 Electromechanical characteristics of a traction motor with series field windings, supplied with constant voltage.

When the link between T, the flux, and current needs to be highlighted, reference is made to (8.27); for the calculation of the net torque (Figure 8.12), calculation (8.30) must be used.

The tractive effort diagram of the motor has the trend indicated in Figure 8.13 (curve a), in which some hyperbolas at constant power are also plotted (that is, at constant current, with $V = \text{const}$).

It can be noted that it falls more than a hyperbole: in fact, the power decreases passing from point A_3 (starting conditions torque, at rotation velocity Ω_a) at point A_1, (nominal torque T_1 at nominal velocity Ω_1) and at point A_2 (maximum rotation velocity Ω_M), corresponding to the decrease of absorbed current.

The form of the tractive effort diagram, particularly suitable to the needs of the traction, is a direct consequence of series field winding connection; because the motor works with emf E, it is basically constant,[5] that is, with $\Omega \cdot \Phi \approx$ cost., at reduction of current when going from point A_3 to point A_2, it corresponds to a reduction of the flux, which leads to an inversely proportional increase in the rotation velocity from Ω_a to Ω_M. The ratio

$$\beta = {}^{\Omega_M}/_{\Omega_1} \tag{8.31}$$

is an elasticity index of the motor operation; in standard-sized DC machines, powered with constant voltage, it is around 2.

[5] Unless the internal voltage drops.

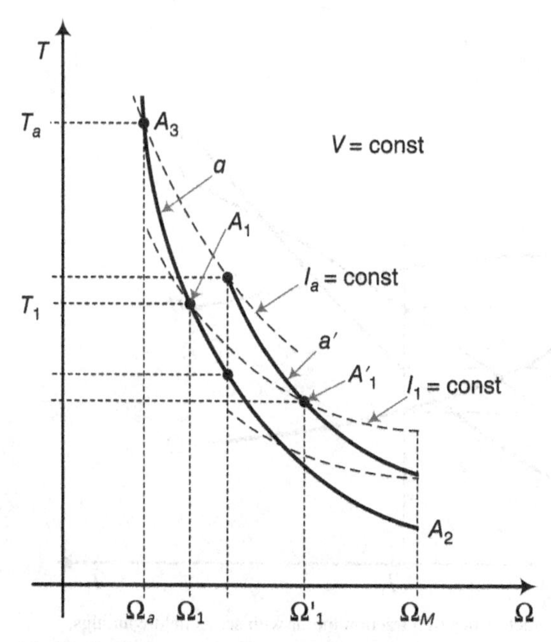

Figure 8.13 Torque–speed diagrams of a traction engine supplied by a constant voltage. a: Characteristic $T(\Omega)$ in a full field; a': characteristics $T'(\Omega)$ in a weak field.

8.12 SPEED REGULATION

From (8.10) we have $\Omega = \frac{E}{k \cdot \Phi}$; if there is a rheostat resistance R (Figure 8.14) in series with the motor, then

$$V = E + r \cdot I + R \cdot I = E + (r + R) \cdot I; \quad E = V - (r + R) \cdot I \tag{8.32}$$

for which the expression of the rotation speed assumes the most general form:

$$\Omega = \frac{V - (r + R) \cdot I}{k \cdot \Phi} \tag{8.33}$$

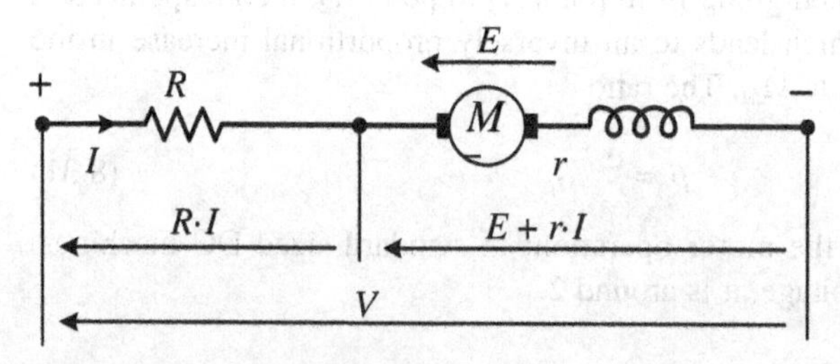

Figure 8.14 Adding of the rheostat starter.

It follows that in order to vary the speed Ω, keeping constant the current I, the following may be done:

1. Vary the power supply voltage V (voltage regulation)
2. Vary the flux by acting on the mmf field (field regulation)
3. Produce an additional voltage drop $R \cdot I$ by a rheostat in series with the armature circuit (rheostatic regulation)

8.12.1 Traditional Drives

Drives that are considered traditional are those where no use is made of an electronic power converter added between line and drive motors: these are powered from the contact line directly, or with the interposition of a rheostat. In equipment comprising two or more motors, all three regulation systems mentioned already are used, in those with a single motor (for example in trolleybuses), only the second and the third.

Section 8.17 will examine rheostatic regulation; although it gives rise to considerable power dissipation $\left(R \cdot I^2\right)$, it is essential to limit the current under starting conditions as we have seen in Section 8.8. The presence of the rheostat starter is typical of this type of drive.

8.12.2 Electronic Drives

If the drive comprises a DC power electronic converter, normally produced in the form of a chopper, the voltage at the motor terminals can be continuously adjusted from a value $V_{\min}$ (8.25) to its nominal value; thus the rheostat becomes useless. Field regulation can be used to improve motor performance.

8.13 VOLTAGE REGULATION

In general, we pass from the power supply voltage V, for which

$$E = V - r \cdot I = k \cdot \Omega \cdot \Phi(I)$$

to the voltage $V' = x \cdot V$, we have (Figure 8.15)

$$E' = V' - r \cdot I = k \cdot \Omega' \cdot \Phi(I)$$

Dividing member to member, the following is obtained,

$$\frac{\Omega'}{\Omega} = \frac{E'}{E} = \frac{V' - r \cdot I}{V - r \cdot I} = \frac{x \cdot V - r \cdot I}{V - r \cdot I}$$

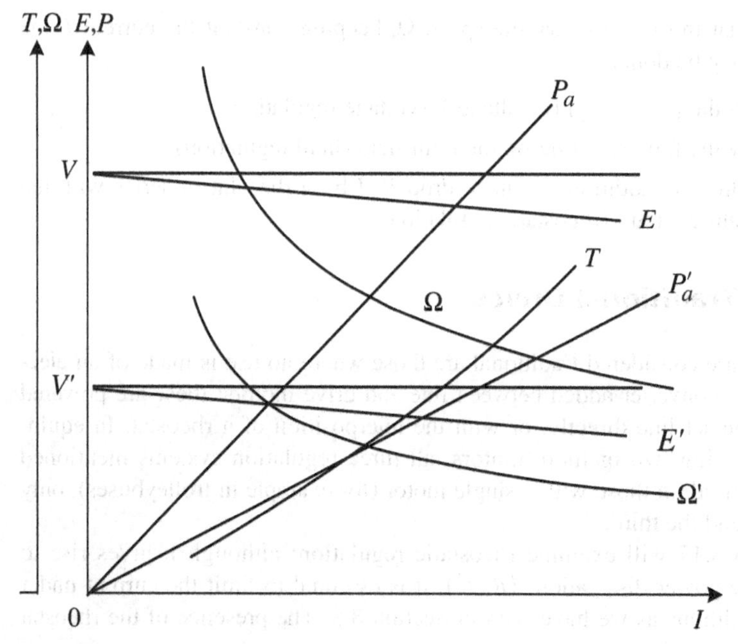

Figure 8.15 Voltage regulation.

and, highlighting the relative internal voltage drop $\varepsilon = r \cdot {}^{I}/_{V}$:

$$\frac{\Omega'}{\Omega} = \frac{x - \varepsilon}{I - \varepsilon} \tag{8.34}$$

Since ε is equal to a few hundredths, by neglecting it compared to x, the following is obtained,

$$^{\Omega'}/_{\Omega} \approx x \tag{8.35}$$

with equal current, therefore, the speed varies approximately in proportion to the voltage applied at the terminals. In fact, for $x < 1$, $\Omega' < x \cdot \Omega$ results. Absorbed power

$$P'_a = V' \cdot I = x \cdot P_a$$

varies in proportion to the voltage.

Per unit p.u. copper losses assume the same absolute value and are multiplied, in relative value, for $1/x$, as the relative voltage drops:

$$p'_1 = \frac{P'_{\mathrm{Cu}}}{P'_a} = \frac{P_{\mathrm{Cu}}}{x \cdot P_a} = \frac{p_1}{x}; \quad \varepsilon' = \frac{r \cdot I}{V'} = \frac{r \cdot I}{x \cdot V} = \frac{\varepsilon}{x}$$

Mechanical losses and core losses reduce as the voltage reduces, the former in a more significant amount than the latter. In practice, the following may be determined:

$$P'_M + P'_C \approx (P_M + P_C) \cdot \left(\frac{\Omega'}{\Omega}\right)^{1.6}$$

from which

$$p'_2 = \frac{P'_M + P'_C}{P_a} \approx \frac{p_2}{x} \cdot \left(\frac{\Omega'}{\Omega}\right)^{1.6}$$

The efficiency is then

$$\eta' = 1 - p'_1 - p'_2 \approx 1 - \frac{p_1}{x} - \frac{p_2}{x} \cdot \left(\frac{\Omega'}{\Omega}\right)^{1.6}$$

For example, if a motor works at its nominal voltage and power and it has

$$\eta = 0.92; \quad p = 0.08; \quad p_1 = \varepsilon = 0.05; \quad p_2 = 0.03$$

at voltage $V' = {}^{V}/_{2}$ it will have

$$\frac{\Omega'}{\Omega} = \frac{0.5 - 0.05}{1 - 0.05} = 0.474$$

$$p'_1 = \frac{0.05}{0.5} = 0.10; \quad p'_2 = \frac{0.03 \cdot (0.474)^{1.6}}{0.5} \approx 0.02$$

$$\eta' \approx 1.00 - 0.10 - 0.02 = 0.88$$

The net power

$$P' = \eta' \cdot P'_a = \eta' \cdot x \cdot P'_a$$

if we neglect the variation of the efficiency, it is proportional to the voltage. The air gap torque $T = k \cdot \Phi \cdot I$ remains unchanged (Figure 8.15) as the current is same.

8.14 FIELD REGULATION

In normal motor operation, the field mmf is equal to (Section 8.3) $M = N_1 \cdot I$ and is the maximum that can be achieved, since all the N_1 winding turns of each pole are crossed by I; when the motor operates under such conditions, the term "full field" is used.

To adjust the speed, the mmf corresponding to the armature current I can be made to be

$$M' = y \cdot M = y \cdot N_1 \cdot I(y < 1) \tag{8.36}$$

in which case the motor operates at a reduced or weak field, with the degree of field weakening z:

$$z = 1 - y = 1 - {}^{M'}/_{M} \tag{8.37}$$

Separate field series motors work similarly; in this case it switches from full field operation to a ($M = K \cdot I$) reduced field operation, placing $M' = K' \cdot I$; with: $K' = y \cdot K$.

The characteristics are always obtained with the same armature current, assuming V=const (Figure 8.17). It must be noted that, in any case, the equivalent internal resistance of the motor is reduced, that is, $r' < r$.

At full field it will have the flux Φ_0 with field current $I = OA$ and mmf $M(I) = N_1 \cdot I$. To obtain flux Φ_0 with a weak field, the mmf M must be the same; for this purpose, the armature current must assume a value $I' = OA'$, such as to satisfy the ratio,

$$M'(I') = M(I)$$

that is,

$$y \cdot N_1 \cdot I' = N_1 \cdot I; \quad I' = {}^{I}/_{y} \tag{8.38}$$

The characteristic of magnetization with a weakened field $\Phi'_0 = f(I)$ is thus derived from $\Phi_0 = f(I)$ at full field, multiplying the abscissae, equal ordinate, by $1/y$. The load flux corresponding to the abscissa $I = OA$ and $\Phi(I) = AB$; the armature reaction is due to the current I. A weak field, at the level of the abscissa: $I' = {}^{1}/_{y} = OA'$, we have flux $\Phi' = A'B'$, which is slightly lower than Φ because, while the field mmf remained unchanged, $M'(I') = M(I)$ armature current increased, thus the armature reaction.

If with equal mmf M, the difference between the two armature reactions is ignored, in full field and weak field, the curve $\Phi' = f(I)$ from $\Phi = f(I)$ multiplying the abscissa by $1/y$ can be derived: then the difference between the segments $A'B'$ e $A'B''$ can be neglected.

Considering Equation 8.36, the filed weakening can be made in two ways:

- By varying the number of active turns of the filed windings → $M' = y \cdot M = (y \cdot N_1) \cdot I$ (Figure 8.16a)
- By inseting a shunt resistive–inductive branch → $M' = y \cdot M = N_1 \cdot (y \cdot I)$ (Figure 8.16b)

8.14.1 Dynamic Behavior of Inductive Shunt Field Regulation

In vehicles powered by a contact line, there may be abrupt changes in voltage at the motor terminals due to the following:

- Sudden variations in the line load, with consequent variations of the voltage drops.

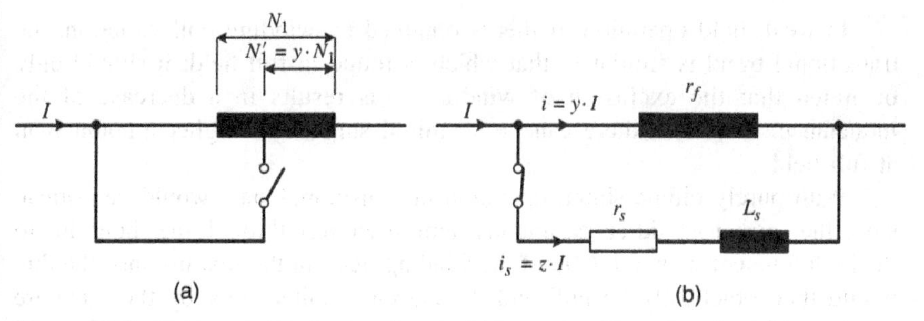

Figure 8.16 Field regulation of motors in series (a) by varying the number of active turns per pole; (b) by resistive–inductive shunt.

- Passage of the vehicle from a segment supplied by a substation to one powered from another substation, with a different voltage.
- Short accidental detachment of the pantograph from the contact line.
- Passage under a neutral zone.

In the last two cases, there is a power supply failure and sudden return of power, without a variation in the configuration of the traction circuit, given the brevity of the phenomenon.

Supposing that, operating in the full field, the motors revolve at the speed Ω and we have

$$I = \frac{V - E}{r}; \quad E = k \cdot \Omega \cdot \Phi \tag{8.39}$$

If, at the very least the power supply is interrupted, the current and the flux become null. When the voltage returns, the current does not immediately reach the steady-state value. Therefore,

$$V = e + r \cdot i + L \cdot di/dt$$

where

- e is the instantaneous value of the electromotive force (emf) induced in the armature windings; since it can be assumed that the speed of the vehicle, given the short interval of time, did not vary, it is found that e is proportional to the magnetic flux Φ.
- L is the total inductance of the motor.

During the first instants, the flux and emf are quite reduced, whereby i is limited almost exclusively by the internal resistance and by the inductance of the motor. There may be an initial surge current, which has the effect of increasing the armature reaction and distortion, therefore, the maximum difference of potential between the segments of the commutator.

In weak field operation, if this is obtained by winding coil variation, the transitional trend is similar to that which is found in full field; it should only be noted that the exclusion of winding coils results in a decrease of the inductance, therefore, there can be an initial surge of a higher amount than at full field.

With purely ohmic shunt, operation in transitional state would be critical, since the current would cross, initially, almost entirely through the shunt due to the high time constant $\tau = L/r_e$ of the winding field. In the first instants, the flux would then practically be null and the current, limited only by the armature winding inductance, would reach unacceptable values. Consequently, it is necessary to provide an inductive shunt, whose inductance L_s is such as to ensure a proper distribution of the current between the two branches of the field circuit at all times. The shunt comprises an inductor and possibly a resistor in series to obtain the desired value of r_s.

The magnetization characteristic of the shunt must be an appropriate inductance value $L_s = d\Phi_s/di_s$; experience suggests the most suitable forms of such characteristics, which can be obtained with an appropriate rating of the magnetic circuit of the inductor.

In electric or diesel vehicles or vehicles with batteries, as there are no sudden variations of the typical power supply voltage of a contact line system, it is possible to weaken the field with simple ohmic shunts. This also applies to traction vehicles powered from the line through auxiliary power converters: the latter are in fact self-protected, in addition to the network filter, by suitable devices that act promptly on regulation, in or for failure of power supply.

The two weak field systems, variating the number of winding coils or shunts, are functionally equivalent in steady state. The second has the advantage of not involving any changes to the stator of the motor connections, but for direct power supply systems from the contact line, it requires an inductive shunt of increasing mass and a footprint with the increase of motor power; conversely, the first system greatly complicates the windings and stator connections. It can be advantageous to vary the number of winding coils in motors with greater power, limiting the field of application of the inductive shunts to small and medium power, but it should be taken into account that, precisely because of the higher power, the necessity to have numerous economic speed control positions for weak field complicates the stator connections of the motors.

In Figure 8.17, it can be presumed that the scale of the emf E is such that the curves $\Phi_0(I)$ and $\Phi(I)$ also represent the curves E_{10} and E_1, for $\Omega_1 = \text{const.}$

At full field, for the abscissa $I = OA$, we have

$$E(I) = V - r \cdot I = k \cdot \Omega \cdot \Phi(I) = AD; \quad E_1(I) = k \cdot \Omega_1 \cdot \Phi(I) = AB$$

$$\frac{\Omega(I)}{\Omega_1} = \frac{E(I)}{E_1(I)} = \frac{AD}{AB} \tag{8.40}$$

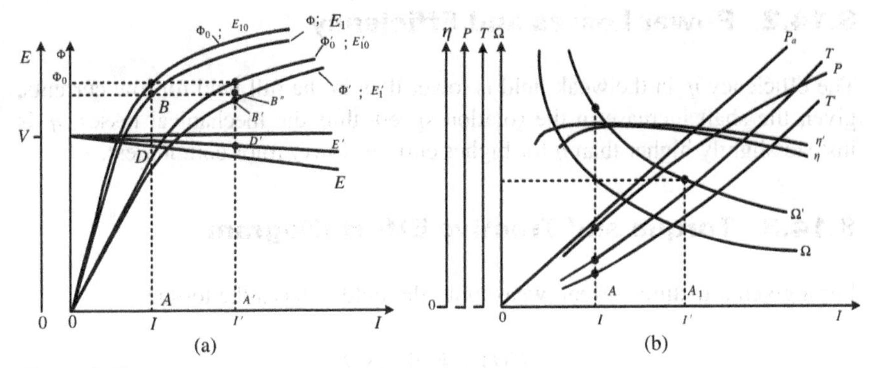

Figure 8.17 Field regulation (V=const.).

At weak field, remembering that $r' < r$, for the abscissa $I' = {}^{I}/_{y} = OA'$ we have

$$
\begin{aligned}
E'(I') &= V - r' \cdot I = k \cdot \Omega' \cdot \Phi'(I') = A'D'; \\
E'(I') &= k \cdot \Omega_1 \cdot \Phi'(I') = A'B' \\
\frac{\Omega'(I')}{\Omega_1} &= \frac{E'(I')}{E_1'(I')} = \frac{A'D'}{A'B'}
\end{aligned}
\tag{8.41}
$$

Dividing member to member (8.41) and (8.40), the following is obtained:

$$
\frac{\Omega'(I')}{\Omega(I)} = \frac{E'(I')}{E(I)} \cdot \frac{E_1(I)}{E_1'(I')} = \frac{A'D'}{AD} \cdot \frac{AB}{A'B'} \tag{8.42}
$$

$E(I)$ and $E'(I')$ differ for the different internal voltage drops of the motor ($r \cdot I$ instead $r' \cdot I'$). The current I' greater than I in the ratio I/y, for which, although there is decreased internal resistance, $A'D'$ is slightly less than AD; neglecting this small difference, it can be considered:

$$
\frac{A'D'}{AD} \approx 1, \text{ that is, } E(I) \approx E'(I')
$$

$E_1(I)$ and $E_1'(I')$ differ only in the different armature core or stack reaction; also in this case it may be assumed

$$
A'B' \approx A'B'' = AB, \text{ therefore, } E_1(I) \approx E_1'(I')
$$

Ultimately, (8.42) can be simplified as follows:

$$
\Omega'(I') \approx \Omega(I) \tag{8.43}
$$

meaning that the speed curve in a reduced field, such as that of magnetization, can also be derived from full field curve by multiplying the abscissae, equal ordinate, by $1/y$ (Figure 8.17b).

The speed increases in a weak field condition with the same current because the flux is reduced, with the emf remaining virtually unchanged.

8.14.2 Power Losses and Efficiency

The efficiency η' in the weak field is lower than in the full field for low currents, given the shaft increase in the rotation speed, thus the mechanical losses; η' is instead slightly higher than η for higher current (lower rotor core losses).

8.14.3 Torque and Tractive Effort Diagram

For a given armature current, weakening the field reduces the torque

$$T'(I) = k \cdot \Phi'(I) \cdot I$$

due to the decrease of the flux. Since the product $T \cdot \Omega$ is proportional to the net power and it is presumed that this does not vary, we have

$$T' \cdot \Omega' = T \cdot \Omega, \rightarrow T'(I) = T(I) \cdot \frac{\Omega(I)}{\Omega'(I)} \tag{8.44}$$

which allows the curve $T' = f(I)$ to be derived from $T = f(I)$ (Figure 8.17b).

In Figure 8.13, the tractive effort diagram shows $T' = f(\Omega)$ (curve a') with weak field, next to the full field $T = f(\Omega)$ (curve a).

8.14.4 Coefficient of Elasticity

If the motor is powered at constant voltage, the field regulation allows for extending the speed range to where the full power of the motor can be used. In this respect, the coefficient of elasticity is defined as the ratio (Figure 8.13)

$$\gamma = {}^{\Omega'_1}/_{\Omega_1} \tag{8.45}$$

between the speed obtained with the maximum weak field and the full field, both corresponding to the continuous power of the motor (points A_1 and A'_1).

The degree of weakening z increases the distortion of the field for uncompensated motors due to the increase of the ratio between the armature mmf and the main pole mmf.

Due to the effect of this distortion, the ratio B_M/B_m between the maximum and average induction increases, and the distribution of the voltage on the commutator worsens. It is, therefore, not possible for these motors to exceed values of z greater than 40–50%.

If higher degrees of weakening must be obtained, up to 70–75% (y_{min} = 25–30%), such as in great power locomotive, it is essential to add the compensating winding (Figure 8.18).

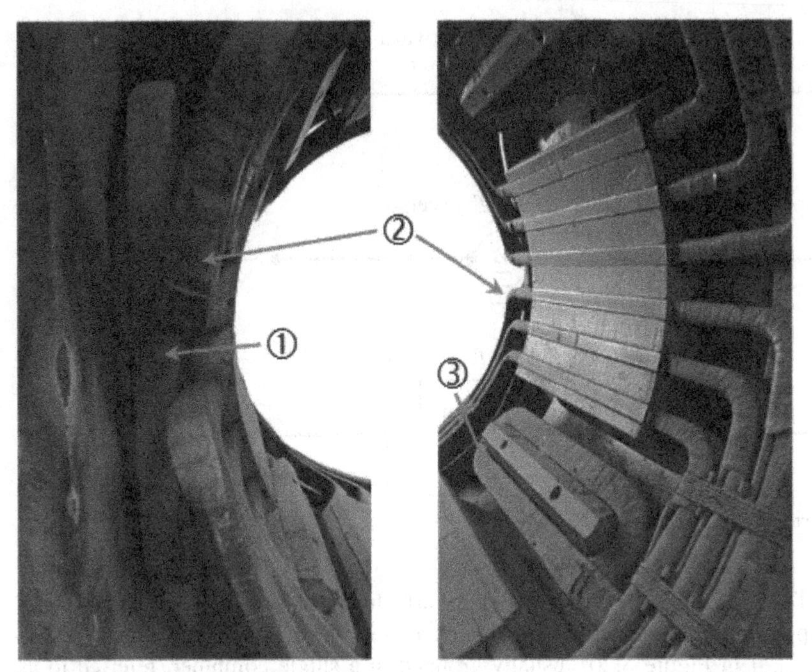

Figure 8.18 Details of the section of a compensated four-pole traction motor. (1) Field winding; (2) compensating windings; and (3) interpole winding.

8.15 FORWARD/REVERSE DRIVE

Since for DC motors the ratios

$$E = k \cdot \Omega \cdot \Phi \quad \text{and} \quad T = k \cdot \Phi \cdot I \tag{8.46}$$

are also valid as regard the signs, to reverse the forward/reverse drive, that is, to change the sign simultaneously to Ω and to T, it is necessary to invert E and I or Φ.

Considering a motor field in series; if, in the forward drive, the direction of various magnitudes is the one represented in Figure 8.19a, to make the reverse drive, it is possible

- to reverse the connections of the field winding (Figure 8.19b), so as to change the sign of Φ and
- to reverse the armature winding connections (Figure 8.19c), so as to change the sign of E and I.

In plane Ω (abscissae)–T (ordinate) with the forward/reverse drive, the tractive effort diagram is transferred from the first to the third quadrant and the braking characteristic from the fourth to the second quadrant.

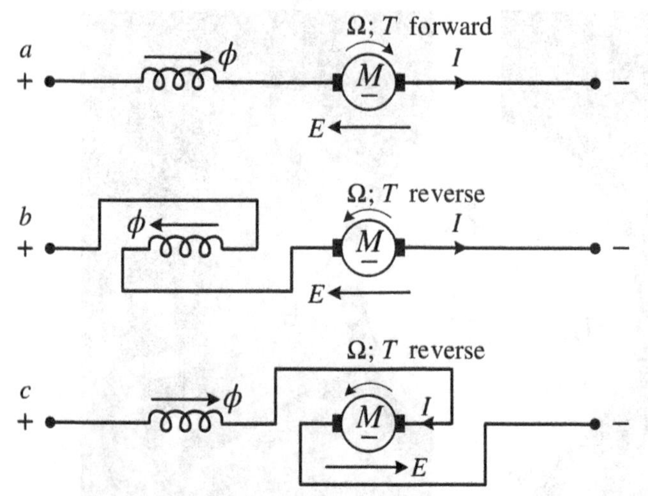

Figure 8.19 Forward/reverse drive.

To make the reversal of the field or armature connections, for each motor or group of motors connected permanently in series,[6] four disconnectors are used (1, 2, 3, 4 in Figure 8.20), usually grouped in a single combiner, referred to as forward/reverse drive switch. It must always be operated with an open circuit configuration; contacts can be made through simple brushes, sliding on suitably shaped segments.

Of course, a mechanical lock is necessary to prevent the forward/reverse drive when the drive circuit is crossed by current. In practice, a lock system links the command of the forward/reverse drive switch to the main combiner forward/reverse drive switch, in order to avoid the latter operating improperly.

The forward/reverse drive command can be direct or indirect.

8.15.1 Direct Command Forward/Reverse Drives

The drive shaft is maneuvered by hand using a lever, removable in an intermediate locked position ("zero" position). The mechanical lock between the shaft of the drive and that of the main combiner causes the following:

- When the master controller is in the "zero" position (open circuit position), it is also possible to bring the forward/reverse drive lever to a "zero" position. In this manner, the main combiner remains locked. In a

[6] In the latter case the field windings are connected permanently in series, as are the armature windings; from the circuit point of view, the motor group is equivalent to the single motor M shown in Figure 8.19.

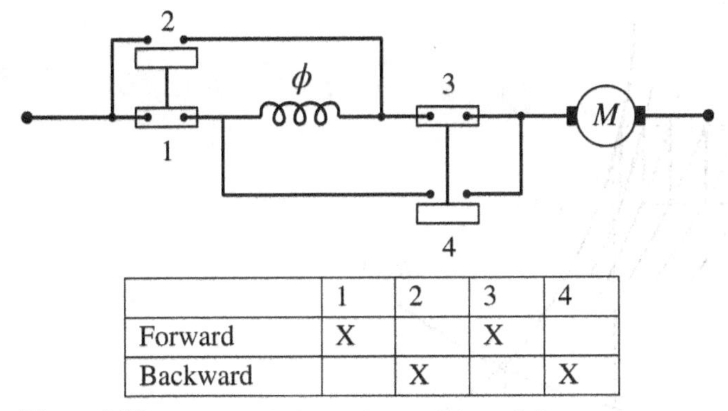

	1	2	3	4
Forward	X		X	
Backward		X		X

Figure 8.20 Diagram of a forward/reverse drive switch.

"zero" position, the forward/reverse drive handle can be removed from the command post[7] to prevent undesired operations.

- With the main combiner in the "zero" position, the forward/reverse drive handle may be shifted forward or back. The main combiner is thus unlocked and can be operated.
- If the main combiner is in any forward/reverse position, with the traction circuit crossed by the current, it is no longer possible to operate the forward/reverse drive, which remains locked in the forward or reverse position already set.

8.15.2 Indirect Command Forward/Reverse Drives

An electropneumatic control is normally used. The shaft is driven by a pneumatic cylinder with two positions (forward, reverse), powered by two solenoid valves.

The command is carried out electrically by an auxiliary combiner (CIM) whose low-voltage contacts command the solenoid valves. The lock in this case is achieved between the positions of the main combiner and the auxiliary CIM, in a manner entirely similar to that described for direct control.

8.15.3 Separate Field Motors

When there are separate field series motors (Section 8.3), the forward/reverse drive can be carried out either by inverting the supply polarity of the field windings (two-quadrant converter), or even by resorting to an electromechanical forward/reverse drive device, however, acting on low-voltage circuits and with limited current, therefore, lighter and less bulky than high-voltage drives.

[7] Thus, it acts as a locking key switch.

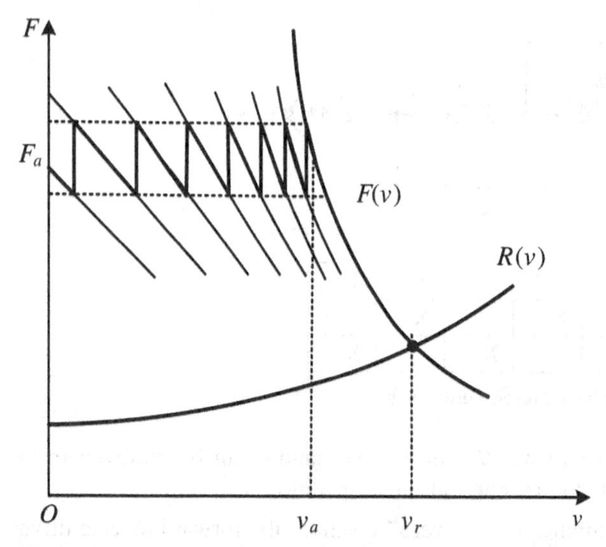

Figure 8.21 Torque–speed diagrams of traction and resistance to motion.

8.16 SPEED CONTROL

Consider a vehicle of weight G that runs with the resistance to motion $R(v)$ (Figure 8.21). If $F(v)$ is its tractive effort diagram, the steady speed v_r corresponds to the intersection of the curves $R(v)$ and $F(v)$.

One tractive effort diagram $F(v)$ is not sufficient; it is first necessary to start the vehicle without the current absorption and the flux exceeding the allowable values and the strength of traction exceeding the adhesion limit.

We shall consider DC traction vehicle equipped with commutator motors: as we have seen in Section 8.12, if the drive is traditional, a rheostat must be added at the inrush to limit the current and then gradually reduce the resistance R, so that the tractive effort is maintained on average equal to F_a (Figure 8.21), until reaching the tractive effort diagram at full voltage $F(v)$, at the speed v_a.

In general, if the vehicle is supplied by the line voltage V_L and is equipped with m drive motors, speed regulation can be performed (Figure 8.22)

1. by varying the voltage at the motor terminals, which are connected in series, in series–parallel, or in parallel. Theoretically, for each motor the voltage may be varied from the minimum value ${}^{V_L}/_m$ (series) to the maximum value V_L (parallel); however, all the possible combinations are not always used. Figure 8.22 shows two characteristics, in series and in parallel and
2. by means of field regulation of the motors.

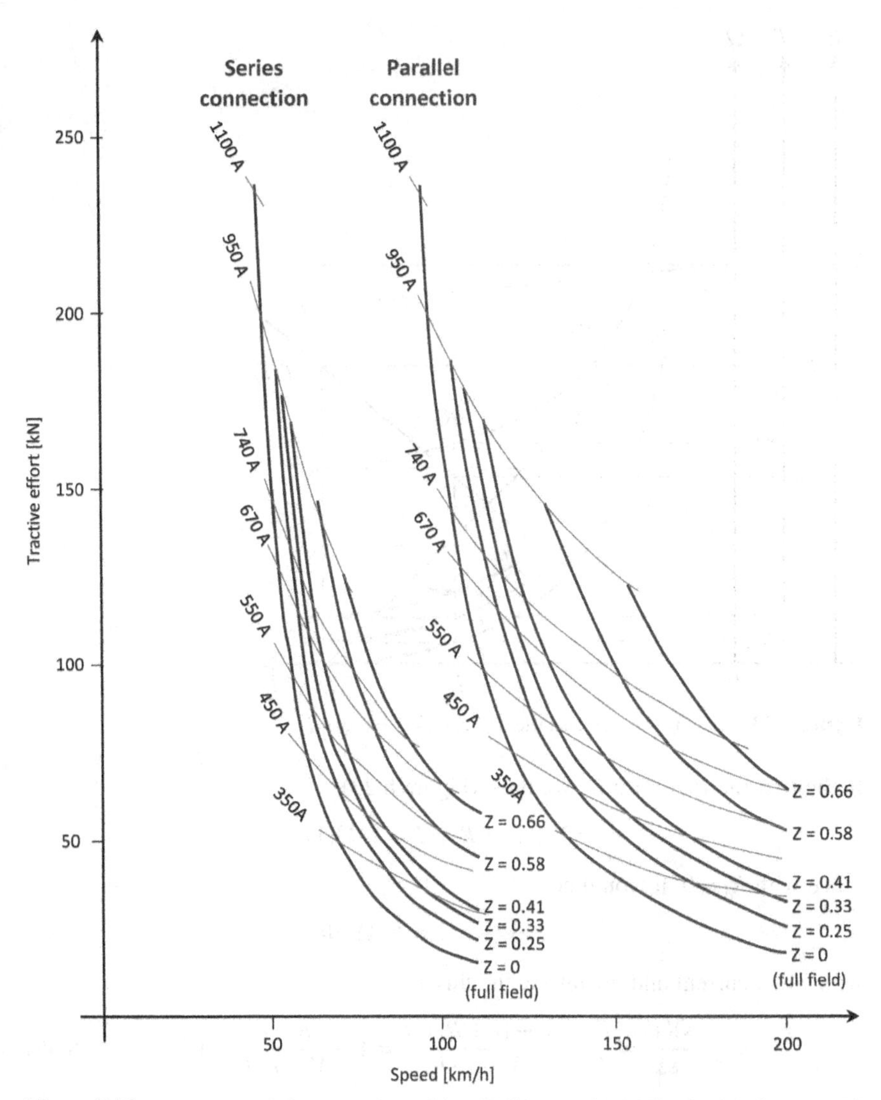

Figure 8.22 Torque–speed diagrams of a traditional DC locomotive with two pairs of motors, powered at half voltage.

8.17 RHEOSTATIC REGULATION

Consider for simplicity's sake only the starting conditions of a traction series field motor supplied at voltage V.

Setting the value I_A of the inrush current from (8.24), the total resistance of the rheostat is obtained:

$$R_1 = \frac{V}{I_A} - r \tag{8.47}$$

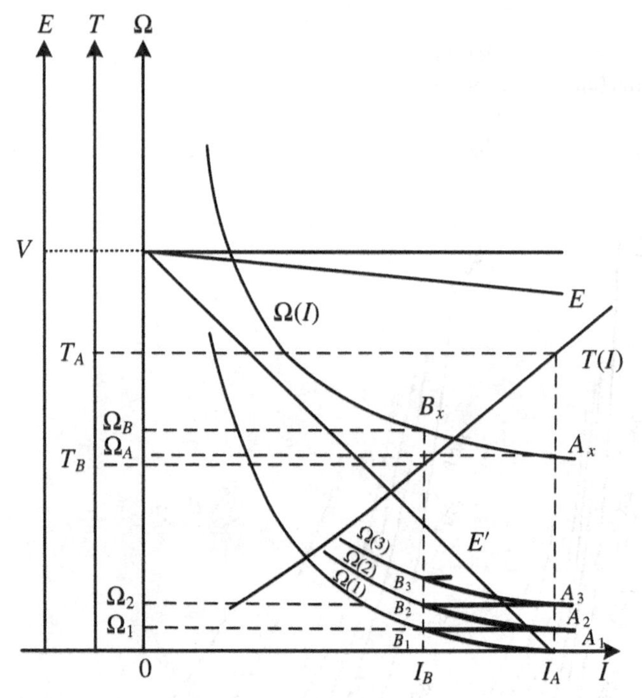

Figure 8.23 Starting rheostat: characteristic curves ($V = \text{const.}$).

In the first rheostat position we have (Figure 8.23)

$$E = V - (r + R_1) \cdot I = k \cdot \Omega(1) \cdot \Phi$$

Since with $R_1 = 0$, it would be

$$E = V - r \cdot I = k \cdot \Omega \cdot \Phi$$

with equal current and, therefore, the flux is

$$\frac{\Omega(1)}{\Omega} = \frac{E'}{E} = \frac{V - (r + R_1) \cdot I}{V - r \cdot I} = 1 - \frac{R_1 \cdot I}{V - r \cdot I} < 1 \tag{8.48}$$

The torque $T = k \cdot \Phi \cdot I$ does not vary.

When the rheostat is added, a part of the power $P_a = V \cdot I$ absorbed is dissipated within it, whereby the net power at the shaft is reduced to

$$P' = \eta \cdot \left(P_a - R_1 \cdot I^2\right)$$

where η is the efficiency of the motor; the total efficiency of the motor + rheostat is

$$\eta' = \frac{P'}{P_a} = \frac{\eta \cdot \left(P_a - R_1 \cdot I^2\right)}{P_a} = \eta \cdot \left(\frac{R_1 \cdot I}{V}\right)$$

From (8.48), neglecting the internal voltage drop r relating to V, we have

$$\frac{\Omega(1)}{\Omega} \approx 1 - \frac{R_1 \cdot I}{V}$$

therefore,

$$\eta' \approx \eta \cdot \frac{\Omega(1)}{\Omega} \tag{8.49}$$

The reduction of speed using the rheostat results in a decrease in the overall efficiency, which is roughly proportional to said reduction.

8.17.1 Rheostat Sections

During starting conditions, to maintain $I = I_A =$ const, it would be necessary to continuously diminish the resistance of the rheostat from R_1 to zero, so as to follow the ordinates A_1 A_2, A_3, . . . , A_x of Figure 8.23.

The rheostat is, in fact, made up of elements in series, of resistance r_1 r_2, r_3, . . . , which are progressively short-circuited by shutting the contactors I, II, III, . . . (Figure 8.24).

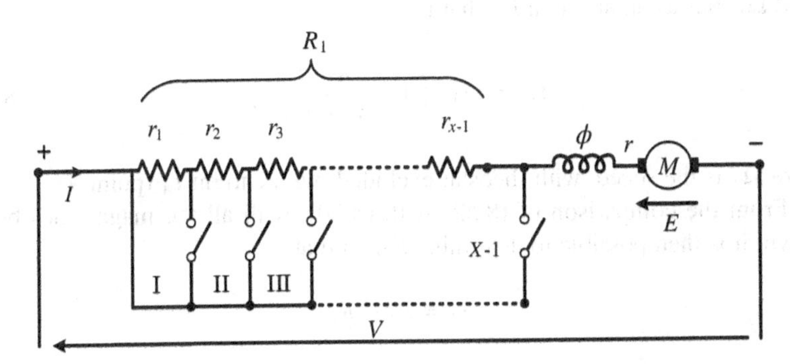

Position	Contactors					
	I	II	III	…	x–1	R
1						$R_1 = r_1 + r_2 + r_3 + \cdots$
2	X					$R_2 = r_2 + r_3 + \cdots$
3	X	X				$R_3 = r_3 + \cdots$
⋮						
x	X	X	X		X	$R_x = 0$

Figure 8.24 Starting conditions rheostat main block diagram.

The decrease of the starter conditions' resistance, therefore, occurs in a discontinuous manner. The current is not maintained constant, but varies, for each position, from the maximum value I_A to the minimum value I_B (Figure 8.23).

In the first position, while the current decreases from I_A to I_B, rotation speed increases from zero (point A_1) to Ω_1 (point B_1). Therefore,

$$\Omega_1 = \Omega_B \cdot \left(1 - \frac{R_1 \cdot I_B}{V - r \cdot I_B}\right) \tag{8.50}$$

where Ω_B is the speed which would correspond, with rheostat excluded, to the I_B current (point B_x). The instant the current I_B is reached, the contactor I is shut, therefore, the resistance of the rheostat is reduced to

$$R_2 = r_2 + r_3 + \cdots < R_1$$

The characteristic of the motor speed in this second position is then

$$\Omega(2) = \Omega \cdot \left(1 - \frac{R_2 \cdot I}{V - r \cdot I}\right) \tag{8.51}$$

It can be assumed, given the speed of the transition from the first to the second position, that the motor speed remains equal to Ω_1; to not exceed the current I_A, it is then necessary that point A_2 of the characteristic (8.51), corresponding to the speed Ω_1, has as an abscissa I_A; that is,

$$\Omega_1 = \Omega_A \cdot \left(1 - \frac{R_2 \cdot I_A}{V - r \cdot I_A}\right) \tag{8.52}$$

where Ω_A is the speed, with rheostat excluded, with current I_A (point A).

From the comparison of (8.52) with (8.50), with all the magnitudes being known, it is then possible to determine R_2, so that

$$r_1 = R_1 - R_2$$

Similarly, in the second position, once reaching point B_2 with speed Ω_2 and current I_B, contactor II shuts, so as to short circuit the resistance segment r_2 and move to the third position

$$R_3 = r_3 + r_4 + \cdots$$

in which the initial operating point A_3 corresponds to the revolutions Ω_2 and the current I_A.

The current thus oscillates between the values I_A and I_B, the torque between T_A and T_B. It is evident that the closer the ratio $^{I_A}/_{I_B}$ is to the unit, the more numerous the rheostat section will be. To obtain x starting conditions, including the final one with rheostat excluded (Figure 8.24), requires $x - 1$ sections and $x - 1$ contactors. Therefore, high regulation finesse requires many contactors; in practice, the solution that allows the current deviation to be contained within acceptable limits, and with a reasonable cost, is selected.

For the starting conditions diagram $A_1 - B_1 - A_2 - B_2 - A_3 - B_3 \ldots$, there is a corresponding similar diagram of the traction forces as a function of speed (Figure 8.21); normally, the average value F_a of the traction force is simplified, constant between zero and the speed v_a.

8.17.2 Approaching Positions

For an EMU, the acceleration corresponding to the traction force F_a must be in the order of 1.0–1.4 m/s^2 in urban services and a little lower in suburban services, and must be reached gradually so as to not cause undue discomfort to passengers: for this purpose, the permissible acceleration gradient is prescribed during the inrush phase at variable acceleration and controlled current. It follows that some approaching rheostatic sections are necessary, corresponding to low currents, starting from the minimum value (8.23):

$$I_{A\,\mathrm{min}} = K_{\mathrm{min}} \cdot I_n$$

This need is even more felt in locomotives for several reasons: here, in fact, F_a is such as to ensure starting conditions of trains with the toughest composition conditions (maximum towed mass) and track (maximum gradient). In the starting conditions, the traction force must gradually reach the value F_a to prevent any tearing in the couplers between the vehicles; it is also necessary to graduate F_a when the towed mass is reduced or at least the locomotive is isolated (maneuvers and approach to trains).

A minimum value $I_{A\,\mathrm{min}}$ inrush current corresponding to a total resistance of the rheostat:

$$R^* = \frac{V}{I_{A\,\mathrm{min}}} - r \tag{8.53}$$

significantly greater than the R_1 given by (8.47). The resistance sections

$$r^* = R^* - R_1$$

must be sized for current values and the inner insertion times, compared to those that constitute R_1; however, their strong ohmic value results in a mass and a footprint that often, especially in locomotives, for which $K_{\mathrm{min}} = 0.2 - 0.4$ "gentleness" of maneuvers would be required, pose significant problems. Therefore, from time to time, a compromise must be found between the conflicting demands.

8.18 AUTOMATIC STARTING CONDITIONS

The transition from one rheostat position to the next can be controlled directly by the driver; in such a case, the opportunity to follow the theoretical starting conditions diagram examined in the previous section depends on the driver's abilities.

Often this involves automatic starting conditions. In this case, the overall devices and control circuits that are given the shorter name, starter, control the contactors of the individual sections of the rheostat, so as to exclude them progressively according to the predetermined sequence, as the passage from one position to the next is subject to certain conditions.

In the past, the starter consisted of electromechanical devices, and subsequently of a system of electronic circuits constituting a static logic supplied at low voltage by the on-board battery. The fundamental control is that of current: a complex of the minimum current relays (acceleration relays) locks the starting conditions in the individual positions as long as the current exceeds the value I_B, and allows passing to subsequent positions only at the points B in Figure 8.23.

With the simple amperometric detection implemented by acceleration relays, the mean value F_a of the traction force is determined; the acceleration of the train,

$$a = \frac{F_a - R}{m_e}$$

therefore, depends on its mass and the gradient of the line. It is necessary to keep in mind that mass can vary, even significantly, due to the load or changes in the number of vehicles driven; consequently, it is possible to have significant acceleration deviation as well as the danger of exceeding the limit of adhesion when the load on the motor wheelsets is reduced.

To obviate these drawbacks, two or more calibration values of the acceleration relays may be implemented so as to implement starting conditions with different average currents. In EMUs, devices called variable loading are also used, which regulate the calibration of the acceleration relays, therefore, the average value of the starting conditions current, as a function of the load; such devices, particularly employed in metropolitan areas, measure the load through the pressure value of the suspensions of the carriages.[8]

With electronic starters it is possible to control not only the starting current, but also the acceleration directly.

8.19 SERIES–PARALLEL CONNECTION OF THE MOTORS

As we know, the voltage regulation in DC vehicles with traditional type equipment and multiple motors consists of connecting the motors by different

[8] In vehicles with secondary pneumatic suspension, the variable loading device is driven by the pressure from the air springs, which is precisely a function of load.

schematics. In equipment with four drive motors (wheel arrangement B_0B_0 or equivalent), it is possible to have the following combinations:

- Series (S)
- Series–parallel (SP), with the motors divided into two groups in parallel, each consisting of two series motors
- Parallel (P)

If the line voltage is $V_L = 3000$ V, the parallel connection is never used, because (Section 8.1) the motors are at half voltage, that is, winding at 3000/2 V. Even with the inner line voltage, however, the first two combinations are usually used in order to not excessively complicate the circuits; in fact, in this manner, each couple of motors permanently in series has a circuit equivalent to a single motor and the diagram does not differ from equipments with two drive motors (BB wheel arrangement or equivalent).

In equipment with six drive motors, frequently used in networks at 3 kV, the possible connections are as follows (considering mid voltage motors):

- Series (S)
- Series–parallel 1 ($SP1$), with two groups in parallel, each formed by three motors in series
- Series–parallel 2 ($SP2$), with three groups in parallel, each formed by two motors in series

There are also locomotives of older conception that have double motors, namely, two motors for each driving wheelset. The locomotives $B_0B_0B_0$, for example, have twelve connectable motors: in series (voltage 3000/12 = 250 V/motor), in two parallel groups of six in series motors (500 V/motor), in three parallel groups of four in series motors (750 V/motor), in four parallel groups of three in series motors (1000 V/motor).

The motors deliver their nominal power in the connection that provides higher voltage to the terminals, that is, normally at the nominal voltage $V_n = V_L$ or $V_n = {}^{V_L}/_2$; in the particular case of the locomotives just mentioned, we have $V_n = {}^{V_L}/_3$.

With regard to the permissible current, it should be noted that in the series combinations, the thermal equilibrium of the motor is modified due to the following causes:

- Copper losses, at equal current, are unchanged in absolute terms.
- Rotor core losses, even in absolute terms, are reduced (Section 8.13), for which the overall losses in the active parts, thus the heat produced, undergo a reduction, which is however limited.
- The ventilation is less effective, particularly for self-ventilated motors, since the voltage is reduced as well as the speed (8.34).

8.20 SERIES–PARALLEL TRANSITION

The transition from one connection of the motors to the next is carried out without interrupting the traction current and requires a number of modifications of the same circuit, which follow one another in a certain order and constitute the transition positions.

The transition from a series to a parallel connection will be considered, with equipment formed by two drive motors, assuming that the transition is carried out under starting conditions with average motor current I_a and traction force F_a.

8.20.1 Short Circuit Transition

The simplified diagram of the drive circuit is shown in Figure 8.25. In the last position, the two series motors, each supplied with voltage $V' = {}^{V}\!/_{2}$, have the speed Ω_0, corresponding to point S of the speed characteristic Ω' (Figure 8.26).

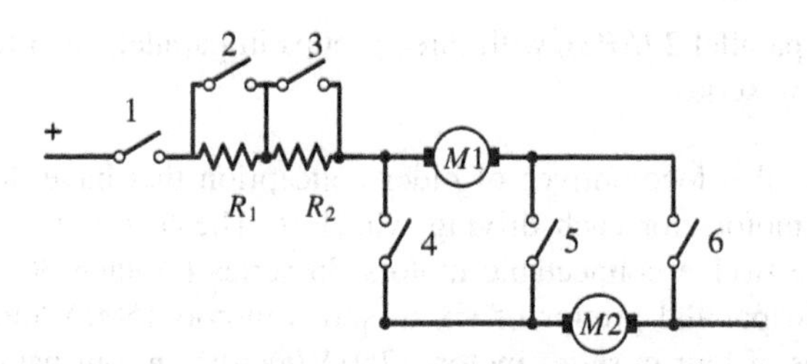

Position		Contactors					
		1	2	3	4	5	6
Zero	0						
Series	1	X				X	
Series	⋮	X				X	
Series	s	X	X	X		X	
Transition	1	X	X			X	
Transition	2	X	X			X	X
Transition	3	X	X				X
Transition	4	X	X		X		X
Parallel	1	X	X	*a*	X		X
Parallel	⋮	X	X		X		X

Figure 8.25 Transition in short circuit. *a*: In the first position in parallel the resistance R_2 must be reduced from R to $R/2$.

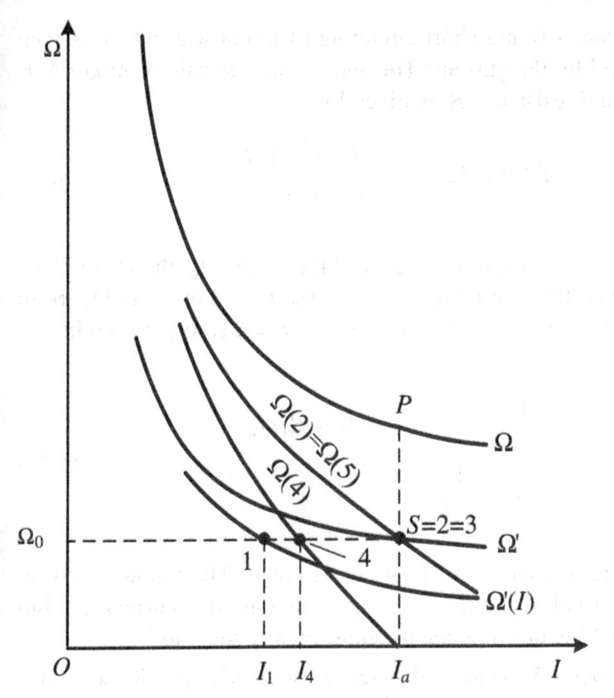

Figure 8.26 Characteristic curves in the short-circuit transition.

Since the transition phases jointly have a very short duration, we can assume that the motor speed throughout the process remains constant and equal to Ω_0. The different phases are the following[9]:

1. Insertion of a part of the rheostat,[10] of resistance R, by the opening of contactor 3. The speed characteristic $\Omega'(1)$ corresponds to operation with voltage $V' = {}^{V}\!/_{2}$ and resistance ${}^{R}\!/_{2}$ for each motor, equivalent to

$$\Omega'(1) = \Omega' \cdot \frac{{}^{V}\!/_{2} - (r + {}^{R}\!/_{2}) \cdot I}{{}^{V}\!/_{2} - r \cdot I}$$

where Ω' is the speed that the motor would have, at equal current, with voltage V' and rheostat excluded. The current absorbed by each motor drops to the value $I_1 < I_a$ (point 1 in Figure 8.26), for which there is a strong decrease of the driving torque and the tractive effort.

[9] These correspond to the transition positions indicated in the table of Figure 8.25.

[10] Corresponding to the section labeled R_2 in Figure 8.25.

2. Shutting of contactor 6 and short circuiting of motor M2, which is then no longer crossed by the current. The speed characteristic of motor M1, with voltage V and resistance R, is given by

$$\Omega(2) = \Omega' \cdot \frac{{}^{V}\!/_{2} - \left(r + {}^{R}\!/_{2}\right) \cdot I}{{}^{V}\!/_{2} - r \cdot I} \tag{8.54}$$

If, as is logical, we do not want to exceed the current I_a, the characteristic $\Omega(2)$ must pass through point S of abscissa I_a and ordinate Ω_a: from (8.54), keeping in mind that for $I = I_a$, $\Omega' = \Omega_0$ the following is obtained,

$$\begin{aligned} \frac{V}{2} - r \cdot I_a &= V - (r + R) \cdot I_a \\ R &= \frac{V}{2 \cdot I_a} \end{aligned} \tag{8.55}$$

which allows the resistance R[11] to be determined. The torque of motor M1 rises to normal value T_a, corresponding to the current I_a, but because motor M2 is inactive, the traction force is equal to ${}^{F_a}\!/_{2}$.

3. Opening of contactor 5, so as to disengage motor M2, previously short-circuited. The different magnitudes are unchanged from the position 2.

4. Insertion of motor M2 in parallel to M1, by shutting contactor 4. If I is the current absorbed by each motor, the rheostat R is crossed by the current $2 \cdot I$, resulting in the voltage drop $2 \cdot R \cdot I$; the speed characteristic $\Omega(4)$ thus refers to the voltage V and the resistance for each motor $2 \cdot R$, that is,

$$\Omega(4) = \Omega' \cdot \frac{{}^{V}\!/_{2} - \left(r + {}^{R}\!/_{2}\right) \cdot I}{{}^{V}\!/_{2} - r \cdot I} \tag{8.56}$$

Given the condition (8.55), this characteristic intersects the abscissae axes at a point corresponding to a current slightly different than I_a; the current I_4, corresponding to point 4 of Ω_0, the ordinate of Figure 8.26, is thus significantly lower than I_a, therefore, the traction force still remains very low.

The transition is thus completed; the value of the series resistance may now be rapidly reduced to the two parallel motors group, so as to have a characteristic $\Omega(5)$ that crosses point S. Then we can proceed to the starting conditions from point S to point P, with average current I_a.

[11] Note that to obtain at inrush ($\Omega = 0$) with two motors connected in series, the current I_a, the rheostat should have resistance $R_t = \frac{V}{I_a} - 2 \cdot r \approx \frac{V}{I_a} = 2 \cdot R$. Therefore, the section indicated with R_1 in the diagram of Figure 8.25 also has a resistance approximately equal to R.

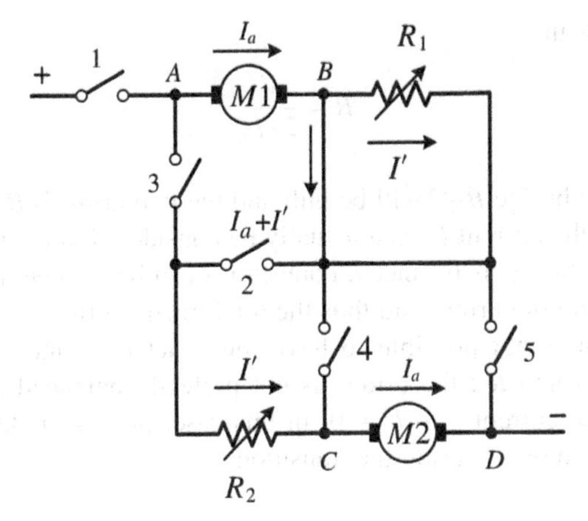

Position		Contactors				
		1	2	3	4	5
Zero	0					
Series	1	X	X			
Series	⋮	X	X			
Series	s	X				
Transition	1	X		X	X	X
Transition	2	X		X		X
Parallel	1	X		X		X
Parallel	⋮	X		X		X

Figure 8.27 Transition to bridge.

8.20.2 Bridge Transition

The principal diagram is shown in Figure 8.27. The rheostat is divided into two groups R_1 and R_2, which will both be assumed to have resistance R.

In the last series position, the rheostats R_1 and R_2 are excluded from contactor 4; the current I_a flows through the loop A–M1–B–C–M2–D. The operating conditions of each of the two motors correspond to point S of Figure 8.26.

If contactors 3 and 5 are shut, R_2 is then connected in parallel to motor M1 and R_1 in parallel to M2. Because the voltage between A and B and between C and D are equal to $V/2$, both rheostats are crossed by the current $I' = (V/2)/R$, that can be assumed to follow the segment A–R_2–C–B–R_1–D. The connection B–C, called bridge, is crossed by the current $I_a - I'$.

The parallel connection of the two motors, each with a rheostat in series, is thus already prepared. If the resistances R are such that

$$I' = I_a$$

Therefore, results in

$$R = \frac{V}{2 \cdot I_a} \tag{8.57}$$

the current in the bridge B–C will be null and the two arms A–B–D and A–C–D, both crossed by the current I_a, can actually be considered in parallel. It is said in this case that the bridge is balanced; contactor 4 can be opened at no-load, without varying the motor current and thus the total tractive effort.

Generally, it is not possible to have the exact resistance value given by (8.57) in both R_1 and R_2; the bridge is not perfectly balanced and contactor 4 must interrupt the current $I_a - I' \neq 0$. In this case, there is a slight variation of the current of the motors during the transition.

8.20.3 Comparison of the Two Systems

Functionally, the bridge transition involves very small or even null changes in the tractive effort; this is beneficial, particularly for vehicles that have to make frequent starts with high accelerations, and leads to a preference for the bridge system in equipment with two and four motors. In equipment with six motors, however, for simplicity of the traction diagram, the short circuit transition must be used.

8.21 ENERGY LOSS IN THE STARTING RHEOSTAT

We shall again consider equipment consisting of two drive motors powered by a voltage V in a contact line. During starting conditions it can be assumed that the current is constant on average and equal to I_a (Figure 8.28).

To go from zero speed (point A) to speed $v_a \sim \Omega_a$ (point D), the following are possible:

- To connect the two motors in parallel from the outset and gradually exclude the rheostat, to go from A to D.
- To connect the two motors in series and exclude the rheostat, so as to go from A to B (speed Ω_a'); make a series–parallel transition, reinsert the rheostat, and continue with the parallel motors, from B to D (speed Ω_a).

Let us see how the work absorbed by the line in the two cases is used, assuming, for simplicity's case, that the resistances to motion are constant, thus the motion can be considered naturally accelerated, with acceleration $a = \text{const}$. The rotation speed is proportional to the time

$$\Omega \sim v = a \cdot t \sim t$$

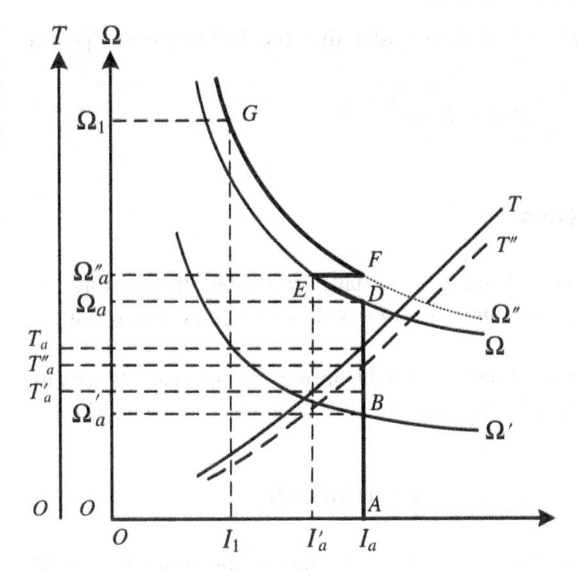

Figure 8.28 Starting conditions sequence.

and also the emf of the motor, since the flux Φ_a is constant corresponding to I_a (Figure 8.29):

$$e \sim \Omega \cdot \Phi_a \sim \Omega \sim t$$

The duration of starting conditions is denoted by t_a, that is,

$$t_a = {}^{v_a}/_a$$

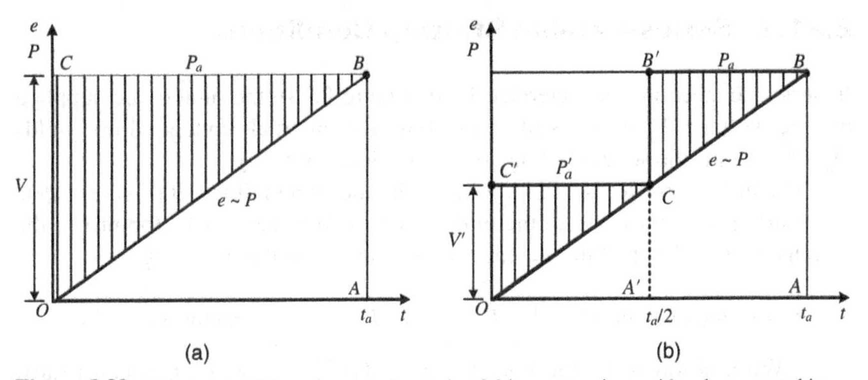

Figure 8.29 Starting conditions rheostat. A couple of drive motors is considered, connected in parallel (a), or in series in the first stage and in parallel in the second (b).

Neglecting the motor losses, we also consider that the instantaneous power developed by each motor is

$$P = e \cdot I_a \sim e \sim t$$

8.21.1 Parallel Motors

The emf increases linearly over time in accordance with the straight line OB (Figure 8.29a); if we neglect the internal motor voltage drop, at that instant t_a will be $e = V$.

The power absorbed by the line is constant during the interval $O - t_a$ and equal, for each motor, to $P_a = V \cdot I_a$ (straight line CB); the work absorbed is given by

$$W_a = V \cdot I_a \cdot t_a \qquad \text{(area } OABCO\text{)}$$

Since the power of the motor $P = e \cdot I_a$ varies according to the straight line OB, the work developed is given by the area $OABO$, that is,

$$W = \frac{1}{2} \cdot V \cdot I_a \cdot t_a = \frac{1}{2} \cdot W_a; \quad \eta = \frac{W}{W_a} = \frac{1}{2} \tag{8.58}$$

Half the energy absorbed from the line during the rheostat starting conditions is equivalent to the $OBCO$ area dashed in the figure,

$$W_r = \frac{1}{2} \cdot V \cdot I_a \cdot t_a \tag{8.59}$$

which is dissipated in the rheostat.

8.21.2 Series–Parallel Starting Conditions

In series connection (characteristic Ω' of Figure 8.28), the motors are supplied by the voltage $V' = {}^{V}/_{2}$; still neglecting the internal voltage drop yields $\Omega'_a = {}^{\Omega_a}/_{2}$, thus the series starting conditions duration is ${}^{t_a}/_{2}$.

The motor's power is still represented by the straight line OB (Figure 8.29).

During the series phase, the emf increases from zero to V' (point C); the various magnitudes, still in reference to one motor, are the following:

- Absorbed power: $P'_a = V' \cdot I_a = \frac{1}{2} \cdot V \cdot I_a = \text{const}$ (straight line $C'C$)
- Work absorbed by the line: $W'_a = V' \cdot I_a \cdot \frac{t_a}{2} = \frac{1}{4} \cdot V \cdot I_a \cdot t_a = \text{const}$ (area $OA'\ CC'\ O$).

In the parallel phase, presuming the duration of the transition to be negligible, the emf increases from $V/2$ to V (point B), and we have the following:

- Absorbed power: $P''_a = V \cdot I_a = \text{const}$ (straight line $B'B$)
- Work absorbed: $W''_a = V \cdot I_a \cdot \frac{t_a}{2}$ (area $A'A\ BB'A'$)

Overall, the work absorbed by the line for each motor is

$$W_a = W'_a + W''_a = \frac{3}{4} \cdot V \cdot I_a \cdot t_a \cdots (\text{area } OA\ BB'\ CC'\ O)$$

The work developed by each motor is still given by (8.58); therefore,

$$W = \frac{1}{2} \cdot V \cdot I_a \cdot t_a = \frac{2}{3} \cdot W_a; \quad \eta = \frac{W}{W_a} = \frac{2}{3}$$

The work dissipated in the rheostat is equal to

$$W_r = \frac{1}{4} \cdot V \cdot I_a \cdot t_a \cdots (\text{area } OCC'\ O + \text{area } CBB'\ C)$$

which is equal to one third of that absorbed from the line and half of that dissipated in the previous case (8.59).

8.21.3 Comparison

Parallel–series starting conditions allow considerable energy savings, reducing the consumption for each motor to three quarters, with equal work developed. In addition, the amount of heat that the rheostat must dissipate, corresponding to W_r, is reduced to half; since the duration of the starting conditions is constant, the average power of the rheostat W_r/t_a is also reduced to half, therefore, it is considerably lighter. For these reasons, series–parallel starting conditions are preferable.

8.22 ELECTRONIC DC MOTOR DRIVES

The introduction of power electronics greatly improved the performance of DC motor drives, enabling them to overcome the inherent limitations of electromechanical drives, that is, low efficiency and torque ripple during the start-up phase.

DC/DC converters (or choppers) were used predominantly for revamping vehicles with electromechanical DC motor drives, albeit only for a brief period, since the superiority of AC motors over their DC counterparts was already becoming apparent. On their own, choppers are used as DC/DC converters on vehicles that run on DC power supplies and are equipped with commutator traction motors, or as the first stage of induction and synchronous three-phase drives.

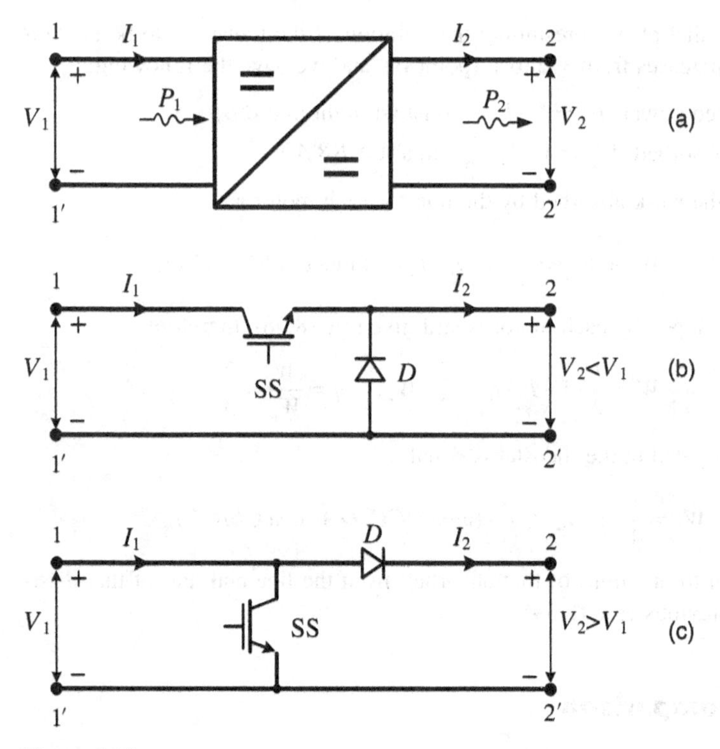

Figure 8.30 DC/DC converter. (a) Sign conventions; (b) step-down chopper, and (c) step-up chopper.

8.22.1 Chopper Description

The following is a functional description of a chopper used to drive a DC motor.

If V_1 represents the DC power supply voltage and V_2 the output voltage (Figure 8.30a), then either

$$V_2 < V_1 \quad \text{or} \quad V_2 > V_1$$

In the first case the chopper is referred to as "step-down", whereas in the second case it is referred to as "step-up". A chopper consists of two basic branches:

- An electronic switch (static switch) SS that operates at the switching frequency f_{sw}
- A freewheeling or blocking diode D.

The arrangement of the two branches with respect to the input terminals 1-1′ and the output terminals 2-2′ is as indicated in Figure 8.30b in the case of a

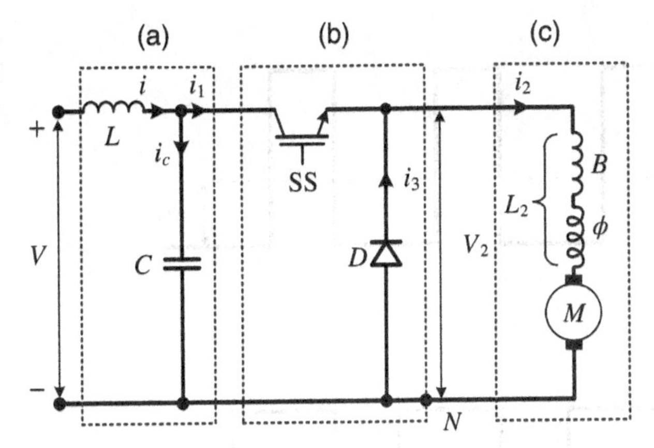

Figure 8.31 DC motor drive diagram. (a) Input filter; (b) power module: SS static switch, *D* freewheeling diode; and (c) load circuit: *M:* traction motor; *B:* leveling inductance; L_2: leveling inductance and motor inductance.

step-down chopper, and in Figure 8.30c in the case of a step-up chopper. These arrangements ensure that power flow is unidirectional, as in Figure 8.30a. Hence, in the case of DC-powered vehicles with commutator motors, a step-down chopper is used during traction, while a step-up chopper is used during regenerative braking.

As shown in Figure 8.31, DC motor drives basically consist of three stages: an input filter, a power module, and a DC traction motor.

In the step-down version of the chopper, the static switch SS connected between the source (contact line or accumulator battery) and the load (traction motors) periodically establishes a connection between the source and the drive circuit, for very short time intervals. Thus, the static switch SS operating period T_{sw} can be divided into two distinct cycles:

- Duty time of duration T_{on}, during which the static switch SS is closed. Since the DC voltage drop across SS is negligible, it can be assumed that the voltage applied to the load is the same as the source voltage (Figure 8.32), for all intents and purposes equivalent to the constant value V_1, that is, $v_2 = V_1$.
- Dead time of duration T_{off}, during which the switch is open.

Therefore, the switching frequency is given by the expression:

$$f_{sw} = \frac{1}{T_{sw}} \text{ in which } T_{sw} = T_{on} + T_{off} \tag{8.60}$$

The chopper has extremely short opening times, which means that it is suitable for use at frequencies in the order of 10^2–10^3 Hz.

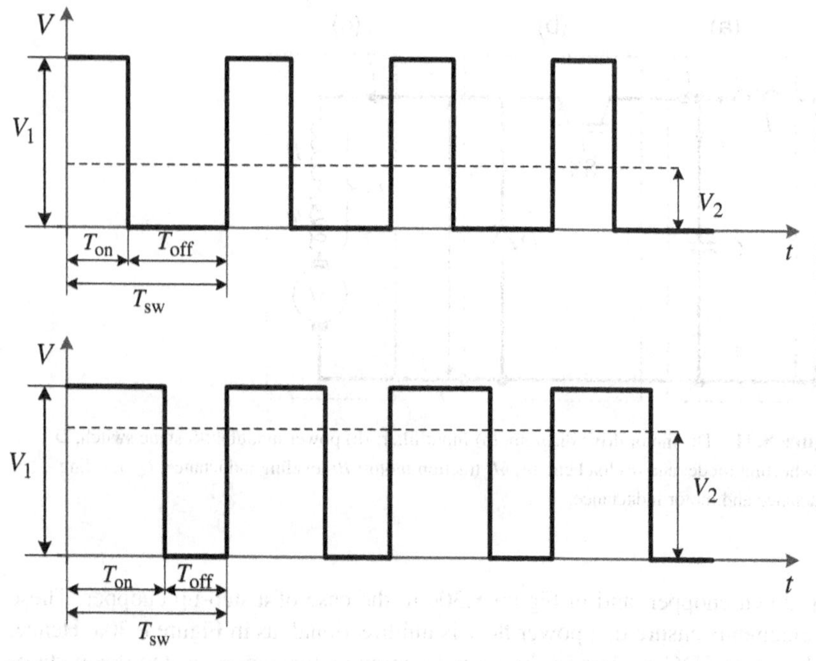

Figure 8.32 Pulse regulation (T_{sw} and f_{sw} are constant; T_{on} variable).

The duty cycle δ is given by the expression

$$\delta = \frac{T_{on}}{T_{sw}}, \text{ therefore, } 0 < \delta < 1 \tag{8.61}$$

Since it is neither possible nor convenient to interrupt the current i_2 drawn by the motors during the dead time interval T_{off}, the chopper circuit includes a freewheeling diode D, connected in parallel with the motor circuit between terminals M and N, which permits the current i_2 to continue flowing during the dead time intervals T_{off}.

Hence, the potential difference v_2 between terminals M and N during the interval T_{off} corresponds to voltage drop across the diode D, which, since this value is negligible, means that the voltage applied to the load is equivalent to zero:

$$v_2 = 0$$

Ignoring, for the moment, the chopper semiconductor switch-on and switch-off phenomena, which respectively occur at the beginning and end of the duty time, it can be assumed that the voltage applied to the load assumes the following values:

$$v_2 = V_1 \text{ in the interval } T_{on} = \delta \cdot T_{sw}$$
$$v_2 = 0 \text{ in the interval } T_{off} = T_{sw} - T_{on} = (1 - \delta) \cdot T_{sw}$$

Thus, the average value of the voltage applied to the load is

$$V_2 = \frac{T_{on}}{T_{sw}} \cdot V_1 = \delta \cdot V_1 \tag{8.62}$$

Hence, by varying the duty cycle δ between 0 and 1, it is possible to vary the average value of V_2 continuously between 0 and V_1, which is basically how the electronic device works.

The voltage V_2 can be regulated in two ways:

- By maintaining the frequency f_{sw}, and hence the operating period T_{sw}, constant and modifying the duration of the duty time T_{on}, as shown in Figure 8.32 (pulse regulation, standard method).
- By maintaining the duration of the duty time T_{on} constant, and modifying the operating period T_{sw}, that is, the frequency f_{sw} (frequency regulation).

In order to ensure that the converter functions correctly, the inductance of the load circuit L_2 must be large enough to reduce the current oscillation, and hence the losses in the motor.

The inductance L_2 is formed partly by the magnetic circuit of the traction motors, and partly by a smoothing inductance, connected in series with the motors.

If the vehicle is powered by a contact line having a significant inductance value, an input filter LC must be installed upstream of the converter (Figure 8.31), consisting of an inductance L connected in series and a capacitor C connected in parallel.

8.22.2 Operating Principle of an Ideal Chopper

In order to simplify the explanation of a chopper, but without overgeneralizing, it is assumed that the source inductance is negligible.

This means that the source can be represented as an ideal voltage generator (Figure 8.33) with $V_1 = \text{const}$, able to absorb the harmonics of the current produced by the chopper, without the need for an input filter LC. Thus,

$$v_1 = V_1 = \text{const}$$

The load circuit always has a very large inductance value L_2; this means that the oscillations of the current i_2 absorbed by the motors are limited. Consequently, during the period T_{sw}, i_2 remains close to its average value I_2:

$$I_2 = \frac{1}{T_{sw}} \cdot \int_0^{T_{sw}} i_2 \, dt$$

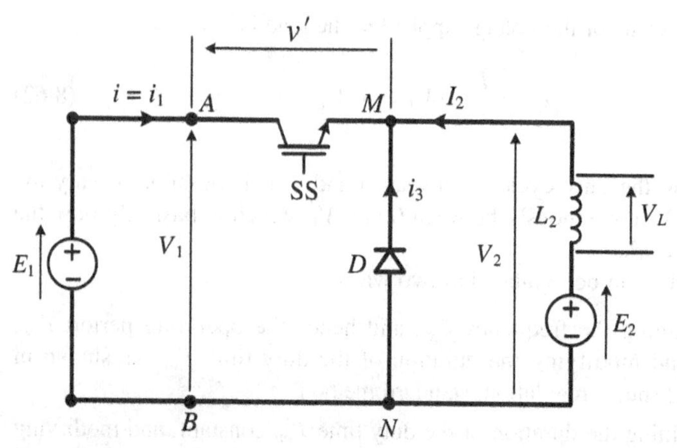

Figure 8.33 Equivalent simplified diagram of a chopper (negligible source inductance; $L_2 = \infty$).

Similarly, when analyzing polyphase rectifiers, it can be assumed that the inductance of the drive circuit is infinite, and that as a consequence the i_2 is perfectly smooth, that is,

$$L_2 = \infty \Rightarrow i_2 = I_2 = \text{const} \tag{8.63}$$

The motors, which are energized in series, are powered by a constant flux so that the induced emf $e_2 = k \cdot \Omega \cdot \Phi$ can be considered constant, provided that the rotation speed Ω does not vary appreciably during the period T_{sw}:

$$E_2 = k \cdot \Omega \cdot \Phi_2 = \text{const} \tag{8.64}$$

where Φ_2 represents the value of the flux corresponding to I_2.

By ignoring dissipation phenomena, it is also possible to represent the load circuit as an ideal voltage generator E_2, connected in series with an inductance $L_2 = \infty$ (Figure 8.33). This means that the motor will absorb power at a constant rate:

$$P_2 = E_2 \cdot I_2 = \text{const} \tag{8.65}$$

8.22.2.1 Analysis of the different states of a chopper

The equations relating to the duty time are derived by applying Kirchoff's laws to the circuit nodes and loops in Figure 8.33, ignoring any voltage drops across the semiconductor elements. The resulting equations are represented graphically in Figure 8.34:

$$v_2 = v_1 = V_1; \quad v_L = v_2 - E_2 = V_1 - E_2$$
$$i_3 = 0; \quad i_1 = i_2 = I_2$$

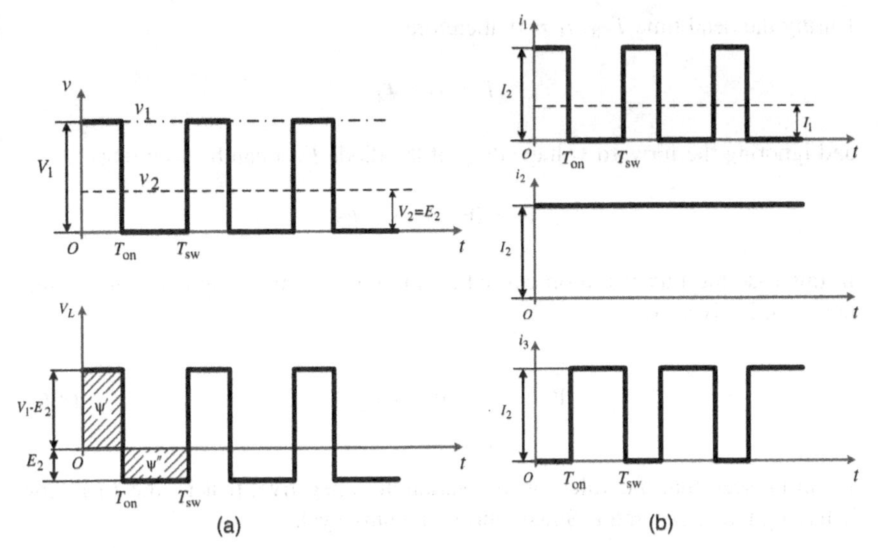

Figure 8.34 Chopper with $L = 0$; $L_2 = \infty$. (a) Voltages; b) currents.

The voltage pulse applied to L_2 causes a variation in the linked flux equivalent to its integral, according to the following formula,

$$\Psi' = \int_0^{T_{on}} v_L \cdot dt = \int_0^{T_{on}} (V_1 - E_2) \cdot dt = (V_1 - E_2) \cdot T_{on} \tag{8.66}$$

which corresponds to the cross-hatched area in Figure 8.34. Hence, it can be seen that, in order for the converter to function correctly, the inductance value of the load circuit must be sufficiently high.

During this phase, the work W'_1 transferred from the source corresponds to the sum of the magnetic energy W'_L stored in the L_2 and the work W'_M absorbed by the motor:

$$W'_1 = W'_M + W'_L$$

Where, on the basis of the assumptions described already

$$W'_1 = \int_0^{T_{on}} V_1 \cdot i_t \cdot dt = V_1 \cdot I_2 \cdot T_{on}$$

$$W'_M = \int_0^{T_{on}} E_2 \cdot i_2 \cdot dt = E_2 \cdot I_2 \cdot T_{on}$$

$$W'_L = \int_0^{T_{on}} v_L \cdot i_2 \cdot dt = I_2 \int_0^{T_{on}} v_L \cdot dt = I_2 \cdot \Psi' = (V_1 - E_2) \cdot I_2 \cdot T_{on}$$

During the dead time T_{off}, $i_1 = 0$, therefore,

$$i_3 = i_2 = I_2$$

and ignoring the forward voltage drop of the diode D, it can be seen that

$$v_2 = 0; \quad v_L = -E_2$$

In this case the flux variation caused by the voltage pulse applied to the inductance L_2, is given by

$$\Psi'' = \int_{T_{\text{on}}}^{T_{\text{sw}}} v_L dt = -E_2 \cdot T_{\text{off}} \tag{8.67}$$

It can be seen that the value of this variation is negative, that is, the L_2 is now behaving like a generator. Since, during the duty cycle

$$\Psi' + \Psi'' = 0$$

on the basis of (8.66) and (8.67):

$$(E_1 - E_2) \cdot T_{\text{on}} - E_2 \cdot T_{\text{off}} = 0; \quad V_1 \cdot T_{\text{on}} = E_2 \cdot T_{\text{sw}}; \quad E_2 = \delta \cdot V_1$$

Therefore, the induced emf of the motor corresponds to the average value V_2 (8.62) of the voltage generated at the output of the chopper. In reality, if r represents the resistance of the load circuit (motor + smoothing inductance), which has been ignored up to this point, then $V_2 > E_2$, that is, $V_2 - E_2 = \delta \cdot V_1 - E_2 = r \cdot I_2$.

During this phase, the work associated with the inductance is equivalent to

$$W_L'' = \int_{T_{\text{on}}}^{T_{\text{sw}}} v_L \cdot i_2 \cdot dt = -I_2 \cdot E_2 \cdot T_{\text{off}} = -W_M''$$

The motor is powered entirely by the energy stored by the inductance L_2 during the duty time.

Thus, the energy balance over the entire operating period T_{sw} is such that the energy produced by the power supply is in part transferred to the motor and in part exchanged with the load inductance:

$$W_1 = W_M + W_L$$

Since the source is connected only during the duty time T_{on}, it can be seen that

$$W_1 = W_1' = V_1 \cdot I_2 \cdot T_{\text{on}} = \delta \cdot V_1 \cdot I_2 \cdot T_{\text{sw}}$$

However, as has already been stated, the motor operates under constant power, therefore,

$$W_M = W'_M + W''_M = E_2 \cdot I_2 \cdot T_{sw} = P_2 \cdot T_{sw}$$

As far as the energy absorbed by the inductance L_2 is concerned, it may be seen that, during the period T_{sw}, its average value is zero. In fact

$$W_L = W'_L + W''_L = I_2 \cdot (\Psi' + \Psi'') = 0$$

Hence,

$$W_1 = W_M \tag{8.68}$$

Since I_1 corresponds to the average value of the absorbed current at terminals AB

$$I_1 = \frac{1}{T_{sw}} \cdot \int_0^T i_t \cdot dt = \frac{T_{on}}{T_{sw}} \cdot I_2 = \delta \cdot I_2 \tag{8.69}$$

furthermore,

$$W_1 = V_1 \cdot I_1 \cdot T_{sw} = P_1 \cdot T_{sw}$$

where the average power delivered by the source to the load corresponds to

$$P_1 = V_1 \cdot I_1 \tag{8.70}$$

Taking the equation indicated in (8.68) into account, it can be seen that

$$P_1 = P_2 \tag{8.71}$$

Thus, the chopper then behaves like a step-down transformer, since the average power values upstream and downstream of the converter are the same, excluding the very limited losses that occur in the converter itself. Unlike an electromechanical transformer, in this case the transformation ratio δ may vary continuously.

Since the relationship in (8.71) is valid for any duty cycle value, the motors always use the power delivered by the source, which means that choppers provide efficient, nondissipative regulation compared to traditional, rheostatic transition devices.

8.22.3 Real Chopper Operation

Since the load circuit inductance, while high, is not infinite, the drive circuit current i_2 is not perfectly smooth. Also, the induced emf e_2 of the motors is not rigorously constant, even if the flux variations during the period T are more limited than the current variations due to the saturation of the magnetic circuit. This means that e_2 differs very little with respect to the average value (8.64).

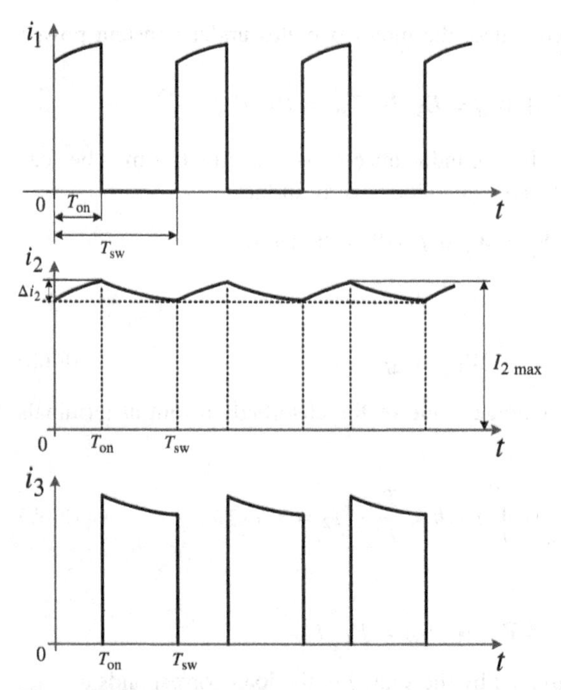

Figure 8.35 Chopper supplying an active load with finite inductance.

During the duty time T_{on}, the current i_2 tends to increase (Figure 8.35); its time derivative is limited by the inductance L_2. Ignoring the dissipation phenomena, it can be seen that

$$v_L = V_1 - e_2 = L_2 \cdot \frac{di_2}{dt}$$
$$\frac{di_2}{dt} = \frac{v_L}{L_2} = \frac{V_1 - e_2}{L_2} \tag{8.72}$$

The total current variation Δi_2 during the period T_{on} depends on the mains voltage rating V_1 and, above all, on L_2. Supposing, on the basis of the above considerations, that $e_2 = E_2$, it can also be assumed, approximately, that $v_L = V_1 - E_2$, and hence, that

$$\frac{di_2}{dt} = \frac{V_1 - E_2}{L_2} = \text{const} \tag{8.73}$$

from which it is possible to derive the current variation by integrating (8.73)

$$\Delta i_2 = \int_0^{T_{on}} \frac{v_L}{L_2} \cdot dt = \frac{\delta \cdot (1 - \delta)}{f_{sw} \cdot L_2} \cdot V_1 \tag{8.74}$$

Since the resistance r of the load circuit is not zero, as assumed in the case of (8.72), the time constant

$$\tau_2 = {}^{L_2}/_{r}$$

is finite, which means that i_2 increases exponentially during the duty cycle, as illustrated in Figure 8.35.

During the dead time T_{off}, if the same assumptions are made as in (8.73), i_2 decreases linearly according to the following relationship:

$$\frac{di_2}{dt} = -\frac{E_2}{L_2} = \text{const}$$

The resulting current variation Δi_2 must coincide with the value given by (8.74) (with the exception of the polarity) in order to maintain the average value of i_2 constant.

Under real operating conditions, the above observations regarding i_2 apply, which follow an exponential time constant τ_2,

$$I_2 = I_{2\,\max} \cdot e^{-\frac{R}{L_2}(t-T_{\text{on}})} = I_{2\,\max} \cdot e^{-\frac{t-T_{\text{on}}}{\tau_2}}$$

where $I_{2\,\max}$ corresponds to the maximum value assumed by i_2 at the end of the duty time.

Although an approximation, (8.74) provides a measurement of the current ripple, which is a function of the duty cycle δ. The amplitude of the ripple is

$$\Delta I = \frac{\Delta i_2}{2} = \frac{\delta \cdot (1-\delta)}{2 \cdot f_{\text{sw}} \cdot L_2} \cdot V_1 \tag{8.75}$$

and the ripple factor is

$$\mu = \frac{\Delta I}{I_2} = \frac{\delta \cdot (1-\delta)}{2 \cdot f_{\text{sw}} \cdot L_2} \cdot \frac{V_1}{I_2} \tag{8.76}$$

For $\overline{\delta} = 0.5$, the maximum ripple value is

$$\Delta \overline{I} = \frac{V_1}{8 \cdot f_{\text{sw}} \cdot L_2}; \quad \overline{\mu} = \frac{V_1}{2 \cdot f_{\text{sw}} \cdot L_2 \cdot I_2} \tag{8.77}$$

Figure 8.36 shows the diagram of (8.75) and (8.76).

As can be noted, in order to reduce the ripple factor, it is possible to increase the rating of the smoothing inductance L_2, which would increase the weight and cost of the vehicle, or increase the switching frequency, which would increase the losses in the converter.

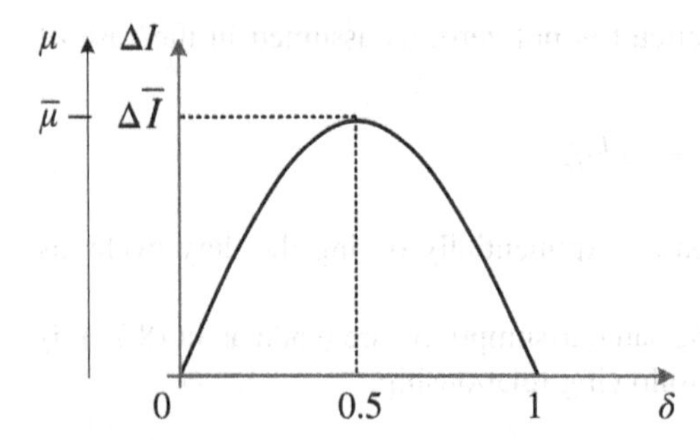

Figure 8.36 Motor current ripple.

8.22.4 Chopper Regulation During Vehicle Operation Phases

Thanks to the transformative behavior of the chopper at start-up when it reduces by δ the applied voltage at the motor terminals, the line current is reduced by the same ratio δ while maintaining a high current in the motors.

Figure 8.37 shows a comparison of the power and current consumption performances of a rheostatic drive (Figure 8.37 a,b) and a chopper (Figure 8.37c,d), ignoring the internal motor voltage drop (and thus assuming $V_2 = E_2$) and the corresponding losses for the sake of simplicity.

On the basis of these assumptions, if $I_2 = \text{const}$ between zero and the base speed v_b, then

$$E_2 = k \cdot \Omega \cdot \Phi(I_2) \sim \Omega \sim v$$

$$\delta = \frac{V_2}{V_1} = \frac{E_2}{V_1} \sim v$$

$$I_1 = \delta \cdot I_2 \sim v$$

In theory, I_1 would reach the value I_2 if, at the speed v_b, $V_2 = V_1$, and $\delta = 1$. Where $v > v_b$, the currents I_1 and I_2 decrease in accordance with the natural characteristics of the motor if no additional measures, such as field regulation (Figure 8.38), are adopted.

The reduction of the line current with respect to the motor current, down to $\delta < 1$, is of considerable importance to TPSSs, contact lines, and other power supply equipments (less heating of the contact line and lower voltage drops).

In the start-up diagram (Figure 8.38), the hatched area expressed in $A \cdot s$ and multiplied by the power supply voltage V_1 depicts the work that would be dissipated by the rheostat in a traditional device. This work is now saved, resulting in an overall decrease in the quantity of energy consumed, and the elimination of the corresponding quantity of heat. This advantage is particularly significant in the case of stretches of track in tunnels, as well as underground sections, where

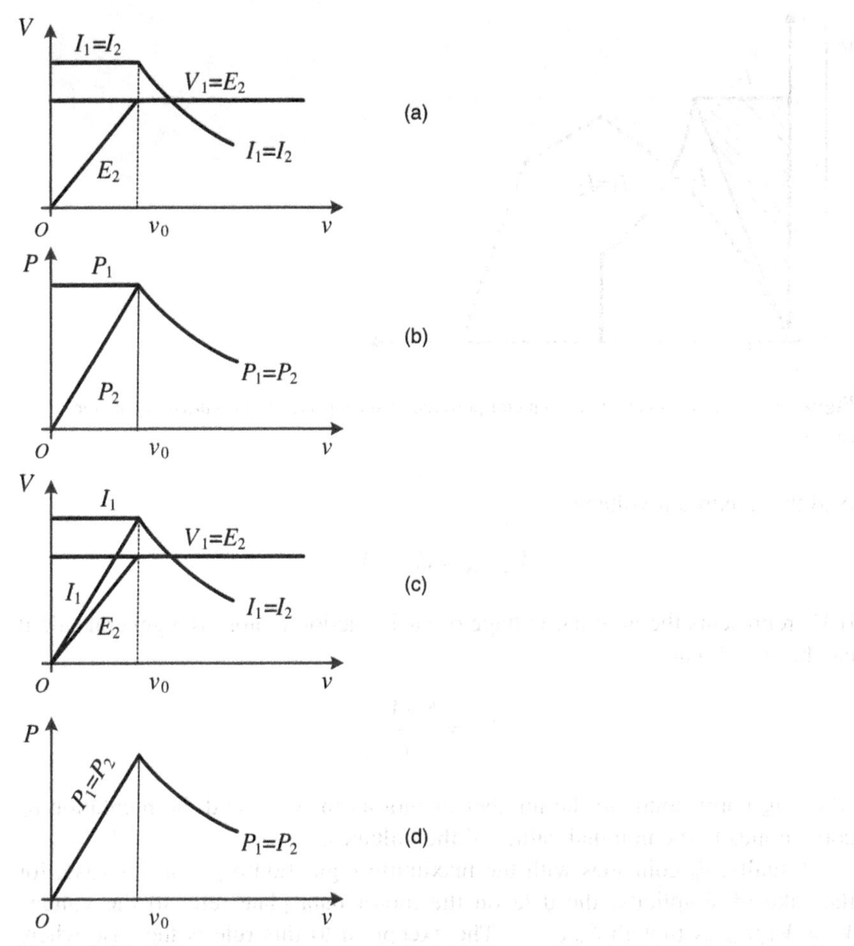

Figure 8.37 DC traction motor power supply. (a,b) Rheostat drive; (c,d) chopper power supply; I_2: line current; I_2: motor current; V_1: line voltage; E_2: motor induced emf; P_1: power drawn from the line; P_2: power drawn by the motor.

traditional drive devices cause an increase in the ambient temperature and require expensive ventilation systems.

In Figure 8.37 it has been assumed that the voltage applied to the motor in the starting phase varies linearly from 0 to V_1 and hence that regulation varies from 0 to 1. As we have seen, the adjustment range is actually defined by the limits δ_{min} and δ_{max}, which correspond, respectively, to the minimum voltage:

$$V_{2\,min} = \delta_{min} \cdot V_1 = f_{sw} \cdot T_{on\,min} \cdot V_1 \tag{8.78}$$

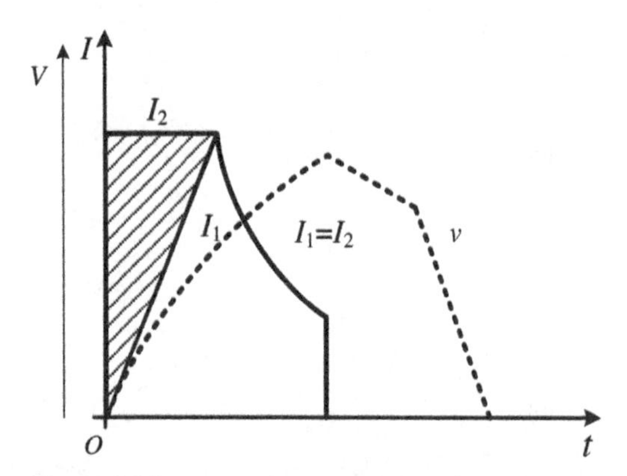

Figure 8.38 Diagram of a traction motor powered by a chopper. I_1: Line current; I_2: motor current.

And the maximum voltage

$$V_{2\,\max} = \delta_{\max} \cdot V_1$$

If V_n represents the nominal voltage of each traction motor, as a general rule it can be stated that

$$V_n = \frac{\delta_n \cdot V_1}{n_s}$$

where n_s corresponds to the number of motors in series and the regulation δ_n corresponds to the nominal values of the voltages.

Usually, δ_n coincides with the maximum regulation $\delta_{\max}$; in this case, for the sake of simplicity, the data on the motor data plate refer to the voltage $V_n = V_1/n_s$, as though $\delta_{\max} = 1$. The exception to this rule is the case where $V_1 = 3000\frac{V}{n_s} = 1$, in which the regulation $\delta_n = V_n/V_1$ is significantly less than the unit.

If the value R_2 represents the total resistance of a motor branch, the smoothing inductance and the circuit cables, the minimum voltage at start-up (8.78) must be equal to the voltage drop corresponding to the predetermined current $I_{A\,\min}$, that is,

$$V_{2\,\min} = R_2 \cdot I_{A\cdot\min}$$

If, for the sake simplicity, we ignore the resistance of the cables and the inductance with respect to the internal resistance of each motor, so that $R_2 = n_s \cdot r$, it can be seen that

$$V_{2\,\min} = n_s \cdot r \cdot I_{A\,\min} = n_s \cdot \varepsilon_n \cdot K_{\min} \cdot V_n \tag{8.79}$$

The desired value of the product $\varepsilon_n \cdot K_n$, in particular for locomotives, are in the order of a few hundredths, therefore it is not easy to satisfy (8.78) and (8.79) simultaneously, especially if $n_s \cdot \frac{V_n}{V_1} < 1$. Since it is not possible to employ frequency regulation, given the problems that such a system would create for the track circuits, and assuming a minimum duty time 2×10^{-4} s, with pulse regulation, it would be necessary to select a frequency (below 100 Hz) that would not be compatible with the rating of the converter passive components and with the reduction of the harmonics. By establishing a nominal frequency f_{swn} between 200 Hz and 500 Hz, it is possible to

- operate at reduced frequencies during the start-up phase f_{sws}, selecting submultiples of f_{swn}, which are compatible with the characteristics of the track circuits;
- insert a resistor R in series with the motors, so that the minimum required voltage

$$V_{2\,\min} = (n_s \cdot r + R) \cdot I_{A\,\min}$$

is greater. The resistor is automatically short-circuited by one or more contactor at the end of the start-up phase. This system can be combined with frequency reduction.

8.22.5 Harmonic Currents Generated by the Chopper

8.22.5.1 Limits of current harmonics in traction motors

Since the inductance L_2 of the load circuit is finite, the current i_2 will present a ripple that must be maintained within the limits tolerated by the traction motors. As an initial approximation, the ripple amplitude ΔI can be expressed by (8.75), which is a function of the duty cycle δ and assumes its maximum value (8.77) when $\overline{\delta} = 0.5$.

Once the line voltage V_1, the operating frequency f_{sw}, and the maximum permissible ripple have been determined, the expression in (8.77) can be used to determine the value of the required inductance L.

Since L_2 is formed by the traction motors (L_m) and the smoothing coil (B, in Figure 8.31)

$$L_2 = L_m + L_B$$

it is possible to establish the rating of the latter. The term L_B is prevalent with respect to L_m.

Equation (8.77) highlights the advantage of operating at high frequencies for limiting the mass and the dimensions of the smoothing chokes.

It may be convenient to adopt traction motors with fully laminated magnetic circuits that accept higher ripple factors μ; such motors also feature better switching properties, but are more complicated and expensive than standard devices with solid stator cores.

8.22.5.2 Line current harmonic limits

The current i_1 drawn by the chopper is shown in Figure 8.34 (assumptions: $L_2 = \infty$; $i_2 = I_2 = \text{const}$) and Figure 8.35 (L_2 finite inductance). This current includes a strong harmonic content that is not, in general, compatible with the characteristics of the source and power supply line.

In systems that are supplied by contact lines, where it is not possible to ignore the inductance, it is necessary to establish limits for both the effective value of the harmonic currents and their frequency spectrum in order to maintain disturbances affecting the track circuits and telecommunications within acceptable limits. For this reason, traction vehicles are always fitted with a low-pass input filter (Figure 8.31). Since the h-order harmonic frequencies f_h depend on the operating frequency of the chopper, it is convenient to adopt pulse regulation in order to obtain constant values for the various f_h under all operating conditions.

8.22.5.3 Input filter specifications

The inductance value L_r of the power supply line may vary significantly, depending on the distance of the rolling stock from power substations, in proximity to which L_r assumes very low values.

When paired with the capacitance of the input filter capacitor (C, in Figure 8.31), the resonant frequency of L_r is given by

$$f'_0 = \frac{1}{2 \cdot \pi \cdot \sqrt{L_r \cdot C}}$$

In order to avoid falling within the spectrum of the harmonics produced by the chopper, it must be less than the minimum working frequency $f_{\min}$. Given the variability of L_r, in order to ensure that this condition is satisfied it is necessary to complete the input filter by adding a branch L_0 in series, so that the total inductance

$$L = L_r + L_0$$

corresponds to a resonant frequency f_0 that satisfies the condition

$$f_0 = \frac{1}{2\pi\sqrt{L \cdot C}} < f_{\min} \tag{8.80}$$

Introducing the relationship

$$K = \frac{f_0}{f_{sw}} = \frac{\omega_0}{\omega_{sw}} = \frac{1}{\omega_{sw} \cdot \sqrt{L \cdot C}} \tag{8.81}$$

It follows that it is always necessary that $K < 1$.

The relationship K is not constant, but varies with f_0 as a function of the line inductance, that is, the distance of the vehicle from the substations. In order to allow for the most demanding conditions, the maximum expected value of f_0 must be considered, that is, the minimum inductance L value (vehicle in the vicinity of a substation).

8.22.5.4 Harmonic analysis of the line current

The following is an analysis of the current i_1 drawn by the chopper (Figure 8.31), maintaining the assumption: $L_2 = \infty; i_2 = I_2 = \text{const.}$

The trend of the current i_1 is illustrated in Figure 8.34, and has an average value of $I = \delta \cdot I_2$. It is periodic, with a period T_{sw}, a frequency $f_{sw} = 1/T_{sw}$, and an rms value given by

$$I_{rms} = \sqrt{\frac{1}{T_{sw}} \cdot \int_0^{T_{on}} i_1^2 \cdot dt} = \sqrt{\frac{1}{T_{sw}} \cdot \int_0^{T_{on}} I_1^2 \cdot dt} = \sqrt{\delta} \cdot I_2 \tag{8.82}$$

The rms value of the ripple current I_R is

$$I_R = \sqrt{I_{rms}^2 - I^2} = \sqrt{\delta \cdot (1 - \delta)} \cdot I_2 \tag{8.83}$$

and assumes its maximum value $\overline{I}_R = I_2/2$ when $\overline{\delta} = 0.5$.

By developing the function $i_1 = f(t)$ in a Fourier series, it can be seen that

$$i_1 = I + \sum_1^{\infty} A_h \cdot \cos h\omega t + \sum_1^{\infty} B_h \cdot \sin h\omega t$$

where the terms A_h and B_h relating to the harmonic order h are given by

$$A_h = \frac{I_2}{\pi \cdot h} \cdot \sin 2\pi\delta h$$

$$B_h = \frac{I_2}{\pi \cdot h} \cdot (1 - \cos 2\pi\delta h) = \frac{2 \cdot I_2}{\pi \cdot h} \cdot \sin^2 \pi\delta h$$

The frequency and rms value of the h-th harmonic i_h are given by

$$f_h = h \cdot f_{sw}; \quad I_h = \frac{\sqrt{A_h^2 + B_h^2}}{\sqrt{2}} = \sqrt{2} \cdot I_2 \cdot \frac{\sin \pi\delta h}{\pi \cdot h} \tag{8.84}$$

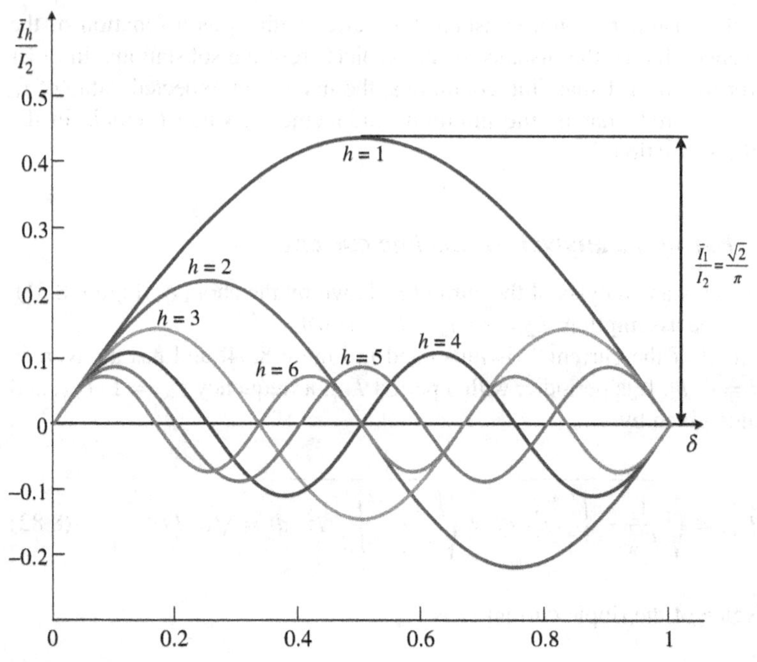

Figure 8.39 Current harmonics generated by the DC/DC converter.

The trend of this harmonic is shown in Figure 8.39. It assumes the maximum (absolute) effective value that decreases with the harmonic order:

$$\bar{I}_h = \frac{\sqrt{2} \cdot I_2}{\pi \cdot h}$$

In correspondence with the duty cycles $\bar{\delta}$

$$\pi \cdot \bar{\delta} \cdot h = \frac{\pi}{2} + k \cdot \pi \ (k = 0,\ 1,\ 2,\ 3, \ldots); \quad \bar{\delta} = \frac{2 \cdot k + 1}{2 \cdot h}$$

In particular, for the first harmonic i_1 of frequency f_{sw}

$$\bar{I}_1 = \frac{\sqrt{2} \cdot I_2}{\pi}; \quad \bar{\delta} = \frac{1}{2}$$

This may be used to derive the following equation:

$$\bar{I}_h = {}^{\bar{I}_1}/_{h}$$

Chapter 9

AC Motor Drives

Thanks to their strength and reliability, induction motors are the preferred solution for modern drives; however, the use of permanent magnets has increased recently, thanks to their high power density capacities.

9.1 DRIVES WITH INDUCTION MOTORS

The use of induction motors onboard locomotives dates back to the early twentieth century; they were fed by a three-phase system at a fixed frequency of 16.7 Hz via a dual-wire contact overhead line, while the third phase was connected to the running rails.

Experiments have been conducted on single-/three-phase conversion systems in the field of industrial frequency single-phase traction since the 1930s: Budapest–Hegyeshalom used rotating converters that supplied a fixed frequency of 50 Hz to its Kando locomotives, as did Krupp on its Hollental line, whereas the Hungarian Ganz locomotives were fitted with induction motors supplied at five different frequencies, thanks to the phase and frequency conversion system. Despite their easy schematic, all the applications required the use of highly complex solutions in order to obtain a certain number of economic fixed speed. From a conceptual point of view, conversion from single phase to three phase at continuously variable frequencies and voltages, which was introduced on a small series of French locomotives at 25 kV–50 Hz in the 1950s, represented remarkable progress. This system used an initial rotating unit to convert the phase, and a second unit to convert the frequency. It could be used to provide continuous regulation with reversible power flow. However, just like the preceding solutions, it was penalized by the presence of rotating units and the need for delicate regulation systems.

Three-phase drives in their current form have only been developed recently, thanks to the evolution of power electronics and the availability of electronic inverters capable of supplying power to induction or synchronous traction motors at variable voltages and frequencies.

Electrical Railway Transportation Systems, First Edition. Morris Brenna, Federica Foiadelli, and Dario Zaninelli.

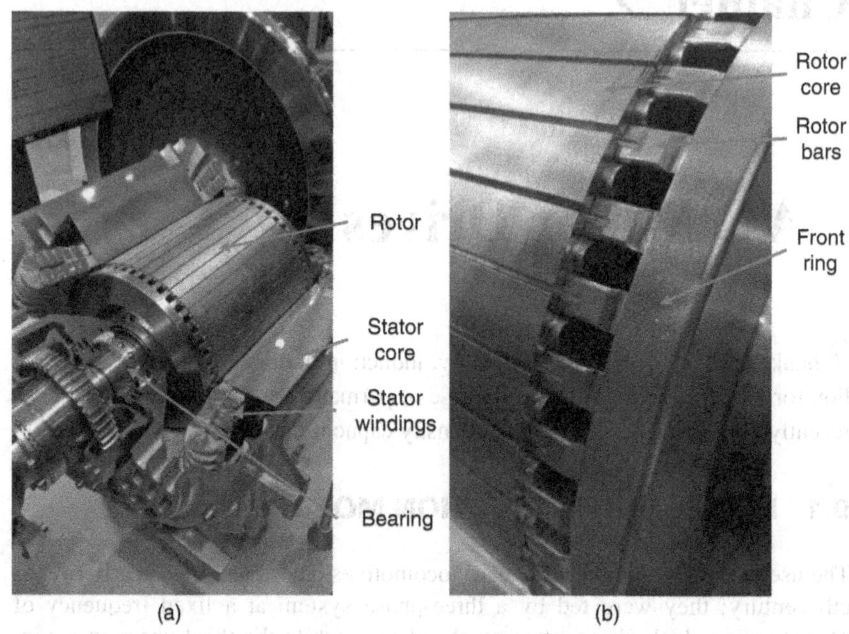

Figure 9.1 Cut-away of an induction traction motor (a) and details of the squirrel cage rotor (b).

Induction motors with electronic drives first began to be used on important applications in the field of electrical traction in the 1970s.

The induction or asynchronous machines as traction motors are almost exclusively three-phase devices with squirrel-cage rotors. This configuration gives the machine considerable robustness and operational flexibility. See Figure 9.1 for a cut-away view of an induction motor used for railway traction applications.

A modern three-phase drive uses induction or synchronous three-phase traction motors supplied at variable voltages and variable frequencies by electronic power converters.

9.1.1 The Advantages of Induction Machines

The advantages of an electronic three-phase drive mainly derive from the use of the induction motor with squirrel cage rotor that offers the following advantages when compared to a commutator motor:

- Reduced dimensions and mass with the same power, since the maximum rotation speed is greater because it is not limited by the presence of a commutator.
- Since it does not have a commutator or brushes, it requires less maintenance.

- Intrinsically performs well for the adhesion exploitation.
- Natural transition from traction to braking, without the need to reconfigure the drive, and develops effective braking even at low speeds.
- Rotation speed can be reversed via the inverter control, without the need for any electromechanical devices.

Thanks to the high power per mass and volume unit values, it is possible to reduce the mass of rail transport motor bogies significantly with respect to the car-body, resulting in improved ride quality.

The variable frequency of the power supply means that, even during the phase when the train starts moving, the slip values, and hence rotor losses, are kept to a minimum.

Both synchronous and induction three-phase drives offer a very high level of elasticity, so under normal operating conditions, it is possible to fully exploit the available power at a wide range of speeds. This has made it possible to produce "universal" locomotives that are suitable for operation at low speeds and high traction forces (e.g., freight trains; high-gradient railways) and for high-speed passenger services.

9.1.2 Operating Principle of an Induction Motor

In order to understand the operating principle of an induction machine, it is necessary to consider a motor in which the stator windings are fed by a three-phase current at a frequency f that generates a rotating magnetic field in the air gap. This field invests the rotor conductors and is capable of generating an electromotive force that is induced only if the speed of the rotor is different from that of the rotating magnetic field, whence the name "asynchronous." The electrodynamic effect of the interaction between the rotating magnetic field and rotor currents generates a torque, which causes the rotor to start rotating.

Thus, a three-phase induction motor wound for p poles, supplied at the phase voltage E at a frequency f and an angular frequency $\omega = 2\pi f$, houses a rotating magnetic field with flux Φ and speed of rotation given by the following expression:

$$\Omega_0 = \frac{2\omega}{p} = \frac{4\pi}{p} \cdot f\,(\text{rad}/s), \quad n_0 = \frac{2f}{p}\,(s^{-1})$$

The rotor rotates at the speed Ω, with a slip, with respect to the synchronous speed Ω_0, of

- absolute : $\Delta\Omega = \Omega_0 - \Omega$
- relative : $s = \dfrac{\Omega_0 - \Omega}{\Omega_0} = \dfrac{\Delta\Omega_0}{\Omega_0}$

Therefore, its angular velocity is given by

$$\Omega = (1 - s)\Omega_0$$

This means that a three-phase system of electromotive forces is induced in the rotor at the slip frequency:

$$f_r = s \cdot f = \frac{\Delta\Omega}{\Omega_0} \cdot f = \frac{p}{4\pi} \cdot \Delta\Omega \tag{9.1}$$

having an rms value of

$$E_r = K_r N_r f_r \Phi$$

where K_r represents the Kapp factor of the rotor winding and N_r the number of conductors per phase.

Note how the rotor frequency f_r is proportional to the absolute slip value, that is, to the rotation speed difference $\Delta\Omega$ between the rotating magnetic field in the air gap and rotor.

The emf induced when the rotor is stationary, that is, $s = 1, f_r = f$, is given by

$$E_{r0} = K_r N_r f \Phi$$

From this it can be deduced that

$$E_r = sE_{r0}$$

Figure 9.2 illustrates the equivalent circuit of a motor phase, where the rotor magnitudes are all referred to the primary frequency f and the load is represented by the equivalent resistance:

$$R_m = R_r \frac{1-s}{s}$$

from which it may be deduced that $R_m = \infty$ in synchronism, that is, when $\Omega = \Omega_0$, whereas $R_m = 0$ when the rotor is stationary, that is, at start-up.

Since the total secondary resistance

$$R_r + R_m = R_r + ((1/s) - 1)R_r = {}^{R_r}/_{s}$$

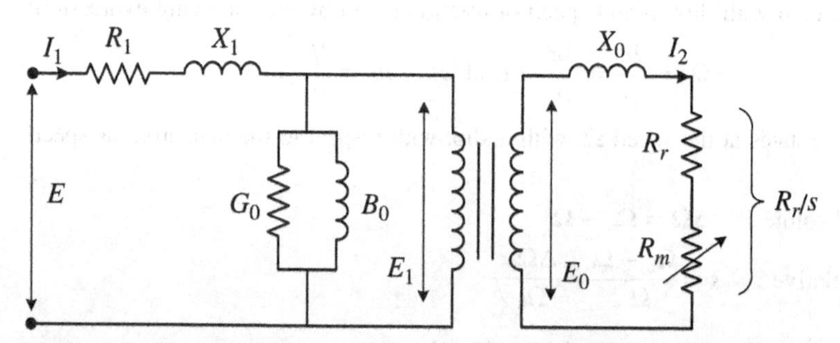

Figure 9.2 Equivalent circuit of a phase of an induction machine, in which the rotor magnitudes are referred to the stator frequency.

it is possible to derive the value of the rotor current:

$$I_r = \frac{E_{r0}}{\sqrt{R_r^2 + s^2 X_{r0}^2}}$$

and the active power transmitted to the rotor:

$$P_t = 3\frac{R_r}{s} I_r^2 = 3E_{r0}^2 \frac{sR_r}{R_r^2 + s^2 X_{r0}^2} \tag{9.2}$$

The power transmitted from stator to rotor is given by the sum of the rotor Joule effect losses P_{Jr} and the mechanical power P_m delivered by the machine:

$$P_{Jr} = 3R_r I_r^2 = s \cdot P_t$$
$$P_{Jr} = 3R_m I_r^2 = (1 - s) \cdot P_t$$

The power P_m is given by the sum of the net power at the shaft P and the mechanical losses; ignoring the latter, it may be stated that $P \cong P_m$.

The losses in an induction machine are concentrated in the following elements of the equivalent circuit in Figure 9.2:

- in the resistance R_s (stator Joule effect losses);
- in the conductance G_0 (losses in the stator ferromagnetic core);
- in the resistance R_r (rotor Joule effect losses);
- the mechanical losses are included in the mechanical power P_m.

9.1.3 Tractive Effort Diagram of the Motor

The tractive effort diagram illustrates the trend of the drive torque as a function of speed. The electromagnetic torque T_e generated at the machine air gap is given by the ratio between the mechanical power P_m and the rotation speed Ω:

$$T_e = \frac{P_m}{\Omega} = \frac{(1 - s)P_t}{(1 - s)\Omega_0} = \frac{P_t}{\Omega_0}$$

By introducing (9.2), the following expression is obtained:

$$T_e = \frac{3E_{r0}^2}{\Omega_0} \frac{sR_r}{R_r^2 + s^2 X_{r0}^2} = Kf^2 \Phi \frac{sR_r}{R_r^2 + s^2 X_{r0}^2}$$

which has the well-known trend illustrated in Figure 9.3.

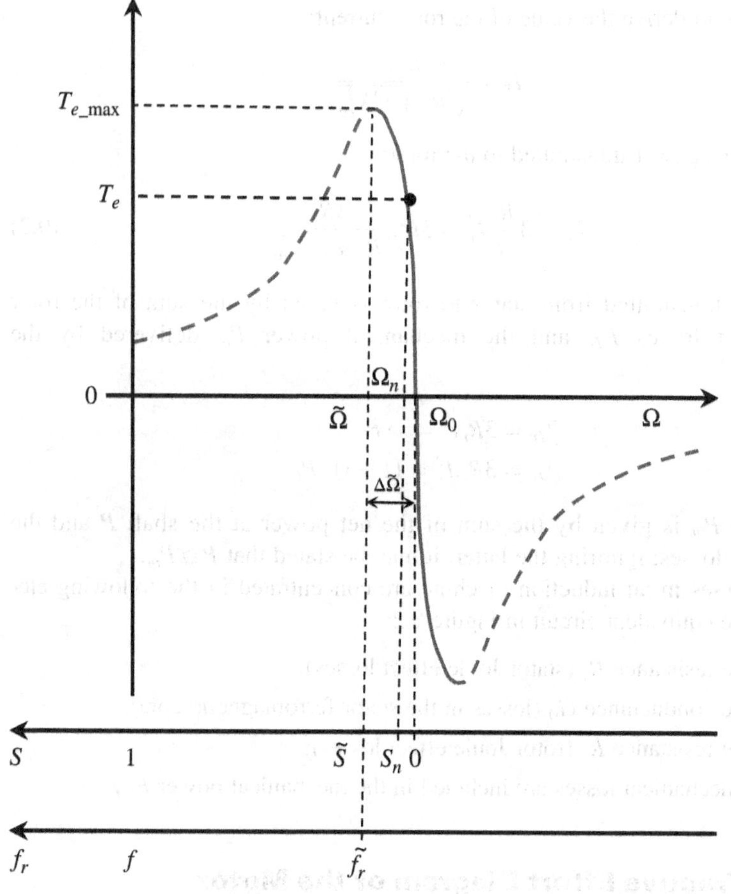

Figure 9.3 Tractive effort diagram of the induction machine with constant voltage and frequency supply.

In traction, that is, where $\Omega < \Omega_0$, the torque reaches its maximum value $T_{e_{\max}}$ in correspondence with the slip $\tilde{s}$ and the rotor frequency $\tilde{f}_r$:

$$T_{e_{\max}} = \frac{Kf\Phi^2}{2X_{r0}} = \frac{K\Phi^2}{4\pi L_r}$$

$$\tilde{s} = \frac{R_r}{X_{r0}}$$

$$\tilde{f}_r = \tilde{s}f = \frac{R_r}{2\pi L_r} \rightarrow \Delta\tilde{\Omega} = \frac{4\pi}{p} \cdot \tilde{f}_r = \frac{2R_r}{p \cdot L_r}$$

As can be seen, both the maximum torque value $T_{e_{\max}}$ and the corresponding rotor frequency value $\tilde{f}_r$, and hence also the absolute slip value $\Delta\tilde{\Omega}$, are independent of the stator power supply frequency f, whereas they are dependent on the construction parameters of the machine; $T_{e_{max}}$ also depends on the square of the magnetic flux.

The actual value of the working torque T_e is less than the maximum torque $T_{e_{max}}$:

$$T_e = \xi \cdot T_{e_{max}}$$

In order to guarantee a sufficient overload margin, ξ is generally less than 1, while it tends to unity either during high acceleration phases or at high speeds where there is significant weakening of the field.

In correspondence with a working frequency f, the characteristic $T_e(\Omega)$ is very steep since the absolute slip $\widetilde{\Delta\Omega} = \tilde{s} \cdot \Omega_0$ required for the transition between the synchronous speed and the maximum torque is very small.

9.1.4 Operation of the Induction Motor at Variable Speeds

When an induction machine is working under normal operating conditions, the rotor rotation speed slip value is such that the rotation speed of the rotor field summed with its rotation is always equal to the synchronous speed:

$$\Omega_0 = \Omega + \Delta\Omega$$

which means that, in terms of frequency:

$$f = \frac{p}{2} \cdot \frac{\Omega}{2\pi} + f_r$$

If the machine rotates at the speed Ω and the converter supplies it at the frequency

$$f > \frac{p}{2} \cdot \frac{\Omega}{2\pi}$$

to which $s > 0$ and $\Omega_0 > \Omega$ corresponds, it means that it is operating as a motor, that is, the vehicle is in the traction phase and the motor is absorbing the active power:

$$P_1 = 3 \cdot E \cdot I_1 \cdot \cos\varphi = \sqrt{3} \cdot V \cdot I_1 \cdot \cos\varphi$$

where E represents the phase voltage and V represents the corresponding line voltage. The active component $I_1 \cdot \cos\varphi$ of the absorbed current is in phase with the power supply voltage E.

During the braking phase, the machine must reverse the sign of the torque with respect to the speed, that is, it operates as a generator. In this phase, the converter has to supply the machine at a frequency of

$$f < \frac{p}{2} \cdot \frac{\Omega}{2\pi}$$

to which $s < 0$ and $\Omega_0 < \Omega$ correspond. In this case, the active component $I_1 \cdot \cos\varphi$ of the current has an opposite phase with respect to the power supply voltage E. The power P_1 is injected into the power supply inverter *DC link* and transferred to the contact line only if the system is completely reversible; otherwise, it is dissipated in the braking rheostats installed in the vehicles.

During traction and braking phases, the machine absorbs the reactive power:

$$Q_1 = 3 \cdot E \cdot I_1 \cdot \sin\varphi = \sqrt{3} \cdot V \cdot I_1 \cdot \sin\varphi$$

which must be generated directly by the power supply inverter. This constitutes one of the limits of the use of induction motors in traction, as it implies the necessity to oversize the traction inverter.

9.1.4.1 Variable voltage and frequency power supply

Modern power supply inverters, as we have seen in the preceding chapters, are capable of generating output voltages with variable amplitudes and frequencies. The relationship between these two quantities is defined by the basic expression used to define electrical machines:

$$E = k_W \cdot f \cdot \Phi$$

which expresses the electromotive force induced in the conductors as a function of frequency and the magnetic flux by means of a coefficient of proportionality k_W that depends on the constructional characteristics of the stator windings.

When supplying an induction machine at voltage $V' = k_V \cdot V$, provided the frequency remains constant, the flux varies approximately in proportion to the supply voltage, that is,

$$\Phi' = k_V \cdot \Phi$$

Since the maximum electromagnetic torque $T_{e_{\max}}$ depends on the square of the magnetic flux, the new torque value is given by the following expression:

$$T'_{e_{\max}} = k_V^2 \cdot T_{e_{\max}} \tag{9.3}$$

However, at the power supply voltage V, if the machine is supplied at a frequency $f' = k_f \cdot f$, the magnetic flux varies according to the following relationship:

$$\Phi' = \frac{\Phi}{k_f}$$

so that the electromagnetic torque is modified as follows:

$$T'_{e_{\max}} = \frac{T_{e_{\max}}}{k_f^2} \tag{9.4}$$

Thus, it can be asserted that the torque generated by an induction machine is highly sensitive to both voltage and frequency variations.

9.1.5 Generation of the Ideal Tractive Effort Diagram

As we have seen, the ideal tractive effort diagram of a given traction vehicle has two characteristic sections:

- A first section where the tractive force remains constant or declines slightly due to such factors as the available adhesion between 0 and the base speed v_b.
- A second section at constant power between the base speed v_b and the maximum speed v_{max}.

9.1.5.1 Operation of the induction machine in the constant force section

In the first section, where the tractive force is constant, the machine must develop a high constant torque in order to transmit the starting acceleration to the vehicle. Since the rotation speed is, in fact, proportional to the power supply frequency due to the small slip value, increasing the frequency alone would result in a substantial reduction of the mechanical torque generated at the shaft as given by (9.4). This means that during this phase, in order to maintain an acceptable torque value, it is necessary to increase the voltage value proportionally with the increase in the frequency in order to maintain a constant flow of power supply. In fact, if $k_V = k_f$, by applying (9.3) and (9.4), it can be seen that

$$T'_{e_{max}} = \frac{T_{e_{max}}}{k_f^2} \cdot k_V^2 = T_{e_{max}}$$

Since the maximum torque remains constant, in order to obtain constant torque during the first section, the parameter γ, that is, the ratio between the working torque and maximum torque, must remain constant. It follows that the tractive effort diagram must translate parallel to itself as the power supply increases, as illustrated in Figure 9.4.

Under these conditions, the absolute slip value $\widetilde{\Delta\Omega}$ remains constant with respect to maximum torque and the absolute working slip value $\overline{\overline{\Delta\Omega}}$, so that according to (9.1), the corresponding rotor frequency also remains constant:

$$\tilde{f}_r = \frac{p}{4\pi} \cdot \widetilde{\Delta\Omega}$$

$$\overline{\overline{f}}_r = \frac{p}{4\pi} \cdot \overline{\overline{\Delta\Omega}}$$

At the starting point, the motor must be supplied at a frequency at least equal to $\overline{\overline{f}}_r$ and at a minimum voltage such as to compensate for the internal voltage drop.

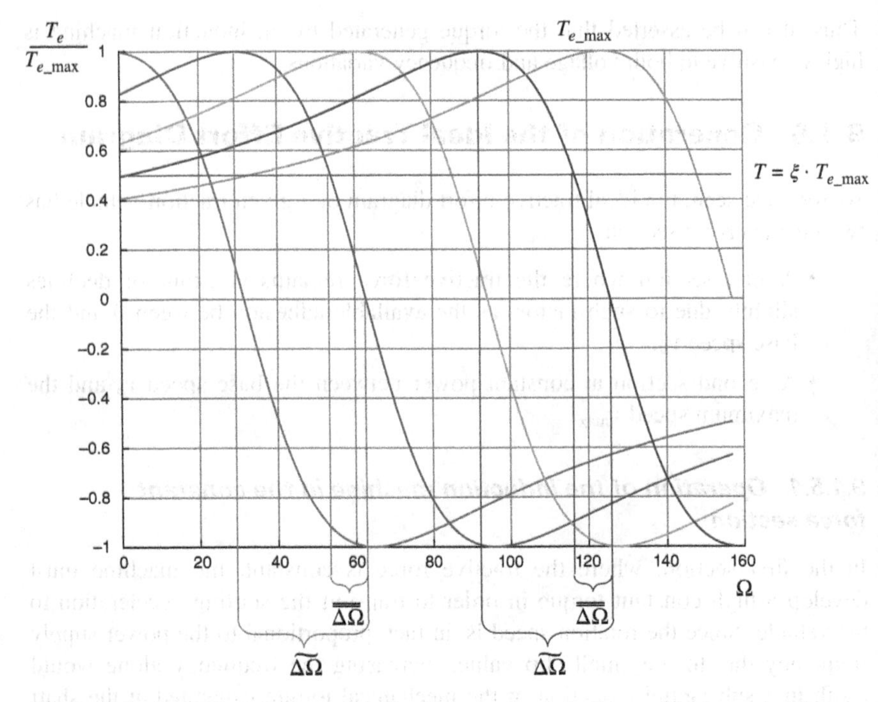

Figure 9.4 Translation of the tractive effort diagram of an induction machine in constant flux operation.

9.1.5.2 Operation of the induction machine in the constant power section

Upon reaching the base speed Ω_b, the torque must decrease proportionally to the speed according to a rectangular hyperbola so as to maintain the power constant.

When the vehicle reaches the base speed, the motor has reached its rated power, or maximum in overload conditions, equal to

$$P_n = T_e(\Omega_b) \cdot \Omega_b = \xi \cdot T_{e_{\max}} \cdot \Omega_b$$

If, in correspondence with Ω_b, the nominal power supply voltage V_n has also been reached, then only the frequency may be increased between the base speed and the maximum speed. It follows from (9.4) that the torque decreases by the square of the increase in frequency, resulting in a decrease in the output power:

$$P(\Omega) = T_e(\Omega) \cdot \Omega = \left[\frac{T_e(\Omega_b)}{k_f^2}\right] \cdot [k_f \cdot \Omega_b] = \frac{\xi \cdot T_{e_{\max}} \cdot \Omega_b}{k_f}$$

If the ratio ξ between working torque and maximum torque is maintained constant, then the power would decrease as $1/k_f$. In particular, at the maximum speed Ω_{max}, $k_f = \beta$; therefore, the power would assume the value:

$$P(\Omega_{max}) = \frac{P_n}{\beta}$$

High elasticity is required in high-speed passenger trains, that is, a wide operating range at constant power, which means that β may reach values close to 3, or higher in some cases. In this case, the power at the maximum speed would be reduced by $1/\sqrt{3}$ of the nominal, representing an unacceptable operating condition. Therefore, as the speed increases, it is necessary to ensure that the value of ξ also increases, that is, the operating torque approaches the maximum torque as shown in Figure 9.5.

As a result, there is an increase in the absolute slip value $\overline{\overline{\Delta\Omega_1}} < \overline{\overline{\Delta\Omega_2}} < \overline{\overline{\Delta\Omega_3}}$ and the corresponding rotor frequency, which in turn causes an increase in losses and a decrease in the power factor.

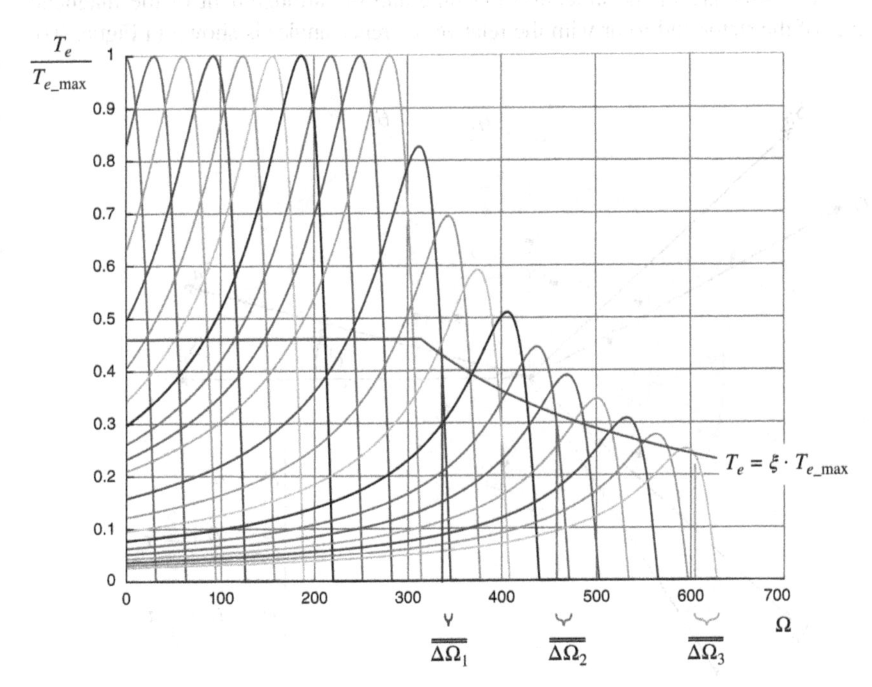

Figure 9.5 Tractive effort diagram of an induction machine in a constant flux and constant power operation.

9.1.6 Torque and Speed Control in an Induction Machine

Once we have understood the behavior of induction machines in static mode, it is necessary to progress to dynamic modeling in order to apply the various control strategies that are currently available. For this reason, the following simplified hypotheses are introduced in order to render this document more comprehensible, but without overgeneralizing:

- The machine is based on an isotropic structure.
- The magnetic permeability of the stator and rotor cores is considered to be of infinite value with respect to the air gap; consequently, the magnetic voltage drops in these sections are null.
- Considering the concentrated windings of N turns, that is, with full concatenation of fluxes, and p poles.
- Considering the machine as a symmetrical and balanced system.
- It is assumed that the electromotive forces in the air gap are perfectly symmetrical.

The structure of the induction machine and the arrangement of the magnetic axes of the stator and rotor with the relative reference angles is shown in Figure 9.6.

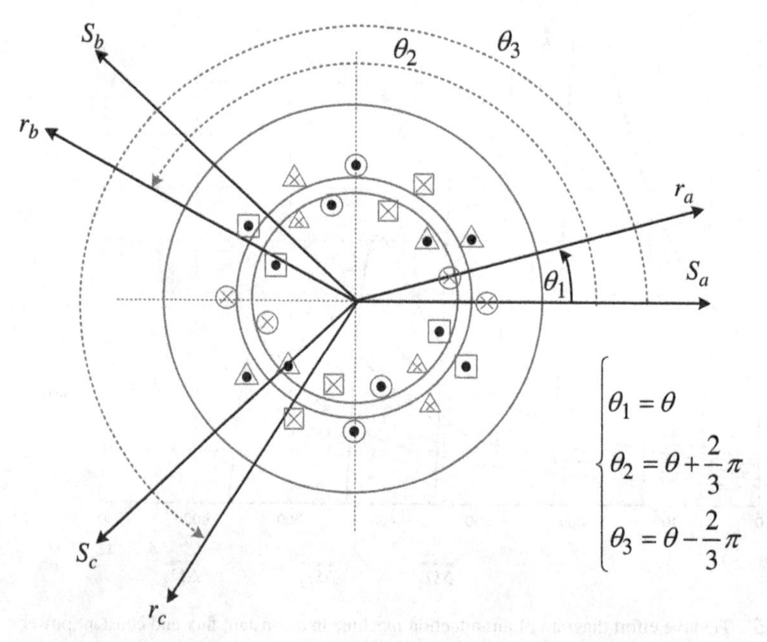

Figure 9.6 Representation of the main variables of the induction machine.

On the basis of these assumptions, the magnetic and electric equations of the machine are as follows:

$$\begin{cases} \left[v_{sr}^{abc}\right] = [R_{sr}]\left[i_{sr}^{abc}\right] + \frac{d}{dt}\left[\psi_{sr}^{abc}\right] \\ \left[\psi_{sr}^{abc}\right] = [L_{sr}(\theta)]\left[i_{sr}^{abc}\right] \end{cases}$$

where

V_{sr} is the column vector of the stator and rotor voltages on the a, b, and c axes;

R_{sr} is the (diagonal) matrix of the stator and rotor resistances on the a, b, and c axes;

ψ_{sr} is the column vector of the magnetic fluxes of the stator and the rotor on the a, b, and c axes;

i_{sr} is the column vector of the stator and rotor currents on the a, b, and c axes;

L_{sr} is the matrix of self and mutual inductances of the stator and rotor on the a, b, c axes.

The following mechanical equations must be added to the magnetic and electric equations:

$$T_e(t) = T_r + \frac{J}{p} \cdot \frac{d^2\theta}{dt^2}$$

Here T_e represents the electromagnetic torque, T_r represents the drag torque, J represents the moment of inertia of the rotor, p represents the number of poles, $d\theta/dt$ represents the angular velocity, and $d^2\theta/dt^2$ represents the angular acceleration.

The set of equations constitutes the differential system of six equations for the six unknowns that describe the machine; for the assumptions that have been made for the purposes of this study, the unknowns correspond to the four currents, the angle θ, and the speed $d\theta/dt$.

In order to control the machine more easily, it may be useful to apply *Park's transformation* to the magnetic and electrical variables. For the sake of simplicity, the orientation and arrangement of the Park α and β axes are selected, so that the α axis is integral with the stator a axis, while the d and q axes form the angle θ with the sa axis. The angle θ is chosen so that $\theta = \theta_s - \theta_r$, where θ represents the phase difference between the α axis and the magnetic axis of the rotor ra, θ_s represents the phase shift between the axis sa and the axis d, θ_r represents the phase shift between the axis ra and the axis d as shown in Figure 9.7.

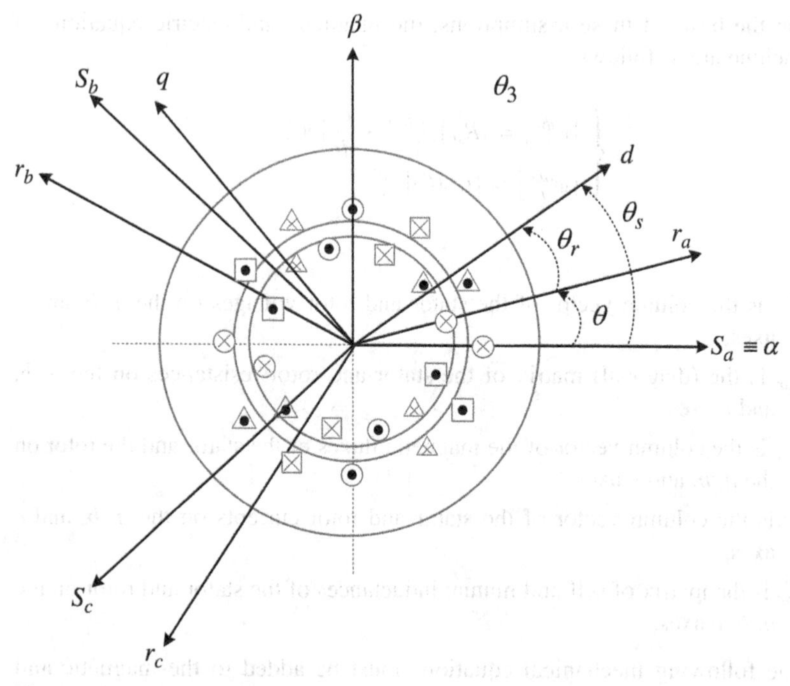

Figure 9.7 Representation of the main variables of the induction machine on the Park axes.

When the Park transformation is applied, the equations plotted on the d,q axes become

$$\begin{cases} \left[v_{sr}^{dqo}\right] = [R_{sr}]\left[i_{sr}^{dqo}\right] + [B]\begin{bmatrix}\theta_s \\ \theta_r\end{bmatrix}\left[\psi_{sr}^{dqo}\right] + \dfrac{d}{dt}\left[\psi_{sr}^{dqo}\right] \\ \left[\psi_{sr}^{dqo}\right] = [L']\left[i_{sr}^{dqo}\right] \end{cases}$$

These are relationships where the matrix of resistance remains unchanged with respect to the original domain, whereas the inductance matrix becomes independent of θ and contains diagonal submatrices, so it is possible to develop a generalized discussion based on the simplified model of the single-phase machine.

Based on the results obtained, it becomes possible to summarize the resolving system using Park's vectors obtained by vectorially summing the various calculated values plotted along the direct d axis and in the quadrature q: $\left[\left(v_{sd}, v_{sq}\right); \left(i_{sd}, i_{sq}\right); \left(\psi_{sd}, \psi_{sq}\right); \left(v_{rd}, v_{rq}\right); \left(i_{rd}, i_{rq}\right); \left(\psi_{rd}, \psi_{rq}\right)\right]$; furthermore, by choosing to place the Gaussian plane rigidly on the rotor of the induction machine so that the real axis always coincides with the d axis, the system becomes

$$\begin{cases} \overline{V}_s = R_s\overline{I}_s + \dfrac{d\overline{\psi}_s}{dt} + j\dfrac{d\theta_s}{dt} \cdot \overline{\psi}_s \\ \overline{V}_r = R_r\overline{I}_r + \dfrac{d\psi_T}{dt} + j\dfrac{d\theta_r}{dt} \cdot \overline{\psi}_r \\ \overline{\psi}_s = L_s\overline{I}_s + M\overline{I}_r \\ \overline{\psi}_r = M\overline{I}_s + L_r\overline{I}_r \end{cases}$$

to which the mechanical equations must be added. Thus, the torque expression derived from the energy balance is as follows:

$$T_e(t) = \frac{P_m}{d\theta/dt} = \frac{p}{2}\left(\psi_{rd}i_{rd} - \psi_{rd}i_{rq}\right) = \frac{p}{2}M\left(i_{sd}i_{rd} - i_{sd}i_{rq}\right)$$

$$= \frac{p \cdot M}{2 \cdot L_r}\left(\psi_{rd}i_{sq} - \psi_{rq}i_{sd}\right) = \frac{p \cdot M}{2 \cdot (L_sL_r - M^2)}\left(\psi_{sq}i_{rd} - \psi_{sd}i_{rq}\right)$$

Also, bearing in mind that $\theta = \theta_s - \theta_r$, the mechanical equilibrium equation can be defined as

$$2\frac{J}{p}\frac{d^2\theta}{dt^2} = \frac{p \cdot M}{2(L_sL_r - M^2)}\left(\psi_{sq}\psi_{rd} - \psi_{sd}\psi_{rq}\right) - T_r$$

Considering the definition of relative slip and being aware that the rotor speed is $\Omega = d\theta/dt$, it can be seen that

$$s = \frac{2\pi f - d\theta/dt}{2\pi f}$$

At this point, it is possible to derive the "equivalent single-phase" network that represents the system. It is assumed that the d,q axes are rotating synchronously with the rotating magnetic field of the stator, so the speed of the rotating field is $\omega = d\theta_s/dt$, and therefore $d\theta_r/dt = s \cdot \omega$. Introducing $K = N_s/N_r$ as the ratio between the number of turns on the stator and rotor windings, the rotor parameters are applied to the stator; furthermore, it is preferable to derive the equivalent four-parameter model by introducing a fictitious ideal transformer with an appropriate transformation ratio. The result is the model mentioned above:

$$\begin{cases} \overline{V}_s = R_s\overline{I}_s + \dfrac{\overline{\psi}_s}{dt} + j\omega\overline{\psi}_s \\ 0 = \dfrac{R_r}{s}\overline{I}_r + \dfrac{1}{s} \cdot \dfrac{\overline{\psi}_r}{dt} + j\omega\overline{\psi}_r \\ \overline{\psi}_s = \overline{\psi}_r + L_{ks}\overline{I}_s \\ \overline{\psi}_r = M\left(\overline{I}_s + \overline{I}_r\right) \end{cases}$$

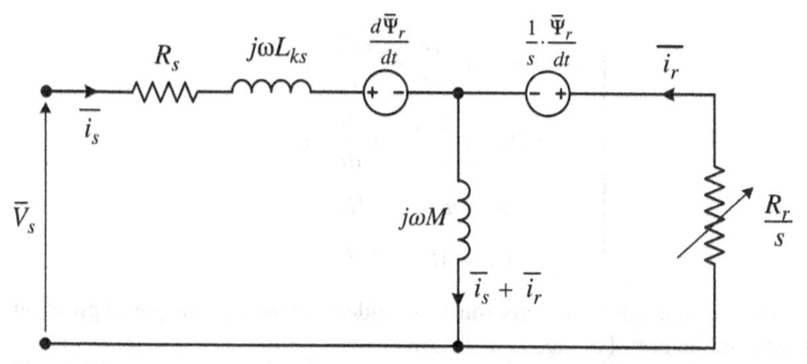

Figure 9.8 Dynamic, four-parameter model of the three-phase induction machine.

to which must be added the mechanical equation:

$$\frac{2J}{p} \cdot \frac{d^2\theta}{dt^2} = \frac{p \cdot M}{2(L_s \cdot L_r - M^2)} (\psi_{sq}\Psi_{rd} - \psi_{sd}\Psi_{rq}) - T_r$$

The circuit model shown in Figure 9.8 corresponds to these equations.

9.1.6.1 Induction machine drive control strategies

In the field of electric traction, it is necessary to use drives with variable dynamics, and as a response to this requirement the electromagnetic torque and the flux of the electrical machine must be controlled as a function of the different speed profiles assumed by the rolling stock.

The most common control strategies for drives with voltage source inverters (VSIs) and three-phase induction motors are represented in Figure 9.9.

A control is defined as *scalar* when the control architecture of the inverter that supplies power to the induction machine makes use of scalar

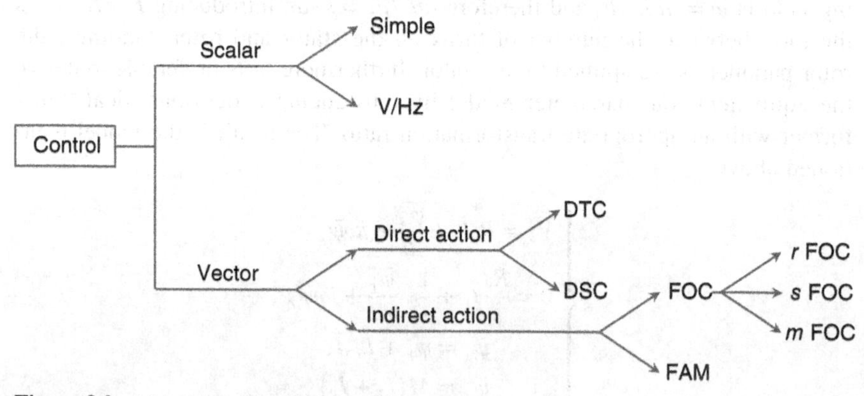

Figure 9.9 Main three-phase induction machine control strategies.

electromechanical values (calculated or measured) only, and not spatial phasors. This is an open-loop torque and flux control method that is widely used when good results are required under stable operating conditions, without stringent dynamic characteristics. However, it is not possible to control the electromagnetic transient, which results in significant oscillations in the current and flux, and hence the torque generated at the shaft.

Under highly dynamic conditions, these oscillations affect the quality of operations and therefore must be eliminated. Moreover, due to the current overshoots that are produced when using scalar controls, especially during the starting phase, it is necessary to oversize the inverter.

In order to govern the electromagnetic transient and obtain the best possible dynamic characteristics, it is necessary to resort to more sophisticated control strategies that provide fast, accurate, and independent control over both flux and torque. Such solutions must be able to control not only both the amplitude of the stator variable but also its phase, or in other words, vector controls. These controls are based on a suitable choice of the reference axes used to determine the components of the stator current vectors.

In particular, it is necessary to select a specific two-phase reference system such that the component of the vector of the stator currents on an axis acts exclusively on the flux, and the component on the other axis acts exclusively on the drive torque.

The vector control system is then further subdivided into two types (Figure 9.9): "direct action" and "indirect action."

The direct action control system provides very satisfactory mechanical transients, while the electromagnetic transitional presents high current overshoots, especially in the starting phase.

In order to obtain a high dynamic, it is necessary to maximize the speed of response of the machine's electromagnetic torque and this can only be achieved by minimizing the variations of the electromagnetic energy of the motor, which is substantially correlated to the magnetic field, and hence to the flux. Thus, it is possible to use this class of controls, known as field-oriented or indirect action, based on the concept that considers a commutator motor as equivalent to an induction motor. To do this, it is assumed that one of the rotating axes d,q is integral with the rotor flux, for example, the d axis. An observer integral with this axis will then see a current that is integral with the flux (similar to that circulating in the excitation circuit of the direct current machine) responsible for producing the fluxing effect in the equivalent machine. It will then observe a current in quadrature with respect to the flux (similar to that flowing in the armature winding), responsible for generating the torque.

Depending on which flux is aligned with the d axis, various types of vector control are possible:

- FOC (field-oriented control)
- FAM (flux acceleration method)

9.1.6.1.1 Simple Scalar Control This is the control technique adopted in the case of medium-level dynamic specifications; it is based on the performance of the operating range seen previously and, in particular until the vehicle reaches the base speed, the ratio between the modulus (amplitude) of the stator voltage and the frequency (related to speed) is maintained constant, thereby maintaining the linked flux constant and equal to the nominal value. In order to accelerate the machine beyond the base speed, it is necessary to apply the flux weakening, maintaining the stator voltage constant at its nominal value and increasing the power supply frequency.

The equation that demonstrates the balance of the three-phase induction machine stator voltages is as follows:

$$\overline{V}_s = R_s\overline{I}_s + \frac{d\overline{\psi}_s}{dt}$$

Under sinusoidal conditions:

$$\overline{V}_s = R_s\overline{I}_s + j\omega\overline{\psi}_s$$

In addition, if ω is different from zero, as a first approximation it may be considered that $R_s \cong 0$. Therefore,

$$\frac{|\overline{V}_s|}{2\pi f} \cong |\overline{\psi}_s|$$

It is immediately obvious from this equation that the machine stator flux is proportional to the ratio between the modulus of the Park vector referred to the power supply voltage and the angular velocity at which the latter rotates; if Ψ_s is maintained constant at the nominal value, then the machine is well fluxed, and hence it is being operated efficiently. The above observations introduce the concept of sensorless scalar machine control, which is illustrated in Figure 9.10.

The "phase voltages" block generates the appropriate stator voltages, starting from the stator voltage modulus and the angle between the d,q axes and the fixed axes. These signals are then used as the modulating PWM control for the inverter that supplies the power to the induction motor.

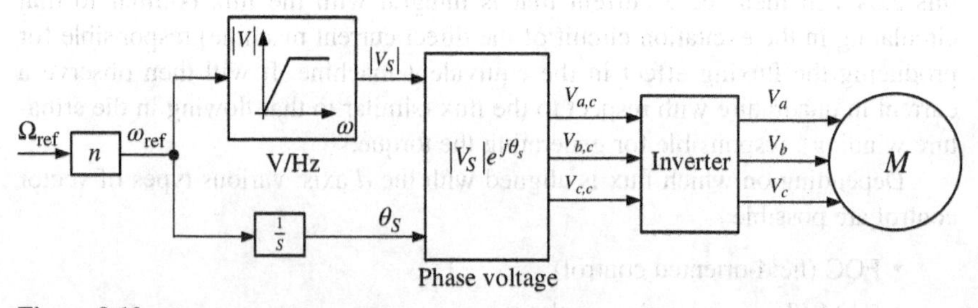

Figure 9.10 Sensorless scalar control for induction machines.

The "V/Hz" block output supplies the modulus of the stator voltage according to the angular velocity of the rotating stator field, compatibly with the operating range. This block compensates for the voltage drop across the resistance R_s, which assumes significant proportions during the start-up phase by means of an offset. As can be seen from the diagram, this is an open-loop arrangement, without any feedback control over the actual speed of the motor; this translates to reduced costs, but also limits its use to situations with low to medium dynamic specifications.

9.1.6.1.2 Scalar Volt/Hertz Control The V/Hz control is constructed on the basis of the simple scalar control; by implementing speed Ω measurement and feedback, it can be used to obtain null speed errors under steady-state conditions.

It can be observed that the only exploitable section of the induction machine static tractive effort diagram is that with the negative slope, that is, the section where stability is verified. In fact, in this section, placing the direct Park axis so that it is aligned with the space phasor of the rotor flux, it can be noted that the torque is directly proportional to the angular velocity and hence the slip, as shown by the following relationship:

$$T_e = \frac{p}{2}\left(\psi_r i_{sq}\right) = \frac{p}{2}\cdot\frac{\psi_r^2}{R_r}\left(\frac{d_s\theta}{dt}-\frac{d\theta}{dt}\right) = \frac{p}{2}\cdot\frac{\psi_r^2}{R_r}(\omega-\omega_m) = \frac{p}{2}\cdot\frac{\psi_r^2}{R_r}\cdot s\cdot\omega$$
$$= K\cdot s\cdot\omega$$

where K represents the *torque constant*, which indicates that if the flux is constant, then the machine tractive effort diagram (T_e, Ω) corresponds to a straight line proportional to $s\cdot\omega$ and thus a function of the slip; the product $s\cdot\omega$ is called ω_{slip}.

For the V/Hz control block diagram, see Figure 9.11.

By measuring the angular speed Ω, the system creates a feedback loop that is used to obtain a null error under steady-state conditions.

9.1.6.1.3 Indirect Action Vectorial Control We will now examine the FAM (flux acceleration method) technique in depth. This method involves aligning the rotor flux with the d axis, in exactly the same way as the "Volt/

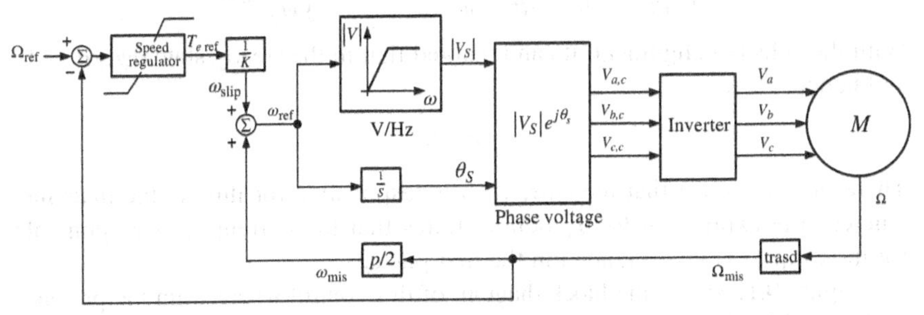

Figure 9.11 Scalar V/Hz control for induction machines.

Hertz" control described previously. The FAM control technique is an open-loop system and is the simplest type of vector control. In particular, it is assumed that the control system "functions correctly" and that it is capable of causing the motor to reach the desired values of i_{sd} and i_{sq} instantaneously, that is,

$$\begin{cases} i_{sd} = i_{sdref} \\ i_{sq} = i_{sqref} \end{cases}$$

for each operating instant. This technique uses the current $\bar{i}_s$ and flux $\bar{\psi}_r$, respectively, as the state variables using them to reelaborate the equations that describe the equivalent, four-parameter model in order to obtain

$$\begin{cases} \bar{v}_s = R_s\bar{i}_s + L_{ks}\dfrac{d\bar{i}_s}{dt} + R_s\bar{i}_s - R_s\dfrac{\bar{\psi}_r}{M} - j\dfrac{d_r}{dt}\bar{\psi}_r + j\dfrac{d_s}{dt}L_{ks}\bar{i}_s + j\dfrac{d_s}{dt}\bar{\psi}_r \\ \dfrac{d\bar{\psi}_r}{dt} = R_s\bar{i}_s - \dfrac{R_s}{M}\bar{\psi}_r - j\dfrac{d_r}{dt}\bar{\psi}_r \end{cases}$$

Projecting them onto the d and q axes, having assumed that $\bar{\psi}_r$ is integral with the d axis, the following expression is obtained:

$$d : v_{sd} = (R_s + R_r)I_{sd} + L_{ks}\frac{dI_{sd}}{dt} - \frac{R_s}{M}\psi_r - \frac{d\theta_s}{dt}L_{ks}i_{sq}$$

$$q : v_{sd} = (R_s + R_r)I_{sq} + L_{ks}\frac{dI_{sq}}{dt} - \frac{d\theta_r}{dt}\psi_r + \frac{d\theta_s}{dt}L_{ks}i_{sd} + \frac{d\theta_s}{dt}\psi_r$$

$$d : 0 = \frac{R_r}{M}\psi_r - R_r i_{sd} + \frac{d\psi_r}{dt}$$

$$q : 0 = -R_r i_{sq} + \theta_r\frac{d\psi_r}{dt}$$

In addition,

$$\begin{cases} T_e = \dfrac{p}{2} \cdot \psi_r i_{sq} \\ \dfrac{d\theta_s}{dt} = \dfrac{d\theta_r}{dt} + \dfrac{d\theta}{dt} = \dfrac{R_r}{\psi_r}i_{sq} \Rightarrow \theta_s = \displaystyle\int \dfrac{R_r}{\psi_r}i_{sq}dt \end{cases}$$

With the selected alignment, it can be noted that in the steady state $d\psi_r/dt = 0$, and hence

$$\psi_r = Mi_{sd}$$

Therefore, it is clear that the current i_{sd} is responsible for fluxing the machine, whereas the expression for T_e demonstrates that the current i_{sq} is responsible for the torque, as was intended in the first place.

Figure 9.12 shows the block diagram of this control where, with the presence of the "FAM" block, the conditions $i_{sd} = i_{sdref}$ and $i_{sq} = i_{sqref}$ are imposed.

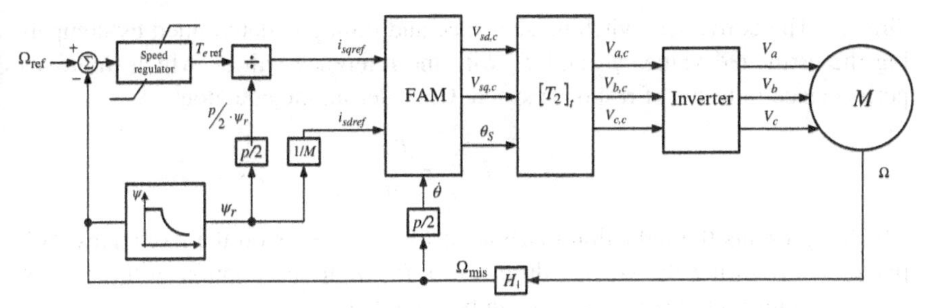

Figure 9.12 FAM scalar control for induction machines.

This control has the following features:

- High performance in dynamic state
- High motor speed measurement accuracy is not necessary
- Separate flux and torque controls

The system does, however, present two main drawbacks:

- The FAM block depends heavily on the induction motor parameters.
- There is no control over the currents that actually circulate in the motor stator: In fact, according to the equations obtained above, it is necessary to "blindly trust" that the drive is operating correctly, and assume that the these currents coincide with the reference values at all times. In practice, only the speed is controlled by a closed feedback loop, while the current and flux loops are open.

9.1.6.1.4 Direct Action Vector Control: DSC and DTC The direct action controls have been designed to drive the induction machine by acting directly on the magnitudes that influence the flux or the torque, driving the traction converters in an unconventional way, and hence without using PWM or SVM techniques.

These systems are inspired by an approach that differs significantly from those covered up to this point, and have established themselves principally in the automotive sector. The direct action controls owe their origins to the studies undertaken by Depenbrock: More specifically, DSC (direct self-control) has been the most advanced solution in the field of heavy trains for many years, thanks to its ability to make the most of even limited switching of the high-power-capacity semiconductors (thyristors or GTO), while DTC (direct torque control) is, in a sense, its evolution. An important advantage of direct action controls is that they can be used to manage fast dynamics in a precise manner, providing high system response speeds without placing excess stress on the converter semiconductors.

Despite the field-oriented and scalar controls, the direct action regulators can be used to regulate the machine flux, $\overline{\psi_s}$ and hence torque T_e, current directly. The converter switching sequence and timing is determined by comparing the estimated values ψ_s and T_e with the reference values, which improves performance in terms of response speed. Considering the equation

$$\overline{V}_s = R_s\overline{I}_s + \frac{d\overline{\psi}_s}{dt}$$

which represents the induction machine stator parameters on the fixed-axes Park plane, and ignoring the voltage drop across R_s (valid assumption in the case of medium–high-power machines), it can be seen that

$$\overline{V}_s \cong \frac{d\overline{\psi}_s}{dt}$$

The above equation demonstrates that $\overline{V}_s$ can be interpreted as the instantaneous speed at which the extremity of the stator flux increases its position. In particular, since $\Delta\overline{\psi}_s = \overline{V}_s \cdot \Delta t$, if we apply $\overline{V}_s \neq 0$, the machine is fluxed or defluxed depending on the direction of the vector $\overline{V}_s$, while if we apply $\overline{V}_s = 0$, the flux $\overline{\psi}_s$ remains constant in magnitude and phase, as though it were "frozen" in a given position.

Ultimately, taking the Park plane into account, the flux tends to align with the direction of the applied vector $\overline{V}_s$ (Figure 9.13).

As far as the torque T_e delivered by the induction machine is concerned, in order to determine its expression in relation to the position of the flux $\overline{\psi}_s$ compared to the flux $\overline{\psi}_r$, it is necessary to refer to the four-parameter model, that is,

$$\begin{cases} \overline{\psi}_s = L_{ks}\overline{\imath}_s + \overline{\psi}_r \\ \overline{\psi}_r = M(\overline{\imath}_s + \overline{\imath}_r) \end{cases}$$

From this it is possible to derive the expression for the current $\overline{\imath}_s$:

$$\overline{\imath}_s = \frac{\overline{\psi}_s - \overline{\psi}_r}{L_{ks}}$$

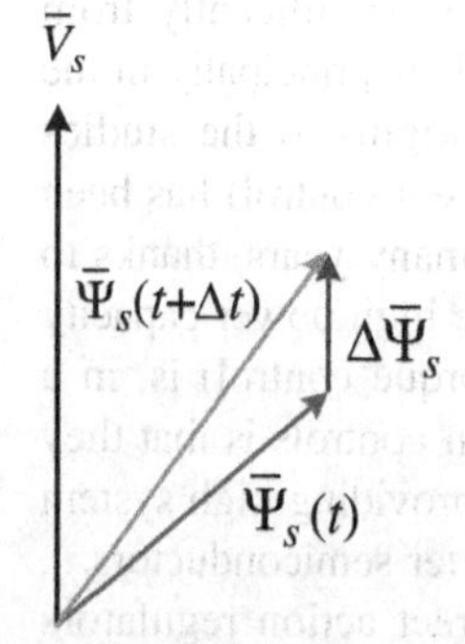

Figure 9.13 Effect of the voltage $\overline{V}_s$ on the flux $\overline{\psi}_s$.

Thus, the expression for the torque becomes

$$T_e = \frac{p}{2} \cdot \mathrm{Im}\left(\overline{\imath_s}\underline{\psi}_r\right) = \frac{p}{2L_{ks}} \cdot \mathrm{Im}(\overline{\psi}_s \underline{\psi}_r) = \frac{p}{2L_{ks}} |\overline{\psi}_s| |\underline{\psi}_r| \cdot \sin\delta$$

where δ represents the phase shift between the two fluxes $\delta = (\measuredangle\overline{\psi}_s - \measuredangle\overline{\psi}_r)$; note that the fluxes $\overline{\psi}_s$ and $\overline{\psi}_r$ are synchronous. Therefore, it can be stated that, when considering the section where the machine stability condition is verified, the torque T_e increases as the angle δ increases.

Finally, it can be noted that direct action controls are generally sensorless types, meaning that they do not require measurement values for the mechanical parameters they are used to control (the torque T_e and the speed Ω); on the other hand, Ω is linked to T_e, which may be estimated; for this reason, the *V–I* state estimator is used: by measuring $\overline{V}_s$ and $\overline{\imath}_s$, it is possible to estimate the values $\overline{\psi}_s$ and T_e, which are used in the control logic feedback chain.

$$V{-}I \text{ estimator} \begin{cases} \overline{\psi}_{s_{\text{estimate}}} = \int \left(\overline{V}_s - R_s\overline{\imath_s}\right) dt \\ T_{e_{\text{estimate}}} = \frac{p}{2} \cdot \mathrm{Im}\left(\overline{\imath_s}\underline{\psi}_s\right) \end{cases}$$

9.1.6.1.5 Direct Self-Control Regulator (DSC) The DSC was the first direct controller to be designed for use with induction and synchronous machines.

The operating principle is based on the revolutionary static switch commutation management method, governed directly on the basis of a comparison between the values to be regulated. After estimating the values $\overline{\psi}_s$ and T_e, they are compared with the corresponding set of reference values. By processing the resulting data, the internal converter switching sequences are handled appropriately. The regulator is also implemented so that it performs controls, depending on whether the requested speed is lower or higher than the base speed.

It should be pointed out that the induction machine operating range is divided into two zones: In the first zone, which is valid for speeds lower than the base speed, the voltage increases more or less linearly with the speed, while the flux $\overline{\psi}_s$ and torque T_e remain constant; in the second zone, known as the flux-weakening zone, the voltage $\overline{V}_s$ remains constant, while $\overline{\psi}_s$ decreases hyperbolically with the speed.

Assuming that the converters are the commonly used VSI type, and hence that there are a total of eight permitted semiconductor state combinations (including six active and two inactive), it can be concluded that the voltage vector $\overline{V}_s$ can assume no more than eight different positions on the Park plane, six of which are positioned at the vertices of a hexagon; this implies that the flux $\overline{\psi}_s$ may follow a hexagonal trajectory, but not a circular one, selecting the eight possible voltage vectors $\overline{V}_s$ as necessary and in sequence. As already explained, the controller adapts the control by implementing two different strategies depending on the current speed.

Low-Speed Regulator ($0 < \Omega < \Omega_b$)

1. *Flux control*

   In this configuration, the modulus $|\psi_s|$ is maintained constant so that the modulus $|V_s|$ varies proportionally with Ω. The flux tends to align with the direction of the applied vector $\overline{V_s}$. Thus, it is clear that in order to conduct the flux along the trajectory of any of the sides of the hexagon, it is necessary to apply the corresponding vector $\overline{V_s}$ to the furthest corner in the direction of the next segment. The inverter switching sequence and timing are determined by comparing the projected flux $\overline{\psi_s}$ (i.e., ψ_s^*) with ψ_b. The latter is determined by considering the direction corresponding to that of the vector $\overline{V_s}$ at the same instant, but placed by 30° ahead. As long as the condition $\psi_b > \psi_s^*$ is true, the static switch is prevented from switching; as soon as this condition is no longer true, the flux regulator causes the converter to switch so that it corresponds to the subsequent vector $\overline{V_s}$.

   Figure 9.14 illustrates the flux regulator switching logic and block diagram.

   The direct control over the flux extends to control over the torque and slip, which means that there is a risk that the absorbed currents are too high; for this reason, it is both useful and necessary to complement this control system by including a torque regulator.

2. *Torque control*

   As has already been stated, the torque increases as the angle δ increases. Moreover, the dynamics of $\overline{\psi_s}$ are considered to be fast, since they are associated with the electrical values; while the dynamics of $\overline{\psi_r}$ are slow, since they are associated with the rotation speed of the rotor and the inertia of the machine.

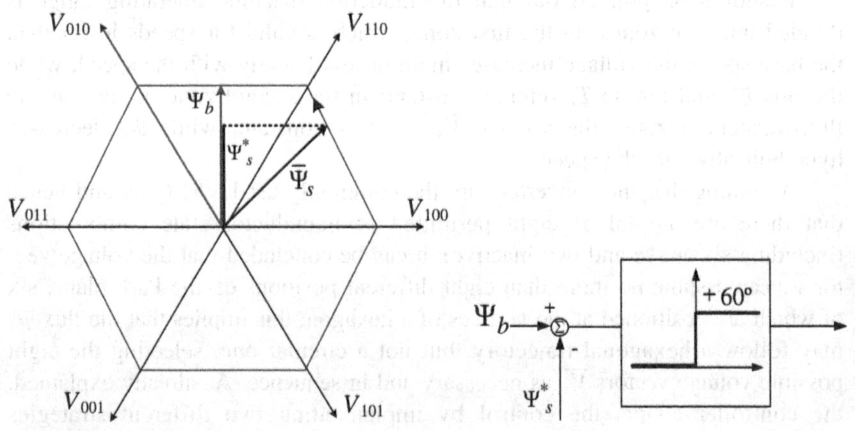

Figure 9.14 Switching logic in the DSC (a) and flux regulator (b).

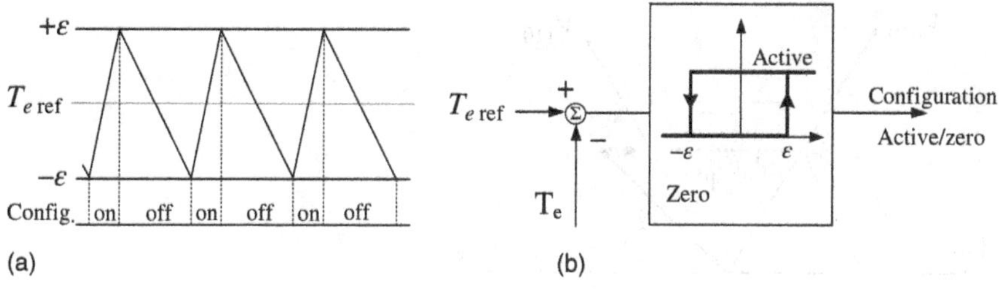

Figure 9.15 $\overline{V_s}$ modulation logic (a) and torque regulator (b).

This means that by applying an appropriate voltage $\overline{V_s}$, the flux $\overline{\psi_s}$ accelerates and moves quickly while $\overline{\psi_r}$ tends to maintain its initial speed: The angle δ increases and therefore so does the torque T_e. Furthermore, by not applying any $\overline{V_s}$, the flux $\overline{\psi_s}$ freezes its position while $\overline{\psi_r}$ tends to rotate regardless: The angle δ decreases and therefore so does the torque T_e delivered by the machine.

Thus, the torque control logic is implemented on the basis of those that have been stated above: This results in $\overline{V_s}$ being fractionated, so the error between $T_{e_{\text{estimate}}}$ and the reference value is maintained within a specific range between $\pm\varepsilon$, as illustrated in Figure 9.15.

The rotation speed can be increased as long as there is a margin for adjusting the voltage V_s; once it reaches its limit value, the machine reaches a speed close to the base speed. In this case, the available torque is balanced by the load-braking torque, so it is not possible to obtain any acceleration. To increase the speed again, the so-called flux weakening must be carried out, which is executed by the high-speed regulator.

High-Speed Regulator ($\Omega > \Omega_b$) In this configuration the modulus $|V_s|$ is maintained constant at its nominal value. To increase the speed Ω, it is necessary to deflux the machine by reducing the modulus $|\psi_s|$ and by increasing its rotation speed. On the basis of the relationship $\overline{V_s} \cong d\overline{\psi_s}/dt$, together with the above affirmation on the modulus $|V_s|$, it can be deduced that, in reality, the derivative of the flux is already at the maximum attainable level, with the inverter operating in square wave mode; therefore, it is necessary to implement a strategy that takes this into account.

The above leads to the adoption of the following control technique: The speed variations are determined by varying the modulus of the stator flux vector, that is,, by having this vector follow the path of hexagons having different apothems. For example, in order to increase the speed, it is sufficient to shorten the edge of the hexagon followed by ψ_s, thus imposing a lower reference flux ψ_b value. In this way it is possible to increase the angular velocity of $\overline{\psi_s}$ but not its peripheral velocity, which is fixed by the relationship $\overline{V_s} \cong d\overline{\psi_s}/dt$.

By applying this logic, the flux ψ_s^* reaches the reference value more rapidly. As a result, the speed of the commands to the static switches is increased, so that

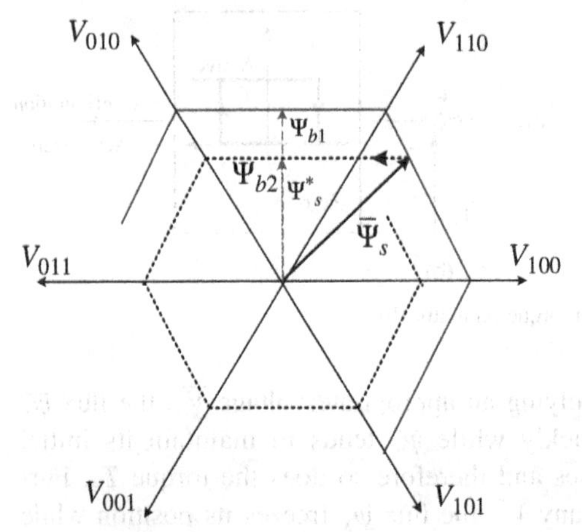

Figure 9.16 DSC control logic for increasing speed above the base speed.

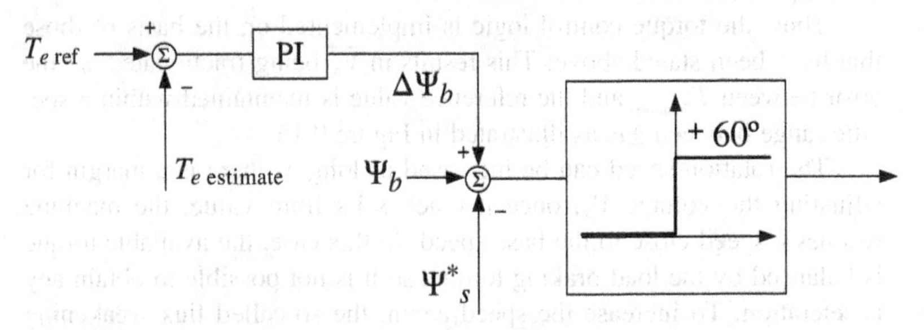

Figure 9.17 DSC torque regulator for increasing the speed above the base speed.

the they anticipate the switching. This determines an increase in the first stator flux harmonic pulse, an increase in the angle δ, and hence an increase in the torque T_e. The logic described above is illustrated in Figure 9.16.

The control, illustrated in Figure 9.17, is implemented by imposing the condition that the electromagnetic torque be equal to a fixed reference torque value, and by modifying the stator flux reference value according to the torque error.

9.1.6.1.6 Direct Torque Control Regulator The direct torque control regulator or DTC can be considered the evolution of the DSC because it retains the basic principles. The major innovations mainly concern the methods used to control stator flux trajectory, and how the inverter semiconductor on and off commands are imparted.

Flux Control As already stated, the stator flux can be manipulated directly by the stator voltage and has a very fast response speed to scaled changes in the latter; it is possible to have the flux vector follow a hexagonal trajectory defined by the six nonnull states that the voltage vector may assume.

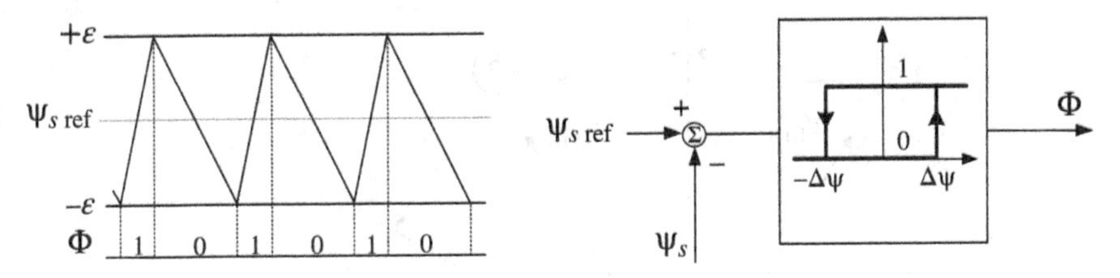

Figure 9.18 The torque comparator in the DTC technique.

In reality, the hexagonal trajectory is not the easiest to follow: In fact, it is possible to approximate the ideal, circular form of the trajectory more closely by confining the extremity of the stator flux vector within a circular crown. The control logic is based on a comparison between the flux, supplied by the state observer, and two threshold values positioned symmetrically with respect to a given reference flux. This ensures that the appropriate voltage vector is selected depending on whether the upper or lower threshold is exceeded, and on the position of the flux $\overline{\psi_s}$ in relation to the reference axes. A level comparator compares the modulus $|\psi_s|$ with the $\psi_{s\ ref}$ reference flux, generating a discrete output equivalent to 0 or 1, depending on whether the flux is touching the upper or lower threshold, respectively. The operating principle is illustrated in Figure 9.18.

At speeds of lower than Ω_b, $\psi_{s\ ref}$ is equal to the nominal value; whereas in the case of speeds exceeding the base speed, the reference $\psi_{s\ ref}$ decreases in line with the operating range.

The output Φ is sent to the control logic, which through suitable processing determines which of the Park voltage vectors to assume as the reference vector. The Park plane is then subdivided into six equally spaced segments, each containing a voltage vector and numbered as shown in Figure 9.19.

The angle θ_s formed by the flux vector and the reference axes is used to determine which of the six segments it is located in and, therefore, which voltage vector must be selected. The logic of selecting the voltages according to the stator flux vector establishes that, in each segment, the voltage vectors belonging to the two segments following the segment where the flux is located are always selected: The adjacent one is used to increase the stator flux modulus, while the subsequent one is used to decrease this modulus. The selection logic is illustrated in Figure 9.20.

The logic described above ensures that, even though a greater number of variations of $\overline{V_s}$ are obtained per unit time, the converter performs the minimum number of main switching operations. The result of this is that only two commutations take place at a time, thus limiting switching and excessive thermal burdens; this prolongs the working life of the static switches and means that it is possible to use smaller heat dissipation devices.

As far as the current drawn by the inverter is concerned, confining ψ_s within a circular crown ensures that the harmonic content is much less significant than in the case of square wave operation.

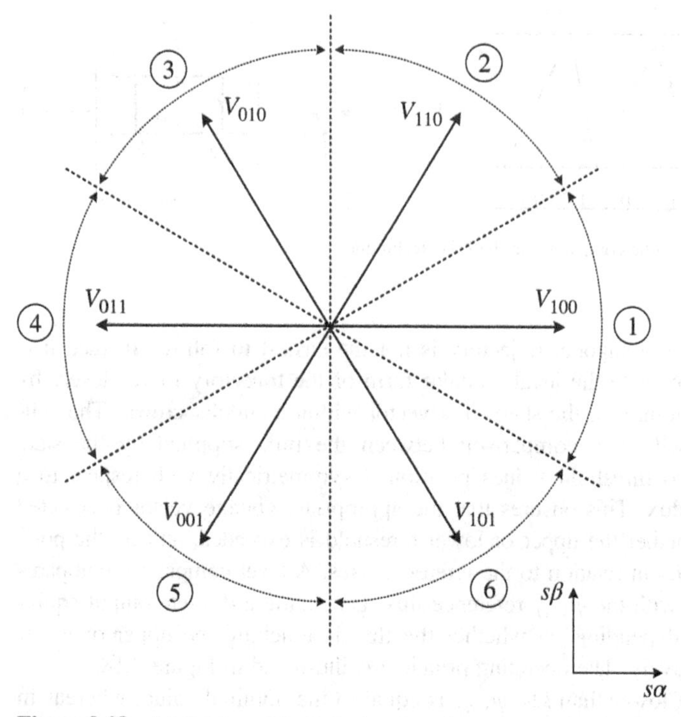

Figure 9.19 Subdivision of the Park plane into six segments when implementing the DTC technique.

Moreover, in the presence of an almost circular rotating magnetic field, the torque delivered by the machine is more constant than in the case of the DSC technique.

Torque Control Torque control is performed using the same methods as the DSC control, arresting or advancing the stator flux vector so as to confine the torque within a given range. Thanks to the level comparator, it is also possible to deliver negative torque if required when braking or reversing the speed; the regulator output τ is 0 when the torque exceeds the upper threshold; 1 when it reaches the lower threshold and the flux vector is rotating counterclockwise; and −1 when the torque reaches the lower threshold and the flux vector is rotating clockwise. The control logic and torque comparison are illustrated in Figure 9.21.

Control Chain Management The techniques described for controlling flux and torque can be managed easily by means of a digital algorithm after implementing the *Switching Table*, a dedicated, three-dimensional [2 × 3 × 6] matrix designed for programming the most suitable switch state combinations at the

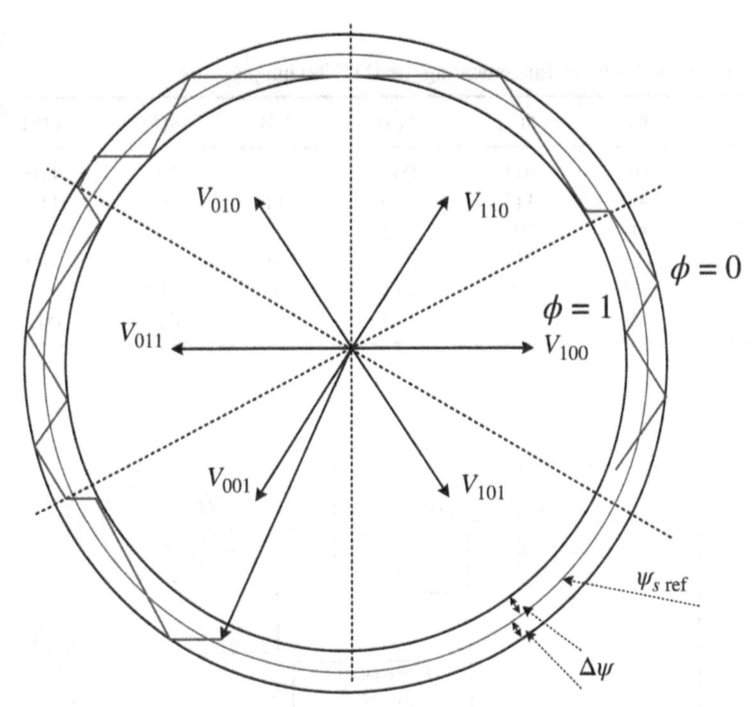

Figure 9.20 Trajectory of the stator flux when implementing DTC.

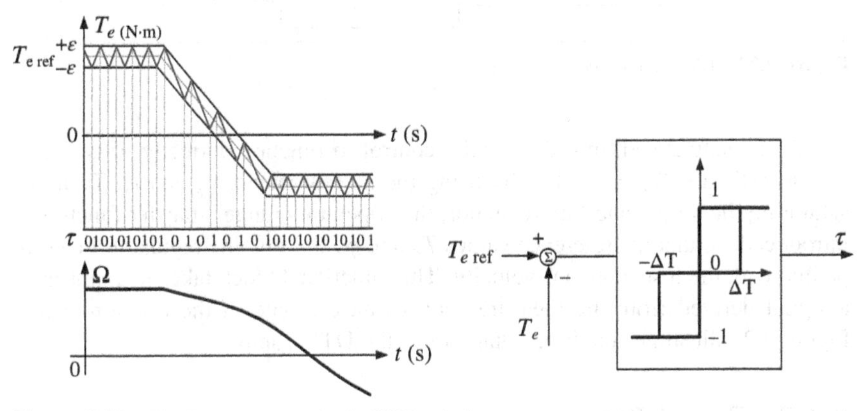

Figure 9.21 The flux comparator in the DTC technique.

instant under consideration, based on the information obtained regarding the modulus ψ_s, its location, and the value of the torque. The Switching Table, which is shown as Table 9.1, has three inputs:Φ, θ, and τ, while the outputs correspond to the value of the voltage vector in the form of commands to the static switches.

Table 9.1 Switching Table for Implementing the DTC Technique

Φ	τ	$\theta(1)$	$\theta(2)$	$\theta(3)$	$\theta(4)$	$\theta(5)$	$\theta(6)$
0	1	010	011	001	101	100	110
	0	000	111	000	111	000	111
	−1	001	101	100	110	010	011
1	1	110	010	011	001	101	100
	0	111	000	111	000	111	000
	−1	101	100	110	010	011	001

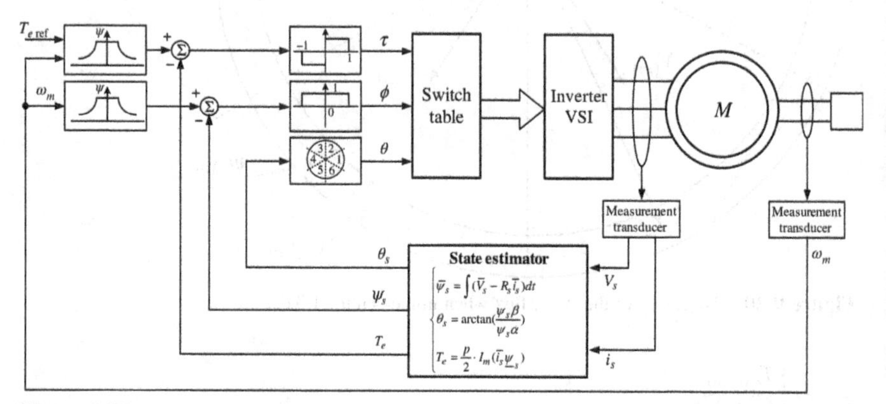

Figure 9.22 DTC control block diagram.

As already stated, in order for the control to function correctly, it is necessary to estimate T_e, ψ_s, and θ_s by using the measurements V_s and I_s. To avoid saturating the torque and flux regulator, the "operating range" transfer function is introduced, so that the reference values T_e and ψ_s used by the regulator are compatible with the instantaneous velocity. This functional block takes its input from a signal derived from the measurement, or an estimate of the rotation speed. Figure 9.22 illustrates the block diagram of the DTC control.

9.1.7 Speed Reverse

An advantage of modern AC motor drives is that the rotational speed can be reversed without changing the circuit configuration, that is, without the need of heavy contactors.

In fact, speed reverse can be easily obtained by changing the phase sequence of the motor drive, by exchanging the firing pulses between two semiconductor modules.

9.2 DRIVES WITH PERMANENT MAGNET MOTORS

9.2.1 Use of Permanent Magnets

Choosing to use permanent magnets implies significant challenges; in fact, multiple complex factors that are more or less simple can affect a project's success. Hence, it is not surprising that these factors are frequently neglected, with unpleasant consequences in the transition from the design phase to the actual construction of the required equipment.

It is rather appropriate therefore that we should review the general aspects concerning the design of electrical machines employing permanent magnets.

Currently, the most common hard magnetic materials are as follows:

AlNiCo: This is a category of alloys composed of aluminum, nickel, and cobalt. Since their discovery, energy production related to their use has been steadily increasing, especially with the introduction of the anisotropic forms. The main characteristics of these alloys are a high flux density (greater than 1 tesla), excellent thermal stability, and reduced manufacturing costs. However, their extreme hardness also implies brittleness and a low coercivity that increasingly limit their use in applications requiring high remanence and very low cost, thus favoring recourse to other materials where higher quality solutions are required.

Ferrites: These stand out from other materials because they are ceramic in nature and nonmetallic, but this does not qualify them as having the highest energy product performance available. They have poor remanence and they are fragile and difficult to work with. Nevertheless, they have had considerable success, with coverage of more than half of the world market for magnets, due to their coercivity and minimal cost. The relevant demagnetization characteristic is essentially linear and this facilitates their use in electrical machines and in several other applications. To contrast their demagnetization tendency at low temperatures (a particularly recurrent condition in electric vehicles), while simultaneously improving the magnetic properties that are typical of ferrites, some manufacturers add powders elements such as cobalt and lanthanum, or mixtures of both, to the base.

Rare Earth Magnets

SmCo: These are obtained from combinations of samarium and cobalt. Their discovery in the 1960s made it possible to combine the advantages of AlNiCo and ferrite magnets, that is, high remanence and coercivity. They also have remarkable thermal stability. Because of the growing demand for permanent magnetic materials and the strategic nature of cobalt materials, SmCo-based material is subject to significant price fluctuations.

NdFeB: Magnetic material composed of neodymium, iron, and boron. In their sintered form, they currently offer the highest performance on the market with BH energy products on the order of 183–414 kJ/m^3. In the sintering process, a crystalline NdFeB alloy is reduced to submicrometer particle powder in a rolling mill; this powder is then subjected to an intense magnetic field and pressed in a mold to obtain its basic shape. The resulting block is then sintered by fusing the material into a solid metal. NdFeB sintered alloys are mechanically more durable and less fragile than other magnets. The Nd and Fe raw materials are very abundant in nature and, therefore, the cost is much lower than for SmCo.

While the demand for soft magnetic materials has not changed significantly over the years, global demand for permanent magnets has grown by an average of 9% annually over the last two decades. In particular, demand for NdFeB magnets (alloyed and sintered) is growing at rates of 12–15% per year, mainly eroding the market share held by the ferrites. Thus, NdFeB permanent magnets are tending to replace conventional ferrite, AlNiCo, and SmCo magnets in many application fields such as electric motors, electroacoustic devices, measuring instruments, and equipment related to the automotive, petrochemical, and medical sectors. A strong influence in the development of NdFeB magnets has been exerted by the expansion of consumer electronic equipment such as computers and mobile phones.

9.2.2 Main Properties of a Magnet

The basic parameters that characterize a PM are essentially its coercivity HC, retentivity Br, and the BH energy product. The importance attributed to these properties depends on the magnet's operational context, taking typical and extreme operating conditions into account. Tables typically provide data for the properties of magnets as a function of ambient temperature; however, if elevated operational temperatures are envisaged, then the reduced flux density and lower resistance to demagnetization must be taken into account. Some manufacturers indicate maximum working temperatures but, at times, these can be confusing and unreliable. Moreover, some specific design factors, such as magnetic circuit geometry or the presence of external demagnetizing fields, can significantly affect the thermal limits.

9.2.2.1 Geometry, dimensions, and tolerances

The size and shape of PMs vary greatly depending on their intended applications. PMs may be shaped as disks, blocks, arcs, rings, and so on. The dimensions range from small magnets for clocks to those of large size for industrial motors.

Project design constraints usually lead to compromises between desired magnet configuration, tolerance limits, and costs. For example, in an ideal motor design, it may be desirable to fit oblique or helical (skewed) magnets to reduce the cogging torque, but the machining cost to create such magnets could be prohibitive in the context of manufacturing budgets.

Other choice conflicts may arise, for example, when comparing arc and ring magnet implementations. For brushless motors, it is easier to assemble rings on the rotor; however, producing rings with the required tight tolerances, also in this case, requires considerable expenditure. Nevertheless, it must be said that ring magnets are more suited to a layout that reduces the cogging effect. As a general principle, achieving reduced tolerances requires additional processes that involve extra costs for both the material itself and the relevant machining operations.

9.2.2.2 Mechanical considerations

Permanent magnets are quite fragile at various levels. Ferrites and SmCo sintered alloys are extremely delicate, while sintered NdFeB is more robust. It is therefore important to keep in mind that magnets should never be fitted in structures where they may be subjected to mechanical stress, nor should their mechanical integrity be compromised by threading or securing them by means of hooks or eyelets: The only task of a magnet is to produce a magnetic flux.

9.2.2.3 Coatings and corrosion resistance

Depending on the type of material selected, several types of corrosion proofing layers may be applied. Ferrite, AlNiCo, and SmCo are generally stable in this regard and rarely require special protective coatings. The high-density NdFeB alloys are very reactive and therefore must be protected. However, it is worth mentioning that even the most stable magnets may benefit from a coating should there be concerns about their aesthetic finish, protection against the loss of magnetic particles, or simply handling them without damage.

Coatings are grouped into two categories: *metallic* and *organic*.

Metallic coatings are implemented with materials such as nickel, copper, tin (iron sheeting with a thin surface layer of tin), or a laminated combination of all three. This category also includes metal ion deposition techniques, for example, exploiting aluminum or cadmium; this approach is particularly suitable where tolerance precision is a major concern.

Organic coatings involve the application of special films or e-coating treatments capable of ensuring remarkable results for NdFeB alloys, even in the presence of saline mists.

The poor corrosion resistance of NdFeB magnets greatly limited their widespread adoption when they were first introduced to the market. The corrosion process in these materials begins with the diffusion of oxygen, hydrogen, or water vapor along the grain boundaries. In particular, hydrogen, which is not naturally present in air, is formed as a by-product of the corrosion reaction when oxygen is stripped from water molecules. Paradoxically, another hydrogen source is associated with the electroplating process that is usually applied as a preparatory surface treatment on the magnet. The oxidation process that leads to the ultimate formation of Nd_2O_3 causes an increase in volume and a loss of the powder material, irreversibly damaging the magnet. Deterioration is also caused by the presence of organic alkaline solvents or acids, conductive liquids (brine), oils, and corrosive gases (Cl, NH_3).

There are two solutions for the protection of NdFeB against corrosion:

Coatings: Table 9.2 shows the most common coating materials and their applications.

Improving the material's intrinsic properties: Initially doping the material with various transition elements such as Co, Ga, Mo, or V, thus limiting the concentration of Nd. The latter aspect was not initially correlated with the oxidation process and there was a tendency to exaggerate the use of Nd.

Usually both of these methods are implemented, taking particular care in the choice of materials used with proper preparation and coating of the surfaces. In other words, the coating alone is not enough; there are cases where the coating remains intact but the underlying magnet material is compromised.

The quality of the coating is usually tested by the magnet manufacturer, but to have a reliable indication of the magnet's structural resistance, the best solution is to test it under operational conditions or in an environment that adequately reproduces them. There are stress tests that subject the magnet to extreme conditions of temperature, pressure, humidity, and saline exposure. It frequently occurs that coatings pass tests on laboratory samples, but in the transition to large-scale production, the results are unsatisfactory, rendering it necessary to reinforce the procedures' quality control.

9.2.2.3.1 Testing There are two approaches to magnet testing. The first is to verify the intrinsic properties of the magnet and compare them with the manufacturer's data. Given the similarity of the methods used, the manufacturer and

Table 9.2 Typical Coatings for NdFeB Permanent Magnets

Coating	Thickness (μm)	Application
Aluminum–chromium	7–19	Motor and sensors
Epoxy resins	40–80	Loudspeakers, magnetic resonance
Electrodeposition	20–30	Industrial motors
Nickel plating	10–20	Loudspeakers, motors, and sensors

the customer will agree when the magnet complies with the specifications. However, the verification of the intrinsic properties cannot totally predict the final performance of the magnet and the sample testing is often destructive. The other approach, therefore, consists of testing the magnet under its actual operating conditions. This does allow for a useful prediction of the magnet's behavior, but it is difficult to relate this information to the intrinsic properties declared by the manufacturer. Both methods have advantages and disadvantages and therefore it is common practice to seek a compromise between the two options.

9.2.2.3.2 Magnetization The purpose of the magnetization process is to fully saturate the magnet, otherwise its magnetic properties would decay in a rather unpredictable manner. Once the magnet is magnetized, it is more difficult to handle; therefore, there is a tendency to delay this step until the last moment or until after the assembly. However, magnetization of the product in its assembled form is more difficult as it is difficult to expose it to sufficiently high fields for full saturation. The general procedure is to apply fields with increasingly stronger intensity to the assembled product until the escaping flux no longer increases; at that point, it is assumed that magnet saturation has been achieved.

9.2.2.3.3 Engineering Considerations In addition to the aspects concerning magnetization, additional considerations must be made with regard to magnet assembly. If magnetization occurs after installation, there are no major concerns regarding the magnet assembly work; diversely, if a magnet is premagnetized before being fitted to its housing (e.g., a large rare earth type), then considerable attraction forces will act on it and these must be taken into consideration. Care must be taken to avoid the magnets attracting each other or other ferromagnetic materials (e.g., metal tools such as wrenches or screwdrivers) until they are properly positioned. This is also relevant from the perspective of the working safety of the personnel involved.

9.2.2.3.4 Adhesives Many types of adhesives are used in magnet assembly, from cyanoacrylates to structural epoxies. The choice depends on the materials to be bonded and the applicable operating atmospheric conditions of the finished unit. In addition to mechanical strength parameters and chemical compatibility between the various materials, the thickness and related tolerance of the adhesive should also be considered among the design parameters.

9.2.3 Magnet Stability

The ability of a magnet to sustain a magnetic field is due to the magnetic domains that are blocked in their orientation by the anisotropic properties of the material. When the initial magnetizing field is removed, the domains retain their uniform orientation until they are subjected to external forces superior to those that tend to hold them fixed. The energy required to alter the field produced by magnets varies greatly for the various types of material. Stability can be described as *the extent to which a magnet can deliver constant optimal performance throughout its entire life*.

Factors that influence stability include time, temperature, variations in reluctance, demagnetizing fields, radiation, stresses, and vibrations.

9.2.3.1 Time

The effect of time on modern permanent magnets is minimal; the greatest changes occur immediately after the magnetization. These changes, known as magnetic decay, occur because of unstable domains that are affected by temperature fluctuations: If they are reduced in number, then so are the variations. From this point of view, the best performance is ensured by rare earth magnets, while AlNiCo 5 magnets can lose at most 3% of their flux density after 100,000 h of life.

9.2.3.2 Temperature

There are three temperature-related effects:

Reversible losses: These are losses that disappear when a magnet returns to its original temperature and cannot be eliminated by stabilization. The reversible losses are described by the reversible temperature coefficient TC, as shown in Table 9.3, expressed as a percentage per degree centigrade. The significant difference between the Br temperature coefficient with respect to HC causes a sharp knee in the demagnetization curve at high temperatures.

Recoverable irreversible losses: These are defined as a partial demagnetization of the magnet due to the presence of high or low temperatures and, unlike reversible losses, they are only recoverable with remagnetizations. Therefore, when designing magnetic circuits, care must be taken to avoid operation at elevated temperatures beyond the curve's knee, precisely to avoid this kind of loss.

Unrecoverable irreversible losses: These occur when the operating temperatures are so high that the chemical structure of the magnet is changed.

Table 9.3 Reversible Temperature Coefficients

Material	T_C for B_r	T_C for H_C
NdFeB	−0.12	−0.6
SmCo	−0.04	−0.3
AlNiCo	−0.02	0.01
Ferrites	−0.2	0.3

Table 9.4 Maximum Operating Temperature for the Different PMS in Relation to the Curie Temperature

Material	T_{CURIE} (°C)	T_{max} (°C)
NdFeB	310	150
SmCo	750	300
AlNiCo	860	540
Ferrites	460	300

Table 9.4 lists the Curie temperatures T_{CURIE}, beyond which the magnetic domains take on a disorderly arrangement and the material loses its magnetism, and the maximum operating temperature T_{max} that represents the practical limit that must not be exceeded in order to avoid damage.

In many cases it is preferable to partially demagnetize magnets by exposing them to high temperatures in order to stabilize them, thus ensuring a reliably constant flux when they must operate at lower temperatures.

NdFeB magnets employed in electric vehicles are required to operate at maximum temperatures of 180 °C and this is made possible by replacing neodymium with a significant amount of dysprosium Dy, which being considerably rarer than neodymium, tends to increase the cost of the magnet.

9.2.4 Reluctance Variations and Demagnetizing Fields

The relevant effects will be analyzed later in greater detail. For now, it should be noted that to limit the effects of reluctance variations, it is common practice to stabilize magnets by subjecting them to the same reluctance variations as those of the magnetic circuit in which they operate.

9.2.4.1 Radiations

Rare earth magnets are commonly used in applications requiring the deflection of charged particle beams, and the consequent effects of such radiations on their magnetic properties must be taken into account. To avoid directly subjecting the magnets to radiation, they are normally fitted behind protective screens and stabilized at the time of assembly by pre-exposure to the expected operational radiation levels.

9.2.5 Use of Permanent Magnets in Electrical Machines

Since the 1950s, the availability of modern magnetic materials with considerable energy products has sustained the replacement of field excitation windings by permanent magnets, with significant benefits in terms of size and loss reductions. Synchronous machines, originally based on traditional rotor field circuits, have been significantly improved with the introduction of permanent magnets and the consequent obsolescence of rings and brushes. Later, the onset of power transistors and thyristors brought about the electronic replacement of mechanical commutators. All these developments have sustained the current widespread adoption of permanent magnet synchronous machines (PMSM) and brushless DC machines (DCBL). In the case of DCBL machines, the armature windings are arranged on the stator rather than the rotor as in conventional DC machines: This has allowed for improved cooling and insulation of the windings and thus operation at higher voltages.

In Figure 9.23, a typical demagnetization characteristic representing the various magnetic materials employed in electrical machines is shown only for the second quadrant since they are not subjected to external excitation once they have been magnetized, and they must therefore resist the presence of external

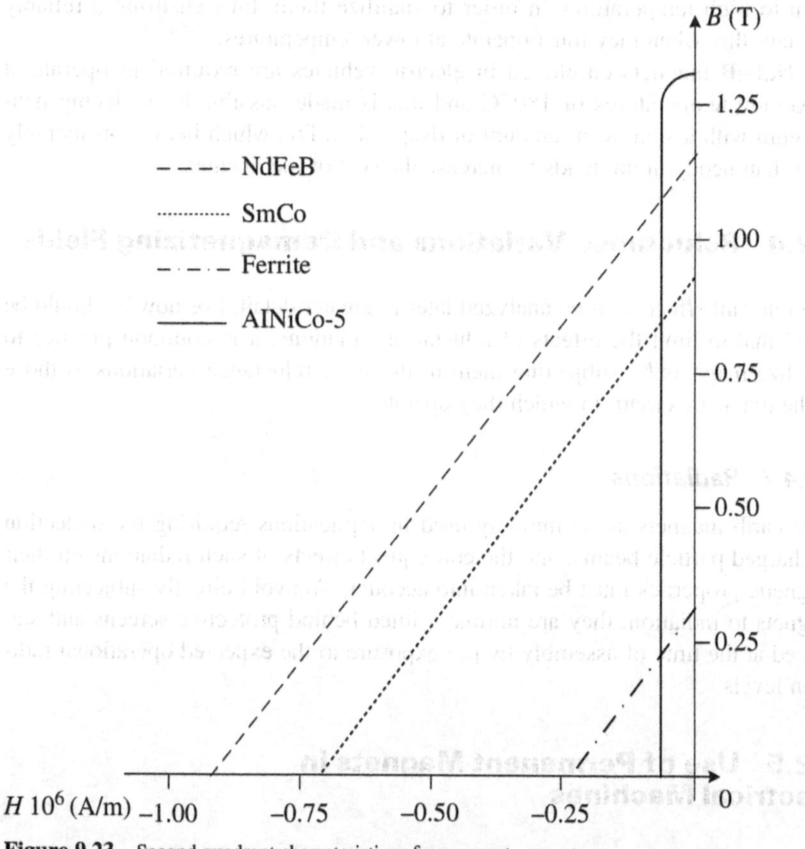

Figure 9.23 Second quadrant characteristics of permanent magnets.

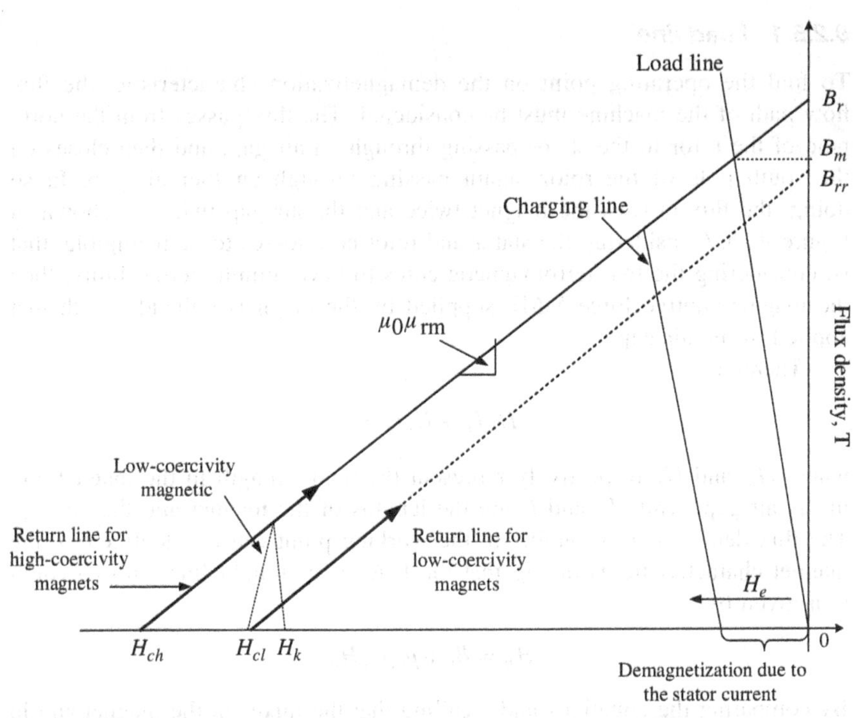

Figure 9.24 Characteristic load line points of permanent magnetic materials.

demagnetizing fields. The third quadrant is theoretically present, but it is rarely reached in normal operation. Except for AlNiCo, which has the highest remanence but its own particular nonlinear characteristic, the other materials have a linear curve in the second quadrant.

With reference to Figure 9.24, the flux density at null excitation is known as the residual flux density B_r. In the case of high-coercivity magnets, the *BH* characteristic is a straight line and the coercivity is expressed as H_{ch}.

Diversely, AlNiCo has a knee characteristic at low flux densities. In the vicinity of this point, it rapidly decreases to null induction and reaches the magnetic field value of H_{cl}. The intensity of the magnetic field at the knee point is H_k. If the external excitation acting on the magnet is removed, it will revert to its intrinsic magnetization by following a curve parallel to the original *BH* characteristic. In this case, a new remanence value *Brr* is reached, which is lower than the initial one and no longer recoverable. Even if the return curve is shown as a straight line, it usually corresponds to a minor cycle (a minor hysteresis cycle loop) whose average value is represented precisely by this straight line. The slope of this line is $\mu_0\mu_{rm}$, where μ_{rm} is the relative return permeability (relative reversible permeability). For SmCo and NdFeB alloys, μ_{rm} has values in the approximate range of 1.03–1.10.

9.2.5.1 Load line

To find the operating point on the demagnetization characteristic, the flux flow path of the machine must be considered. The flux passes from the north pole of the rotor to the stator passing through an air gap, and then closes on the south pole of the rotor, again passing through another air gap. In so doing, the flux crosses the magnet twice and the air gap twice, as shown in Figure 9.25. Considering the stator and rotor core losses to be negligible, that is, considering the two ferromagnetic cores to have infinite permeability, then the magnetomotive force MMF supplied by the magnets coincides with that applied in the air gap.

Therefore,

$$H_m l_m + H_g l_g = 0$$

where H_m and H_g respectively represent the field strength in the magnet and in the air gap, while l_m and l_g are the lengths of the magnet and the air gap. The flux density corresponding to the working point can be identified on the magnet characteristic assuming that the latter is a straight line with an equation given by

$$B_m = B_r + \mu_0 \mu_{rm} H_m$$

By comparing the equations and recalling that the fluxes in the magnet and in the air gap coincide, it can be stated that

$$B_m = \frac{B_r}{\left(1 + \dfrac{\mu_{rm} \cdot l_g \cdot A_m}{l_m \cdot A_g}\right)}$$

It is therefore confirmed that the air gap excitation causes the operational flux density to be always less than the remanence. Recall also that the core reluctances and flux dispersion have been ignored when deriving these equations. The operating flux density may be increased by reducing the air gap or by increasing the thickness of the magnet.

The working point is indicated in Figure 9.24 and the line that connects it to the origin of the axes is called the *load line*. The slope of this line is equal to the permeance coefficient μ_c.

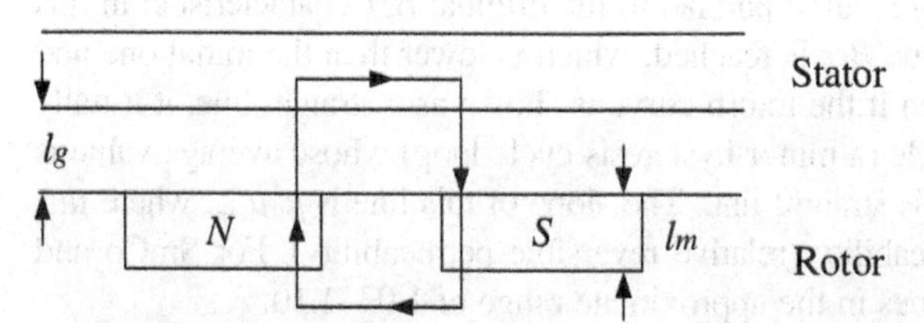

Figure 9.25 Representation of the magnetic flux path in a synchronous permanent magnet motor.

If the stator is powered, producing a demagnetizing field, the load line moves to the left, parallel to the original line, by an amount H_e, as shown again in Figure 9.4.

$$|H_e| = \frac{N \cdot I}{l_m}$$

Note that the intersection with the H axis does not depend on the geometric parameters of the air gap, but on the external field and the size of the permanent magnet. The working point moves again, further reducing the flux density B_m.

Note that for a certain working point defined by B_m and H_m, the permeance coefficient is derived as

$$B_m = B_r + \mu_0 \mu_{rm} H_m = -\mu_o \mu_c H_m$$

which in turn gives

$$\mu_c = \frac{B_r}{-\mu_0 H_m} - \mu_{rm} = \frac{-\mu_0 \mu_{re} H_m}{-\mu_0 H_m} - \mu_{rm} = \mu_{re} - \mu_{rm}$$

where μ_{re} is the *external permeability*.

The variations of the remanence are caused by changes in temperature as well as by the impact of the external fields, both of which are induced by external operating conditions, hence the term external permeability. If, for example, an external demagnetizing field were to be introduced, the permeance coefficient would decrease in the same manner as the external permeance for a given operating point. For a hard magnet, the external permeability varies from 1 to 10 under rated conditions.

9.2.5.2 Energy product

The energy product of a magnet is obtained from the product of the field strength and its flux density. The magnet's optimal working point coincides with the location of the peak of the energy product curve. It is obtained by equating the derivative of the energy product to zero with respect to the field intensity. The maximum energy product for a magnet, as shown in Figure 9.26, can be expressed as

$$E_{\max} = -\frac{B_r^2}{4\mu_0 \mu_{rm}}$$

where the flux density at which the maximum energy is available is equal to $B_r/2$. The load line passing through this induction value gives the corresponding magnetic field intensity. It should be noted that this maximum energy product working point requires a high demagnetizing field to be generated by the machine stator windings. Furthermore, it is not always possible to maintain the same working point in a variable speed machine because the stator currents vary considerably when covering the operating area.

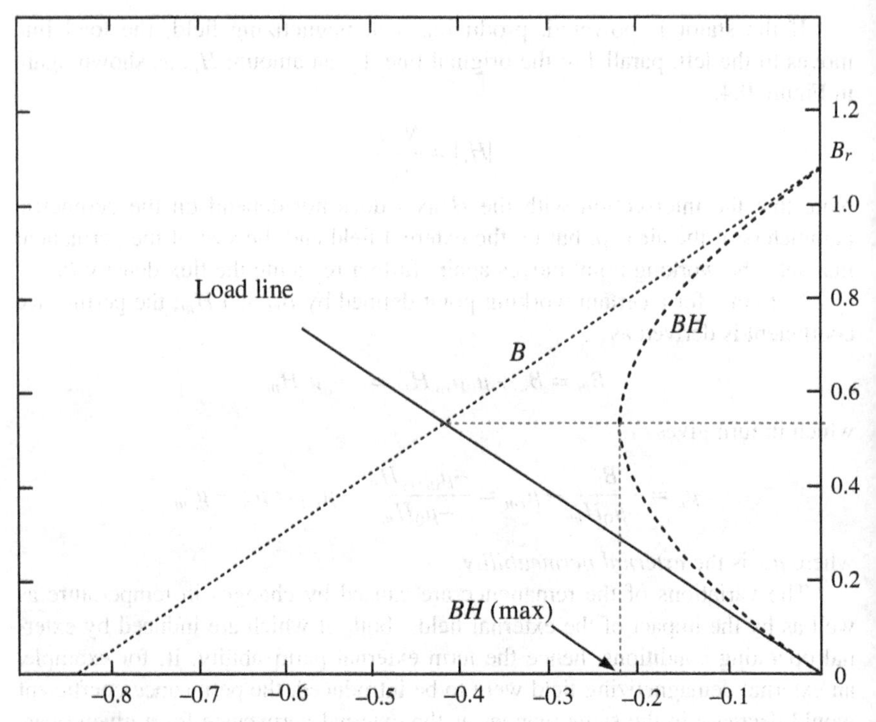

Figure 9.26 H (A/m) and BH (MJ/m³) characteristics.

9.2.5.3 Magnet size

Magnet size can be derived in terms of the working point and the air gap volume; given

$$B_g l_g = \mu_0 H_m l_m$$
$$B_m A_m = B_g A_g$$

the size of the magnet is

$$V_m = A_m l_m = \left(\frac{B_g A_g}{B_m}\right)\left(\frac{B_g l_g}{\mu_0 |H_m|}\right) = \frac{B_g^2 (A_g l_g)}{\mu_0 |B_m H_m|} = \frac{B_g^2 V_g}{\mu_0 |E_m|}$$

where V_g is the volume of the air gap, E_m is the magnet operational energy product, and A_m and A_g are the magnet and air gap surfaces, respectively. From this last relation, it can be deduced that setting the working point close to the maximum energy product allows smaller magnets to be used and hence reduced material costs.

In other words, given that the energy stored in the air gap can be expressed by

$$W_g = \frac{1}{2} \cdot \frac{B_g^2}{\mu_0} \cdot A_g \cdot l_g = -\frac{1}{2} \cdot B_m \cdot H_m \cdot V_m$$

for a given magnet size, the greatest air gap energy is achieved by selecting magnets with the highest $B_m H_m$ energy products. However, these considerations are valid when the magnet operates under static conditions. In dynamic machine applications, the magnet is subjected to external demagnetizing fields, air gap variations, or temperature changes. Consequently, the working point may shift to a location below the knee, causing irreversible magnet demagnetization. Therefore, with conventional magnets, it is better not to work exactly at the point of maximum energy product, but further to the right on the demagnetization curve.

9.2.5.4 The magnet equivalent circuit

Consider a permanent magnet of length l_m, cross section A_m, and demagnetization characteristic as in Figure 9.27.

Let B_r equal the remanence, H_c the normal coercive force, and H_0 the extrapolated value of the straight line demagnetization (linear portion of the demagnetization curve) for $B = 0$.

Defining the reversible relative permeability as

$$\mu_{rm} = \frac{B_r}{\mu_0 H_0}$$

then

$$B = B_r + \mu_0 \mu_{rm} H$$

Recalling the passive sign convention to the magnetic quantities U and Φ at the equivalent magnetic circuit terminals (Figure 9.28), the following expressions apply:

$$U = H \cdot l_m = \frac{B - B_r}{\mu_0 \mu_{rm}} \cdot l_m = \frac{B}{\mu_0 \mu_{rm}} \cdot l_m - \frac{B_r}{\mu_0 \mu_{rm}} \cdot l_m = B \cdot A_m \frac{l_m}{\mu_0 \mu_{rm} \cdot A_m} \cdot - H_0 \cdot l_m$$
$$= \Phi \cdot \theta_m - M_m$$

where $\theta_m = \frac{l_m}{\mu_0 \mu_{rm} \cdot A_m}$ and $M_m = H_0 \cdot l_m$.

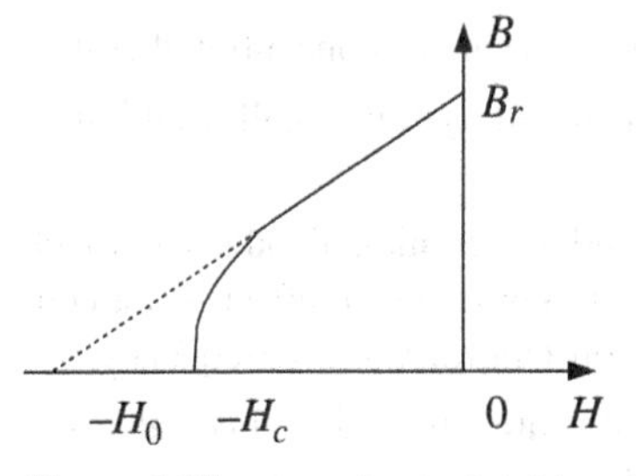

Figure 9.27 Curve for the definition of the equivalent circuit of a magnet.

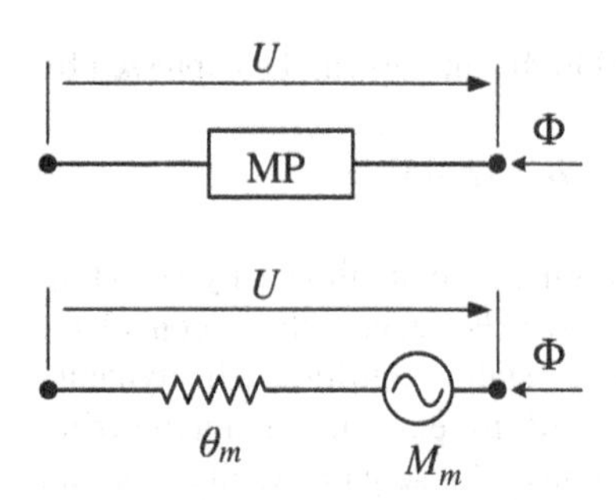

Figure 9.28 Equivalent Thèvenin circuit representation of a permanent magnet.

Thus, the $U - \Phi$ relationship in a permanent magnet is analogous to current in an active bipole.

9.2.6 Model of a Synchronous Machine with Permanent Magnets

The model of a synchronous machine with internal permanent magnets will now be introduced. With an appropriate choice of parameters, this model can also be extended to machines with surface magnets. Furthermore, the model to be described is applicable only to radial-flux machines, as these are more widespread than the axial and transverse flux counterparts. A classification of permanent magnet synchronous machines shall be outlined later according to the principle of torque generation.It is assumed that the machine is used as a motor, although in certain operating conditions (e.g., during vehicle braking) functionality may invert to that of a generator.

In order to derive a simple and practically usable model, the following simplifications are introduced:

- Magnetic saturation is not considered.
- The permanent magnet is assumed as having a constant total flux with a linear demagnetization curve.
- The flux of the magnets, the density of the windings, and the resultant stator phase inductance have a sinusoidal angular distribution.
- The distributed windings are represented by means of concentrated coils.
- The three-stator phases are identical, arranged symmetrically, and star-connected.
- The electromagnetic field wavelengths and penetration depths are much greater than the physical dimensions of the motor (the skin effect is not considered), so resistance and inductance are constant for varying frequency.
- Core losses (hysteresis and eddy currents) are ignored; hence, also the rotor-induced currents are not considered.

- Mechanical time constants are much larger than the electrical ones.
- Parameter drift over temperature is not considered, which is equivalent to the hypothesis of constant ambient temperature conditions.

9.2.6.1 Space vector modeling

As seen in previous chapters, in the theory and analysis of AC systems it is common practice to express time-variant functions by means of a complex number notation; for example, a sinusoidal current $i(t)$ can be expressed as

$$i(t) = I(\cos\phi + j\sin\phi) = Ie^{j\phi}$$

where I is the peak current value and $\phi = \omega t + \varphi$ is the phase angle of the current. Either of the complex components can be chosen to represent the instantaneous value of the current, although it is more common to chose the imaginary part $i(t) = \mathrm{Im}(\bar{i}) = I \cdot \sin\phi$.

The instantaneous values of the currents in a three-phase system can therefore be expressed as follows:

$$\begin{aligned}
i_a(t) &= I \cdot \cos(\omega t + \theta) \\
i_b(t) &= I \cdot \cos\left(\omega t + \theta - \frac{2}{3}\pi\right) \\
i_c(t) &= I \cdot \cos\left(\omega t + \theta + \frac{2}{3}\pi\right)
\end{aligned}$$

Consider the stator of an AC machine with three-phase windings. For simplicity, suppose that each winding consists of a single coil with a corresponding sinusoidal magnetomotive force. The spatial harmonics are ignored. The magnetomotive force distribution m_S created by the three-phase currents is therefore expressed as

$$m_S(\theta, t) = N_{SE}\left[i_a(t) \cdot \cos(\theta) + i_b(t) \cdot \cos\left(\theta - \frac{2}{3}\pi\right) + i_c(t) \cdot \cos\left(\theta + \frac{2}{3}\pi\right)\right]$$

where θ is the angle between the reference frame axis and N_{SE} is the equivalent number of turns. The equation can be rewritten in another form

$$m_S(\theta, t) = \frac{1}{c} \cdot N_{SE} \cdot \mathrm{Re}\left\{c \cdot \left[i_a(t) + \alpha \cdot i_b(t) + \alpha^2 \cdot i_c(t)\right] \cdot e^{-j\theta}\right\}$$

having defined the operator $\alpha = e^{j(2/3)\pi}$. This leads to the definition of the stator current space vector as

$$\bar{i}_s(t) = c \cdot \left[i_a(t) + \alpha \cdot i_b(t) + \alpha^2 \cdot i_c(t)\right] = \left|\bar{i}_s\right| \cdot e^{j\alpha_s}$$

where c is a scale factor.

Similarly, the space vectors of flux and voltage are as follows:

$$\overline{\psi}_s(t) = c \cdot \left[\Psi_a(t) + \alpha \cdot \Psi_b(t) + \alpha^2 \cdot \Psi_c(t)\right]$$
$$\overline{v}_s(t) = c \cdot \left[v_a(t) + \alpha \cdot v_b(t) + \alpha^2 \cdot v_c(t)\right]$$

The choice of c is arbitrary, but its definition does influence some related identities, such as those for torque and power. The three-phase power P is given by

$$P = 3 \cdot \text{Re}\left\{\overline{V} \cdot \overline{I}^*\right\} = \frac{3}{2} \cdot VI \cos\varphi$$

where $\overline{V}$ is the phasor of the phase voltages, $\overline{I}^*$ is the complex conjugate of the phasor of the phase currents, and V and I are the respective peak values of the same quantities. As space vectors are used to represent the entire three-phase system, the power should be expressed without considering the number of the three phases, namely,

$$P = \text{Re}\left\{\overline{v} \cdot \overline{i}^*\right\} = c^2 \cdot VI \cos\varphi$$

The choice of $c = \sqrt{3/2}$ makes these two equations equivalent; that is, the space vectors are invariant with respect to power. If $c = 3/2$ is chosen, then the transformation is not invariant with respect to power. In the following steps, noninvariant form will be adopted, except for the calculation of the relative values. Assuming the absence of zero-sequence current components, then

$$i_a(t) + i_b(t) + i_c(t) = 0$$

One of the currents can be eliminated, thus reducing the degrees of freedom. This allows for the space vectors to be expressed as an equivalent two-phase system comprising a real and an imaginary part:

$$\overline{i}_s(t) = \text{Re}\left\{\overline{i}_s\right\} + j\,\text{Im}\left\{\overline{i}_s\right\} = i_{s\alpha}(t) + j i_{s\beta}(t)$$

9.2.6.2 Three-phase electric model

Taking the assumptions made so far into account, it is reasonable to consider the electrical equivalent of the internal magnet motor as seen from the terminals of the stator phases, for given stationary values of angle θ_r and angular velocity ω_r. It is important to note that the electrical angular quantities θ_r and ω_r are related to the periodicity with which the angular position has an electromagnetic effect on the stator phases, while the mechanical variables that characterize motion are as follows:

$$\theta_r = \frac{p}{2}\theta_m, \quad \omega_r = \frac{p}{2}\omega_m$$

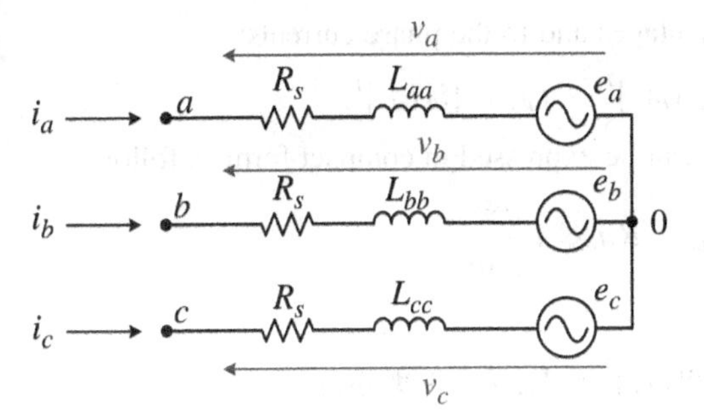

Figure 9.29 Equivalent circuit of a PM machine seen from the stator phases.

For a permanent magnet machine, the direction that identifies the rotor's position is conventionally made to coincide with the axis of the magnet; while for the stator, the reference coincides with the phase a. This is only valid for a single pole pair, but the extension to multiple pairs is easily obtained by exploiting periodicity in the definition of the angle; in other words, for the purpose of the model, it is indifferent whether the rotor is in the position θ_r or $\theta_r + \frac{2\pi}{p/2}$. Each of the phases can be represented by the equivalent electric circuit shown in Figure 9.29, composed of an emf generator, a self-inductance, and a resistance. The induced voltage generators include terms related to the mutual inductances and an additional term related to changes of permanent magnet flux, linked to each phase, as a result of position variations.

The self-inductances are given by

$$L_{aa} = L_{s\sigma} + L_{s0} + L_g \cos(2\theta_r)$$

$$L_{bb} = L_{s\sigma} + L_{s0} + L_g \cos\left(2\theta_r + \frac{2}{3}\pi\right)$$

$$L_{cc} = L_{s\sigma} + L_{s0} + L_g \cos\left(2\theta_r - \frac{2}{3}\pi\right)$$

where L_{s0} is the average inductance, $L_{s\sigma}$ is the leakage inductance, and L_g is the maximum value of the term that depends on the rotor angle, which in this case has a negative value. The mutual inductances can be expressed as follows:

$$L_{ab} = L_{ba} = -\frac{1}{2}L_{s0} + L_g \cos\left(2\theta_r - \frac{2}{3}\pi\right)$$

$$L_{bc} = L_{cb} = -\frac{1}{2}L_{s0} + L_g \cos(2\theta_r)$$

$$L_{ca} = L_{ac} = -\frac{1}{2}L_{s0} + L_g \cos\left(2\theta_r + \frac{2}{3}\pi\right)$$

Assigning vectors to the phase voltages and to the phase currents:

$$\boldsymbol{v}_{abc} = [v_a\ v_b v_c]^T, \quad \boldsymbol{i}_{abc} = [i_a\ i_b i_c]^T$$

equations for the three branches can be expressed in compact form as follows:

$$\boldsymbol{v}_{abc} = \boldsymbol{R}_s \boldsymbol{i}_{abc} + \frac{d\boldsymbol{\Psi}_{abc}}{dt}$$

where

$$\boldsymbol{\Psi}_{abc} = [\Psi_a\ \Psi_b \Psi_c]^T = \boldsymbol{L}_{abc} \boldsymbol{i}_{abc} + \boldsymbol{\Psi}_{MP\ abc}$$

for the total linked fluxes of each winding.

The inductance matrix $\boldsymbol{L}_{abc}$ has the form

$$\boldsymbol{L}_{abc}(2\theta_r) = \begin{bmatrix} L_{aa}(2\theta_r) & L_{ab}(2\theta_r) & L_{ac}(2\theta_r) \\ L_{ba}(2\theta_r) & L_{bb}(2\theta_r) & L_{bc}(2\theta_r) \\ L_{ca}(2\theta_r) & L_{cb}(2\theta_r) & L_{cc}(2\theta_r) \end{bmatrix}$$

while the vector of the linked fluxes due to the permanent magnet is

$$\boldsymbol{\Psi}_{MP\ abc} = \Psi_{MP} \begin{bmatrix} \cos(\theta_r) \\ \cos\left(\theta_r + \frac{2}{3}\pi\right) \\ \cos\left(\theta_r - \frac{2}{3}\pi\right) \end{bmatrix}$$

Therefore, distinguishing the self-induced voltages from those related to the mutual inductances leads to

$$\boldsymbol{e}_{abc} = \frac{d}{dt}(\boldsymbol{M}_{abc}(\theta_r) \boldsymbol{i}_{abc} + \boldsymbol{\Psi}_{MP\ abc})$$

with

$$\boldsymbol{M}_{abc}(2\theta_r) = \begin{bmatrix} 0 & L_{ab}(2\theta_r) & L_{ac}(2\theta_r) \\ L_{ba}(2\theta_r) & 0 & L_{bc}(2\theta_r) \\ L_{ca}(2\theta_r) & L_{cb}(2\theta_r) & 0 \end{bmatrix}$$

$$\boldsymbol{e}_{abc} = [e_a\ e_b e_c]^T$$

It can be noted that the three voltage equations are not linearly independent, since the phase voltages and currents are bound by additional constraints imposed by Kirchhoff's laws:

$$i_a + i_b + i_c = 0, \quad v_a + v_b + v_c = 0$$

It is therefore possible to reduce the formulation of the electrical model to two dimensions in order to be able to exploit an orthogonal reference frame anchored, for example, to the stator or to the rotor.

9.2.6.3 A two-phase model referenced to stator coordinates

Consider the system shown in Figure 9.30 illustrating an $\alpha\beta$ reference system anchored to the stator with the α abscissa aligned with the a phase, the stationary reference frame Park transform can then be invoked.

The model then simplifies as follows:

$$\begin{bmatrix} v_{s\alpha} \\ v_{s\beta} \end{bmatrix} = R_s \begin{bmatrix} i_{s\alpha} \\ i_{s\beta} \end{bmatrix} + \begin{bmatrix} {}^{d}/_{dt} & 0 \\ 0 & {}^{d}/_{dt} \end{bmatrix} \begin{bmatrix} \Psi_{s\alpha} \\ \Psi_{s\beta} \end{bmatrix}$$

or in matrix form:

$$\boldsymbol{v}_{s\,\alpha\beta} = \boldsymbol{R}_s \boldsymbol{i}_{s\,\alpha\beta} + \begin{bmatrix} {}^{d}/_{dt} & 0 \\ 0 & {}^{d}/_{dt} \end{bmatrix} \boldsymbol{\Psi}_{s\,\alpha\beta}$$

while the linked fluxes of the two orthogonal phases are expressed as

$$\begin{bmatrix} \Psi_{s\alpha} \\ \Psi_{s\beta} \end{bmatrix} = \begin{bmatrix} L_{s\sigma} + \frac{3}{2}L_{s0} + \frac{3}{2}L_g\cos(2\theta_r) & -\frac{3}{2}L_g\cos(2\theta_r) \\ -\frac{3}{2}L_g\cos(2\theta_r) & L_{s\sigma} + \frac{3}{2}L_{s0} - \frac{3}{2}L_g\cos(2\theta_r) \end{bmatrix} \begin{bmatrix} i_{s\alpha} \\ i_{s\beta} \end{bmatrix} + \Psi_{MP} \begin{bmatrix} \cos(\theta_r) \\ \sin(\theta_r) \end{bmatrix}$$

$$\boldsymbol{\Psi}_{s\,\alpha\beta} = \boldsymbol{L}_{\alpha\beta} \boldsymbol{i}_{s\,\alpha\beta} + \boldsymbol{\Psi}_{MP\,\alpha\beta}$$

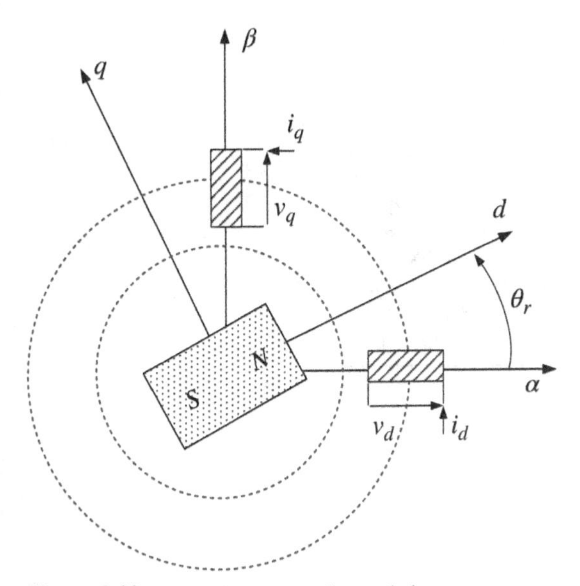

Figure 9.30 Equivalent stator-referenced phases.

Alternatively, the following space vector representation can be adopted:

$$\overline{v}_s^s = R_s \cdot \overline{i}_s^s + \frac{d\overline{\psi}_s^s}{dt}, \quad \overline{\Psi}_s^s = L_s \cdot \overline{i}_s^s + \Psi_{MP} \cdot e^{j\theta_r}$$

9.2.6.4 A two-phase model referenced to synchronous rotor coordinates

The analysis undertaken for the stator-referenced case can repeated by moving the reference system in a direction identified by the rotor angle, that is, by applying a rotating reference frame Park transform as shown in Figure 9.31.

The electrical model of a three-phase synchronous machine is often constructed by projecting the three-phase variables on two axes that are usually anchored to the stator ($\alpha\beta$ axes) or to the rotor (dq axes). Generally, the control system acts in the rotor reference frame.

It is possible to switch from the $\alpha\beta$ reference frame to the rotor-referenced frame by means of the expressions:

$$\overline{v}_s^r = \overline{v}_s^s \cdot e^{-j\theta_r}$$
$$\overline{i}_s^r = \overline{i}_s^s \cdot e^{-j\theta_r}$$

The voltage equations thus become

$$\overline{v}_s^r = R_s \cdot \overline{i}_s^r + \frac{d}{dt}\left(L_s \cdot \overline{i}_s^r\right) + j\omega_r \cdot \left(L_s \cdot \overline{i}_s^r + \Psi_{MP}\right)$$

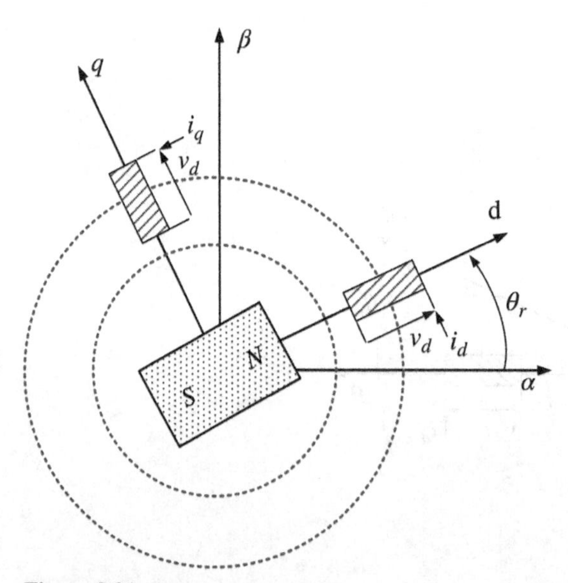

Figure 9.31 Equivalent rotor-referenced phases.

By adopting a Cartesian representation, voltages and currents can be expressed in the following form:

$$\overline{v}_s^r = v_{sd} + jv_{sq} \overline{i}_s^r = i_{sd} + ji_{sq}$$

leading to the following equations:

$$v_{sd} = R_s i_{sd} + \frac{d}{dt}\Psi_{sd} - \omega_r \Psi_{sq}$$
$$v_{sq} = R_s i_{sq} + \frac{d}{dt}\Psi_{sq} + \omega_r \Psi_{sd}$$

or in matrix form:

$$\boldsymbol{v}_{s\,dq} = R_s \boldsymbol{i}_{s\,dq} + \begin{bmatrix} d/dt & -\omega_r \\ \omega_r & d/dt \end{bmatrix} \boldsymbol{\Psi}_{s\,dq}$$

On the other hand, flux is modified as follows:

$$\begin{bmatrix} \Psi_{sd} \\ \Psi_{sq} \end{bmatrix} = \begin{bmatrix} L_{sd} & 0 \\ 0 & L_{sq} \end{bmatrix} \begin{bmatrix} i_{sd} \\ i_{sq} \end{bmatrix} + \begin{bmatrix} \Psi_{MP} \\ 0 \end{bmatrix}$$
$$\boldsymbol{\psi}_{s\,dq} = \boldsymbol{L}_{s\,dq} \boldsymbol{i}_{s\,dq} + \boldsymbol{\Psi}_{MP\,dq}$$

having applied the replacements:

$$L_{sq} = \frac{3}{2} L_{s0} + L_{s\sigma} + \frac{3}{2} L_g$$
$$L_{sd} = \frac{3}{2} L_{s0} + L_{s\sigma} - \frac{3}{2} L_g$$

Note how the dependence on the rotor angle θ_r has decoupled and it no longer appears in the model equations. Therefore, in the rotating reference frame formulation, the flux vector due to the magnet will have a null q component given its alignment with the rotor-referenced frame d axis.

9.2.6.4.1 The Torque Equation Considering only the fundamental of the stator magnetomotive force, the electromagnetic torque T_e of an AC machine is expressed as a vector that, given the noninvariant power transformation, is obtainable by the following vector product:

$$\overline{T}_e = \frac{3}{2} \cdot \frac{p}{2} \overline{\Psi}_s \times \overline{i}_s$$

where p is the number of poles. With fluxes and currents considered as vectors in the $\alpha\beta$ plane,

$$\overline{\Psi}_s = \overline{\Psi}_{s\alpha} \overline{i} + \overline{\Psi}_{s\beta} \overline{j}$$
$$\overline{i}_s = i_{s\alpha} \overline{i} + i_{s\beta} \overline{j}$$

Hence, the torque is orthogonal to the plane, that is,

$$\overline{T}_e = \frac{3}{2} \cdot \frac{p}{2} \left(\Psi_{s\alpha} i_{s\beta} - \Psi_{s\beta} i_{s\alpha}\right) \overline{k}$$

Usually, since $\overline{\Psi}_s$ and T_e are complex values, the z axis has no meaning. The torque can then be considered as a scalar quantity T_e that implies considering only the z axis component of the cross-product. In other words, the torque $\overline{T}_e$ is subject to a scalar projection onto the unit vector $\overline{k}$:

$$T_e = \overline{T}_e \cdot \overline{k} = \frac{3}{2} \cdot \frac{p}{2} \left(\Psi_{s\alpha} i_{s\beta} - \Psi_{s\beta} i_{s\alpha}\right)$$

The cross-product of this equation shows that the torque does not depend on the coordinate system used, but only on the angle between the vectors: Therefore, the torque can be calculated from quantities aligned to either the stator or rotor reference frames. In the rotor reference frame, the torque can be expressed as follows:

$$T_e = \frac{3}{2} \cdot \frac{p}{2} \left(\Psi_{sd} i_{sq} - \Psi_{sq} i_{sd}\right) = \frac{3}{2} \cdot \frac{p}{2} \left[\Psi_{MP} i_{sq} - \left(L_{sq} - L_{sd}\right) i_{sd} i_{sq}\right]$$

It is frequently useful to express the reluctance torque differently, that is, by invoking the *salience ratio* defined as

$$\xi = \frac{L_{sq}}{L_{sd}}$$

An expression emerges for the torque that allows its dependence on the difference between the inductances to be analyzed. The salience ratio best describes the range of the inductances compared to the difference $L_{sq} - L_{sd}$:

$$T_e = \frac{3}{2} \cdot \frac{p}{2} \left[\Psi_{MP} i_{sq} - L_{sd}(\xi - 1) i_{sd} i_{sq}\right]$$

9.2.6.5 Mechanical equations

To complete the model of the machine, besides the electrical equations, it is also necessary to take account of the mechanical equations and the relationship between them. This relationship is based on the expression for the electromagnetic torque and the presence of θ_r and ω_r in the equations. Thus far, these two mechanical entities were considered as parameters, but in reality they are mechanical state variables; hence, they are not given values, rather they evolve over time. Consequently, the model should also incorporate the following two equations:

$$\omega_r = \frac{d_r}{dt}$$

$$T_e - T_r = \frac{J}{p/2} \frac{d\omega_r}{dt}$$

These equations take into account the electromagnetic torque T_e produced by the interaction of fluxes and currents. Note, therefore, how electromechanical power conversion occurs: Power delivered to the input terminals of an electrical port is transformed into a torque that, in turn, causes a mechanical acceleration and a variation in kinetic energy. Power transformation does not occur with perfect efficiency since there are losses such as those related to the Joule effect, modeled as part of the resistive drops, and mechanical losses, considered in the load torque T_r. There are also other losses (eddy currents in the rotor, magnetic hysteresis, dispersion flux, etc.) that are normally ignored because they are usually of minor importance compared to other types of losses.

9.2.6.6 Equations based on relative values

It is frequently convenient to express AC system quantities in a dimensionless form in order to compare machines with different performance. First consider Faraday's law:

$$\frac{d\Psi}{dt} = v$$

Then, it is possible to define relative values for voltage and flux as follows:

$$v_{pu} = \frac{v}{V_b} \Psi_{pu} = \frac{\Psi}{\Psi_b}$$

where the value of the base voltage is the rated value of the phase voltage, while the base value of the flux is given by the ratio between the base voltage V_b and the base frequency ω_b:

$$V_b = \sqrt{2}\, V_{n_{\text{phase}}} = \sqrt{\frac{2}{3}}\, V_n \; \Psi_b = \frac{V_b}{\omega_b}$$

Dividing both terms of the expression $d\Psi/dt = v$ by Ψ_b gives $d\Psi_{pu}/dt = \omega_b v_{pu}$.

The base current is defined as

$$I_b = \frac{2T_b\omega_b}{3^p/_2 V_b}$$

From this the basic inductance is derived:

$$L_b = \frac{\Psi_b}{I_b} = \frac{3^p/_2 V_b^2}{2T_b\omega_b^2}$$

Finally, the base torque is given by

$$T_b = \frac{3}{2} \cdot \frac{p}{2} \Psi_b I_b = \frac{3^p/_2 I_b V_b}{\omega_b}$$

The machine model for *sinusoidal steady-state conditions* expressed in relative values, disregarding the voltage drops, becomes

$$\Psi_{sd\ pu} = \Psi_{MP\ pu} + L_{sd\ pu} i_{sd\ pu}$$
$$\Psi_{sq\ pu} = L_{sq\ pu} i_{sq\ pu}$$
$$v_{sd\ pu} \cong -\omega L_{sq\ pu} i_{sq\ pu}$$
$$v_{sq\ pu} \cong \omega \Psi_{MP\ pu} + \omega L_{sq\ pu} i_{sq\ pu}$$
$$T_{e\ pu} = \Psi_{MP\ pu}\, i_{sq\ pu} - L_{sd\ pu}(\xi - 1) i_{sd\ pu} i_{sq\ pu}$$

9.2.6.7 Classification and characteristics of PMSMS

Recalling the expression for torque T_e, note that it consists of two terms, one related to the alignment of the magnet and the other given by the difference between the d and q axes inductances:

$$T_e = \frac{3}{2} \cdot \frac{p}{2} \left[\Psi_{MP} i_{sq} - \left(L_{sq} - L_{sd}\right) i_{sd} i_{sq}\right]$$

On the electrical side, the terminal voltage is given by

$$V_a = \sqrt{\left(R_s i_{sd} - \omega L_{sq} i_{sq}\right)^2 + \left(R_s i_{sq} + \omega L_{sd} i_{sd} + \omega \Psi_{MP}\right)^2}$$

The latter equation is used to define the maximum speed and the amplitude of the constant power area. Figure 9.32 shows the relationship between reluctance torque and magnetic torque for different configurations of synchronous machines, classified on the basis of the torque generation principle. The synchronous machine with surface magnets (SPMSM), in which arch-shaped magnets are mounted on the surface of the rotor core, is a pure permanent magnet machine because only the magnet torque is present. The synchronous reluctance machine (SynRM), on the other hand, is not a permanent magnet machine and

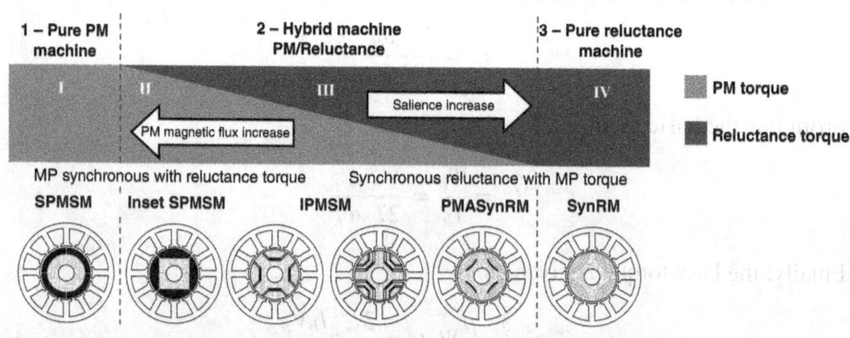

Figure 9.32 Classification of synchronous machines according to the principle of torque generation.

the torque is generated only by the rotor anisotropy. The category of surface magnet machines also includes the so-called InsetSPMSM, but it is actually a hybrid machine with magnetic saliences. The internal magnet machines (IPMSMs) are also hybrid units, but the magnets are arranged inside the rotor. The latter category, in turn, can be divided into PM motors with reluctance torque (zone II) and reluctance machines with magnetic torque (zone III) on the basis of the prevailing torque generating principle.

From a magnetic point of view, surface magnet motors can be considered as isotropic motors with high air gap thicknesses, because the magnets have a relative permeability on the order of approximately 1.02–1.2. The magnetization inductances aligned on the direct axis (the direct axis coincides with the axis of a north pole) and those in quadrature with the rotor (the quadrature axis is at 90 electrical degrees, since the electrical angles are equal to the mechanical angles multiplied by the number of air gap poles) are therefore equal (Figure 9.33). Furthermore, due to the substantial air gap, the magnetization inductance is small and, therefore, the armature reaction effects are negligible. The stator electric time constant is also small.

PM motors have a more robust mechanical structure, which makes them particularly suitable for high-speed applications, since the magnets are physically protected within the rotor.

Although these motors are based on rotors with geometric isotropic structures, they cannot be considered isotropic from a magnetic circuit point of view. In fact, since the permeability of ceramic or rare earth magnetic materials is close to that of empty space, the thickness of the magnet constitutes a large air gap along the magnetic flux path aligned to the direct axis. Diversely, the

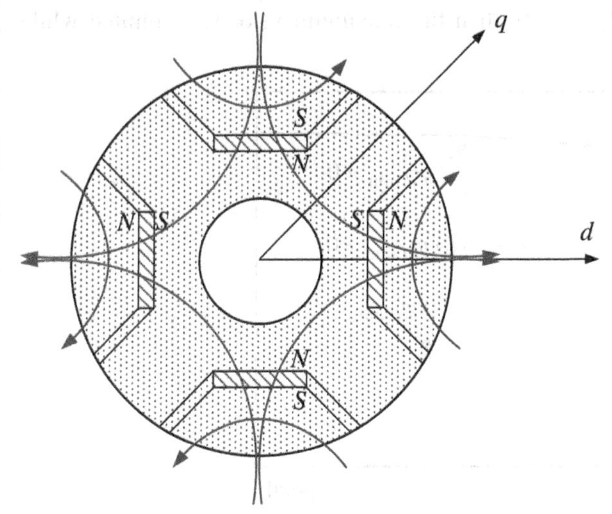

Figure 9.33 *d* axis (dark gray) and *q* axis (light gray) rotor magnetic flux paths in an IPMSM.

magnetic flux path aligned to the quadrature axis is not subject to appreciable permeance variation because each magnet is encased in a high-permeability soft steel pole shoe (Figure 9.33).

It is important to emphasize that, in a PMSM with embedded magnets, the magnetization inductance aligned to the quadrature axis is greater than that aligned to the direct axis, contrary to what occurs in conventional salient-pole synchronous machines.

Machines with internal magnets (IPMSMs) are preferred to those with superficial magnets, thanks to a significant number of other advantages.

Square shapes may be adopted instead of the arch-shaped magnets that are more costly, delicate, and difficult to fit on the rotor.

The presence of sufficiently strong magnetic anisotropy encourages its use for the production of torque and for operation under weak field conditions, enabling the achievement of a constant power operating range extended to high speeds. Moreover, most IPMSM machines have a very versatile structure that leaves several degrees of freedom open to the designer, regarding both the mechanical aspects and the speed-torque characteristics.

The benefits associated with PMSM designs depend on machine parameters and the control system. The torque and power profiles as a function of angular velocity, as well as operation at constant power, depend on the minimum d-axis flux defined as

$$\Psi_{d_{\min}} = \Psi_{MP} - L_{sd} I_C$$

where I_C is the so-called ceiling current.

Figure 9.34 shows a typical power profile as a function of velocity with varying minimum flux $\Psi_{d\,\min}$. When $\Psi_{d\,\min} < 0$, the maximum achievable velocity is theoretically infinite but the maximum power decreases with decreasing $\Psi_{d\,\min}$; if $\Psi_{d\,\min} > 0$, then the maximum velocity is limited while

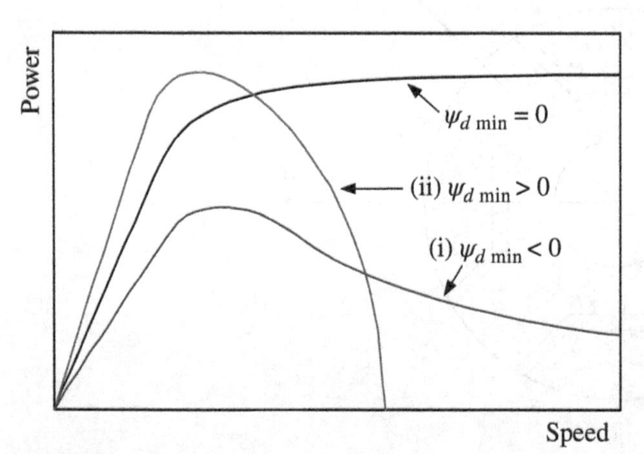

Figure 9.34 Power characteristics of a PMSM with varying $\Psi_{d\,\min}$.

the maximum power is greater compared to the case in which $\Psi_{d\,\min} < 0$. For traction applications, the optimal case is $\Psi_{d\,\min} = 0$, since a constant power zone is established that extends to infinite velocity without significantly sacrificing maximum power at the same time. If a PMSM is intended for a traction drive application, the parameter $\Psi_{d\,\min}$ should be chosen as close to zero as possible. As will be demonstrated more clearly later, the flux produced by the magnet must meet certain requirements in order to prevent inverted operation as a generator and consequent damage to the inverter. Hence, it is preferable to limit the alignment torque and rely on the contribution of the reluctance torque with a suitable field-weakening control technique, which is equivalent to choosing a rotor that falls between zone II and zone III in Figure 9.32. A choice of this type also facilitates control when $\Psi_{d\,\min} \approx 0$ as well as the reduction of core losses at high velocities.

9.2.7 Control Techniques for PM Synchronous Machines

Some of the more important control techniques for permanent magnet machines are now presented, with particular consideration for those with internal magnets that, as stated, are the preferred choice in traction applications.

There are essentially two methods for torque production in a PMSM: vector control, which is usually based on the rotor flux (RFOC), and direct torque control.

9.2.7.1 Vector control

The previous definition of a machine model based on a *dq* rotating reference frame allows for considerable simplification, also from the point of view of control requirements, since dispensing with position-related variables in the motor equations eliminates the need to treat sinusoidal signals and only constant reference variables may be considered. It is therefore possible to produce torque by controlling current amplitude and direction with respect to flux in a manner similar to the techniques applied in the control of DC machines. In general, this last statement is valid for isotropic machines that are controlled under conditions with $i_{sq} = i_{s\,\mathrm{MAX}}$ and maximum torque $(\varepsilon = \pi/2)$ up to the base velocity. Diversely, internal magnet machines have a further degree of freedom in the choice of the current space vector, thus allowing optimized control strategies to be defined.

In fact, by observing Figure 9.35, the following identities are evident:

$$\begin{aligned} i_{sd} &= i_s \cos\varepsilon \\ i_{sq} &= i_s \sin\varepsilon \end{aligned}$$

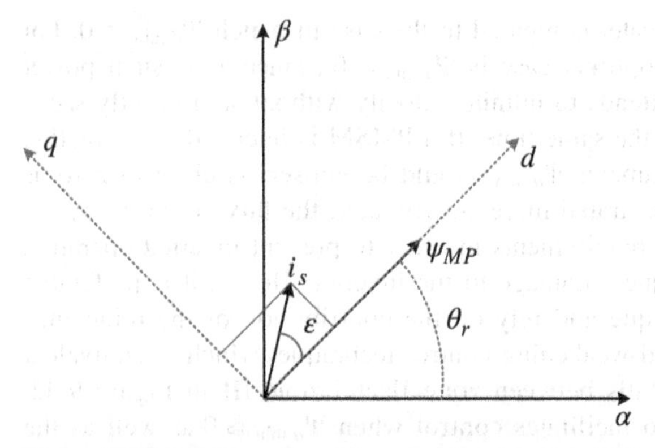

Figure 9.35 Representation of the vector diagram of a PMSM on stationary and rotating axes.

and replacement in the torque equation T_e yields

$$T_e = \frac{3}{2} \cdot \frac{p}{2} \left[\Psi_{MP} i_s \sin \varepsilon - \frac{1}{2} L_{sd}(\xi - 1) i_s^2 \sin 2\varepsilon \right]$$

Since internal magnet machines have $\xi \gg 1$, in order to take advantage of reluctance torque it is always necessary to impose a negative i_{sd}, that is, $\varepsilon > (\pi/2)$. To deliver maximum torque, once the dependence of torque on i_s^2 is known, then the angle ε will extend beyond $\pi/2$ to the extent by which the current magnitude increases (Figure 9.36).

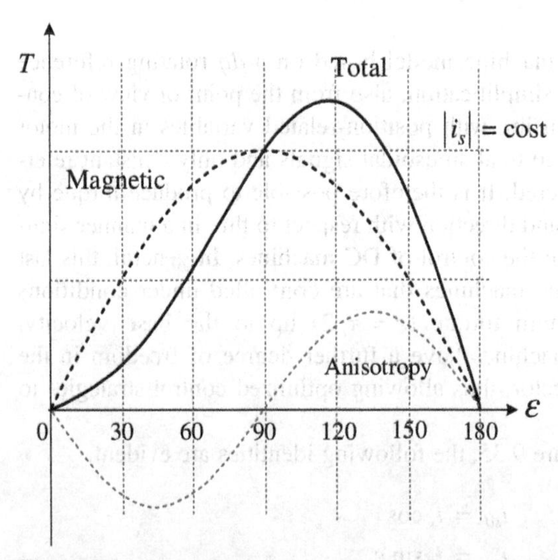

Figure 9.36 Torque behavior as a function of the angle ε.

The control approach for internal magnet machines can follow different criteria, the most popular being the following:

- Control with $i_{sd} = 0$, as for isotropic motors, it has the advantage of linearizing the torque–current relationship, but it does not exploit the reluctance torque, thereby preventing optimization of the torque/current ratio.
- Control with constant angle $\varepsilon > (\pi/2)$, chosen so as to optimize torque in certain prevalent requirements; however, this technique is not suitable for use in traction applications.
- Maximizing the torque produced per ampere (maximum torque per ampere), which reduces ohmic losses, but has the disadvantage of being limited in velocity.
- Maximizing the torque produced per volt (maximum torque per voltage), suitable for operation under voltage limited conditions with field weakening.

Recourse to the steady-state machine model with relative values allows analytical derivation of the so-called *machine operating characteristics*, that is, the constant voltage, torque, and current locations in the (i_{sd}, i_{sq}) plane. Figure 9.37 illustrates the constant torque hyperbolas, the constant voltage ellipses, decreasing in value with increasing speed and centered on point $C\left(-\Psi_{MP\,pu}/L_{sd\,pu}, 0\right)$, and the maximum current circumference curve. Point B represents the *base working point at* which the motor works at rated current, voltage, velocity, and torque.

Figure 9.37 also shows the trajectories for the *maximum torque/current* ratio and *maximum torque/voltage* ratio, which represent the loci of the points of

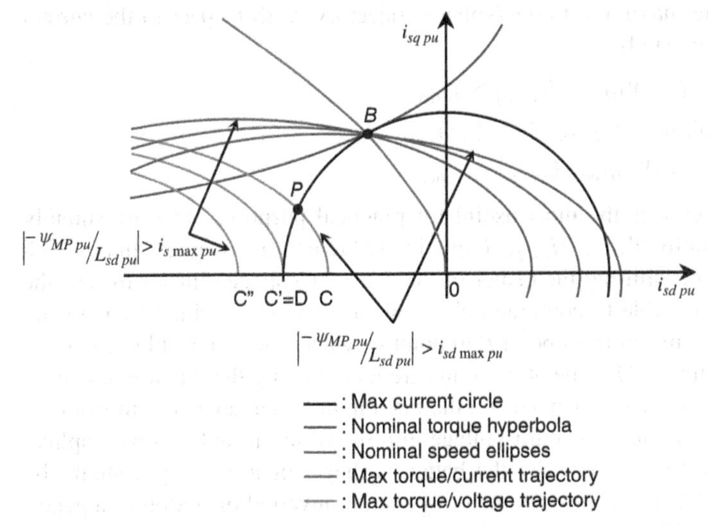

Figure 9.37 Operating characteristics of a PMSM under steady-state conditions.

tangency between the constant torque hyperbolas and, respectively, the constant current circumferences and the constant voltage ellipses. These allow optimal motor operation range to be identified.

By defining,

$$i_{s\,pu} = \left|\bar{i}_{s\,pu}\right| = \sqrt{i_{sd\,pu}^2 + i_{sq\,pu}^2}$$

$$\text{ang}\left(\bar{i}_{s\,pu}\right) = \gamma = \arctan\left(\frac{i_{sd\,pu}}{i_{sq\,pu}}\right)$$

$$v_{s\,pu} = \left|\bar{v}_{s\,pu}\right| = \omega\sqrt{\left(\psi_{MP\,pu} + L_{sd} i_{sd\,pu}\right)^2 + \left(L_{sq} i_{sq\,pu}\right)^2}$$

$$\theta = \arctan\left(\frac{\Psi_{MP\,pu} + L_{sd\,pu} i_{sd\,pu}}{\xi L_{sd\,pu} i_{sq\,pu}}\right)$$

where the phase γ of the current is measured in a counterclockwise sense from the $i_{sq\,pu}$ axis. It is a simple matter to analytically derive these trajectories by imposing

$$\frac{d}{d\gamma}\left(\frac{T_{e\,pu}}{i_{s\,pu}}\right) = i_{s\,pu} L_{sd\,pu}(1-\xi)\left(\sin^2\gamma - \cos^2\gamma\right) + \Psi_{MP\,pu}\sin\gamma$$

$$= L_{sd\,pu}(1-\xi)\left(i_{sd\,pu}^2 - i_{sq\,pu}^2\right) + \Psi_{MP\,pu} i_{sd\,pu} = 0$$

thus obtaining the ellipse that passes through OB and

$$\frac{d}{d\theta}\left(\frac{T_{e\,pu}}{v_{s\,pu}}\right) = \frac{v_{s\,pu}}{\omega\xi}(1-\xi)\left(\sin^2\theta - \cos^2\theta\right) + \Psi_{MP\,pu}\sin\theta = 0$$

obtaining the characteristics that pass through the C points. As seen in Figure 9.37, the maximum torque/voltage trajectory, with respect to the current maximum curve, can be

- external, if $\left|-\Psi_{MP\,pu}/L_{sd\,pu}\right| > i_{s\,\max}$
- tangential, if $\left|-\Psi_{MP\,pu}/L_{sd\,pu}\right| = i_{s\,\max}$
- internal, if $\left|-\Psi_{MP\,pu}/L_{sd\,pu}\right| < i_{s\,\max}$

The latter case is the most useful for practical purposes since by suitably choosing the ratio $\Psi_{MP\,pu}/L_{sd\,pu}$ (e.g., by reducing the magnet flux, which graphically means shifting the center of the constant voltage ellipses toward the origin O) it is possible to continue field weakening in the machine by reducing the value of the current from point P to point C, where the achievable velocity is theoretically infinite. The operating limits are identified by the maximum current circle (the value of which depends on the maximum sustainable current from the converter) and by the maximum voltage ellipse (which in automotive applications is limited, for example, by the battery voltage, or in railway systems, by line voltage or DC link). Besides motor operation, inverted operation as a generator should also be possible during braking: This is achievable in a symmetrical manner by forcing both i_{sd} and i_{sq} to be negative. Given these constraints and adopting strategies that maximize torque, *the motor operating range* for traction is bounded by the area OBPC. In particular, the control paradigm should follow these steps:

- Segment *OB*: This is indicated for operation from standstill up to rated velocity with torques from zero up to the maximum.
- Segment *BP*: This is indicated for operation with field weakening beyond the rated velocity with maximum current and decreasing torque.
- Segment *PC*: This allows for field weakening to be extended up to infinite velocity and zero torque ($\varepsilon = \pi$ in Figure 9.36).

A vector control approach that takes account of these limits has the advantage of being very robust and is capable of ensuring high operating efficiency over the entire operational drive range. Furthermore, in the event of sudden power loss, the intrinsic high stability ensures better behavior, compared to DTC.

A representative vector control architecture, based on velocity and current control loops, is illustrated in Figure 9.38. The inner control loop bandwidth must be greater than that of the external loop so that the velocity algorithm sees the current control process as an ideal amplifier.

The motor power supply is provided by a space vector PWM-controlled voltage source inverter, stimulated by the $v^*_{s\alpha}$ and $v^*_{s\beta}$ space vector components of the reference voltage.

The voltage reference vector is generated from the current loop that regulates on the basis of the error between the reference currents $\left(i^*_{sd} i^*_{sq}\right)$ and the

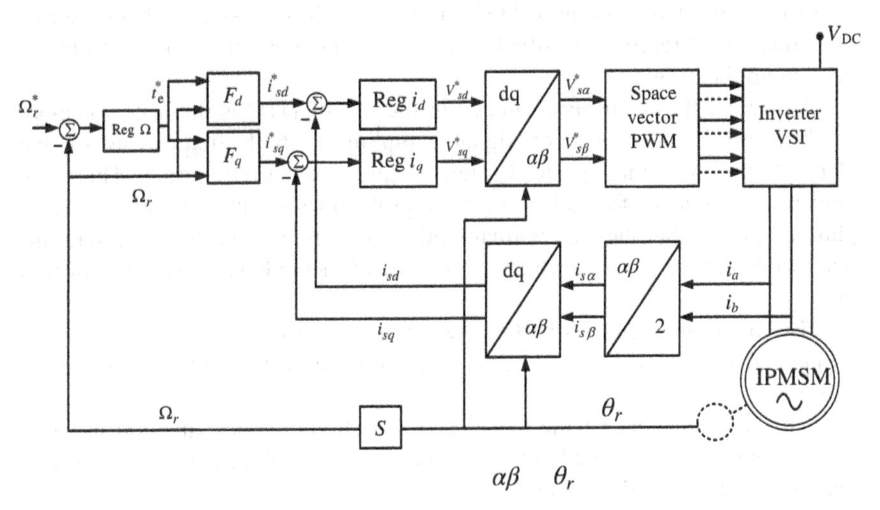

Figure 9.38 Diagram of a vector control system with position sensor.

currents $(i_{sd} i_{sq})$ (obtained following an axis transformation enabled by the rotor position feedback signal). As can be seen, only two of the motor current measurements are necessary since the windings are assumed to be star-connected and hence the sum of the phase currents is zero.

The current references i^*_{sd} and i^*_{sq} are, respectively, given by the functions $F_d(T^*_e)$ and $F_q(T^*_e)$; these allow the optimal torque current/ratio trajectory (segment OB) to be followed for each value of the required torque input T^*_e.

When a sign change is required for the torque during braking, it is possible to operate the same control loops by changing the sign of the reference i^*_{sq}.

To run in the optimal torque mode (segment BPC) with a weakened field, information is also needed regarding the velocity Ω_r at the input to the blocks F_d and F_q.

The velocity controller is usually of the proportional-integral type (PI) so as to have zero steady-state error, that is, when $\Omega_r = \Omega^*_r$. Normally, the output of the speed regulator is saturated at the maximum allowable torque value so as to be able to fully exploit the capacity of the machine overload, for example, during transients or accelerations.

9.2.7.2 Direct torque control

Direct torque control was initially proposed as a new control method for induction motors powered by voltage source inverters, and still represents the most viable alternative to vector control. The main advantages of DTC are the good torque response and the simplicity of the control scheme that does not require a position sensor for flux and torque control; essentially, it can be considered as a sensorless control technique. These advantages frequently render it preferable to FOC in all those drive systems where the torque must be directly controlled rather than the velocity, for example, in vehicle applications.

However, the main disadvantage in the use of DTC is the need for more costly controllers capable of signal sampling at high frequencies (above 40 kHz), thus minimizing steady-state torque and current ripple. There are numerous techniques for addressing these problems without increasing the sampling frequency, but they are normally difficult to implement. It is therefore difficult to determine *a priori* which of the control methods is better for a traction drive.

The basic diagram of a DTC drive, illustrated in Figure 9.39, consists of a torque and flux estimator, hysteresis comparators, a switching table, and a voltage source inverter.

The basic idea underlying the DTC approach is to choose the best instantaneous voltage vector capable of simultaneously controlling both the stator flux and the electromagnetic torque.

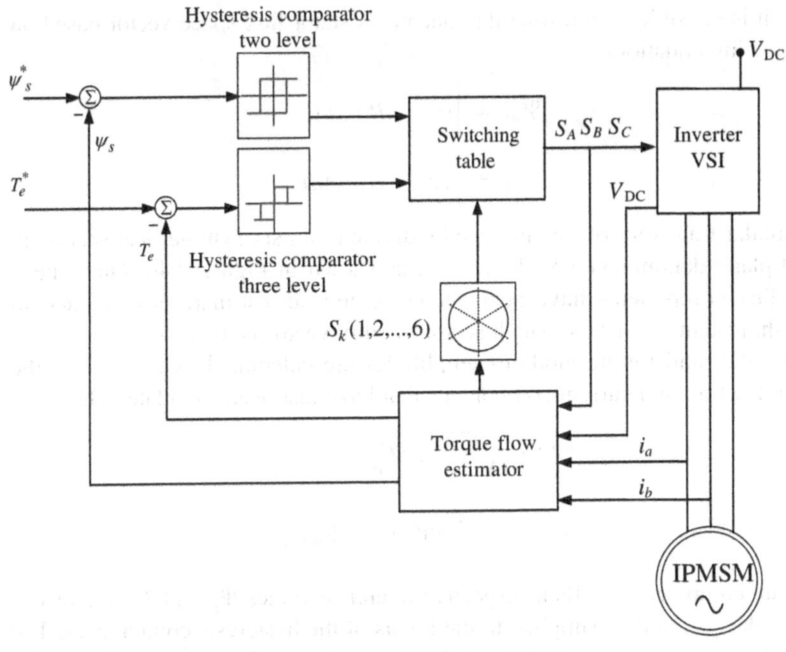

Figure 9.39 A direct torque control system (DTC) for a PMSM.

The two-phase currents and the inverter's DC link voltage are measured at each sampling instant to estimate the α and β voltage components:

$$v_{s\alpha} = \frac{2}{3}V_{DC}\left(S_A - \frac{S_B - S_C}{2}\right)$$

$$v_{s\beta} = \frac{2}{3}V_{DC}\left(\frac{S_B - S_C}{\sqrt{3}}\right)$$

where S_A, S_B, and S_C represent the states of the inverter semiconductors ($S_i = 1$ if the upper semiconductor of ith branch is *on*, $S_i = 0$ if the upper semiconductor of ith branch is *off*).

Invoking again the star-connected hypothesis, the stator-referenced current components are calculated as follows:

$$i_{s\alpha} = i_a$$

$$i_{s\beta} = \frac{i_a + 2i_b}{\sqrt{3}}$$

Hence, it is possible to construct the machine's stator flux space vector based on the following equations:

$$\Psi_{s\alpha} = \int (v_{s\alpha} - R_s i_{s\alpha}) dt$$

$$\Psi_{s\beta} = \int (v_{s\beta} - R_s i_{s\beta}) dt$$

The circular trajectory of the flux can be divided into six symmetrical sectors in the $\alpha\beta$ plane denominated $S_1, S_2, \ldots, S_6$, as shown in Figure 9.40. Once the α and β flux components have been reconstructed, an estimate is generated to establish, instant by instant, which sector the flux vector is located in.

The flux and torque modulus amplitudes are calculated as follows on the basis of the flux and current components that have just been calculated above:

$$|\Psi_s| = \sqrt{\Psi_{s\alpha}^2 + \Psi_{s\beta}^2}$$

$$T_e = \frac{3}{2} \cdot \frac{p}{2} \left[\Psi_{s\alpha} i_{s\beta} - \Psi_{s\beta} i_{s\alpha} \right]$$

These are compared with their respective reference values Ψ_s^* and T_e^*, the resulting error levels are then supplied to the inputs of the hysteresis comparators. The flux and torque comparators are similar; however, the latter also provides for

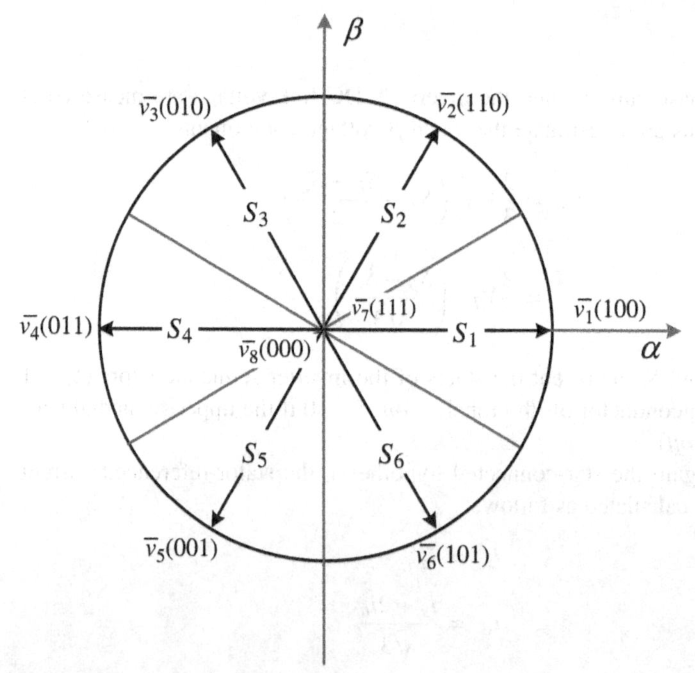

Figure 9.40 Representation of inverter space vectors and stator flux segments.

possible negative torques in the event of motor rotation inversion or braking. The logic signals output from the comparators and the information on the sector hosting the flux space vector control the switching table, which has the task of selecting a suitable inverter space vector.

9.2.7.3 Sensorless control systems

The widespread tendency to eliminate position sensor transducers associated with the governance of drive systems in industrial applications is well rewarded in terms of reduced cost, weight, and system complexity as well as improved noise immunity and reliability. Particularly for the traction requirements in the context of mass production of low-cost reliable vehicles, there is considerable motivated interest in exploiting the machine's intrinsic self-sensing properties in order to estimate the rotor position.

In general, sensorless estimation techniques are diversified according to the following:

High-speed operation: Based on the measurement of the induced emf, this quantity however achieves little or no significant amplitude at low rotational velocity or with a stationary rotor.

Low speed operation: Exploiting detectable magnetic saliency effects following the injection of known tracer signals superimposed on the motor controls. The state of the art in this area currently adopts two main injection methods: one is based on step switching the voltage at particular time instants and the other on the superposition of sinusoidal signals. However, as generally implemented, these solutions are incapable of defining polarity; therefore, additional precautions must be incorporated to prevent motor rotation from starting in an uncontrolled direction.

From this point of view, machines with high salience such as IPMSMs or SRMs offer further advantages, since it is possible to successfully implement interesting position prediction techniques. The following discussion provides an outline of the fundamental characteristics of these techniques.

9.2.7.3.1 State Observers Belonging to the first category, state observers employ a dynamic model of the machine and exploit measurement of the stator currents and the power supply voltages to calculate the flux by means of an integration:

$$\Psi_{s\,\alpha\beta} = \int \left(v_{s\,\alpha\beta} - R_s i_{s\,\alpha\beta}\right) dt$$

and the equations for the overall linked flux provide the expression:

$$\Psi_{MP\,\alpha\beta} = \psi_{s\,\alpha\beta} - L_{\alpha\beta} i_{s\,\alpha\beta} = \psi_{MP} \begin{bmatrix} \cos(\theta_r) \\ \sin(\theta_r) \end{bmatrix}$$

It is thus possible to calculate the position of the rotor as

$$\theta_r = \arctan\left(\frac{\Psi_{MP\,\beta}}{\Psi_{MP\,\alpha}}\right)$$

In this case, the estimate is not corrected based on the measurements, since the system runs open loop and inaccuracies in computation or knowledge of the parameters (inductances vary in the presence of saturation and resistances are affected by temperature fluctuations) can cause the estimates to diverge. The most common solution is to adopt a closed-loop configuration. This enables the estimate to be corrected based on a comparison between the actually measured quantities and those obtained by inserting the position estimates for the previous sampling period into the electrical system model. There are several possible observer algorithms: deterministic (Luenberger), nonlinear (sliding mode), stochastic (extended Kalman filter (EKF)), and solutions based on adaptive systems with a reference model (MRAS). In any case, this approach still has the major shortcoming of relying on the measurement of voltage, which at low speeds has modest levels that can be corrupted by inverter voltage drops or the effects of dead times.

9.2.7.4 Methods based on magnetic saliency

In a machine with a certain degree of magnetic saliency, the inductance seen by the stator windings varies naturally according to the angle between the flux and rotor directions; this can also be seen by observing how the matrix of the machine inductances is a function of the rotor position. Thus, the injection of a known electrical signal into the stator phase will trigger effects that reveal position-related information. This is similar to what occurs with rotating magnetic position sensors (resolvers), but in this case it is the machine itself that functions as a "sensor." However, it must be considered that unlike a resolver, these electrical signals are superimposed on the driving voltages required for powering the machine and this implies noise disturbances, nonlinearities, and generic errors that degrade position-sensing accuracy. These problems may be contrasted by again resorting to a closed-loop observer capable of processing the perceived error between the estimated and actual position data (Figure 9.41). This is

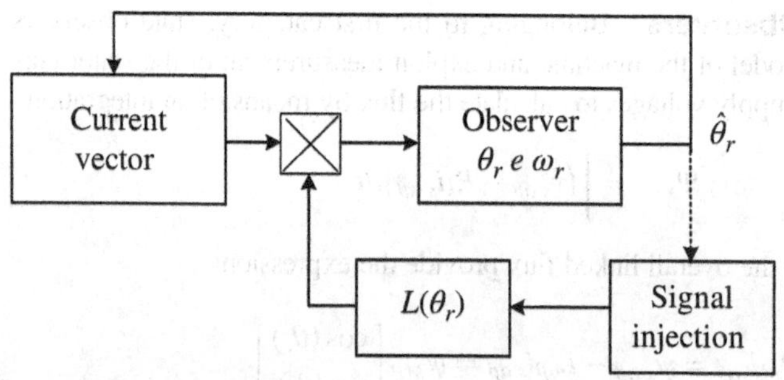

Figure 9.41 Basic architecture of a position estimator based on saliency techniques.

performed by calculating the vectorial cross-product of the actually measured current space vector (linked to position through the inductor matrix) and expected space vector (obtained by considering the injected flux and the position estimate).

The injected signals can be sharp transients (*transient excitation*) or *high-frequency sinusoids* (high-frequency injection). In both cases, the common goal is to track the rotor saliency.

9.2.7.5 Transient excitation

This method consists of injecting a zero mean value square wave through the inverter by exploiting the instants in which the PWM fundamental signal drives the zero-voltage vector. This generates current variations around the basic value imposed by the current control, but without giving rise to mechanical oscillations. Proceeding to the measurement of current in a given time interval, it is possible to derive the slope of the current waveform that, neglecting resistive drops, is linear in nature. The inductance, which is inversely proportional to the slope, is then compared with the distribution that would be expected if the estimate were correct, and applied to the input of the position observer.

However, this technique requires high sampling accuracy with a considerable increase in control hardware costs; furthermore, the task of generating the square wave signal increases the inverter's switching rate. Compromised solutions are thus required in order to overcome these drawbacks, such as reducing the switching frequency of the drive system or applying the excitation at lower repetition frequencies, but the accuracy is nevertheless compromised.

9.2.7.6 High-frequency injection

These techniques are the most widespread because they enable estimation at very high speeds. They are simple in construction but some of the available drive control voltage must be subtracted to provide the injection signal. The signal may generate a rotating field vector or a pulsating one.

In the first case, the injected sinusoidal voltages generate a high-frequency rotating field. With a known generated flux direction and by deriving the direction of the resultant current vector from the measurements, the relative displacement between them can be calculated and this provides the rotor angle information. Comparison with the estimated angle provides the required error signal. The expression for the imposed flux on stationary axes has the following form:

$$\Psi_{i\,\alpha\beta} = \frac{V_i}{\omega_i}\begin{bmatrix}\cos(\omega_i t)\\ \sin(\omega_i t)\end{bmatrix}$$

where ω_i is the signal angular frequency and V_i is the amplitude of excitation voltage.Knowing the inductance matrix thus yields an expression for the current generated by this flux:

$$\boldsymbol{i}_{i\,\alpha\beta} = \boldsymbol{L}_{\alpha\beta}^{-1} \cdot \boldsymbol{\Psi}_{i\,\alpha\beta} = K\left[f_1(\omega_i t) + f_2(2\theta_r - \omega_i t)\right]$$

It is observed that the signal is a function of two high-frequency components: one a constant amplitude rotating in the same direction as $\boldsymbol{\Psi}_{i\,\alpha\beta}$, the other rotating in the opposite direction but with phase linked to the position of the rotor. Once the frequency components that do not contribute to the estimate have been filtered out, the currents are transformed from high-frequency stationary $\alpha\beta$ axes to rotating dq axes as follows:

$$\boldsymbol{i}_{i\,dq} = \boldsymbol{T}\left(2\theta_r - \omega_i t + \frac{\pi}{2}\right) \cdot \boldsymbol{i}_{i\,\alpha\beta} = \begin{bmatrix} f_d\left(sen\left(2\theta_r - \omega_i t + \frac{\pi}{2}\right), \sin\left[2\left(\theta_r - \hat{\theta}_r\right)\right]\right) \\ f_q\left(\cos\left(2\theta_r - \omega_i t + \frac{\pi}{2}\right), \cos\left[2\left(\theta_r - \hat{\theta}_r\right)\right]\right) \end{bmatrix}$$

The direct component has low-frequency content proportional to the sine of the error estimate $\left(\theta_r - \hat{\theta}_r\right)$, which can be approximated by the angle itself for small errors and this is then used to correct the estimate in an observer.

Diversely, a pulsating vector imposes a flux, whose intensity this time varies sinusoidally with a known frequency aligned to the estimated $\hat{q}$ axis direction, from which it is then possible to extract an error estimation in a manner similar to that discussed previously. In this case, the injected flux is expressed as

$$\boldsymbol{\Psi}_{i\,\alpha\beta} = \frac{V_i}{\omega_i}\sin\left(\omega_i t\right)\begin{bmatrix} \cos\left(\hat{\theta}_r\right) \\ -\sin\left(\hat{\theta}_r\right) \end{bmatrix}$$

whose amplitude, as can be seen, is pulsating.

Recourse to high frequencies makes it possible to derive the injected currents in the $\hat{d}\hat{q}$ rotating reference frame aligned with the estimated direction. Hence,

$$\boldsymbol{i}_{i\,\hat{d}\hat{q}} = \boldsymbol{T}\left(\hat{\theta}_r\right) \cdot \boldsymbol{L}_{\alpha\beta}^{-1} \cdot \boldsymbol{\Psi}_{i\,\alpha\beta} = \sin\left(\omega_i t\right)\begin{bmatrix} f_d\left(\sin\left[2\left(\theta_r - \hat{\theta}_r\right)\right]\right) \\ f_q\left(\cos\left[2\left(\theta_r - \hat{\theta}_r\right)\right]\right) \end{bmatrix}$$

By demodulating the current $i_{i\,\hat{d}}$ to remove the sinusoidal variation $\sin\left(\omega_i t\right)$, an estimation error signal suitable for the position and velocity observer is again recovered. While in the case of rotating vectors, a delay between the injection and the measurement of the signal causes a static error, in this case there is an error proportional to the velocity that must be compensated in order to avoid a loss of accuracy.

9.2.8 Use of PMSMS in Electric Traction

The propulsion system is the main part of an electric vehicle and consists of the motor, the power converter, the electronic controller, the transmission, and the wheels. In direct-drive mechanisms, there is no real transmission system because the motor is directly coupled to the wheels.

Essentially, a traction drive must ensure

- high instantaneous power and power density;
- high torque at low speeds for starting and high power at high speeds for cruising;
- a broad operating range, including the constant torque and constant power zones;
- fast torque response;
- high efficiency over the entire operating range;
- the possibility of high-efficiency regenerative braking;
- reliability and robustness;
- fault tolerance; and
- reasonable cost.

Permanent magnet machines are the only ones able to compete directly with their induction motor counterparts.

The main advantage of the PMSM with respect to the induction motor is the absence of rotor windings and the consequent reduction of Joule effect and rotor core losses. This leads to improved efficiencies on the order of 95–97% compared to 90–92% for asynchronous machines.

Machines with relatively low efficiencies must necessarily have large dimensions in order to properly dissipate a greater amount of generated heat; conversely, PM machines dissipate little heat and therefore are more compact in construction. For a given size, the rated power of a PM motor will exceed that of an asynchronous machine by up to 50%. Thus, it is commonly held that PM machines ensure a greater power/weight advantage.

PMSM cooling requirements are simplified. Since it is not necessary to dissipate heat generated in the rotor, machine housing construction can be fully sealed to the benefit of silent operation and reliability. Fully enclosed motors avoid the need for cleaning and removal of internally deposited dust. In any case, the cooling efficiency of a closed motor is lower than that of a traditional one and, consequently, it must operate with lower losses than a ventilated motor of the same power and size. Alternatively, to reduce the weight and size, an external forced ventilation system could be adopted, but at the expense of greater noise. Drive motors usually employ a cooling system to achieve high power but with limited structural dimensions.

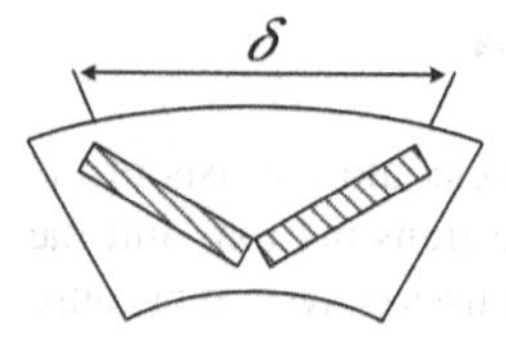

Figure 9.42 A PMSM rotor with magnets set in the V-configuration.

The mechanical time constants are also low and driving accelerations for a given input power can be higher. The rotor has a practically smooth surface without bars or hollow sections, with a consequent reduction of noise during rotation.

A traction motor is subjected to intense vibrations and shocks when running, which is not the case for PMSMs employed in industrial applications. Motors directly mounted on the wheels are even more exposed to this kind of problem compared to the case where they are suspension-mounted and protected by the vehicle car-body. Hence, some concerns remain about the use of rare earth magnets because of their relative fragility. Usually prototypes are subjected to stress tests that try to simulate the harsh actual deployment conditions.

The state-of-the-art practice is to set magnets in a V configuration in 6–12 pole formats (Figure 9.42). Compared to the radial structure, the V-shaped arrangement increases magnetic torque by sustaining a greater flux concentration. Or for a given torque requirement, lower flux density magnets may be fitted compared to the radial type. A wide aperture angle δ maximizes the magnet's torque, while a small δ reduces the rotor core losses and the open-circuit voltage.

9.2.8.1 Problems related to induced EMF

In a PMSM, the permanent magnets cause flux linkage to be generated between the stator and rotor even in the absence of an external power source. When employing PMSMs for electric vehicles' traction, induced emf may be generated during coasting, which is a motion phase, especially for rail vehicles. If the peak value of the induced emf is greater than the inverter DC link voltage, then the motor acts as a generator, returning energy to the power supply with the current passing through the inverter switch antiparallel diodes. Power semiconductors can be destroyed when the peak value of the induced emf is greater than reverse voltage limits.

In order to ensure that the maximum induced emf is lower than the inverter DC link voltage, the following condition must apply:

$$E_{0\,\mathrm{MAX}} = \frac{p}{2}\Omega_M \Psi_{MP} < E_{\mathrm{DC}\,link}$$

where Ω_M is the maximum motor velocity and, $E_{\mathrm{DC\ link}}$ is the inverter power supply voltage. This phenomenon is of significant impact, particularly for high-speed vehicles.

Generally, as the speed increases, the field weakening also caters for limiting the induced voltage. However, in the event of a malfunction of the electronic controller, there is a risk that this limiting mechanism will fail.

Even if the power supply circuit is usually protected against this eventuality, the issue must be addressed at the design level so that the generated open-circuit voltage is never excessively high at maximum velocity.

Recalling the equation for electromagnetic torque,

$$T_e = \frac{3}{2} \cdot \frac{p}{2} \left[\Psi_{MP} i_{sq} - L_{sd}(\xi - 1) i_{sd} i_{sq} \right]$$

it can be seen that $E_{0\,\mathrm{MAX}}$ may be reduced by lowering Ψ_{MP}, but this would also compromise the torque. Hence, to meet the high starting torque requirement while simultaneously limiting the induced emf, the solution is to properly choose the machine's saliency ratio of ξ.

Reducing the induced emf, however, does imply some further disadvantages, as seen in Figure 9.43, where $E_1 < E_2 < E_3$; an increase of the phase currents during the constant torque start-up phase can force the inverter into an overload condition.

The power factor also changes as induced emf varies; in fact, Figure 9.44 shows that the machine with lower induced emf E_1 has a very low power factor during the constant torque phase, while it grows rapidly during the transition to weakened field operation.

By reducing Ψ_{MP}, the efficiency degrades at low velocities while it is higher during the constant power phase (Figure 9.45). Predicting the effect of reduced

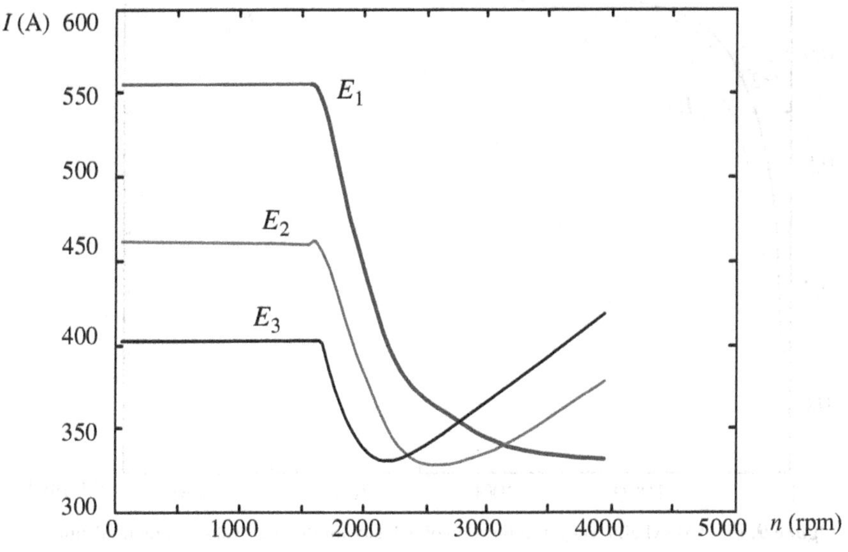

Figure 9.43 PMSM phase currents as a function of velocity for different values of induced emf ($E_1 < E_2 < E_3$).

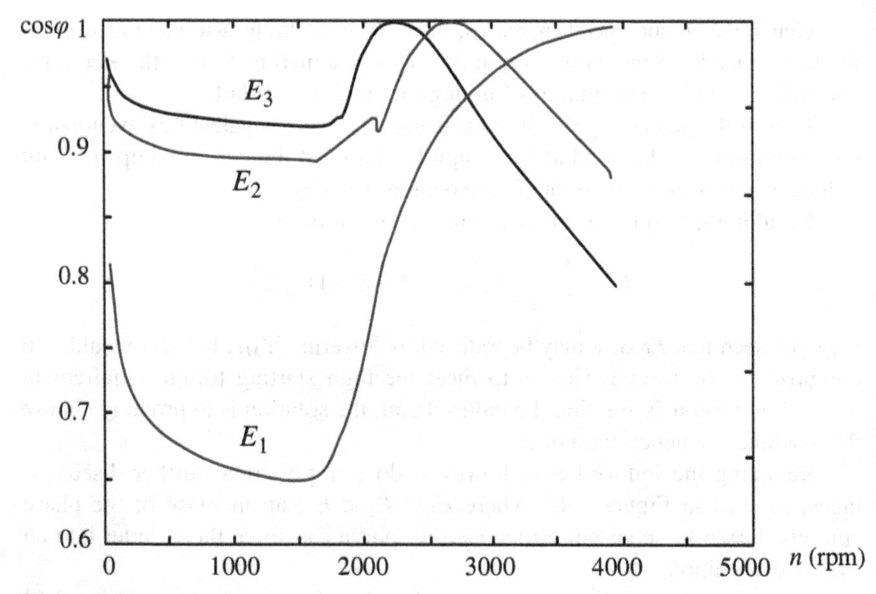

Figure 9.44 PMSM power factor as a function of velocity for different values of induced emf ($E_1 < E_2 < E_3$).

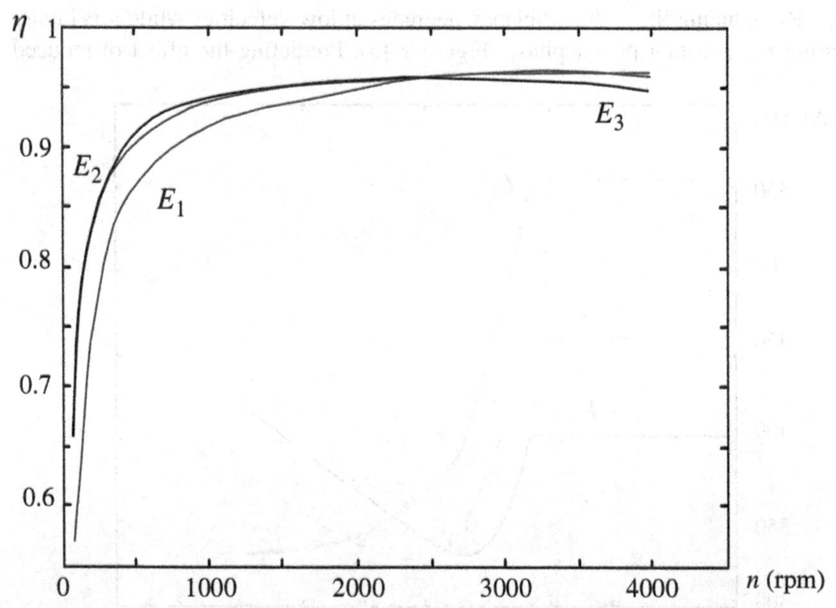

Figure 9.45 PMSM efficiency as a function of velocity for different values of induced emf ($E_1 < E_2 < E_3$)

flux on efficiency is difficult because several factors come into play. To reduce the induced emf at high velocities during coasting, it is necessary to weaken the field by introducing a negative demagnetizing current component in a quantity proportional to the induced emf and this increases the ohmic losses.

As has been shown, limiting the induced emf by acting on Ψ_{MP} and compensating the loss of torque by increasing the salience leads to a deterioration of the current, the power factor, and the low-velocity efficiency; hence, the disadvantage of using PMSMs.

9.2.9 Design Criteria for Limiting Fault Conditions

One of the major concerns in the use of brushless PM drives in traction is their behavior under fault conditions. In the event of a short circuit, there is a risk of irreversibly or partially demagnetizing the motor magnets due to intense opposing magnetomotive force caused by the fault currents. SPMSMs (surface magnet machines) are more sensitive to this problem than IPMSMs (internal magnet machines).

Since the excitation cannot be removed, induced currents will continue to circulate in the windings as long as the rotor is in motion; on detecting a fault, it is therefore not sufficient to simply shut down the machine by turning off the inverter, but other measures must also be activated to cope with the situation.

One of the main requirements is that a fault must not affect the maneuverability of the vehicle, which may be compromised, for example, by a sudden loss of braking torque. In any case, the peak torque associated with a short circuit can be so high that it can damage the machine or the transmission. Another example could be the uncontrolled behavior of a fixed or direct drive transmission system in which the motor is not mechanically separable from the wheels, even in case of failure.

It is also desirable that a fault in an inverter element should not damage the traction motor and, vice versa, faults in the motor or power supply equipment should not affect the inverter.

9.2.9.1 Fault-tolerant machines

The grievous fault implications associated with permanent magnet devices has led to the development of *fault-tolerant machines* whose design criteria differ significantly from that of conventional machines.

Nevertheless, some of the measures taken to achieve fault tolerance are sometimes contradictory when compared with the standard rules of good design: A fault-tolerant machine, besides being more expensive, is generally heavier and less efficient. Added to this is an increased risk of torque ripple as well as greater levels of vibration, noise, and magnetomotive force harmonics.

Fault-tolerant machines are classified according to two classes:

- Motors that have been engineered for low failure probability.
- Motors capable of operating even after a fault, in a provisional or permanent manner.

The first class refers to the reduction of the physical, electrical, and magnetic couplings between the active parts of a PM motor. The second class is in turn divided into two other subgroups:

- PM motors with a number of phases that can operate independently. Should one phase fail, the other functional phases take over.
- Redundant systems in which some or all of the electric drive components are duplicated. During normal operation, redundant elements are partially or totally excluded from operation, whereas they are fully activated only under fault conditions.

9.2.9.2 Configurations adopted for reducing the probability of failure

To avoid the occurrence of a fault and its propagation to other parts of a motor, a PM machine must have the following characteristics:

Electrical separation between the phases: To prevent their interaction, both connection terminals of each phase are accessible and each is powered by an H-bridge as shown in Figure 9.46. The advantage lies in the fact that the motor can also operate with one of its phases short-circuited because it is no longer powered by the corresponding converter bridge. The converter, in turn, must be able to withstand the short circuit for a time sufficient for the fault to be detected. The rated power of an inverter of this type, however, is 15% greater than that of a three-phase inverter with six switches.

Physical separation between the phases: To prevent a phase-to-phase fault, the coils of one phase are wound on a single tooth so that each slot contains coils of the same phase (Figure 9.47). The possibility of contact between coils of different phases is negligible, making failure very unlikely. Furthermore, thermal isolation between the phases and a

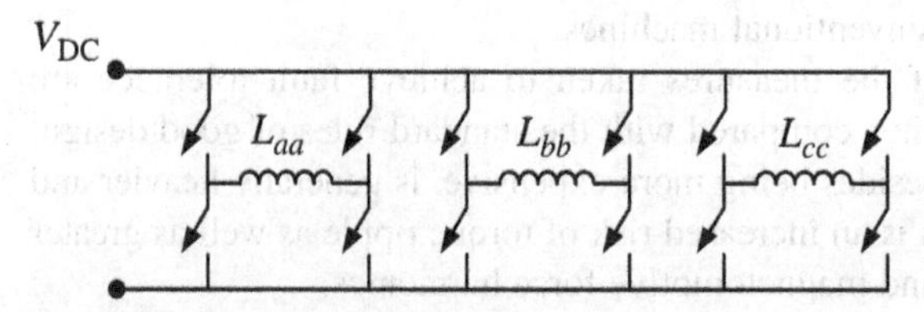

Figure 9.46 Three-phase converter with H-bridges.

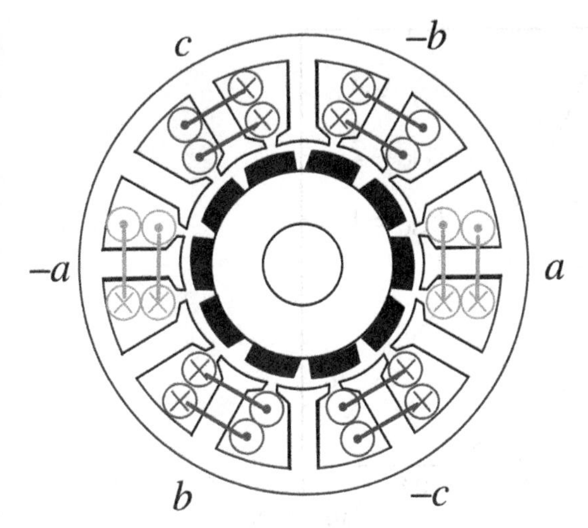

Figure 9.47 Separation of the windings to prevent faults between the phases.

simplified construction process are achieved at low cost since only half of the teeth need to be wound. However, the distribution of magnetomotive force in the air gap is affected by a higher harmonic content that causes greater rotor losses, increased torque ripple and noise, and the risk of magnet demagnetization. These must be chosen with increased thickness to cater for the greater load caused by a distorted waveform.

Sufficiently large-phase inductance: To limit short-circuit current amplitudes, thus avoiding the propagation of excessive heat to functional phases. To increase inductance, the machines are made with high leakage inductance by an appropriate choice of slot geometry.

From the increased leakage inductance, an increase in the apparent sizing-related power is derived that can be extended by a factor of $\sqrt{2}$ when the machine is designed so that the inductive voltage drop, under rated conditions, is equal to the induced emf. In this manner, the machine can sustain an armature reaction capable of completely opposing the excitation provided by the magnets. This characteristic is very much appreciated in traction applications because it also extends the ability to function in weakened field conditions.

The steady state short-circuit current may be derived from the dq-referenced equations as follows:

$$I_{sSC} = \frac{\sqrt{\left(\omega^2 L_{sq} \Psi_{MP}\right)^2 + \left(\omega R_s \Psi_{MP}\right)^2}}{R_s^2 + \omega^2 L_{sd} L_{sq}}$$

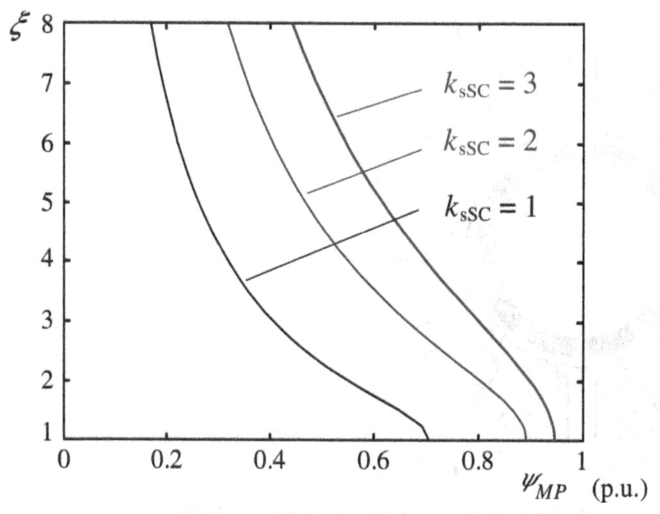

Figure 9.48 Operation areas as a function of ξ and Ψ_{MP}.

Under the worst conditions, ignoring the resistance R_s gives

$$I_{sSC} = \frac{\Psi_{MP}}{L_{sd}}$$

$$\Psi_{MP} = k_{sSC} L_{sd} I_n$$

Having set the short-circuit current to its relative value k_{sSC}, the machine parameters can be identified analytically.

Figure 9.48 shows the possible operating areas eligible for a given combination of ξ and Ψ_{MP}. The maximum short-circuit currents increase with increasing Ψ_{MP}. Simultaneously, if the saliency ratio increases, then the magnet flux decreases. Conversely, if Ψ_{MP} is fixed, then ξ must increase so that the machine can withstand faults of greater intensity.

Almost null value mutual inductance between the windings: To prevent the functional currents from inducing currents in the faulty section. Each coil must be wound around a single tooth and the number of slots should be appropriately chosen as a function of the number of poles.

9.2.9.3 Polyphase drive configurations

The main advantage of increasing the number of phases is to be able to divide the power across a number of inverter branches, thus reducing the stress on the switches (Figure 9.49). Being mutually independent, the functional phases are capable of compensating for the loss of the faulty ones. Some polyphase solutions can change their control structure in the event of failure without particular requirements for the redundancy of the elements or for the neutral connection.

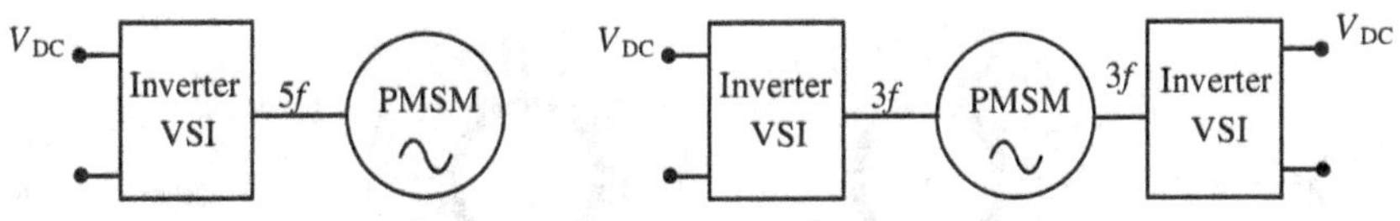

Figure 9.49 Five- and six-phase drives.

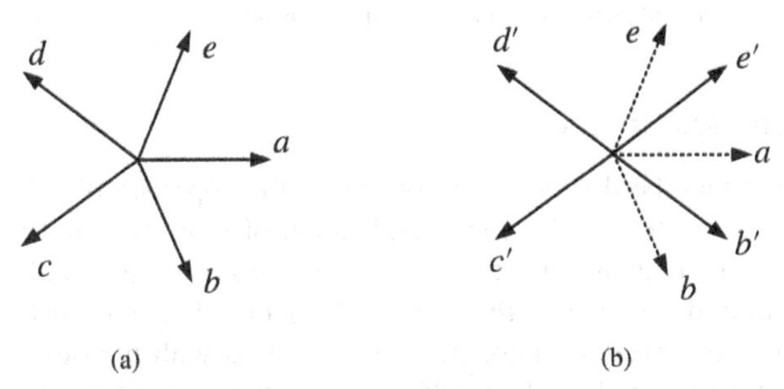

Figure 9.50 Reference current phasors before and after a fault for a five-phase drive.

In the case of a five-phase motor, for a given delivered power the current amplitude is reduced by 1/3 with respect to the three-phase case, hence also the three-phase inverter sizing-related power grows by 10%. Should one phase fail, the other phases will be overloaded by 40% so that the motor can continue to run at rated power. Once the faulty phase is disconnected, the others continue to be powered, but the power control must be changed.

The current references can thus be "rotated" passing from the case in Figure 9.50a to that in Figure 9.50b so as to simultaneously achieve a rotating field in the air gap and a null zero-sequence current (if zero-sequence currents were to circulate, then a neutral connection would be required).

In the case of a six-phase drive, the configuration is that of a dual three-phase structure, that is, with two separate three-phase inverters. In the event of a fault, one of the two structures is disconnected and the machine operates at half power without switch overload. Conversely, to run at a given power with half the number of phases, the converters must be able to withstand a 100% overload. In the latter case, it is more appropriate to make the converters and motors fully redundant.

The six-phase motor has the structure of a three-phase unit, where the windings are rearranged at no extra cost. A possible geometric solution is to separate the two three-phase windings by $\pi/6$ electric radians and power them with currents that are phase-shifted by 1/12 of the period.

Alternatively, they can be placed in two separate parts of the machine and powered by in-phase currents. In this manner, the coils are completely separate and there is limited mutual coupling between the two three-phase windings. Figure 9.51 highlights just one of the two three-phase windings and the associated phasor diagrams for both cases.

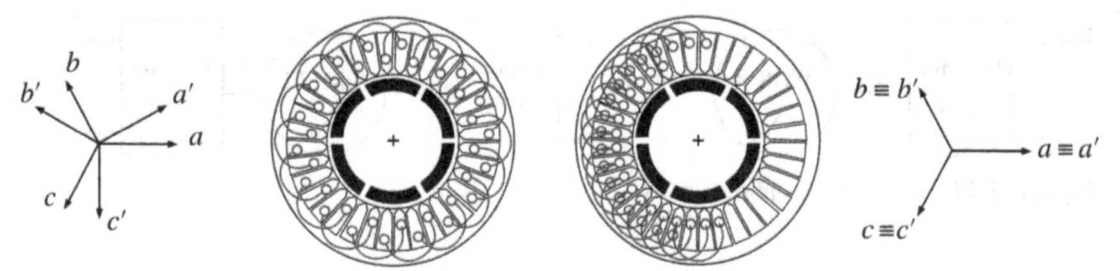

Figure 9.51 Winding arrangements and phasor diagrams for six-phase motors.

9.2.9.4 Redundant configurations

Traction drive redundancy may be diversified according to the approach of *full* and *partial* redundancy (Figure 9.52). Complete duplication of sensors, control circuits, and power devices is frequently necessary for applications where a high level of reliability is required, but it has the drawback of doubling the cost. Under normal conditions, one inverter–motor pair shall be used with the other off, or both may operate together with each at half power. In the event of a malfunction, the functional pair compensates for the failure of the faulty one.

Partial redundancy, on the other hand, simply duplicates the drive components that are exposed to the greatest risk of failure in a manner that optimally satisfies the compromise between cost and reliability. One option could be the duplication of the entire inverter in Figure 9.52, or only a part of it as in Figure 9.53. In this case the inverter is equipped with a fourth leg; in the event

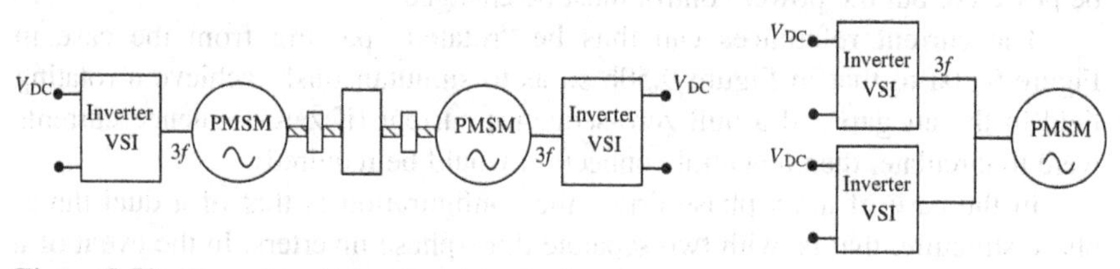

Figure 9.52 Examples of full and partial redundancy.

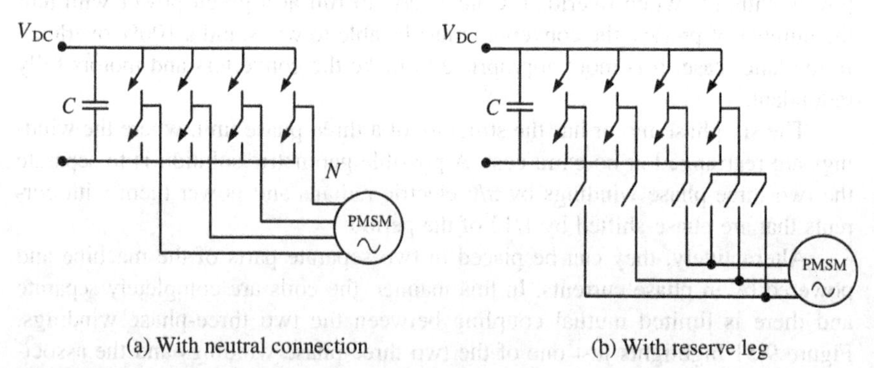

Figure 9.53 Partial redundancy with four inverter branches.

of short circuit, the faulty branch is disconnected and the auxiliary branch is activated.

The auxiliary leg can be connected to the motor star center point by means of a neutral connection; alternatively, it can be activated by operating supplementary contactors in order to replace the defective branch but without changing the control strategy in the post-fault phase.

When duplicating the PM motor, as in Figure 9.52, it is desirable to prevent the faulty machine acting as a brake on the functional one. In fact, if the PM rotor is dragged from the outside, induced emf is created in the faulty machine and this causes a braking torque. An example case is that of a distributed power rail vehicle for which a failure of one motor must not overly compromise the performance of the vehicle, for which it must be possible to drag the faulty motor without creating a braking torque.

This torque is a function of the rotation velocity and its peak value, defined as

$$T_{e_p} = -\frac{3}{2} \cdot \frac{p}{2} \frac{\Psi_{MP}^2}{L_{sq}} f(\xi)$$

and depends only on the machine parameters.

$f(\xi)$ is a function of the salience ratio and is linear $f(\xi) \cong \xi - 1$ for $2 \leq \xi \leq 6$.

In an internal magnet machine, the interaction between the stator currents and the rotor generates a torque ripple, which is rather unpleasant for the comfort of the passengers. In this regard, a redundant drive allows the control of the two motors to be misaligned so as to produce torque harmonics that compensate each other.

By way of example, consider what happens when a hard fault kills the voltage of the inverter's DC link. The behavior is similar in the case of a fault at terminals of a three-phase machine.

The effect on the torque is immediate, as shown in Figure 9.54: It collapses rapidly and then starts oscillating as in Figure 9.55. The first braking peak is greater than the maximum braking torque for operation in motor mode and it can cause loss of wheel adhesion, especially in rail vehicles. Once the oscillations are damped, the machine runs continuously as a brake.

However, the braking torque under fault conditions is minimal at high velocity and grows as the vehicle slows down.

When there is no DC supply voltage, the energy source evolves into the induced emf that maintains the fault currents until the vehicle comes to a stop.

The q axis current follows the same dynamic as the torque. At the end of a damped oscillating transient, it converges to a new steady-state amplitude level with limited values at high and growing velocities as the vehicle slows down (Figure 9.56). From this point of view, the q axis does not provide a significant contribution to the fault current.

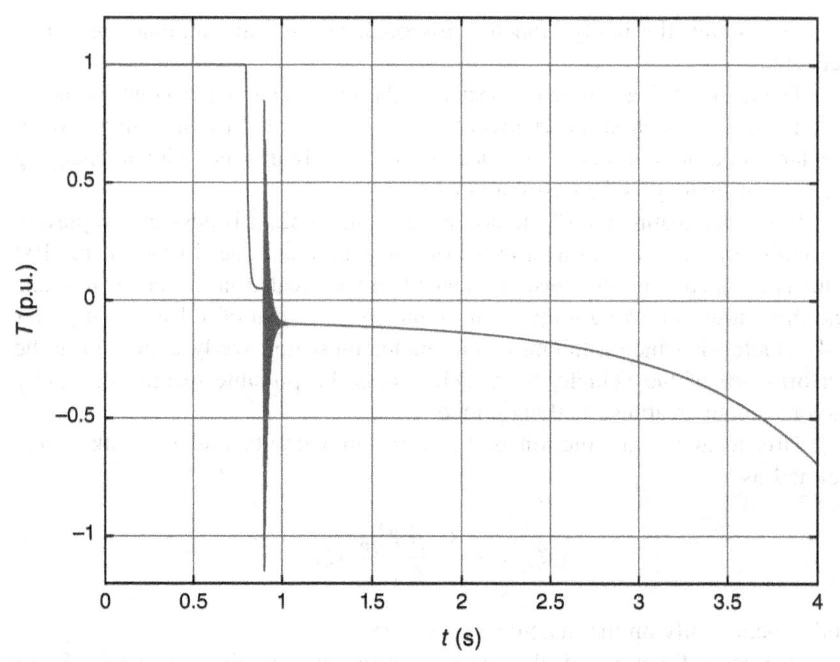

Figure 9.54 Evolution of the electromagnetic torque until the vehicle stops (failure event at 0.8 s).

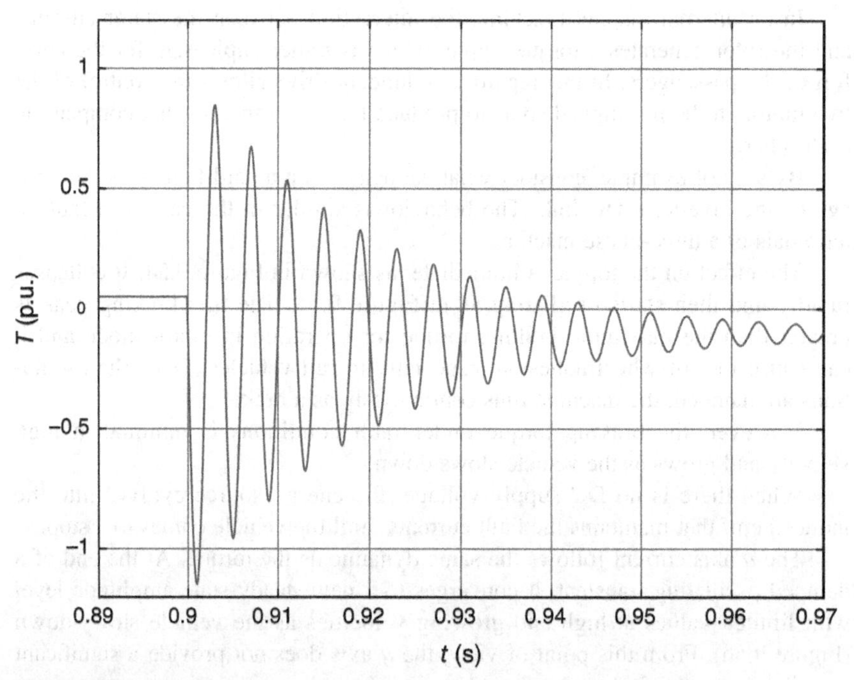

Figure 9.55 Analysis of the torque dynamic in the first instants after the fault event.

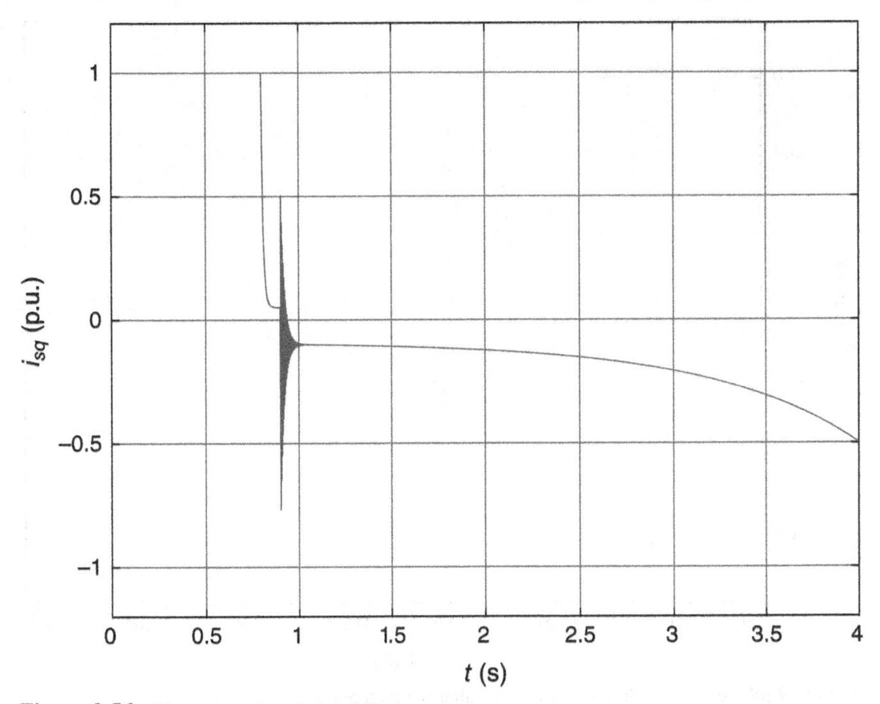

Figure 9.56 The current is q during failure.

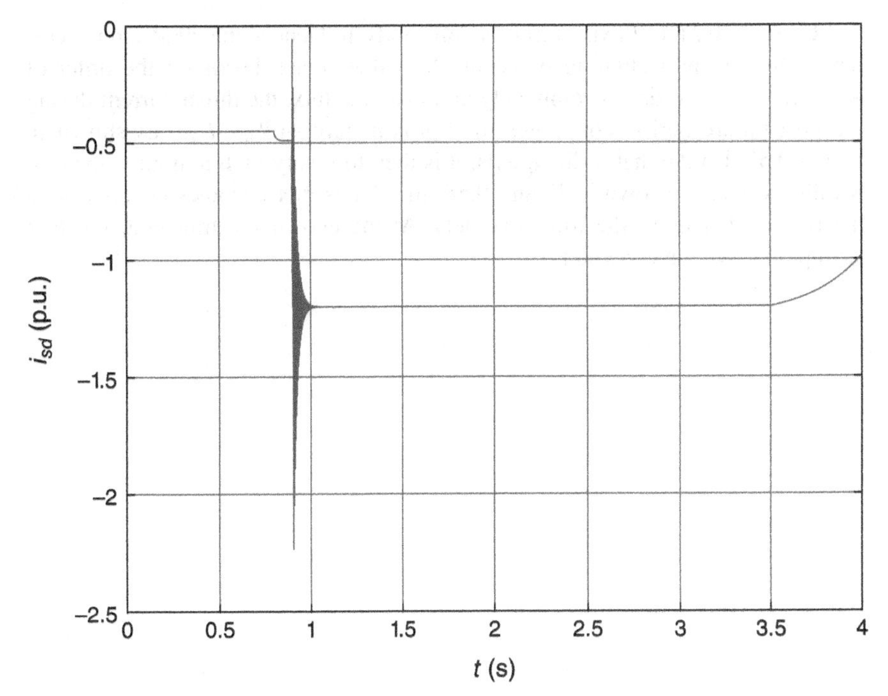

Figure 9.57 The current is d during failure.

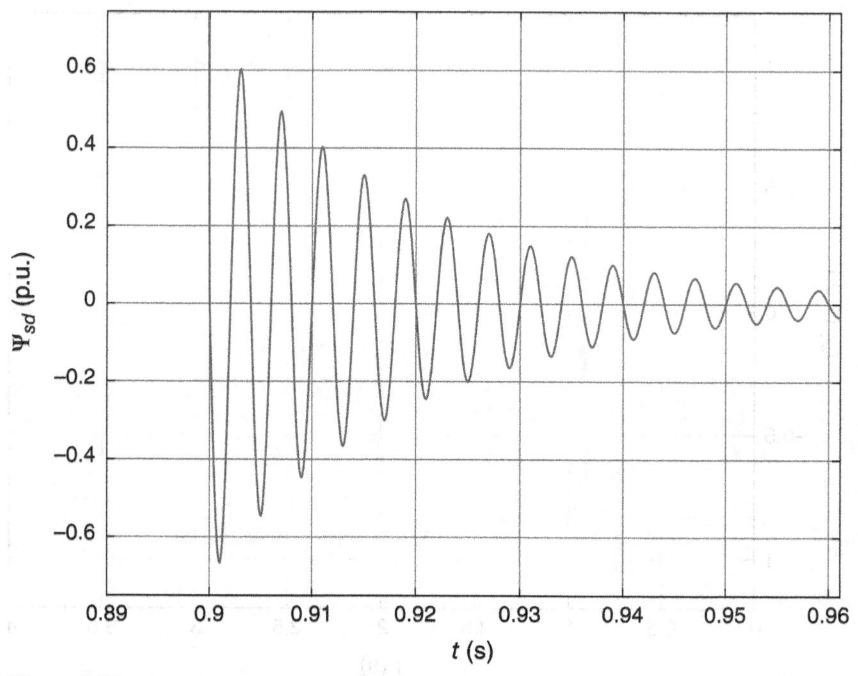

Figure 9.58 *d* axis flux in the first instants after the fault event.

Conversely, the *d* axis current significantly influences the short-circuit currents. During the oscillating transient, the values reach levels on the order of $-2(\Psi_{MP}/L_{sd})$. In order to completely eliminate the flux, the direct current during the post-failure period converges to approximately $-(\Psi_{MP}/L_{sd})$, as shown in Figure 9.57. Unlike that of the *q* axis, it is therefore very high at high velocities. It follows that, as shown in Figure 9.58, the *d* axis flux becomes negative with the risk of damaging the rotor magnets. At the end of the transient, the flux asymptotically converges to zero.

Chapter 10

Current Collecting Systems, Protection Systems, and Auxiliary Services onboard Vehicles

While being a key part for the movement of trains, motor drives alone are not sufficient for the operation of trains.

Additional equipment is of fundamental importance to ensure the collecting, transport, and use of energy to run not only the functionality related to traction and complementary services, but also to support safe driving of the vehicles and the comfort of the passengers.

This chapter will introduce you to the main devices installed onboard to provide these services.

10.1 CURRENT COLLECTING SYSTEM

Current collecting systems are a fundamental part of electrical train traction, as they allow an exchange of energy between the fixed power installations and the vehicle via appropriate sliding contacts that can receive (or collecting) an electric current. They are exposed to considerable mechanical and environmental stresses, and are one of the most vulnerable parts of the vehicle, and must, therefore, be adequately sized, monitored, and maintained during the operation of a railway.

Current collecting systems can be divided into two main families: systems based on overhead lines and pantographs, and systems based on third side rails and slippers.

Electrical Railway Transportation Systems, First Edition. Morris Brenna, Federica Foiadelli, and Dario Zaninelli.

10.1.1 Pantograph

A pantograph is a device on railway vehicles that allows electrical energy to be collected from a overhead line for feeding equipment like the motors, electric heating, air conditioning, and any onboard auxiliary devices.

The quality of collecting is greatly influenced by the dynamic behavior of the pantograph. This, with its own movements, must try to compensate for those of the catenary, maintain a secure and stable contact, with a low number of detachments. Even when using the same catenary it is possible to have completely different dynamic behavior depending on the characteristics and calibration of the pantographs that are passing over it.

A pantograph comprises a double-frame kinematic mechanism of tubular or steel boxes that confers significant flexibility on the collector head (Figure 10.1).

A pantograph can thus adopt the resting position, in which it is lower and not in contact with the air line, or the working position, when it is fully raised to maintain contact with the wire.

The main framework of steel tubes is anchored below the base of the pantograph, which in turn is supported by insulators installed on the roof of the rolling stock. The main frame is subjected to the action of two elastic systems, consisting of upward or work springs, and downward or main springs.

The kinematic mechanism allows the purely vertical translation motion of the rod at the apex of the upper head to be achieved. This rod is usually at the lower end of the suspension, and it is therefore important to prevent it moving laterally for correct operation of the suspension itself. The kinematic structure is completed by the two levels of suspension and the arches that include the carbon strips for the collection of the current. At the base of the lower frame there is a pneumatic actuator that raises the frame itself and

Figure 10.1 The kinematic mechanism of a pantograph.

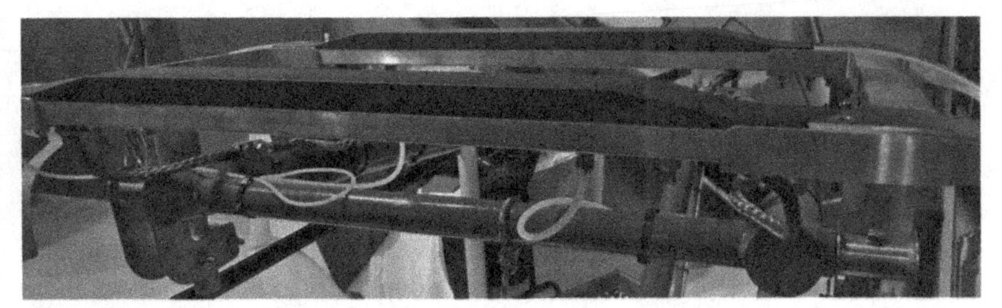

Figure 10.2 Sliding contacts.

provides the static preload, that is, the force required to ensure contact between the overhead line and the pantograph head. The pneumatic actuator is generally parallel to a damper, and thus constitutes the primary suspension. Connection between the frame and the pantograph head is made via three hinges that connect the respective rods. The pantograph head is connected to the arches through the second level of the suspension. The secondary suspension is usually made from helical springs with variable wheelbases, and is intended to follow the higher frequency oscillations in the pantograph-line contact, given the low inertia of the framework itself.

On the arches, which are usually made of steel or a light alloy, the sliding contacts are fixed; these elements are made of materials that are resistant to atmospheric agents, with a low coefficient of friction and good electrical conductivity. Over time, sliding contacts have been made of various types of materials, usually copper-based alloys, but nowadays they try to use sliding contacts that are sintered in graphite and copper, because they have better lubricating properties and tend to reduce wear on the contact wire due to the longer replacement intervals for the sliding contacts themselves (Figure 10.2).

10.1.2 Current Collecting Quality

The quality of electrical current collecting ensures the correct operation of a commercial line, and depends on the coupling between the pantograph and the catenary: if the contact is broken the power of the electric motors is stopped. When the pantograph and wire are separated, an arc is generated; therefore, the total duration of arcs in relation to traveling time T_t is an excellent indicator of collecting quality,

$$q_c = \frac{\sum a_k}{T_t}$$

where a_k is the duration of the k-th arc.

Current collecting quality also depends on the condition and regularity of the catenary. If the catenary or pantograph has a defect, current collecting quality will be significantly degraded. Even the weather can influence current collecting: A strong wind can cause a lateral deviation of the contact wire that disturbs the dynamic interaction between the pantograph and catenary.

To ensure the passage of current takes place correctly, the force with which the pantograph presses the arch against the catenary must remain as constant as possible, and equal to the static preload value allocated in the project phase. However, the vertical stiffness of the catenary is not constant along its horizontal progress, but instead adopts a variable value with the position of the pantograph along the span, which is periodic in the space, with a period equal to the length of a single span. Because of this variable stiffness, passage of the pantograph underneath the catenary, at a speed different from zero, produces oscillations in both the catenary and in the pantograph, with a consequent variation in the contact force. As the traveling speed increases, the variation in the contact force increases due to the dynamic interaction between the pantograph and the catenary. The quality of the electric current transmission decreases until it becomes compromised. In cases where the contact force is cancelled out, with detachment of the pantograph from the line, the electric arc phenomenon occurs, which is seriously detrimental to the line itself. Interruption of the current supply also damages the sliding contacts, causing either an increase in wear, or malfunction of the motor drives. To prevent this, one can optimize the elastic, viscous, and inertial properties of the pantograph and the catenary. In particular, it can be advantageous to lighten the mass of the pantograph. This measure is however, limited by the mechanical strength and wear required for the pantograph materials in the contact area, which is responsible for the conduction of high currents and subject to wear due to sliding. The framework of the pantograph should thus have its own mechanical strength, for reasons connected to its deformability, which limits the ability to lower its mass.

As for the catenary, an increase of structural stiffness would be advantageous. Using a rigid catenary does not allow high speeds to be reached, so these are used in metropolitan railway systems, in tunnels, or where there is insufficient space to mount a conventional catenary.

In studies of the behavior of the pantograph/catenary coupling, the natural frequency f_0 of the catenary assumes notable importance and depends on the length of the span, the force of tension T of the conductors, and their linear mass m,

$$f_0 = \frac{k_{f_0}}{l} \cdot \sqrt{\frac{\sum T}{\sum m}} \tag{10.1}$$

where k_{f_0} is a coefficient between 0.4 and 0.5, depending on the characteristics of the catenary; f_0 decreases with increasing temperature, if the messenger wire is not counterbalanced.

We have particularly unfavorable collecting conditions, with very large pantograph excursions, at the critical speed v_0 for which the time taken to cover a span corresponds to $^1/_{f_0}$:

$$v_0 = l \cdot f_0 = k_{f_0} \cdot \sqrt{\frac{\sum T}{\sum m}} \tag{10.2}$$

Standard, simple catenaries, with $l = 60-65$ m and $f_0 = 0.75-0.8$ Hz, have critical speeds on the order of 160–180 km/h; the increase of the conductors' force of tension increases these values.

Vertical displacements y of a periodic nature also have to be taken into consideration, due to the compliance of the contact line, which is raised due to the total force $F(v)$ exerted by the pantograph.

Such movements are overlaid on the nonperiodic displacements caused by theoretical H height variations of the contact wires with respect to the "normal" amount, at singular points on the line; sections with differing amounts are connected by very limited sloping ramps, so no disturbance to the dynamism of the tap is created.

To these vertical movements are associated accelerations a_y, then inertial forces F_i that overlap with the force $F(v)$ of the pantograph.

This, with the dynamic effects, is characterized by an equivalent or fictitious mass, referred to as the arch, which can cover the whole frame + arch for larger movements (mass m_e) or the arc alone for small changes in level, of a few millimeters.

It is essential that the equivalent mass of the pantograph is kept as low as possible, consistent with geometric requirements (maximum excursion Δh requirement) and electrical (maximum current collecting, which determine the number and the size of the sliding bars).

Compared to the "normal" H height, the contact line experiences, as the pantograph passes, an average lift of

$$y_m = c_m \cdot F(v) \tag{10.3}$$

where F is the total force, c_m the average compliance. Since the compliance c is not uniform, the lift y varies, within each span l, from a minimum at the supports to a maximum at the mid-span. It can be assumed, in the first approximation, that the trajectory of the arch is sinusoidal with a period of l; by counting the vertical amounts y from the horizontal corresponding to the average lift y_m and assuming the origin of the abscissa corresponds to a support, one can write (Figure 10.3)

$$y = -Y \cdot \cos \omega_l t \tag{10.4}$$

$$\omega_l = \frac{2 \cdot \pi \cdot v}{l} \tag{10.5}$$

where, v is the running speed.

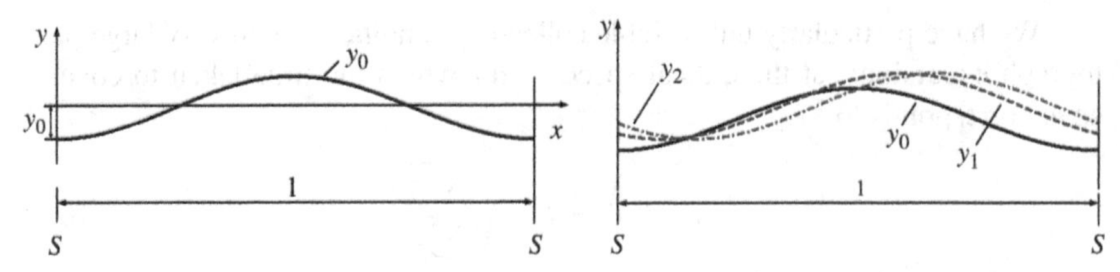

Figure 10.3 Line of contact at catenary: trajectory of arch $y = f(x)$ S: positions of masts; y_0: $v \approx 0$; y_1: $v1 > 0$; y_2: $v2 > v1$.

Having followed the trajectory (10.4), the arch is subject to a vertical acceleration:

$$a_y = \frac{d^2y}{dt^2} = Y \cdot \omega_l^2 \cdot \cos \omega_l t = -\omega_l^2 \cdot y \tag{10.6}$$

Due to the inertial force

$$F_i = -m_e \cdot a_y = m_e \cdot \omega_l^2 \cdot y$$

the total force becomes

$$F_t = F(v) + F_i = F_s + F_d - m_e \cdot a_y = F_s + K \cdot v^2 + m_e \cdot \omega_l^2 \cdot y$$

This is so in all cases: $F_t > 0$ so that detachments are prevented at the lower vertices of the trajectory, where the inertial force reaches the value

$$F_i^* = -m_e \cdot \omega_l^2 \cdot Y \tag{10.7}$$

It is important to respect the conditions:

$$F_{t\,\min} = F_s + k \cdot v^2 - m_e \cdot \omega_l^2 \cdot Y > 0$$

$$F_s + k \cdot v^2 > m_e \cdot \omega_l^2 \cdot Y$$

The dynamic effort F_d, proportional to the square of v as the force of inertia, is therefore useful for preventing detachment of the archs in areas next to the supports.

In full span, where the trajectory touches the upper vertex, there is a considerable increase in the force

$$F_{t\,\max} = F_s + k \cdot v^2 + m_e \cdot \omega^2 \cdot Y \tag{10.8}$$

on which the elastic deformation of the catenary becomes amplified.

The phenomenon of current collecting is much more complex than the previous scheme. First, at high speeds, the periodic movements of the so-called long-wave l due to nonuniform compliance of the catenary no longer obey a sinusoidal law. For

$$v > 0.6 \cdot v_0 \tag{10.2}$$

the vertices of the trajectory move in the direction of motion, as indicated in Figure 10.3, with the result that near the suspension, the trajectory of the sinusoid is much steeper; accelerations, directed downward, increase, with a greater chance of detachments.

As the speed increases, the excursions amplify: It is essential that the geometry of the support and guide devices allow the maximum predicted lift at the points of suspension.

To improve current collecting conditions, it is necessary to reduce the equivalent mass of the pantograph.

For this purpose, dual stage pantographs have been made, which comprise the following:

- A normal framework, with a high, wide working area. The large variations in the height H of the line, in fact, occur with small sloping connectors and impart vertical accelerations and moderate inertia forces onto the archs.
- A small pantograph mounted on the framework of m_0 reduced mass. Its limited size (less than 0.5 m) is sufficient to absorb height variations due to the compliance of the catenary.

Recently "controlled force" pantographs have been tested, in which, thanks to the presence of a controlled spring, it is possible to obtain $F(v)$ more appropriate force values, possibly different to those exerted by a second pantograph, where these are used.

Another useful measure is the increase of the force of tension of the conductors, which leads to a reduction of the quantities e_m, y_m, and Y and an increase in the critical speed v_0. It is also possible to reduce the effective width of the vertical excursions of the arch by using a stitched catenary, positioned using a positive deflection at the mid-span. On high-speed lines in particular, copper–magnesium alloys are being tested that simply allow an increase in the pull of the conductors for catenary lines.

In addition, the pantograph/catenary system is subject to other movements, some of which are periodic, some random and named "short wave."

The former are due to the connection of droppers to the contact wires, the elasticity of which, while following average periodic changes l between c_{max} and c_{min}, undergoes similar localized variations between one dropper and the next, spaced at a distance l_p. The l_p is on the order of 4–5 m; if we bear in mind that the minimum compliance c_{min} of a simple catenary depends on the distance l_p between the droppers located astride the suspension point, this is a distance that should, therefore, not be too small.

The behavior of the system is related to the characteristics of the droppers; solutions for flexing and damping the droppers are possible, in order to improve the quality of the current collecting.

The speed of propagation of the oscillations along the catenary is given by (see the expression for critical speed) (10.2)

$$v_p = \sqrt{\frac{\sum T}{\sum m}} \tag{10.9}$$

where the quantities T and m have the meanings already mentioned. For a conventional catenary, we have

$v_p \approx 108\,\text{m/s} = 389\,\text{km/h}$; the increase of the force of tension of the conductors, on equal sections and thus linear masses, leading to values v_p on the order of 420 km/h. In any case v_p is greater than the actual operating speed, for which the pantograph as it moves, finds the catenary already disturbed by its own action; these vibrations may overlap the reflected waves from singular points, producing additional movements.

In addition to this cause, others contribute to movement disturbances: fluctuations from the arch deriving from the vehicle as well as from the line (track and ballast) and catenary motions arising from fluctuations of the brackets and masts, which should, therefore, be rigid.

At high speed, the problem of the possible presence of a second pantograph exists, which, as already mentioned, is essential in a direct current system when power consumption exceeds a certain level: at 3 kV, taking into account that a pantograph speed of 250 km/h cannot collect more than 2 kA, such limit is on the order of 6 MW.

The second pantograph finds the contact line is already oscillating due to the passage of the first, so is working under very difficult conditions; the worst is where the lifting motion of the line due to the force of the second pantograph is in phase with the free fluctuations produced by the first.

Recent experience has shown that, despite using a heavy catenary and pantographs specifically designed for high speeds, the current collecting of the second pantograph becomes critical above 250–260 km/h

10.1.3 Third Rail

As an alternative to the overhead contact line, the current is collected by means of a sliding contact between one or more slippers placed on the train, and a rail lying alongside, connected to the positive pole of the substation. Current return is normally achieved through the running rails or, more rarely, by a fourth rail located at the center (Figure 10.4).

This solution is normally used on subway lines in systems fed by a direct current, so as to contain the cross-section of the tunnel, and thereby keep infrastructure construction costs down. The global system is more economical because of its greater simplicity, which does not require masts, catenaries, or wire-tensioning systems. On the other hand, problems can arise regarding

Figure 10.4 Third rail power supply.

turnouts in which the third rail has to be interrupted to allow passage of the train. For which the train must be long enough and have a sufficient number of slippers to maintain contact at all times. It is used, therefore, on subway lines, but it is hard to apply to surface railway lines.

Depending on the geometry, current collecting can take place from above, laterally, or from below, as shown in Figure 10.5, and can be protected by a special casing. At present, current collecting from below is preferred, due to safety concerns.

Among the many systems operating with a third rail we can cite the underground systems in New York, Vancouver, Sao Paulo, London, Vienna, and Osaka. Generally, the power supply voltage is equal to 750 V, but there are exceptions, which have power supply voltages, ranging from 600 V to 1200 V, such as Tokyo, Moscow, and Berlin. Exceptional cases, that can be considered limits for power safety with third rails, adopt a voltage of 1500 V, such as in Guangzhou. Then there are the modern driverless lines that use this current collecting system, for example in Copenhagen, Milan, Riyadh, Thessaloniki, and Taipei. Special cases where a fourth rail is used as a current return are the M1 line of Milan and on the London Underground.

Figure 10.5 Current collection from third rail (a) above, (b) laterally, and (c) below.

10.2 ONBOARD PROTECTION SYSTEMS

Onboard protection systems are assigned to a main circuit breaker that is positioned immediately downstream of the current collecting system. This circuit breaker may be a vacuum type (vacuum circuit breaker VCB) for trains fed by an alternating current, or an ultrafast type (High-speed circuit breaker HSCB) for those fed by direct current. The principle of operation for such circuit breakers is

(a) (b)

Figure 10.6 (a) Changeover switches and (b) disconnector for onboard protection.

completely analogous to those used in substations, which has previously been illustrated.

These devices can be mounted inside the vehicle, or in special housing on the roof, or in the underbody of the rolling stock.

The circuit breakers are integrated via suitable isolators that allow reconfiguration of the circuit or earthing during maintenance operations, Figure 10.6. The circuit breakers are controlled via a relay to protect the circuit against overloads and short circuits.

Given the likelihood of over voltages forming on the contact line, either originating internally, due to maneuvers, or externally, caused, for example, by atmospheric discharges, it is necessary to protect the onboard equipment by means of special surge protection devices installed downstream of the pantograph.

10.3 ELECTRICAL POWER SYSTEMS AUXILIARY SERVICES

The auxiliary services onboard, depending on their power, can be supplied at different voltage levels. In particular, a "high voltage" level can be identified, which corresponds to the line voltage in DC systems, while this varies from 1000 to 1500 V in AC systems at the railway frequency and network frequency, respectively. In this case, the power for auxiliary services is provided via a special coil in the main transformer.

In their general composition, there is a feeder that runs along the train feed from the locomotive (HEP) head-end power, which is fed back through the running rails (Figure 10.7).

Alternatively, in electric trains with distributed power, since it is also necessary to supply the traction systems, the feeder can be at the same voltage as the contact line.

Figure 10.7 HEP head-end power.

However, in modern rolling stock, it is preferable not to supply auxiliary services at such high voltage values, but rather to create a three-phase distribution onboard, similar to that in civil systems, which also allows the use of components that are already on the market.

Alternatively, systems that are associated with train safety, such as commands and controls, telecommunications and lighting, are powered by a direct current from buffer batteries, at a voltage that is generally between 24 and 110 V.

The two power supplies are made through electronic converters or directly connected to the high-voltage direct current line or to the intermediate DC link in the AC-powered systems. For security, galvanic isolation is always required between the high-voltage section and the power supply lines of the auxiliary services. Such separation can be obtained by a transformer downstream of the converter that operates at the mains frequency according to the scheme in Figure 10.8.

However, given the low frequency of operation, nowadays solutions are supplied with high-frequency transformers (Figure 10.9), according to the scheme in Figure 10.10, which allows a reduction in the weight and dimensions of the converter itself.

The DC–DC converter that serves as a battery charger and power supply for essential services is normally connected to the three-phase distribution. For reasons of reliability, the distribution systems onboard are redundant. In this case, the return cannot occur through the running rails, as also at the dielectric joints of the rail, like the EBP (electric border points), the power supply must be guaranteed in any event.

10.4 BATTERIES

Electrochemical rechargeable batteries are of great importance in many electrical systems because they can store chemical energy that can be converted and sent whenever and wherever needed, in electrical form.

Especially in railway systems, this kind of battery covers an important role, both for safety and for operations. In case of main low-voltage supply system (LVPS) loss, with their stored energy, the train's batteries are able to supply all operational and safety equipment – during the time and in the modes specified by national and international standards, ensuring a supply voltage level inside any acceptable and safety equipment limits.

Moreover, batteries can be used during normal operations like in "buffer mode," that consists in connecting the battery output terminal directly to the static converter low-voltage output and so to the distribution network. This kind of architecture forces the converter output voltage to be set to a level compatible with the battery charging limits, keeping it always charged. Due to this operational mode, during switching operations or transients, once the converter output current limits are reached, the battery naturally provides for the missing current.

While lead–acid batteries were the leading technology in railway systems in the past, nowadays they are no longer used in passenger transportation – also due to the intrinsic danger of sulfuric acid, such as electrolytes, preferring new and safer technologies.

Today, the main technologies used in railways are nickel and lithium based.

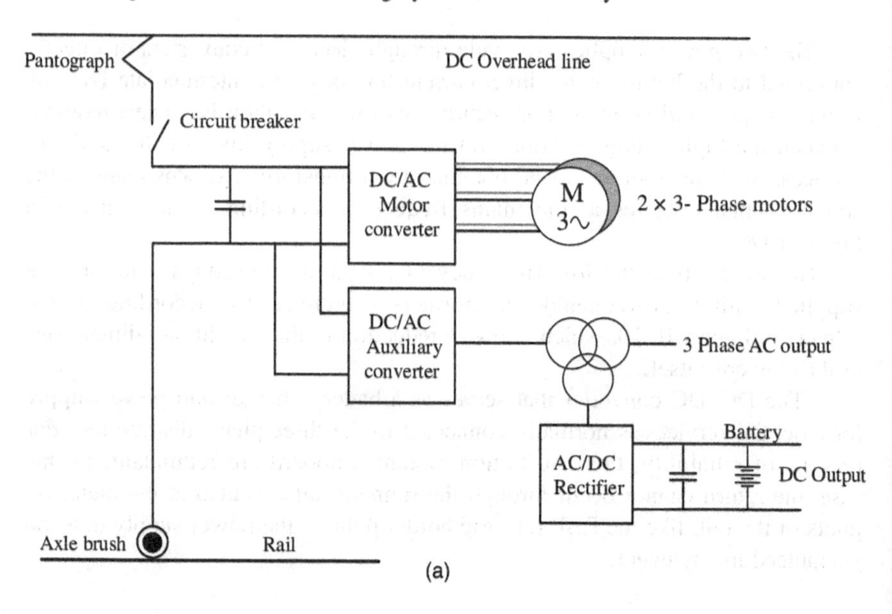

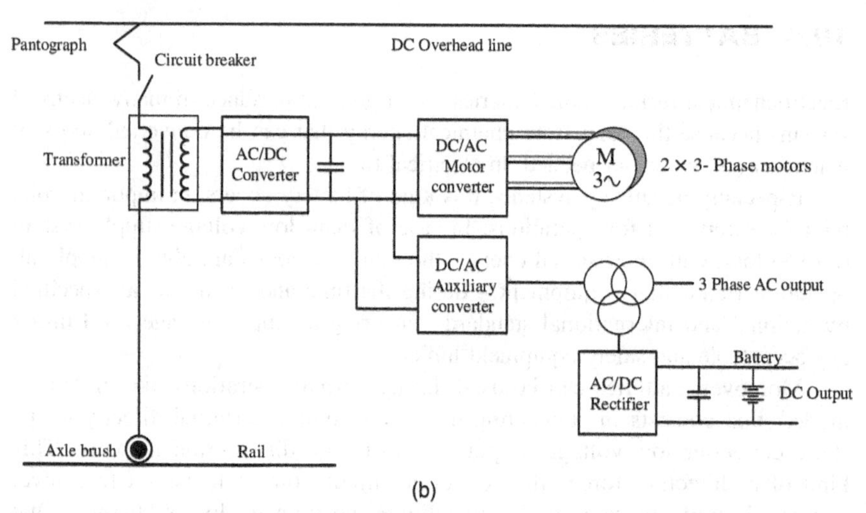

Figure 10.8 Power supply of auxiliary converters.

Since lithium-based technology is not yet considered well proven, special attention will be paid to the nickel-based batteries.

10.4.1 Electrochemical Batteries

Despite differences among technologies, all the electrochemical batteries have similar intrinsic behaviors that is useful to summarize.

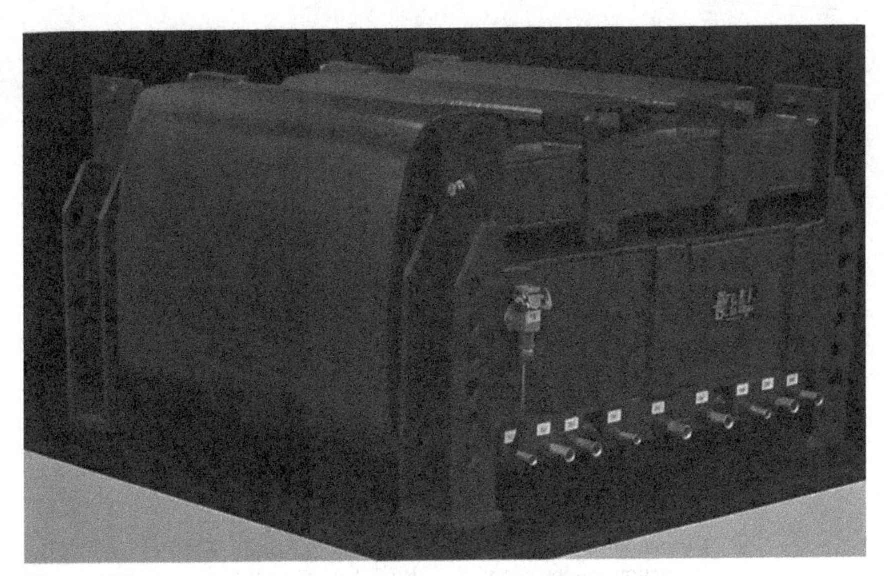

Figure 10.9 High-frequency transformer.

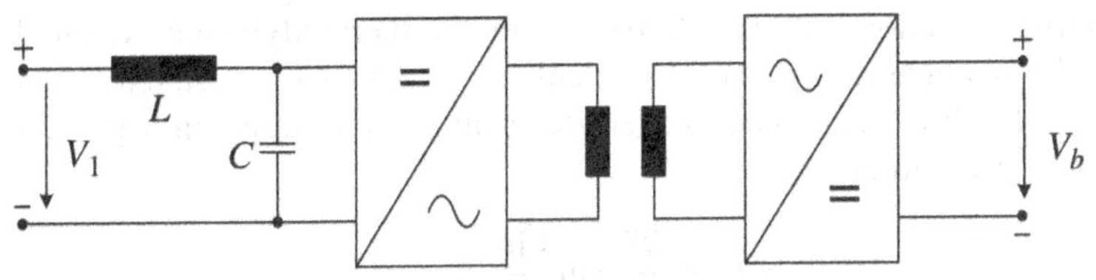

Figure 10.10 Basic diagram of an auxiliary power electronic DC/DC converter with high-frequency transformer.

An electrochemical cell is normally defined through its electric parameters: nominal voltage and charge. While the terminal's voltage is defined by the chemistry of the electrochemical reaction, the storable charge depends on its geometry.

One or more elementary parts of an electrochemical cell form an electrochemical battery. Several cells can be connected through serial/parallel connections in order to achieve the desired battery voltage and charge.

An electrochemical cell, as the name suggest, demonstrates iteration between two different physical systems: chemical and electromagnetic. The chemical behavior of a cell can be described by a redox equation. Two reduction and oxidation semireactions generate a potential differential, as the electromotive force that sustains the reaction. To have a useful electrochemical cell, it is sufficient to separate the semireactions and convey the electronic flow (between the two subsystems, anode and cathode) into an electric circuit (Figure 10.11).

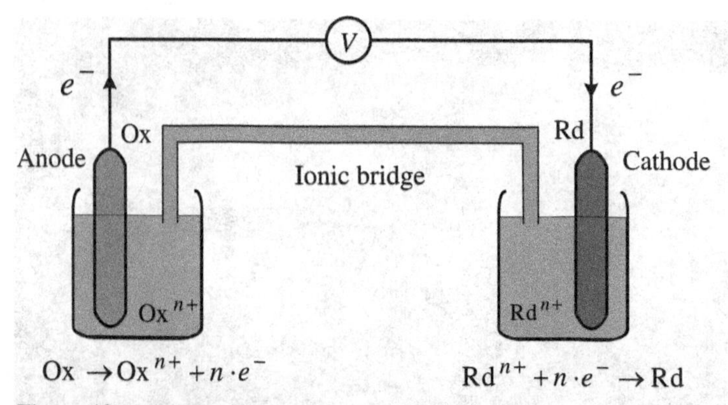

Figure 10.11 Electrochemical cell main diagram.

Oxidation semireaction:	$Ox \rightarrow Ox^{n+} + n \cdot e^-$
Reduction semireaction:	$Rd^{n+} + n \cdot e^- \rightarrow Rd$
Total redox reaction:	$Ox + Rd^{n+} \rightarrow Ox^{n+} + Rd$

The Nernst equation, directly derived from the thermodynamic potential definition (Gibbs Potential), permits to calculate the potential generated by an electrochemical cell in each state of thermodynamic equilibrium. In a general formula, for each electrode

$$E = E_0 + \frac{RT}{nF} \cdot \ln \frac{\prod a_{i,ox}^{\nu ox}}{\prod a_{i,rd}^{\nu rd}}$$

Where

- R is the gas constant (8.3144 J/(K mol))
- T is the absolute temperature (K)
- F is the Faraday constant (98485 C/mol)
- n is the number of electrons transferred in the semireaction
- a are the activities of the reduced and oxidized reagents (for a reduction, right- and left-hand sides of the reaction)
- ν are their stoichiometric coefficients

This formula provides the voltage generated from each semireaction, referring to the standard semireaction of hydrogen reduction $2H^+ + 2e^- \rightarrow H_2(g)$, which gives $E = 0$ V under standard measurement conditions[1].

[1] Standard measurement conditions are defined as: $T = 25$ °C; pressure of gaseous reagents 1 atm; reagents in concentration 1 M (1 mol/l); metallic electrodes.

For aqueous solutions with low reagent concentrations, activities can be expressed as concentrations:

$$E = E_0 + \frac{RT}{nF} \cdot \ln \frac{\prod [\mathrm{Ox}]_i^{\nu\mathrm{ox}}}{\prod [\mathrm{Rd}]_i^{\nu\mathrm{rd}}}$$

For oxidation semireaction,

$$E = E_0 + \frac{RT}{nF} \cdot \ln \frac{[\mathrm{Ox}]^{n+}}{[\mathrm{Ox}]}$$

While for reduction semireaction

$$E = E_0 + \frac{RT}{nF} \cdot \ln \frac{[\mathrm{Rd}]^{n+}}{[\mathrm{Rd}]}$$

So, for the global reaction, taking into account a unitary concentration for solid electrodes, we obtain the cell output voltage:

$$\Delta E = \Delta E_0 + \frac{RT}{nF} \cdot \ln \frac{[\mathrm{Rd}]^{n+}}{[\mathrm{Ox}]^{n+}}$$

This equation is so able to link, in steady-state condition, the external electrical variable voltage with the internal chemical variable concentration of all the reagents acting in the reaction.

10.4.2 Batteries for Railway Applications

During the last two centuries, a lot of different battery technologies have been developed, based on different electrodes and electrolytes and redox reactions.

In battery sizing, the first choice is on technology: since weight is a very important parameter in most applications, energy and power, depending on service requirements, are always referred to in unit of weight and described by the Ragone chart (Figure 10.12).

Regardless of weight, most application often have to meet other requirements. Railways is a field that involves people, so safety requirements apply since various dangers are linked to electrochemical batteries (explosion, burning, loss of acid or toxic electrolyte, etc.).

Currently, most cells used are based on nickel (despite its toxicity):

- Nickel–cadmium cell (Ni-Cd)
- Nickel–metal hydride cell (Ni-MH)

Nevertheless, other technologies, such as lithium-ion and polymer lithium-ion, are being considered.

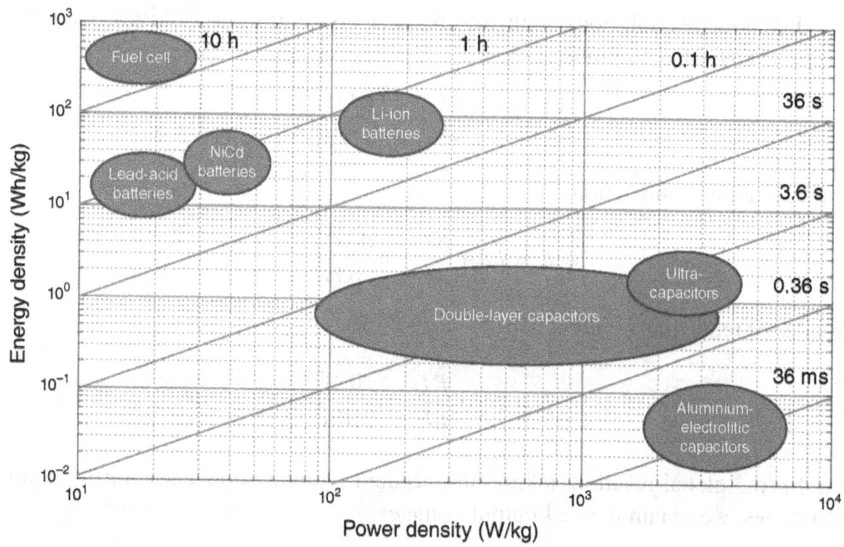

Figure 10.12 Ragone chart for different storage technologies.

Nickel-based cells are very suitable for railway applications since they have the following:

- *Very low internal series resistance*: A low internal resistance ensures low internal voltage drops, so that battery output voltage varies as least possible, varying the applied load. They are so suitable to supply equipment with high insertion current, such as electromagnetic brakes.
- *High robustness*: They require less maintenance and can be used on a non-stationary application. In addition, electrolyte does not take part in the chemical reaction, so there is no need to periodically refill the electrolyte tank. In this way, the battery can be mounted regardless of short-period maintenance.
- *Very long life*: They can be charged and discharged more times than other technologies.
- *Higher specific power and energy*: Compared to lead–acid technologies.

On the contrary, they present some drawbacks:

- *High self-discharge rate*: Ni-Cd cells present self-discharge rates from 15% per month up to 20% per month and Ni-MH cells of about 30% per month, while this value is around 5% for Pb-Acid cells.
- *High cost*: Materials and production cycle are more expensive than for other technologies.
- *Memory effect*: This technology suffers the so-called memory effect, a loss in net storable energy if the battery is charged before it is not completely discharged.

The reactions associated with this type of batteries are as follows:
For the anodic electrode

$$2\cdot Ni(OH)_2 + 2\times OH^- \xrightarrow{\text{charge}} 2\times NiOOH + 2H_2O + 2\times e^-$$

with an anodic potential EOA = 0.38 V.
The cathodic reaction is

$$Cd(OH)_2 + 2\times e^- \xrightarrow{\text{charge}} Cd(s) + 2\times OH^-$$

with a cathodic potential EOC = −0.91 V.
Therefore, the total reaction is the following:

$$2\times Ni(OH)_2 + Cd(OH)_2 \xrightarrow{\text{charge}} 2\times NiOOH + 2H_2O + Cd(s)$$

that provide the theoretical potential

$$E_o = E_{OA} - E_{OC} = 1.29\ V$$

10.4.3 Battery Variables and Parameters

Although the Nernst equation is able to represent the cell's voltage at each equilibrium point, it is not very useful for our purposes. The use of non-linear equations expressed through internal chemical variables constitutes a serious problem for the implementation and the integration of the model. We thus need to begin a modeling process that takes into account the physical phenomena and the dynamics that we are interested in representing. In addition, this process shall lead to a battery model expressed through the external electrical variables.

In a Ni-Cd battery, the main variables are the ones represented in most manufacturers' voltage versus state of charge (SOC) discharge charts (Figure 10.13). We can consider the battery voltage as a state variable, since the conservative chemical behavior acts like a capacitor's electrical field. According to the network model developed, the most suitable battery model will be the one that will provide for the actual battery voltage, depending on the environmental and functional variables and parameters.

Battery manufacturers usually represent several discharge curves measured at different constant discharge current rates. These levels are referred to multiples of C_5 (Ah), that is the capacity declared at 5 h constant current discharge. The reference test current is obtained dividing the reference capacity by 1 h. Starting from the full-charge conditions, the cell voltage is plotted on the *y*-axis while the capacity used is plotted on the *x*-axis. This chart shows the cell internal resistance nonlinearity and time variance (with the SOC).

Under steady-state conditions, output voltage mainly depends on current, state of charge, aging, and temperature.

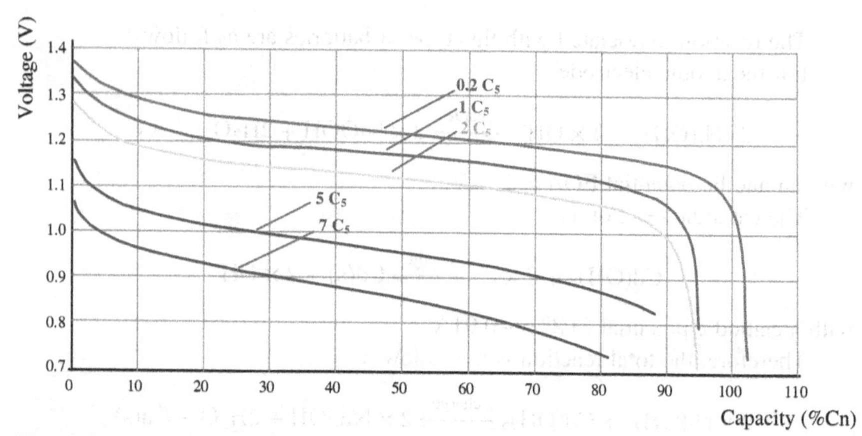

Figure 10.13 Ni-Cd battery, voltage/state of charge plot.

Current Dependence Voltage varies with current mainly due to the internal resistivity of the battery.

In electrochemical batteries, the resistivity property can be seen as the sum of two different contributions: electrical resistivity and electrochemical resistivity. The integral of this property on the cell geometry leads to the equivalent parameter resistance. The sum of electrical and electro chemical resistances make up the so-called cell's internal resistance.

While the electrical resistance can be considered as constant, electrochemical resistance does not refer to a simple electronic flow but to a more complex phenomenon, which makes this kind of resistance nonlinear and time variant.

Indeed, while in a solid metal the current conduction is only carried out by electrons, in a electrolytic solution, charge flow (the current) is maintained by a ionic flow, energized by an electric field. Positive and negative ions have, due to their bigger size, lower mobility compared to electrons so that, at the same charge concentration, they present higher resistivity.

The second effect is more closely linked to the structure of an electrochemical cell. It is stated that cell operations are mainly based on the interaction between two different systems: the electrical one and the chemical one. This interaction occurs in a small layer between them, made by the electrode surface and a little electrolyte film. In fact, here it is where the redox semireactions act, bonding an electronic flow and an ionic flow through a chemical reaction.

It is intuitive to understand that this kind of interface can strongly limit the system charge transport capability.

The redox semireactions work with an exchange of material between anode or cathode and their relative electrolytic solutions. Each elementary reaction needs a place to permit molecules to reach the right orientation, needed to break the old chemical bonds and to form the new ones. In addition, to promote this situation, the reagent concentration in the layer closer to the electrode surface has to be high, or, in another way, the formed product shall be able to quickly diffuse in the rest of the electrolytic solution.

For these reasons the electrochemical resistance cannot be considered constant with the current density σ (A/mm^2), but it presents two saturation zones (anodic and cathodic) where it is higher.

Looking at a constant current discharge curve, it is easy to identify a third-order system, with two nonlinear parts and a quite constant slope in the central part. This curve directly comes from V/σ curve, since $C_{\%} \propto \int i \cdot dt \propto i(t) \propto \sigma(t)$.

State of Charge Dependence Decreasing voltage due to decreasing state of charge presents the same characteristic shown for current dependence.

Decreasing state of charge, first means a decrease in the reagent concentration at disposal of the reaction. Redox, working in lack of reagent, is not able to provide for the theoretical potential anymore. The Nernst equation shows this kind of behavior in detail; the external equivalent effect is a decrease in the open-circuit voltage of the Thevenin equivalent generator or, in the same way, the increase of the output equivalent resistance.

Temperature Temperature strongly affects the mobility of ions in the electrolyte solution. Higher temperature means lower viscosity of the solution and higher thermal excitation of molecules, so higher speed in diffusion and transport processes.

Another effect of the temperature variation is the presence of a charging control system onboard the train that automatically reduces the charging voltage when the temperature in the battery case is too high (overtemperatures due to overcurrents can lead to cell damages). Therefore, for high environment temperature, the full charging voltage will be lower.

Ageing The age of the battery influences its storage capability and the internal resistance.

Assuming that the battery is discharged from an equal charge state to the same end of discharge voltage, the usable capacity declines with cycle number.

With ageing, the available reagents' quantity decreases, mainly due to the fact that electrolyte degrades, for example, forming inert composites, and it could take part in the reaction, so its tank has to be periodically refilled.

Under dynamic conditions other effects have to be taken into account: the main one considers that saturation occurs when a current step is applied. In this event, the battery demonstrates an additional resistance due to the fact that is not possible to reach the new thermodynamic equilibrium point in a null time. From the electrical point of view, this phenomenon is equivalent to the presence of a bulk capacitor that keep the continuity of its state variable, that is, the voltage applied to its terminals.

However, due to the short duration of these transients, due to the slight influence that they have on the rest of the network and on the battery state of

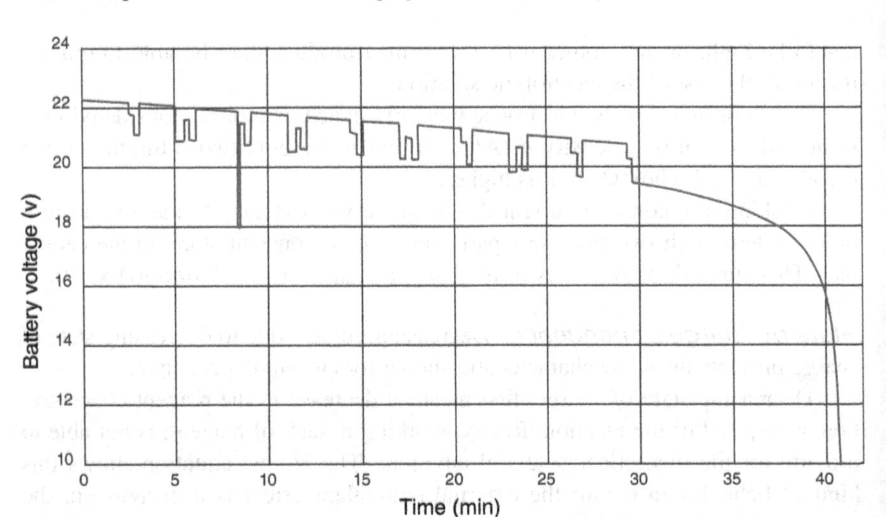

Figure 10.14 Example of battery discharge profile.

charge, we can consider them as a negligible phenomena, thus representing the battery dynamic as a collection of steady-state functioning points. Moreover, we can state that not considering this phenomena constitutes a safety gap, since it should give a higher battery voltage like, for instance, during emergency braking.

10.4.4 Battery Sizing

In order to size the auxiliary batteries onboard, battery discharge profile calculations are used.

It is a dynamical study and consists in analyzing the more critical battery uses in order to choose, once the LV voltage level is defined, the total battery capacity.

Battery behavior analyses, regarding test scenarios, have to ensure respect of equipment's supply voltage limits, to ensure no equipment failures, and the cell's voltage limit, to prevent damage from deep discharge.

Battery sizing consists in making several simulations with different increasing battery sizes, until all the technical requirements are completely met.

A typical battery discharge profile is represented in Figure 10.14.

10.5 COMPRESSED AIR PRODUCTION

The systems for the production of compressed air are made up of compressors and systems for air treatment, at dimensions for the desired air flow rate, depending on the services to be implemented.

Compressed air is the main source of energy for the various auxiliary systems onboard trains. In addition to the actuators of the brakes, it can also be used for opening the doors, controlling the suspension systems, the sand distribution systems, as well as for lifting the pantographs. However, the trend is to make electric actuators, especially for controlling the doors and for small motor drives, to obviate the problems of condensation and occlusion that may occur in compressed air ducts. It does, however, remain essential to control trains, as will be shown further.

The trend toward achieving increasingly more compact and lightweight vehicles with a higher level of safety and performance requires innovative solutions even in this sector. For example, the power supply for motors of the three-phase distribution system is preferred instead of operating at high-traction voltage. Another innovation concerns the use of volumetric rotary screw compressors that reduce vibration and thus increase passenger comfort.

The pressure produced ranges from 700–900 kPa. Since the compressors feed safety systems, such as the brakes, they are always redundant.

10.6 THE BRAKING SYSTEM

The rolling stock needs to have a braking system that can stop the vehicle or to restrain it during downhill driving.

The first rudimentary devices that they implemented for braking were composed of a lever pushing a wooden block against one of the two wheels; as the inertial mass and the speeds reached increased, it became necessary to introduce more efficient braking systems. In particular, toward the end of the nineteenth century, English railway trains were slowed or stopped by manually actuated brakes and, later, by a back-pressure steam braking system on single locomotives. These systems became increasingly less efficient due to the fact that the increase in size of the rolling stock and the technology used did not offer much room for improvement; two innovative devices were therefore developed; the *direct acting air brakes* or *Westinghouse* and the *vacuum brake* or *Hardy*.

The first exploited the force developed by air pressure to bring the brake system into the released position that required the use of ducts, valves, and compressors. The second exploited a permanent negative pressure (vacuum) to obtain the brake release position. In general, in both systems, deceleration of the vehicle during braking was caused by dissipation of its kinetic energy via the friction generated by

- *beams* that press against the *rims* of the wheels
- *brake shoes* that press against *drums* applied to the wheels
- *brake pads* that press onto *discs* integrated to the wheels

These braking mechanisms were governed by *devices called brake cylinders*: the air introduced, compensating for the force exerted by a *return spring*, operated a *piston* placed inside the cylinder that put the *braking device* in motion. Just as today, even at that time, the importance of safety was appreciated, and led to rigorous systems *of passive braking*, which were designed so that, if a fault caused a malfunction of the air circuit duct, the rest position of the cylinder was such that the brake was activated.

The working principles of the systems that are still in use are shown further.

10.6.1 Westinghouse System (Compressed Air Brake)

Expressed more simply as a *pneumatic brake*, its operation is based on compressed air transmission along the brake duct extending from the front to the rear of the train. This signal is generated at the front of the train and spreads toward the rear to reach all the carriages. The pressure wave propagation speed is due to a delay between the instant at which the lead carriage begins to brake with respect to the others. This time interval is the source of serious drawbacks because, while some carriages are braking, others tend to maintain their speed without braking. During the release of the brake, the front part of the train pulls the rear that, however, is still braking, causing mechanical stress on the hooks and the couplers, as well as inefficient braking.

As shown, activation of the braking valve directs the compressed air through the air duct directly to the brake cylinders, activating the basic braking mechanism: the piston, moving backward, exerts a force, which contrasts to that exerted by the tensioned spring, pressing the brake release. The braking valve, called the *triple valve*, is a control device and has the function of allowing or interrupting communication between the *direct air duct* and the *main air reserve*, the *brake cylinder* and the atmosphere, according to the positive or negative pressure variations that take place in the *main duct.* This device can, therefore, be ascribed to the *servomotor devices* category, that is, those mechanical devices which, by acting indirectly and without an appreciable consumption of power, allow the control signal actuated by the operator to be amplified.

However, since the direct ducts do not contain compressed air during normal operating conditions, the brakes would fail if the carriages were uncoupled. To avoid this problem, the direct braking system can be used in combination with braking systems of the electropneumatic type.

10.6.2 Electropneumatic System (EP Brake)

The advent of electromechanical technology has allowed the development of more efficient systems. In particular, the company designed the *electropneumatic brake or EP*: a braking system based the Westinghouse system, which gives a significant improvement in performance. This system is still present in most of the rolling stock. Originally designed for subway lines, where stops and restarts are frequent, the electropneumatic brake was also introduced on electric trains. When compared to the classic direct air system, its main advantage is a considerable increase in the driver's control sensitivity, and its rapid reaction times.

When comparted to a Westinghouse system, the difference is that the brake duct indirectly supplies air to the cylinders; in fact, this task is left to the *solenoid valves* that, at the instant they are commanded, allow its introduction. Since the control of solenoid valves is managed by using electrical signals, it is apparent that the *EP brake* gives its reaction simultaneously and immediately by braking all of the carriages in addition to controlling the *gradualness* of the braking itself. This results in improved performance and greater driving comfort. As regard to safety, the solenoid valve configuration is such that the absence of a power supply (signal) determines the decrease of air pressure in the cylinder and thus the braking of the carriage: in this way, the serious drawbacks of the Westinghouse system are excluded. Development of the EP brake led to previous technology systems being abandoned as obsolete. It should be emphasized that the *electropneumatic brake* is used as a service brake, while *the pneumatic brake* is a backup system.

10.6.3 Electrodynamic Brake (ED Brake)

Referring to the latest *AC motor drives*, the electrodynamic brake exploits the reversibility of the drive motors so that the power flow has a reversed direction with respect to the conditions for acceleration.

The kinetic energy accumulated in the mass of the train is then converted into electrical work by means of the same traction motors that, when properly fed, function as generators.

The electrical power thus obtained can be delivered to the network in AC systems, but can only be injected into DC systems if the line is receptive, that is, if there is the simultaneous presence of other absorbing trains. Since this probability is not very high, the power is often dissipated in appropriate braking rheostats installed onboard and controlled by a chopper that adjusts the power.

It should be noted that electric braking does not work for speeds close to zero; which is an advantage, in that, given its principle of operation, it is not possible to lock the wheels and at the same time, which is a disadvantage, as it cannot be used as a parking or emergency brake.

10.6.4 The Electrohydraulic Brake

This type of brake is widely used on tram vehicles, where the need to build vehicles with low floors involves a significant reduction in available space. The *brake pump* acts on the working fluid, typically oil composed of *polyethylene glycol* that actuates the piston under pressure in the cylinder, which moves backward by exerting a counter force to the spring, and actuates the brake release.

Figure 10.15 Hydraulic-type brake system.

Control of the pump and valves is effected electrically and electronically. In Figure 10.15, an example of the hydraulic-type brake system is shown.

The use of compressed air not being necessary, these systems allowed the development of all-electric vehicles.

However, the inability to restore fluid pressure in case of leakage, and the change in viscosity as a function of temperature, which makes the system non-linear, does not allow their use on transport systems, such as trains and subways, but only on light rail vehicles (LRV).

10.6.5 Eddy Current Brake

In general, the effectiveness of the braking system depends closely on the adhesion of the wheels to the rail.

It is a matter of observation that such adhesion is reduced by increases in speed and by the presence of moisture, ice, leaves, or other matter on the rails. To avoid wheel spin, a reduction in the braking force is necessary, which results in an increase in the stopping distance. To overcome this problem, recourse can be made to auxiliary braking systems that exploit the phenomenon of magnetic repulsion.

In *magnetic brakes*, in fact, the braking force is obtained by means of the magnetic interaction created between the runners' inductors placed onboard the vehicle and the rails, due to eddy currents induced on their upper surfaces. The effectiveness of these brake systems does not depend on the adhesion between the wheel and the rail, and thus operates in any climatic conditions, especially at high speeds.

This braking system consists of runners into which are inserted electromagnets that are electrically connected to each other so as to form an alternating sequence of N and S magnetic polarities.

The pneumatic pistons have the function of lowering the *electromagnetic* runner up to about 8 mm from the track so as to create an optimal operational air gap. After reaching the working position, they are energized by the DC coils. The magnetic runners, passing above a connection taken up in a section of the track with a speed different from zero, therefore, offer alternating magnetic polarities N and S.

According to the laws of electromagnetic induction, a variable magnetic flow path generates electromotive forces along closed paths on normal planes, in the direction of the field; if the medium in which the magnetic field manifests itself is an electrical conductor, the phenomenon gives rise to a flow of current in the material.

Within the rails, therefore, the eddy currents that are induced, give rise to a magnetic flux that tends to oppose the inductor flux.

According to the Lorentz's law, a magnetic circuit is subjected to a force that can be decomposed into two components; a braking force along the horizontal axis, and an attractive force, between the runner and the rail.

We can see therefore, that the magnetic brake cannot be used as a parking or stopping brake because, at speeds near or equal to zero, the braking component is zero, as shown in Figure 10.16.

For this reason, a magnetic brake can only be used at high speeds and is always combined with an electropneumatic or electrohydraulic braking system. It can provide, particularly at high speeds of travel, a significant braking force, considering that for every kilowatt of electric power fed into the runners, we can obtain about 70 kN; the maximum permissible deceleration is thus equal to 2.5 m/s^2.

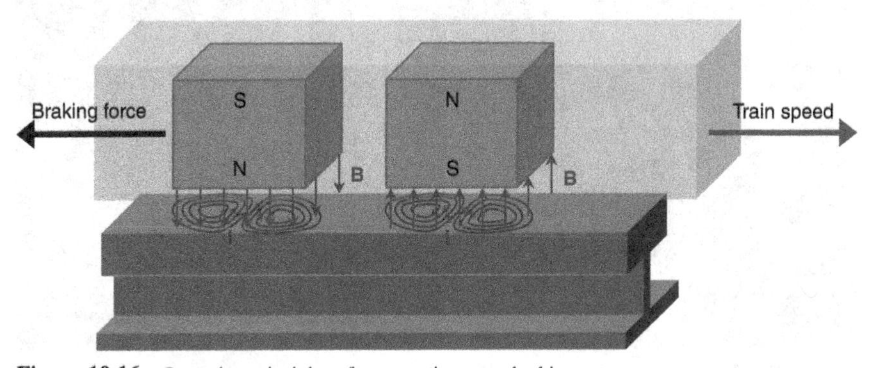

Figure 10.16 Operating principles of a magnetic runner braking system.

However, this can cause the train rails to overheat, because kinetic energy is dissipated within them by the Joule effect due to the induced eddy currents. In high traffic lines, there may not be enough time for them to cool down between one train and another, due to which abnormal expansion might occur, causing deformation of the track.

Another problem that is sometimes due to nonapplicability, is interference caused by eddy currents with the signaling circuits (track circuit current) that can compromise data carried or exchanged with the train.

10.6.6 Electromagnetic Runner Brakes

The electromagnetic runner has a similar structure to the eddy current brake, the difference being that excitation of the coils causes attraction of the runner on the rails. The sliding friction between the runner and the rail is due to the dissipation of power, and therefore, generation of a braking force.

The electromagnetic runner brake is used as an emergency auxiliary brake primarily in trams and subway systems. In comparison to the eddy current brake, its advantage is that it can operate at low speeds, up to the complete stop of the vehicle. As a safety device, it can also be powered by onboard batteries in the absence of power supply from the contact line (Figure 10.17).

10.6.7 Brake Control Unit (BCU)

The brake control unit is the automatic braking management system for rolling stock: It is thus a type of self-regulating system with programmed variation of

Figure 10.17 Electromagnetic runner brake.

the level of response. In addition to the functions needed to implement braking in normal operation, this system ensures the stopping of the train in case of the failure of the main braking system.

As far as construction is concerned, it is composed of a brake manipulator, a EBCU control unit (electronic brake control unit), and electropneumatic actuators.

The system control logic is based on a comparison between the reference signals corresponding to the various positions of the control lever of the manipulator, and response signals from the pressure/current transducers that detect variations in the pressure of the pilot chamber of the solenoid valves and the general brake conduction.

The task of the electronic brake control unit during the braking phase is to activate the solenoid valves by using reference signals relating to the position of the manipulator, the speed of the train, and the efficiency of the electrodynamic braking (ED brake). It also performs electropneumatic brake control functions (EP brake), surveillance, and, depending on the installed programme, can also perform all of the control and diagnostic operations.

Specifically, digital signals encoded in PWM mode are received from the manipulator along with any additional data provided by the vehicle software, so as to control all the actuators for a correct pressure setting.

The EBCU, in addition to providing functionality and surveillance for the braking system, has diagnostics for the inputs, outputs, and self-diagnosis.

The electropneumatic devices in the system are mainly isolation taps, electropneumatic valves, and valves for emergency braking.

The electropneumatic valves can

- limit the flow, if de-energized, open for passing a small fraction of air;
- activate the brake-release, which, once activated, feeds the pilot chamber engaging gradual release of the brake;
- activate the braking, which if activated, discharges the pilot chamber;
- for safety, and if de-energized, quickly discharges the pilot chamber.

The valves for the emergency braking are used by service personnel whenever it is necessary to manually discharge the main brake duct, and thereby cause the train to brake.

The main functions of the BCU system can be grouped into two macro areas: managing the braking and processing signals for guiding the braking and driving other devices onboard the train.

In particular, as regard the managing braking, there are generally four distinct braking modes:

- *Service brake mode*. The service braking mode is used to slow the vehicle; the BCU receives the value of the current speed, brake weight, slope of the line, and target deceleration (force required), and then, using an algorithm, processes the correct braking allocation.

- *Stopping brake mode*. The stop braking mode has the function of reducing the speed of the rolling stock to 0 velocity. This mode can follow service braking, and is divided into the following four phases:
  - Phase 1. This occurs when the vehicle speed is less than the speed of takeover; in this phase the braking force of ED progressively decreases while the braking force of the EP increases.
  - Phase 2. The ED brake function is completely inhibited, to be replaced by the EP function that takes over all the force required
  - Phase 3. At speeds below 1 km/h, the EP function is no longer reduced, but applied at its maximum value.
  - Phase 4. The vehicle is stopped and the brake is applied. The brake hold function is inhibited only if a brake release signal is sent and there is no braking command.
- *Parking brake mode*. Parking keeps the rolling stock in the locked position. When stationary, the brake is activated by a spring and released by pneumatic action. The parking brake command can be made in manual or automatic mode; the latter is made autonomously whenever it is disabled at the control desk by turning the key on the control desk. In cases where the train is in "remote control" mode, it is still possible to deactivate it.
- *Emergency brake mode*. The emergency brake is used to stop the train in a short time when necessary. Generally, it is used to provoke discharge of the main duct and closure of the isolation valve that prevents refeeding of the main duct. The direct consequence is that the train is stopped with maximum braking force. Moreover, there is another mode called quick braking, that allows the same effect to be obtained using a manual, direct discharge of the main duct.

10.6.8 Vehicle Air Conditioning: the HVAC System

The HVAC systems (heating, ventilation, and air conditioning) onboard railway vehicles are the result of continuous research and improvements in heating and cooling systems suitable to give passengers a high level of comfort, and comprehensively treat the air fed into the carriage regardless of changes in external weather conditions during the journey, or the speed of the train.

In particular, the air-conditioning provides a series of services that aim to achieve the optimum temperature and humidity conditions inside the carriages. An ideal balance is achieved when thermohygrometric conditions assume temperature values between 20 and 26 °C with a relative humidity of 40–60%. To keep the microclimate within predetermined limits, the air conditioning system must work on the heating, cooling, and ventilation systems.

The problem of air-conditioning is complex in itself, because a feeling of comfort is subjective, and depends on the individual, or the inhabitants of different geographical areas. In general, in both the heating and cooling phase, it is appropriate that the temperature difference between external and treated air never exceeds a certain limit, which might cause physiological discomfort for passengers (e.g., colds, draughts). In addition, optimum comfort is achieved by taking into due consideration the speed, flux, and purity of the treated air, and providing a minimum volume of air per passenger, depending on the outside temperature.

For railway vehicles, the conditions that air-conditioning systems of carriages must comply with under international traffic arrangements are dictated by specific regulations, such as, at a European level, EN 13129 and UIC Sheet 553.

The main systems installed can be classified according to their locations in a train, the air distribution system, the regulation of the temperature and the electrical and thermal services.

There are various solutions regarding the distribution system and the arrangements, including the roof-mounted HVAC machinery with top-down distribution.

In this solution, the system consists of a single conditioner unit located between the ceiling of a carriage and the roof, while the motor compressor group and the condenser are arranged in the undercarriage.

The air is distributed through a main duct in the cavity under the roof by appropriate distributors: the treated air then filters into the various compartments from the top downward.

This solution can become uncomfortable due to the effect on the floor: in heating mode, when the external temperature becomes freezing, the floor remains cold, causing passenger discomfort. It, therefore, uses additional heaters along the walls of the carriage at the level of the floor. As an advantage, top-down air distribution offers the advantage of cleaner air, as the vents are placed on the roof.

With the induction solution, the motor compressor unit, the condenser, and the centralized air-conditioning units are placed in the undercarriage, while the air distribution duct is placed inside the carriage via the various compartments along the sides.

The treated air is distributed in these compartments by means of suitable distributors, from which the air flow exits from under the window and lightly touches the inner walls. Treated air of a primary type, that is, one that includes all the fresh air, coming in from the distributor nozzles, drags a secondary air flow from within the compartment, which mixes with it; this is returned to the same compartment through the nozzles, causing a continuous recirculation.

The induction systems may have a single or dual air duct.

In dual air ducts, the two channels run parallel, but independently, along the side of the carriage at floor level. The two air currents are separate, and are mixed by the distributors, then sent to the compartment through the nozzles.

The air-conditioning system consists of the following elements:

- The conversion unit, powered by the three-phase onboard network.
- The air treatment unit is the unit containing the air inlet sockets, the filter, the fan motor, the heater assembly, and the cooling evaporator assembly. The air vents are generally made of grills with sufficiently small slits to prevent foreign bodies getting in. Because atmospheric air contains many substances suspended in the form of dust, they can cause problems from a hygiene perspective: this is why proper air filtration is required, to intercept as many harmful or irritating agents as possible. Generally, the filters used most often are of the following two types:
  - *Metallic or frame*: Formed by a rectangular frame of variable thickness; the filter mass is made by superimposed layers of wire mesh with narrow links, to give a high dust holding capacity.
  - *Artificial fiber mat*: Composed of acrylic fibers expanded in three dimensions linked to special synthetic resins. These have the property of being heat resistant up to 120 °C, waterproof, moisture insensitive, and high efficiency, since, due to their fibrous structure, they allow uniform filtration.
- The *compressor/condenser unit* comprises the compressor, the condenser, the collecting tanks of the cooling fluid and filter drier, pressure switches/thermostats, and pressure gauges; it is also called the "refrigerant circuit."

All vehicles are equipped with a low-voltage command and control switchboard that connects the various power supply lines of the various circuits and equipment that are separated each other. Each line is equipped with protection via fuses and/or thermal–magnetic circuit breakers. Generally, the systems are all set up with the control circuit, for which switching of the various circuits takes place automatically as soon as the high voltage in the AT duct is present on the vehicle.

Use of an inverter for controlling the motors allowing it to reduce the absorption peaks in the start-up phase, impacting especially on the onboard distribution system.

Rolling stock temperature is regulated automatically by the automatic temperature regulator, that gathers signals via internal and external probes. During operation, depending on the value of the outside temperature, the system can operate in heating mode, with ventilation alone, in cooling mode, or dehumidification alone.

The ventilation constitutes the heart of the air-conditioning system as it is responsible for the intake of external air and its distribution inside the vehicle after appropriate treatment.

The size of the fans is determined by the air flow required to meet the needs of comfort; the power of the motors must be able to overcome the pressure drop caused by the filters and ducts. Their main features include the following:

- *Q*: air flow rate (m^3/h), the amount of air passing through the fan unit at a time
- *N*: rotation speed (rpm)
- *D*: diameter of the impeller
- *p*: air density (kg/m^3)
- *u*: fluid velocity (m/s)

Effective operation of the system depends only on the right choice: it must be able, that is, to bring a certain amount of air at a given total pressure. This total pressure is the sum of two components, static and dynamic,

$$P_{tot} = P_s + P_d$$

in which the first term (P_s), represents the pressure that the air exerts on the wall of a duct running through it longitudinally; the second (P_d) is due to the speed that is established in the duct to reach the desired flow rate.

In the driver's cab, the air-conditioning system must be separated for safety reasons.

10.6.9 Passengers Information System (PIS)

The passengers information system is an information and multimedia system that offers passengers various types of service information via appropriate terminals, like graphic displays, audio amplifiers, and display panels in the compartments of the rolling stock. From the aesthetic and operational perspective, these elements are similar to equipment for daily use, like TV/LCD monitors or informational displays for sale points. In truth, they have to work on rolling stock, and must comply with very stringent regulations such as the European Standard EN 50155 and ST 306158 for electronic equipment used on railway vehicles related to operating temperatures, vibration, power supply, low power consumption, stresses and compact mechanical dimensions, duration.

This system has the task of informing the passengers, in automatic or semi-automatic mode, through audio/visual messages, about the progress of the journey, broadcasting short animations using graphic displays, video surveillance of the main internal external and system doors, and checking for/reporting fires.

Moreover, also in terms of safety, the system can provide real-time information about the various emergency evacuation methods, in order to control the exit of passengers in a safe and efficient way.

With regard to video surveillance, the system allows images from the internal cameras to be recorded, the viewing of images from external and internal cameras and the viewing of recorded images.

On connection of the power supply and in the presence of the driver's control enabling the signal, the system starts operating automatically, without operator intervention. It can independently switch to viewing the images taken by cameras external to the opening side of the entry doors once the doors' opening signal becomes active.

Chapter 11

Multisystem Rolling Stocks

The services and auxiliary motors of modern traction vehicles are powered by auxiliary power converters driven by microprocessor control systems. This architecture allows important advantages to be obtained compared with traditional electromechanical traction vehicles in terms of efficiency, the reduction of weight and volume, reliability, and maintainability. The versatility of electronic converters has also made the economic development of multisystem rolling stock possible, which allows overcoming the obstacle of different power supply systems and thus the implementation of the railway interoperability.

Achievements in relation to rolling stock differ not only because of the different construction traditions of different countries but also as a result of the different socioeconomic and environmental situations they must adapt to.

Due to the particularities of the rail system, the vehicles must be adapted to different voltage levels and frequencies of the power supply network while seeking to maintain the maximum allowable performance for each system as much as possible.

A multisystem rolling stock is generally equipped with a pair of pantographs for each type of power supply or country in which it must carry out service. In special cases, such as for freight locomotives where there is not enough space, only one pantograph is installed per type of power supply, losing redundancy.

To avoid incorrect maneuvers, the vehicles are equipped with a special sensor that controls the main power switch only when the collected line voltage is consistent with the configuration of the traction circuit.

In modern applications there is a tendency to create a dedicated onboard DC link to which the traction and auxiliary services converters are connected. The DC-link voltage remains constant, regardless of the power supply, in order to avoid changing the configuration of onboard systems. Therefore, there will be special input stages responsible for the transformation and/or power supply conversion present from time to time.

Electrical Railway Transportation Systems, First Edition. Morris Brenna, Federica Foiadelli, and Dario Zaninelli.

Since the space onboard is limited and there are weight constraints, particularly for high-speed vehicles, it is not possible to install a dedicated input stage for each system. It is therefore necessary to reconfigure the circuit in the transition from one system to another, making the most of the same components and also varying their functions. For example, parts of the main transformer can be transformed into switching or filtering inductances, going from AC to DC, or modules of a four-quadrant converter (4Q) may be used as DC–DC converters.

11.1 TRANSFORMER

The electromechanical components onboard the multisystem vehicles feature considerable weight and dimensions. Therefore, their application should be limited only as needed and their installation must be carefully considered in relation to the available space and weight distribution on the wheels of vehicles. For these reasons, wherever possible, we prefer installation on the underbody in a central position.

In multisystem vehicles it is therefore difficult, if not impossible in certain cases, to use dedicated components for each power supply system, particularly if they use mainly electromechanical devices.

Thus, it is necessary to run the main transformer with different voltage levels and different frequencies, as well as use it as an inductor in the systems.

11.1.1 Multivoltage and Multifrequency Transformer Operation

Assume that the main transformer should operate with two systems characterized by primary voltages and frequencies (V'_1,f'_1) and (V''_1,f''_1), and that in both systems the supply comes from nominal power S_n.

Therefore, the current absorbed by the primary side of the two systems will be $I'_1 = \frac{S_n}{V'_1}$ and $I''_1 = \frac{S_n}{V''_1}$, whereby the primary side must be isolated for the higher voltage, but the winding will be sized for the system with the lower voltage and thus for the higher current. These transformers have a single primary side winding and therefore the number of winding turns N_1 remains constant for each power supply system. This affects the rating of the magnetic core that must withstand the maximum flux between the two systems. The magnetic flux Φ is related to the inducted electromotive force e in each individual winding turn by the ratio:

$$e = \sqrt{2}\pi \cdot f \cdot \Phi$$

The electromotive force per turn can be obtained from the relationship between the power supply voltage and the number of primary winding coils. Consequently,

$$e' = \frac{V_1'}{N_1} \Rightarrow \Phi' = \frac{e'}{\sqrt{2} \cdot \pi \cdot f'} = \frac{V_1'}{\sqrt{2} \cdot \pi \cdot N_1 \cdot f'}$$

$$e'' = \frac{V_1''}{N_1} \Rightarrow \Phi'' = \frac{e''}{\sqrt{2} \cdot \pi \cdot f''} = \frac{V_1''}{\sqrt{2} \cdot \pi \cdot N_1 \cdot f''}$$

On the other hand, the secondary side will have to deliver the same voltage E_2 and therefore the same current in both systems. Thus, by going from voltage system V_1' to voltage system V_1'', the number of winding turns must be changed, according to the ratio:

$$N_2' = \frac{E_2}{e'}, \quad N_2'' = \frac{E_2}{e''}$$

For example, when going from a 25 kV, 50 Hz system to a 15 kV, 16.7 Hz system, the primary side will have an overrating equal to 67% due to the increase of current caused by the lower voltage and equal to 80% of the magnetic core due to the reduced frequency. At the secondary side, instead, the number of winding coils will have to be increased by 67%.

11.1.2 Power Electronic Traction Transformer (PETT)

At present, AC and multisystem rolling stocks employ a traditional configuration for the input stage, which is based on the line frequency transformer and four-quadrant converters that are tuned to comply with electromagnetic compatibility standards (Figure 11.1a). Generally, this transformer is a bulky and heavy machine since it can work also at low railway frequencies, as 16.7 or 25 Hz.

A new type of transformer based on power electronic devices is being introduced into the railway system, which achieves voltage transformation, galvanic isolation, and power quality improvements with a single component (Figure 11.1b), instead of the traditional bulk transformer working at contact line frequency.

Several prototypes are under construction, as is testing for railway application by the main train manufacturers.

The research and development of PETT has been encouraged recently because of the appearance of silicon carbide (SiC) power modules. As an aside, thanks to their properties, which allow switching frequencies on the order of tens

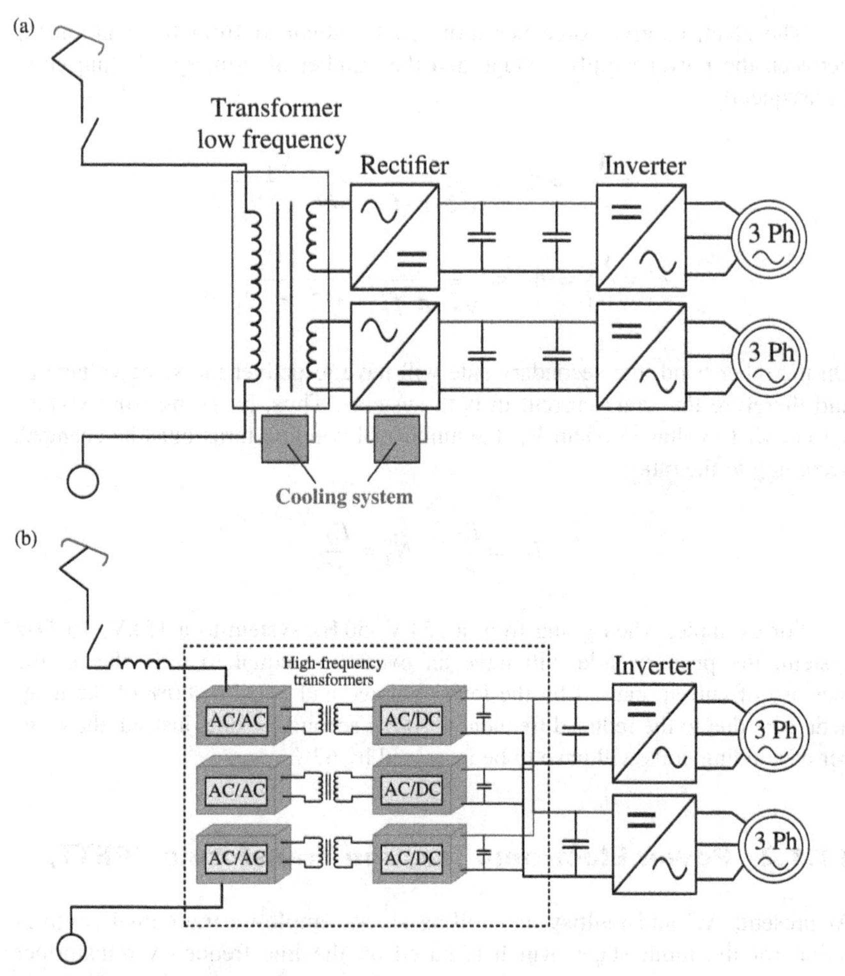

Figure 11.1 (a) Bulky transformer versus (b) power electronic traction transformer.

of kilohertz, it will be possible to downsize the bulky traction transformer to a lighter medium-frequency transformer, which will, additionally, improve the efficiency of the system. On the other hand, high-voltage devices are not fully mature and commercial application will be available fully once the SiC 10 kV components are ready.

In the meantime, several studies and patents are published on this topic, mainly focused on realizing the high insulation conversion AC/AC of Figure 11.1b. As a general remark, two different solutions are actually envisaged:

- Direct structure, which considers matrix converters that are able to realize AC–AC conversion from low (16.67, 25, or 50–60 Hz) to medium frequency (5–10 kHz).

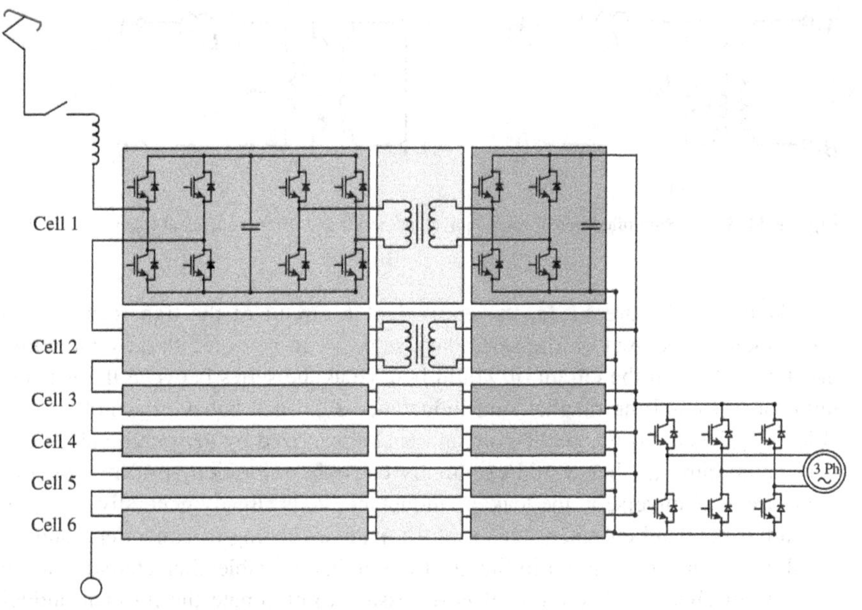

Figure 11.2 Indirect structure for power electronic traction transformer.

- Indirect structure (Figure 11.2), which considers AC–DC conversion realized in the first place, and then DC–AC converter provides alternative current at medium frequency.

In both solutions, the working frequency of the various transformers is always equal to the switching frequency of the power modules, so it is independent from the input frequency of the contact line. In this way, the sizing of each transformer is not affected by the line frequency.

The medium-frequency transformers are then connected to the AC side of such converters in order to scale down voltage and provide galvanic insulation. The final connection, with the four-quadrant converter, supplies the DC link where it is connected to the usual three-phase power inverter.

Among the several AC lines, the particular interest in such development today appears confined to the 15 kV AC line, where 16.7 Hz obliges a train to onboard tens of tons of classical transformer. A global weight reduction of 50% and improved efficiency of almost 7% can be achieved with respect to classic solutions.

11.1.3 Operation as an Inductor

A transformer may also be used as a simple inductor by appropriately reconfiguring the circuit.

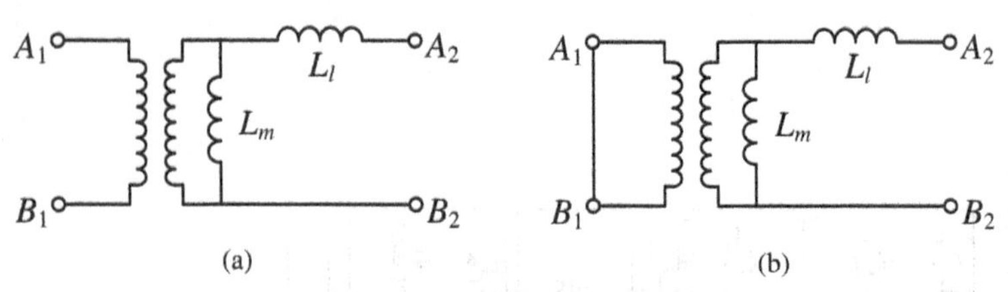

Figure 11.3 (a) Simplified transformer equivalent circuit and (b) its equivalent to an inductor L.

Analyzing Figure 11.3a, the equivalent L circuit of the transformer with series elements shown on the secondary side, it can be noted that by including the dipole A_2B_2 in the circuit, it would behave as the series between the leakage inductance L_l and the magnetizing inductance L_m. Such behavior is not acceptable as L_m has a ferromagnetic core and is characterized by very high values and very small currents that would eventually cause its saturation at normal operating currents. Conversely, the leakage inductance L_l is already generally sized for the inductance and current values useful for the switching of converters, and by involving a magnetic circuit in the air it maintains its stable characteristics when the current changes. Therefore, it is necessary to eliminate the intrinsic inductance L_m to the transformer through the short-circuiting of A_1B_1 at the primary side. In this manner, the dipole A_2B_2 will take on the characteristics of the single inductor L_l (Figure 11.3b).

11.2 FOUR-QUADRANT CONVERTER

In order to evaluate the impact of railway rolling stock on the infrastructure, the AC/DC conversion input stages are considered, consisting of 4Q to IGBT converters.

The 4Q converter is a forced switching bidirectional converter with pulse regulation that can be considered structurally similar to a single-phase voltage source inverter, with an H-shaped bridge structure. Multilevel structures allow higher voltage to be adopted in DC section, achieving the goal of increasing the transmittable power with equal transmission losses. The DC voltage has a maximum limit imposed by the presence of the semiconductors and voltage leveling capacitors. Furthermore, the high switching frequency of the semiconductors introduces the problem of filtering of the harmonics on the alternating current side to only high-frequency interference.

The system, for which the basic diagram is found in Figure 11.4, essentially consists in the secondary side of the transformer, 4 GTO- or IGBT-type semiconductors and 4-freewheeling diodes connected in antiparallel, so that the current can circulate in both directions. Each pair of opposed switches cyclically connects the two input terminals to the positive pole or the negative pole of the DC link. Each terminal can therefore assume two values of the potential, for

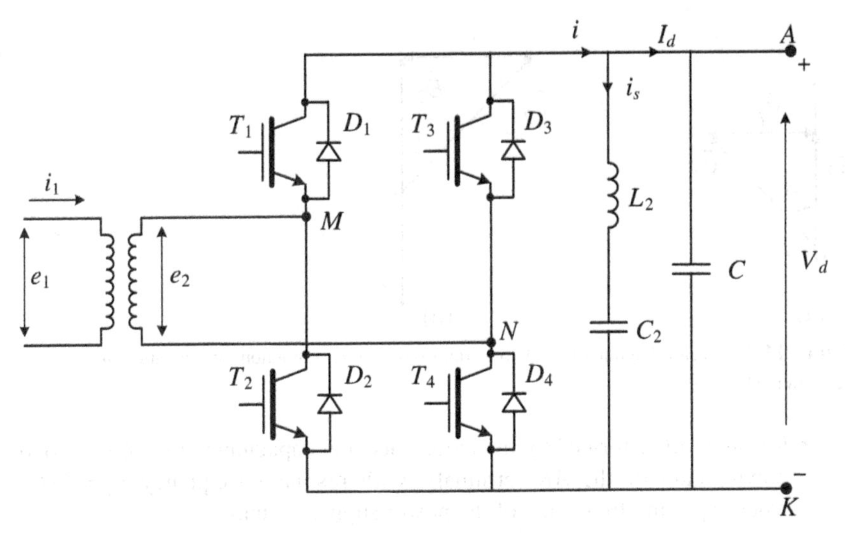

Figure 11.4 Basic diagram of the 4Q converter.

which the converter operates at two levels. In the DC link, a capacitive intermediate circuit C is placed with the aim of imparting a constant voltage.

First, the system has the task of absorbing from the contact line, at voltage e_1 and frequency f_1, a current having the fundamental component i_1 in phase with e_1 and negligible harmonic content, so that the following conditions are met:

$$\text{phase shift factor } \cos\varphi_1 = 1; \quad \text{form factor } \lambda \approx 1 \tag{11.1}$$

Second, it must absorb an average value power P_1 and frequency pulse $2f_1$ from the line and provide the intermediate DC circuit continuous power P_d for traction motors and auxiliary services.

Said i_2 fundamental harmonic of the current supplied by the transformer, v_2 that of the voltage at the AC terminals of the single-phase bridge, the condition (11.1) translates into the vector diagram, as shown in Figure 11.5.

The leakage inductance of the transformer and the switch resistance shifts the $\vec{V}_2$ in a delayed manner compared with $\vec{E}_2$.

With reference to Figure 11.5, the 4Q converter consists of the following elements:

- Secondary side of the main transformer that lowers the line voltage to the level suitable for the operation of the converter.
- Converter with H-shaped bridge that includes four power switches $T_1 - T_4$ and four recirculation diodes $D_1 - D_4$.

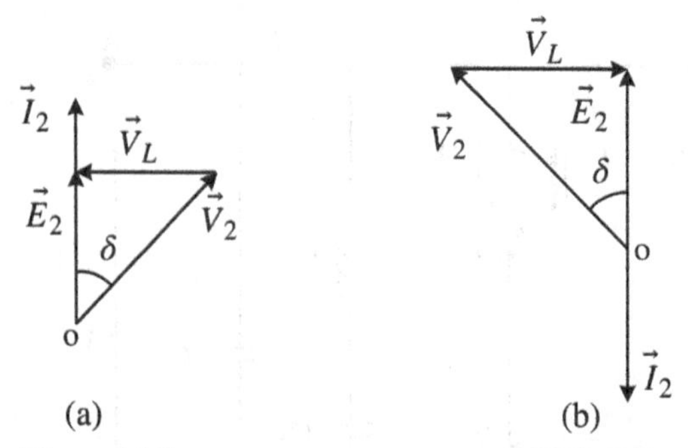

Figure 11.5 Vector diagram of a 4Q converter (a) operating in traction, (b) operating in regenerative braking.

- Resonant filter formed by the inductance and capacitance $L_2 - C_2$ inserted downstream of the AK terminals, with resonant frequency $f_2 = 2 \cdot f_1$, where f_1 is the frequency of the power supply system.
- Capacitor C always connected downstream of the AK terminals and works on the voltage source V_d.

For the study of the 4Q converter's operation, the secondary side of the transformer is replaced with a Thevenin equivalent circuit that considers the no-load circuit voltage $e_2 = \frac{e_1}{k}$ and the only leakage inductance as a series impedance that gives rise to a reactance equal to $X = \omega_1 \cdot L$ (Figure 11.6). The V_d voltage of the DC link is also considered constant, and the magnitudes of the AC side sinusoidal, neglecting the ripple voltage and current because a 4Q converter employs the pulse-width modulation (PWM) modulation.

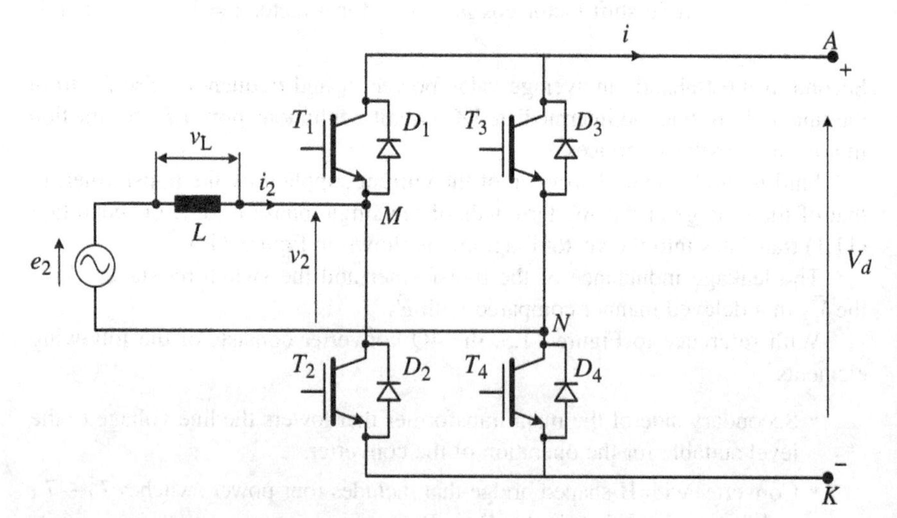

Figure 11.6 Four-quadrant equivalent circuit converter.

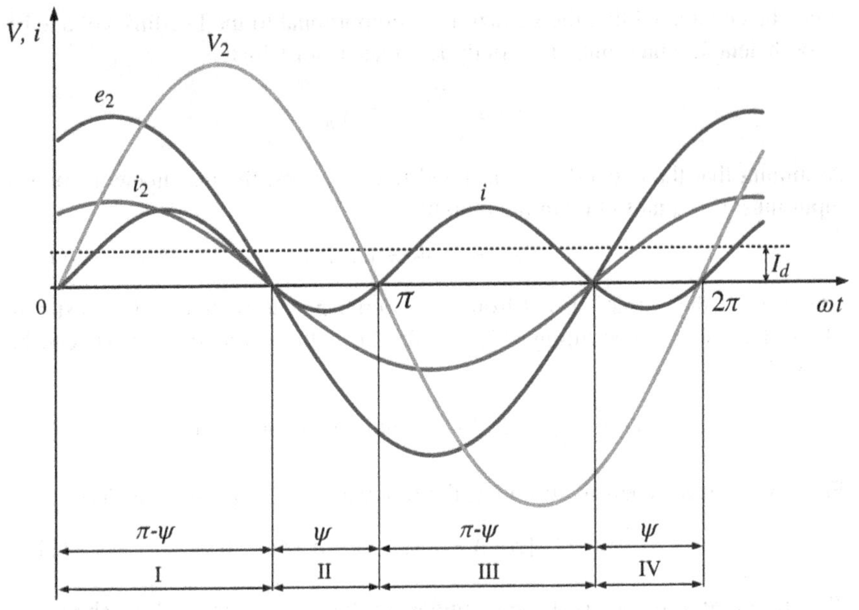

Figure 11.7 Performance of voltage magnitudes and current in traction operation.

The voltage drop on the leakage reactance is such to cause a phase shift between the voltage e_2 output from the transformer and the voltage v_2 generated by the 4Q converter through PWM modulation. This phase shift must be equal to the angle δ of Figure 11.7 in order to have an absorption with unity power factor. Therefore, the equations governing the input electrical quantities are

$$
\begin{aligned}
i_2 &= \sqrt{2} \cdot I_2 \cdot \sin(\omega_{.1} t) \\
e_2 &= \sqrt{2} \cdot E_2 \cdot \sin(\omega_1 \cdot t) \\
v_2 &= \sqrt{2} \cdot I_2 \cdot \sin(\omega_1 \cdot t - \delta)
\end{aligned}
\tag{11.2}
$$

This converter can be considered as a single-phase inverter supplied with impressed voltage V_d and adjusted by means of pulses with the switching frequency f_{sw}; this should be as high as possible, compatibly with the limits imposed by the switching losses that are proportional to f_{sw} and the recovery times of the semiconductors used. With current technologies, this frequency is on the order of 500–1000 Hz.

In addition, in the vector diagram of Figure 11.7, the following conditions must apply:

$$
\mathrm{tg}\delta = \frac{V_L}{E_2} = \frac{X \cdot I_2}{E_2}
$$

It can be considered that the voltage v_2 is proportional to the DC-link voltage by a coefficient k, which must be strictly less than 1, for which

$$V_2 = \frac{E_2}{\cos\delta} = k \cdot V_d$$

Assuming that the converter is ideal and free of losses, the instantaneous power input must be equal to that in the output:

$$p_2 = i_2 \cdot v_2 = V_d \cdot i \tag{11.3}$$

where $i(t)$ is the output current from the inverter power module and is upstream of the LC filter. Substituting (11.2) with (11.3) the expression of $i(t)$ can be derived:

$$i(t) = i_2 \frac{v_2}{V_d} = 2 \cdot k \cdot I_2 \cdot \sin(\omega_1 t) \cdot \sin(\omega_1 t - \delta)$$

From the prosthaphaeresis formulas, the following final expression is derived:

$$i(t) = I_2 \cdot k \cdot [\cos\delta - \cos(2\omega_1 t - \delta)] = I_d + i_{2f} \tag{11.4}$$

The trend of the electrical magnitudes in the converter is thus shown in Figure 11.7.

From (11.4), it can be seen that the current output from the converter is formed by a DC component $I_d = I_2 \cdot k \cdot \cos\delta$, to which the desired continuous power $P_d = V_{DC} \cdot I_d$ transmitted to the DC link is associated, and a sinusoidal component $i_{2f} = I_2 \cdot k \cdot \cos(2\omega_1 t - \delta)$ that causes an oscillation of zero mean value power at double frequency, compared with that of the power supply and to which no useful effect is associated.

The i_{2f} component is typical of single-phase converters and is caused by the oscillation of the instantaneous power on the AC side. Conversely, in balanced three phase it does not appear that the three-phase instantaneous power is constant.

If only the capacitor C is present in the DC link, the situation shown in Figure 11.8 would occur, where the constant current I_d is absorbed by the traction drives and auxiliary services, while the components i_{2f} would affect only the smoothing capacitor. It follows that the voltage on the DC link will have the expression:

$$V_d = V_0 + \frac{1}{C}\int i_{2f}dt = V_0 + v_r$$

Namely, it is formed by a DC component V_0 equal to the desired value plus a ripple at twice the frequency of the network given by

$$v_r = \frac{1}{C}\int i_{2f}dt = \frac{I_2 \cdot k}{2\omega_1 C}\sin(2\omega_1 t - \delta)$$

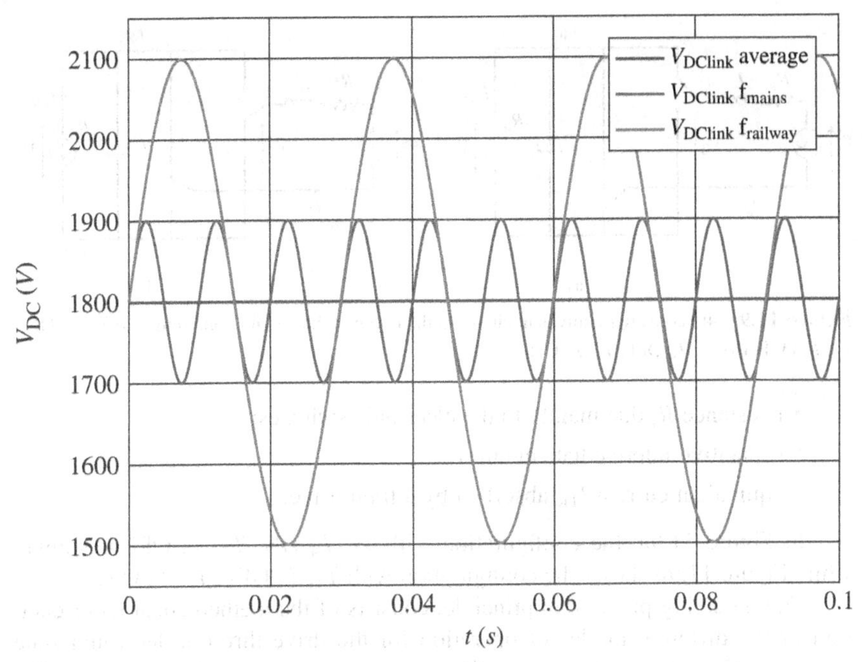

Figure 11.8 Behavior of the DC-link output from 4Q converter.

As can be seen, the ripple increases along with the power and thus the absorbed current i_2, and decreases with the increase of the leveling capacity and input frequency.

Since the ripple causes interference toward the other onboard converters, its value must be as low as possible.

In systems powered at the network frequency, the increase of the leveling capacitor's capacity may be sufficient to contain the ripple within acceptable limits (Figure 11.8). Conversely, in railway frequency systems, this solution would require overly large sizes and would not be compatible with the space onboard, for which the installation of the tuned LC filter is preferable, which behaves like a short circuit at the frequency $2f_1$ and therefore allows the share of i_{2f} that affects the DC link to be reduced.

11.2.1 Stability Analysis of the 4Q Converter

To analyze the converter's stability, the two different circuit configurations it can assume based on the state of the switches must be considered. Therefore, Figure 11.9 shows a representation of the two-level converter considered, in which the circuit is composed of

- ideal voltage generator ($e_2 = e_1/h$, where h is the transformation ratio);
- inductor L, equivalent to the leakage inductance of the transformer;

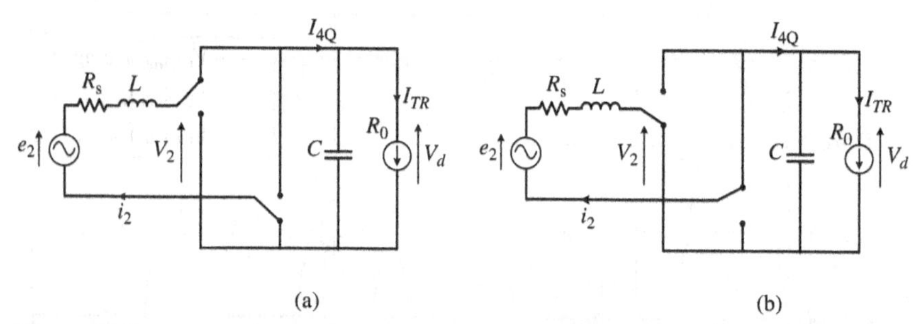

Figure 11.9 4Q converter equivalent circuit in the two possible circuit conditions. (a) T_1, T_4 ON, T_2, T_3 OFF. (b) T_1, T_4 OFF, T_2, T_3 ON.

- resistance R_s due mainly to the electronic switches;
- capacitive intermediate circuit C;
- equivalent current I_{TR} absorbed by a train drive.

In Figure 11.9a, the configuration with T_1, T_4 ON, T_2, T_3 OFF is shown, while Figure 11.9b shows the configuration with T_1, T_4 OFF, T_2, T_3 ON.

The modeling process, in principle, consists of the mathematical representation of the different modes of operation for the drive through dedicated state equations. Considering the magnitudes i_2 and v_d as state variables and with reference to Figure 11.9a, the following equations of state can be written:

$$L\frac{di_2}{dt} = -R_s i_2 - v_d + E_2 \sin(\omega t)$$

$$C\frac{dv_d}{dt} = i_2 - I_{\text{TR}}$$

where ω is the pulse associated with the power supply frequency. Rendering the equations in matrix form, we obtain

$$\begin{bmatrix} L & 0 \\ 0 & C \end{bmatrix} \frac{d}{dt} \begin{bmatrix} i_2 \\ v_d \end{bmatrix} = \begin{bmatrix} -R_s & -1 \\ 1 & 0 \end{bmatrix} \begin{bmatrix} i_2 \\ v_d \end{bmatrix} + \begin{bmatrix} E_2 \sin(\omega t) \\ -I_{\text{TR}} \end{bmatrix}$$

also rewritable as $Z \cdot \frac{dx}{dt} = A \cdot x + F$, where x is the state vector, A is the matrix of coefficients of the state variables that represents the specific mode of operation, and F is the vector of driving force. Indeed, another possible mode of operation is envisaged, since the analyzed converter operates at two levels. Considering Figure 11.9b, the matrix A will therefore take on a different form given by

$$A = \begin{bmatrix} -R_s & 1 \\ -1 & 0 \end{bmatrix}$$

Observing matrices A obtained for the two modes of operation, it can be noted that the elements on the main diagonal do not vary with changing operation. Consequently, the state equation in the generic conditions can be rewritten as

$$\begin{pmatrix} L & 0 \\ 0 & C \end{pmatrix} \frac{dx}{dt} = \begin{bmatrix} -R_s & -\xi \\ \xi & 0 \end{bmatrix} x + \begin{bmatrix} E_2 \sin(\omega t) \\ -I_{TR} \end{bmatrix} \tag{11.5}$$

where ξ depends on the converter's mode of operation and can take the value 1 or -1 ($\xi = \pm 1$). Because there is numerous switching within a period of the network frequency and with PWM modulation, ξ for the purpose of the present study can be represented using a continuous time-varying function given by

$$\xi = m \cdot \sin(\omega t - \delta)$$

in which m is the modulation index and δ is the load angle. The time variance ξ certainly makes the solution of Equation 11.5 more complicated. Changing the reference system into a rotating one is possible for a symmetrical three-phase system, but is not useful for a single-phase system where the power supply amplitude oscillates. In order to then delete the time variance from the equations, instead of transforming the reference system, a substitute method has been developed.

First, the equations are rewritten by introducing ξ:

$$\begin{cases} L\dfrac{di_2}{dt} = -R_s i_2 - \xi v_d + E_2 \sin(\omega t) \\ C\dfrac{dv_d}{dt} = \xi i_2 - I_{TR} \end{cases}$$

from which, substituting the expression of ξ:

$$\begin{cases} L\dfrac{di_2}{dt} + R_s i_2 = -m \sin(\omega t - \delta) v_d + E_2 \sin(\omega t) \\ C\dfrac{dv_d}{dt} = m \sin(\omega t - \delta) i_2 - I_{TR} \end{cases} \tag{11.6}$$

Considering i_2 since it is a driving force, and the coefficients of the state variables are sinusoidal of angular frequency ω, and a driving force is constant, its solution will have the following general form:

$$i_2 = i \sin(\omega t + \varphi) + I_{20}$$

As previously stated, one of the main tasks of the converter is to operate at unity power factor ($\varphi = 0$) while eliminating the DC component I_{20} that would be detrimental to the transformer, the i_2 will assume the expression $i_2 = i \cdot \sin(\omega t)$. Consequently, it will result in

$$\frac{di_2}{dt} = \frac{di}{dt}\sin(\omega t) + i(\omega\cos(\omega t)) \tag{11.7}$$

Substituting (11.7) in system of Equations 11.7, the following is obtained:

$$\begin{cases} L\left[\dfrac{di}{dt}\sin(\omega t) + i\omega\cos(\omega t)\right] + R_s i\sin(\omega t) = -m\sin(\omega t - \delta)v_d + E_2\sin(\omega t) \\ C\dfrac{dv_d}{dt} = mi\sin(\omega t)\sin(\omega t - \delta) - I_{\mathrm{TR}} \end{cases} \tag{11.8}$$

knowing that $\sin(\omega t - \delta) = \sin(\omega t)\cos\delta - \sin\delta\cos(\omega t)$, the first equation of the system (11.8) can be rewritten as follows:

$$\begin{aligned} L\left[\frac{di}{dt}\sin(\omega t) + i\omega\cos(\omega t)\right] + R_s i\sin(\omega t) &= -m\sin(\omega t)\cos(\delta)v_d \\ &\quad + m\sin\delta\cos(\omega t)v_d + E_2\sin(\omega t) \end{aligned}$$

$$\begin{aligned} \sin(\omega t)\left[L\frac{di}{dt} + R_s i\right] + Li\omega\cos(\omega t) &= \sin(\omega t)[E_2 - mv_d\cos(\delta)] \\ &\quad + \cos(\omega t)[mv_d\sin(\delta)] \end{aligned}$$

Since the solution has been inserted in the equation and therefore an identity for each value of t is obtained, the coefficients $\sin(\omega t)$ and $\cos(\omega t)$ must be equal on both sides, respectively. As a result, two equations will be obtained, which must simultaneously have a value of

$$\begin{cases} L\dfrac{di}{dt} + R_s i = E_2 - mv_d\cos\delta \\ Li\omega = mv_d\sin(\delta) \end{cases} \tag{11.9}$$

Unlike the three-phase converters, the single-phase converter has a DC voltage with a high ripple at twice the frequency of that of the network due to the oscillation of the power absorbed. Said ripple v_r is separated from the

DC component v_0 attributing the following expression to the solution of the second state variable:

$$v_d = v_0 + v_r \cos(2\omega t + \vartheta)$$

in which one can therefore note the presence of a unidirectional component (v_0) and a harmonic component 2ω (v_r). Deriving the same with respect to time yields

$$\frac{dv_d}{dt} = \frac{dv_0}{dt} + \frac{dv_r}{dt}\cos(2\omega t - \vartheta) - v_r\left(\sin(\omega t - \vartheta)\left(2\omega - \frac{d\vartheta}{dt}\right)\right)$$

Considering now the second equation of the system (11.8) and knowing that $\sin(\alpha)\sin(\beta) = \frac{1}{2}[\cos(\alpha - \beta) - \cos(\alpha + \beta)]$, the following is obtained:

$$C\frac{dv_d}{dt} = \frac{1}{2}m \cdot i \cdot \cos(-\delta) - \frac{1}{2}m \cdot i \cdot \cos(2\omega t)\cos\delta - \frac{1}{2}m \cdot i \cdot \sin(2\omega t)\sin\delta - I_{TR}$$

Introducing the value of dv_d/dt leads to the following expression:

$$\begin{aligned} &C\frac{dv_0}{dt} + C\frac{dv_r}{dt}\cos(2\omega t - \vartheta) - Cv_r\left(\sin(2\omega t - \vartheta)\left(2\omega - \frac{d\vartheta}{dt}\right)\right) \\ &= \frac{1}{2}mi\cos(-\delta) - \frac{1}{2}mi\cos(2\omega t)\cos\delta - \frac{1}{2}mi\sin(2\omega t)\sin\delta - I_{TR} \end{aligned} \tag{11.10}$$

Expanding the terms of sine and cosine once again, and exactly:

$$\begin{aligned} \sin(2\omega t - \vartheta) &= \sin(2\omega t)\cos\vartheta - \cos(2\omega t)\sin\vartheta \\ \cos(2\omega t - \vartheta) &= \cos(2\omega t)\cos\vartheta + \sin(2\omega t)\sin\vartheta \end{aligned}$$

(11.10) becomes

$$\begin{aligned} &C\frac{dv_0}{dt} + C\frac{dv_r}{dt}\cos(2\omega t)\cos\vartheta + C\frac{dv_r}{dt}\sin(2\omega t)\sin\vartheta - Cv_r\sin(2\omega t)\cos\vartheta\left(2\omega - \frac{d\vartheta}{dt}\right) \\ &+Cv_r\cos(2\omega t)\sin\vartheta\left(2\omega - \frac{d\vartheta}{dt}\right) = \\ &\frac{1}{2}mi\cos(-\delta) - \frac{1}{2}mi\cos(2\omega t)\cos\delta - \frac{1}{2}mi\sin(2\omega t)\sin\delta - I_{TR} \end{aligned}$$

$$\cos(2\omega t)\left[C\frac{dv_r}{dt}\cos\vartheta + Cv_r\sin\vartheta\left(2\omega - \frac{d\vartheta}{dt}\right)\right]$$
$$+\sin(2\omega t)\left[C\frac{dv_r}{dt}\sin\vartheta + Cv_r\cos\vartheta\left(2\omega - \frac{d\vartheta}{dt}\right)\right] + C\frac{dv_0}{dt} =$$
$$\cos(2\omega t)\left[-\frac{1}{2}mi\cos(\delta)\right] + \sin(2\omega t)\left[-\frac{1}{2}mi\sin\delta\right] + \frac{1}{2}m\cos(-\delta) - I_{TR}$$

Since the solution has been inserted into the equation and the equation must hold for all values of t, the terms in sine, cosine, and those known must be, respectively, equal on both sides. As a result, we obtain the following system of three equations that must apply simultaneously:

$$\begin{cases} C\dfrac{dv_r}{dt}\cos\vartheta + Cv_r\sin\vartheta\left(2\omega t - \dfrac{d\vartheta}{dt}\right) = -\dfrac{1}{2}mi\cos\delta \\ C\dfrac{dv_r}{dt}\sin\vartheta - Cv_r\cos\vartheta\left(2\omega t - \dfrac{d\vartheta}{dt}\right) = -\dfrac{1}{2}mi\sin\delta \\ C\dfrac{dv_0}{dt} = \dfrac{1}{2}mi\cos(-\delta) - I_{TR} \end{cases} \tag{11.11}$$

From the analysis carried out so far, we have reached five equations simultaneously given by (11.10) and (11.11) in the five unknowns v_r, v_0, ϑ, m, and δ, based on the known terms i, E_2, and I_{TR}.

These equations are time invariant, but not linear. Because they can be useful for the calibration of the control system and the study of stability, they must be linearized with respect to the steady-state point.

Before proceeding to the linearization, it is possible to simplify the system by eliminating the two variables v_r and ϑ. Considering, in fact, the application onboard a train, in practice for real converters, the ripple component v_r is rather limited, also thanks to the presence of an LC filter between the two direct current terminals, the resonant frequency $f_2 = 2{\cdot}f_1$ double that of the network, and the high capacity and voltage of the DC link. Furthermore, the dynamic of the regulator is slow enough and such as to not be able to follow such quick changes; consequently, the voltage signal of the DC link at regulator input must be filtered anyway. These considerations, therefore, justify being able to neglect the effects of the variables v_r and ϑ from the stability study that follows, thus reaching three equations in three unknowns (v_0, m, δ):

$$\begin{aligned} &L\frac{di}{dt} + R_s i = E_2 - mv_0\cos\delta \\ &Li\omega = mv_0\sin(\delta) \\ &C\frac{dv_0}{dt} = \frac{1}{2}mi\cos(-\delta) - I_{TR} \end{aligned} \tag{11.12}$$

In order to linearize the system, each variable is replaced with the sum of the magnitude at the point of steady state plus the respective deviation:

$$\begin{aligned}
&\text{modulation index} : m = M + m' \\
&\text{DC-link voltage} : v_0 = V_0 + v_0' \\
&\text{load angle} : \delta = \Delta + \delta' \\
&\text{current consumption} : i = I + i'
\end{aligned}$$

where M, V_0, Δ, and I represent the respective magnitudes at the working point, while m', v_0', δ', and i' are the respective deviations. Instead, the value of E_2 is imposed by the power supply originating from the contact line, while I_{TR} is imposed by the power demanded by the train. Replacing the magnitudes thus defined in (11.12), the following is obtained:

$$\begin{aligned}
&L\frac{di'}{dt} + R_s(I + i') = E_2 - (M + m')\left(V_0 + v_0'\right)\cos(\Delta + \delta') \\
&L(I + i')\omega = (M + m')\left(V_0 + v_0'\right)\sin(\Delta + \delta') \\
&C\frac{dv_0'}{dt} = \frac{1}{2}(M + m')(I + i')\cos(-(\Delta + \delta')) - I_{TR}
\end{aligned} \tag{11.13}$$

To continue linearization, the following valid approximation is applied for small angles x, for which we have $\sin x \approx x$ and and $\cos x \approx 1$.

Consequently, for the derivation of x compared with the stationary value X it follows that $\sin(X + x) \approx \sin X + x\cos X$ and $\cos(X + x) \approx \cos X - x\sin X$; therefore, substituting in (11.13), we reach

$$\begin{aligned}
&L\frac{di'}{dt} + R_s(I + i') = E_2 - (M + m')\left(V_0 + v_0'\right)\cos\Delta + (M + m')\left(V_0 + v_0'\right)\delta' \sin\Delta \\
&L(I + i')\omega = (M + m')\left(V_0 + v_0'\right)\sin\Delta + (M + m')\left(V_0 + v_0'\right)\delta' \cos\Delta \\
&C\frac{dv_0'}{dt} = \frac{1}{2}(M + m')(I + i')\cos(-\Delta) - \frac{1}{2}(M + m')(I + i')\delta' \sin\Delta - I_{TR}
\end{aligned}$$

Considering only the constant terms, the model in steady state we started from is obtained:

$$\begin{aligned}
&R_s I = E_2 - M \cdot V_0 \cos\Delta \\
&LI\omega = M \cdot V_0 \sin\Delta \\
&\frac{1}{2}M \cdot I\cos(-\Delta) - I_{TR} = 0
\end{aligned}$$

Considering, instead, only terms of first order, we obtain the small-signal model, useful in order to find the frequency response of the converter:

$$
\begin{aligned}
&L\frac{di'}{dt} + R_s i' = -Mv_0' \cos\Delta - m' V_0 \cos\Delta \\
&Li'\omega = Mv_0' \sin\Delta + m' V_0 \sin\Delta \\
&C\frac{dv_0'}{dt} = \frac{1}{2}Mi' \cos(-\Delta) + \frac{1}{2}m' I \cos(-\Delta)
\end{aligned}
\tag{11.14}
$$

The small signal model is developed using the Laplace operator. Thereby applying the Laplace transform to (11.14) and considering the initial conditions null, we get

$$
\begin{aligned}
&(Ls + R_s)i' = -Mv_0' \cos\Delta - m' V_0 \cos\Delta \\
&Li'\omega = Mv_0' \sin\Delta + m' V_0 \sin\Delta \\
&(C)sv_0' = \frac{1}{2}Mi' \cos(-\Delta) + \frac{1}{2}m' I \cos(-\Delta)
\end{aligned}
\tag{11.15}
$$

As will be explained below, the logical choice for controlling the 4Q converter follows current tracking logic, for which the modulation index m does not appear directly in the control algorithm. From the second equation of the system (11.15), the value of m' can be derived:

$$
m' = -\frac{(sL + R_s)i' + Mv_0' \cos\Delta}{V_0 \cos\Delta} = -\frac{sL + R_s}{V_0 \cos\Delta} i' - \frac{M}{V_0} v_0'
$$

which when replaced in the third equation of (11.16) yields

$$
\begin{aligned}
&sCv_0' = \frac{1}{2}Mi' \cos(-\Delta) - \frac{1}{2}I\cos(-\Delta)\frac{sL + R_s}{V_0 \cos\Delta} i' - \frac{1}{2}I\cos(-\Delta)\frac{M}{V_0} v_0' \\
&\left(sC + \frac{1}{2}I\cos(\Delta)\frac{M}{V_0}\right)v_0' = \left(\frac{1}{2}M\cos(\Delta) - \frac{1}{2}I\frac{sL + R_s}{V_0}\right)i' \\
&v_0' = \frac{\left(\frac{1}{2}M\cos(\Delta) - \frac{1}{2}I\frac{sL + R_s}{V_0}\right)}{\left(sC + \frac{1}{2}I\cos(\Delta)\frac{M}{V_0}\right)} i'
\end{aligned}
\tag{11.16}
$$

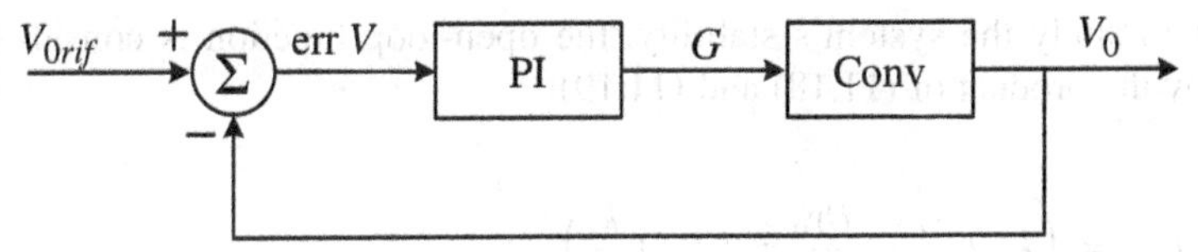

Figure 11.10 Block diagram of the regulation system.

The Equation 11.16 provides the value v_0' as a function of i'. At this point, it is possible to find the variations in the DC-link voltage for small changes of the current absorbed by the converter:

$$\frac{\partial v_0'}{\partial i'} = \frac{\left(\frac{1}{2}M\cos(\Delta) - \frac{1}{2}I\frac{sL+R_s}{V_0}\right)}{\left(sC + \frac{1}{2}I\cos(\Delta)\frac{M}{V_0}\right)} = \frac{-\left(\frac{1}{2}L\frac{I}{V_0}\right)s + \frac{1}{2}M\cos(\Delta) - \frac{1}{2}R_s\frac{I}{V_0}}{(C)s + \frac{1}{2}I\cos(\Delta)\frac{M}{V_0}} \tag{11.17}$$

The current is given by the product between the equivalent conductance G provided by the PI regulator which maintains the DC link and the power supply voltage E_2 constant, where

$$G = \left(k_p + \frac{k_i}{s}\right)(V_{0rif} - V_0) \tag{11.18}$$

Figure 11.10 shows the block diagram of the system considered.

The variations in the voltage of the DC link for small changes of the current $\frac{\partial v_0'}{\partial i'}$ can thus be rewritten as

$$\frac{\partial v_0'}{\partial i'} = \frac{\partial v_0'}{\partial G} \cdot \frac{1}{E_2}$$

Since for the study in question we are interested in evaluating the variation of the DC-link voltage for small changes in the equivalent conductance G provided by the regulator, consider

$$\frac{\partial v_0'}{\partial G} = \frac{\partial v_0'}{\partial i'} \cdot E_2 \tag{11.19}$$

multiplying (11.17) for the input voltage E_2.

In order to study the system's stability, the open-loop function is considered, given by the product of (11.18) and (11.19):

$$TF_{\text{OL}} = \left(k_p + \frac{k_i}{s}\right) \cdot \frac{\partial v_0'}{\partial G} = \left(k_p + \frac{k_i}{s}\right) \cdot \frac{-\left(\frac{1}{2}LI\frac{E_2}{V_0}\right)s + \frac{1}{2}ME_2\cos(\Delta) - \frac{1}{2}R_sI\frac{E_2}{V_0}}{(C)s + \frac{1}{2}I\cos(\Delta)\frac{M}{V_0}} \tag{11.20}$$

The stability has been studied with the classical methods of automatic controls through the use of the Bode diagrams, implemented in a mathematical software environment. Each value considered is one of the following:

- $V_L = X_S I$
- $V_2 = MV_0$
- $E_{2\text{MAX}} = 1000\text{ V}$
- $V_0 = 1800\text{ V}$
- $P_{\text{ass}} = 900\text{ kW}$
- $I_{\text{TR}} = 900/1.8 = 500\text{ A}$
- $I_{\text{MAX}} = 2P_{\text{ass}}/V_{S\text{MAX}} = 1800\text{ A}$
- $X_L = 0.471\ \Omega$
- $V_{L\text{MAX}} = 847.8\text{ V}$
- $V_{2\text{MAX}} = \sqrt{V_{L\text{MAX}}^2 + E_{2\text{MAX}}^2} = 1311\text{ V}$
- $M = V_{2\text{MAX}}/V_0 = 0.7283$
- $\text{tg}\delta = 0.8478$
- $\delta = 0.703\text{ rad}$

Considering the proportional regulator parameter k_p and the integral k_i, a good compromise is such as to obtain a prompt response of the system and thus have a high k_p which at the same time should ensure a reasonable phase margin. In this case the solution found to achieve the above objectives considers the parameters $k_p = 0.1$ and $k_i = 3$, values confirmed by the simulation conducted on the entire system.

Figure 11.11 shows the Bode diagram obtained from (11.20) considering said values.

Figure 11.12 shows the Bode diagram obtained from (11.20) considering $k_p = 0.2$ and $k_i = 3$, values obtained for the condition at the limit of stability.

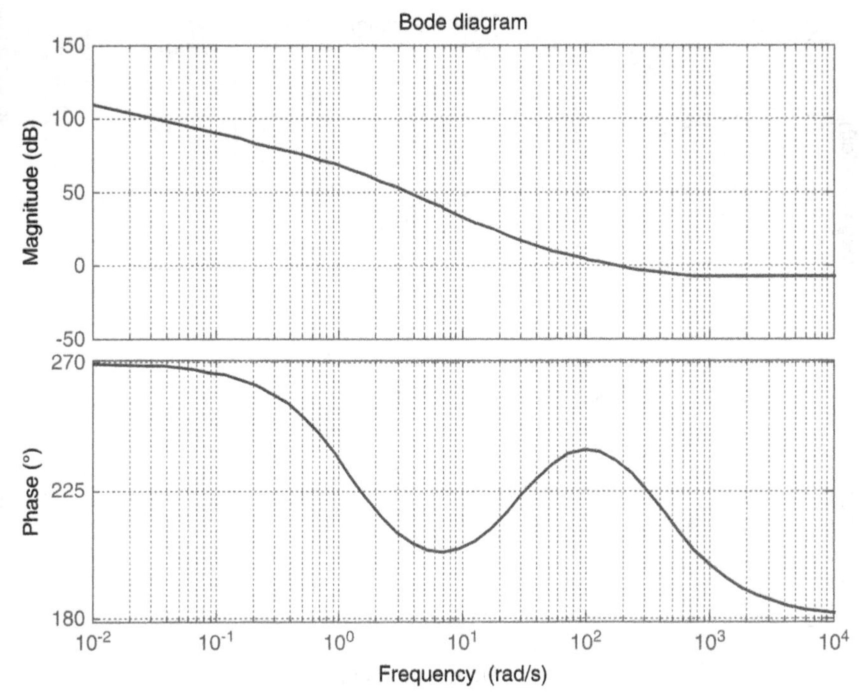

Figure 11.11 Bode diagram of the converter plus regulator for $k_p = 0.1$ and $k_i = 3$.

11.2.2 Interleaving of Multiple 4Q Converters

The parallel operation of several power converters allows the phase of the switching instants to be varied so as to reduce the amplitude of the input current ripple and at the same time increase its frequency in order to decrease the harmonic content and improve its filterability. This practice, known as interleaving, is only possible if the converters are controlled with constant frequency switching techniques in order to synchronize the control pulses of the converters of the various switches between them.

The interleaving is particularly effective when there are several power converters with low switching frequencies, and thus by currents absorbed with high harmonic content which are hardly filterable. In the railway field this has been applied in several series of DC locomotives regulated by multiple interleaved DC/DC converters.

As was presented in the previous paragraphs, modern rolling stocks, such as interoperable locomotives or EMU for high-speed AC lines, have two or four four-quadrant converters as an input stage that can then be interleaved together. However, to obtain effective interleaving, the phase shift of each converter switching frequency must be, respectively, $^1/_2$ or $^1/_4$ of the switching period.

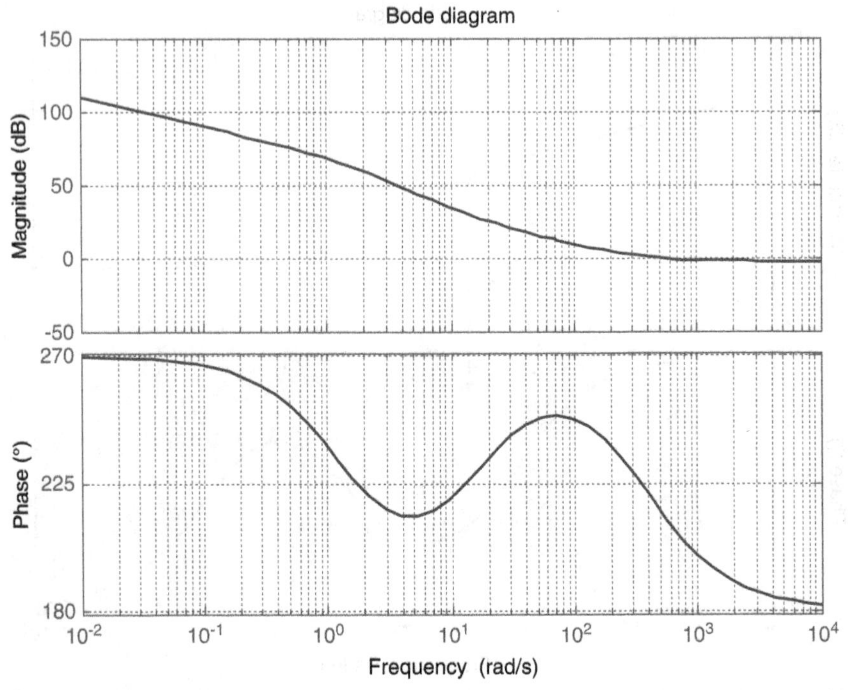

Figure 11.12 Bode diagram of converter plus controller for $k_p = 0.2$ and $k_i = 3$.

In this manner, the frequency of the ripple current absorbed by the transformer due to the set of 4Q converters appears to be double or quadruple that of switching, with the advantage of having a lower and more easily filterable amplitude.

The operation of interleaving leads to a phase shift between the waves of the converters' output current: A phase shift that, among all those possible, must be the one that minimizes the harmonic pollution.

We then consider two boost converters whose absorbed current waves have a phase shift of a generic angle φ, as shown in Figure 11.13.

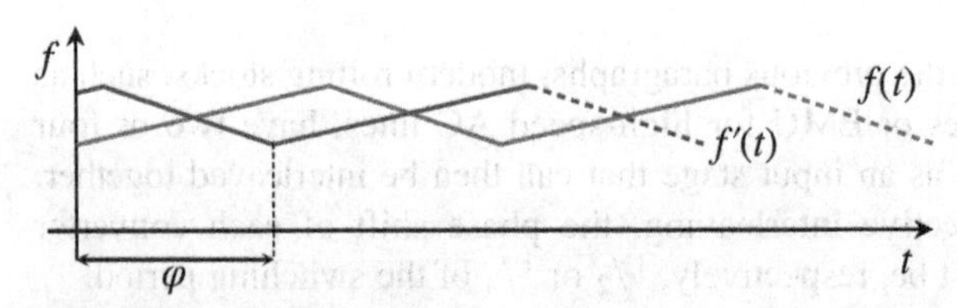

Figure 11.13 Phase shift of two interleaved boost converters' current waves.

The two waves are identical to each other and only have a phase shift of an angle φ, from which it follows that the development in Fourier series of the two functions f and f' is the following:

$$f(t) = A_0 + \sum_{n=1}^{\infty}[A_n \cdot \cos(n\omega t) + B_n \cdot \text{sen}(n\omega t)]$$

$$f'(t) = A_0 + \sum_{n=1}^{\infty}[A_n \cdot \cos(n\omega t + n\varphi) + B_n \cdot \text{sen}(n\omega t + n\varphi)]$$

Applying the rules of trigonometry, for which

$$A_n \cdot \cos(n\omega t + n\varphi) = A_n \cdot [\cos(n\omega t) \cdot \cos(n\varphi) - \text{sen}(n\omega t) \cdot \text{sen}(n\varphi)]$$
$$B_n \cdot \text{sen}(n\omega t + n\varphi) = B_n \cdot [\text{sen}(n\omega t) \cdot \cos(n\varphi) + \cos(n\omega t) \cdot \text{sen}(n\varphi)]$$

f' can be rewritten as

$$f'(t) = A_0 + \sum_{n=1}^{\infty} \begin{bmatrix} A_n[\cos(n\omega t) \cdot \cos(n\varphi) - \text{sen}(n\omega t) \cdot \text{sen}(n\varphi)] + \\ +B_n[\text{sen}(n\omega t) \cdot \cos(n\varphi) + \cos(n\omega t) \cdot \text{sen}(n\varphi)] \end{bmatrix}$$

Amalgamating the terms $\cos(n\omega t)$ and $\sin(n\omega t)$ yields

$$f'(t) = A_0 + \sum_{1}^{\infty} [A'_n \cdot \cos(n\omega t) + B'_n \cdot \text{sen}(n\omega t)]$$

where

$$A'_n = A_n \cdot \cos(n\varphi) + B_n \cdot \text{sen}(n\varphi)$$
$$B'_n = B_n \cdot \cos(n\varphi) - A_n \cdot \text{sen}(n\varphi)$$

Since the current absorbed by the train is the sum of those absorbed by the two converters, the resulting function to be considered for the calculation of the THD will be given by the sum of the two individual functions; therefore,

$$f_t(t) = f(t) + f'(t) = A'_0 + \sum_{1}^{\infty} [A''_n \cdot \cos(n\omega t) + B''_n \cdot \text{sen}(n\omega t)]$$

with $A''_n = A_n + A'_n$ and $B''_n = B_n + B'_n$.

In this case, the absolute THD is defined expressed in the following manner:

$$\text{THD} = \sqrt{\sum_{1}^{\infty} Z_n^2} \tag{11.21}$$

with $Z_n^2 = (A''_n)^2 + (B''_n)^2$.

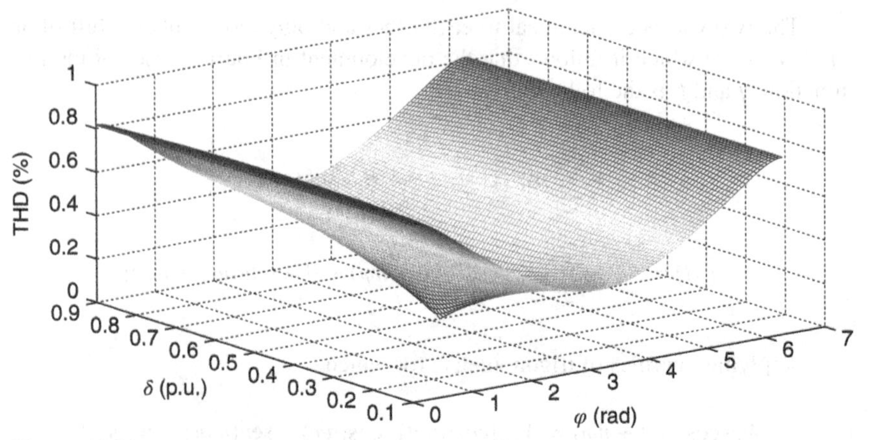

Figure 11.14 Performance of THD to vary the duty cycle δ and the phase shift φ between the currents drawn by two boost converters.

As can be seen, the harmonic residue also contains the first harmonic, referring to the switching frequency, because the entire ripple is an undesired component.

The THD thus calculated is function of the phase shift φ between the two waves and the duty cycle δ, in this case assumed equal for both converters, and its performance is shown in Figure 11.14.

Several considerations can be drawn from Figure 11.14. First of all, the THD is null for $\delta = 0.5$ and $\varphi = \pi$, since it is the only case in which the two triangular waves are symmetric and in antiphase, thus canceling each other out.

From the optimum point, the THD initially tends to grow rapidly when varying both δ and φ, highlighting the fact that even small asymmetries in the switching of the two converters tend to decrease the interleaving advantages. The optimum phase shift φ is always equal to π, and in this case the THD remains substantially lower than in the case of converters with no interleaving ($\varphi = 0$).

Finally, it is worth pointing out that the THD is directly proportional to the amplitude of the ripple with respect to the DC component, for which the actual values that occur in practice are higher than those reported in Figure 11.14 due to the increased current ripple caused by low switching frequency.

In the case of real 4Q converters onboard trains, the low switching frequency makes it so that varying the phase of the switching instants of the two converters, the duty cycle is no longer the same for both, but different from one drive to another while remaining within the same period of switching, as is clearly visible in Figure 11.15.

To then analyze the effect of this variation, the coefficients (A'_n, B'_n) and (A''_n, B''_n) were calculated for different values of δ' and δ'' in the case of optimum phase shift $\varphi = \pi$; the result of said analysis is shown in Figure 11.15.

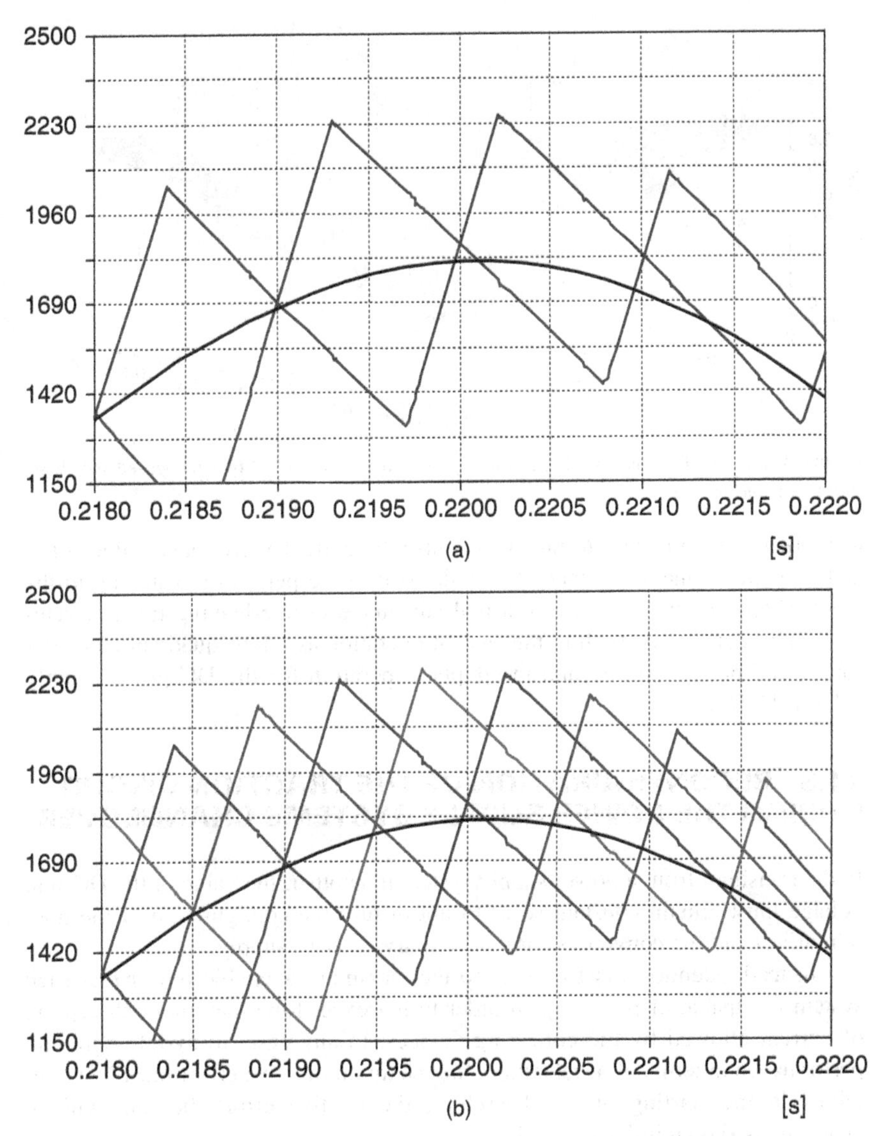

Figure 11.15 The ripple current results in the case of (a) two and (b) four 4Q converters interleaved during the pursuit of a reference sinusoidal current.

Figure 11.15 shows that the most favorable condition is located on the main diagonal, that is, when δ' and δ'' are equal, while the most unfavorable condition is when δ' and δ'' are complementary, or on the secondary diagonal.

The most significant increase of THD is thus obtained for very different values between δ' and δ'', while it stays low for similar values. The latter is the case of 4Q converters which, while varying the duty cycle within a period of the

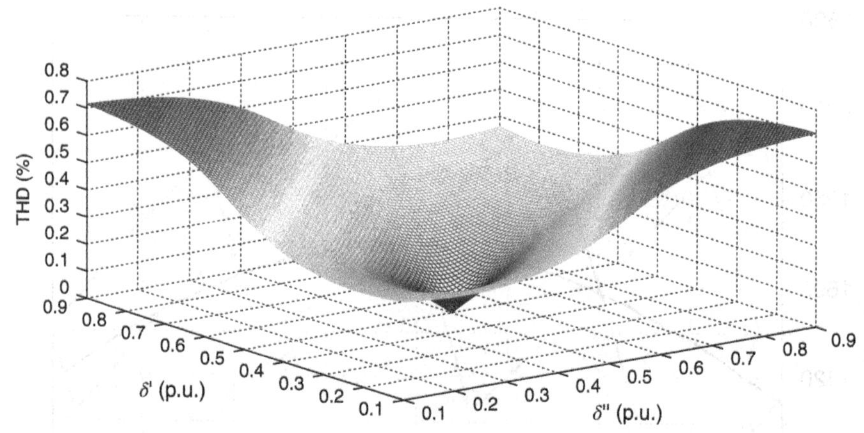

Figure 11.16 Performance of THD at varying duty cycles δ' and δ'' of two interleaved and phase shift boost converters π.

network voltage in order to pursue a sinusoidal current wave, have values of δ' and δ'' which remain substantially similar between a period of switching to the other. From this we can deduce a real advantage in interleaving traction vehicles' input converters, both in the case of operation as a four-quadrant converter for an AC power supply, and in chopper operation for the DC power supply (Figure 11.16).

11.3 RECONFIGURATION OF THE TRACTION CIRCUIT DURING THE POWER SUPPLY SYSTEMS CHANGEOVER

In the transition from a power supply system to another, the value of the DC-link voltage must remain constant so as to not disturb the configuration of the traction drives and the converters for the auxiliary services supply.

Instead, attention must be paid to the maximum available power that each system is capable of providing in order to not exceed the maximum absorption of current allowed by the collecting devices. In any case, to avoid degrading performance in terms of acceleration, the maximum tractive effort must be maintained in the starting phase. Therefore, the tractive effort diagram will be changed, as shown in Figure 11.17.

The input stage must therefore adapt the power supply voltage, DC or AC at different voltages and frequencies, to the value of the voltage of the desired DC link.

11.3.1 Example of Transition between 25 kV AC and 3 kV DC

The first difference is in the pantographs that must be dedicated to each power supply system, even if in emergency conditions, which for higher voltages may

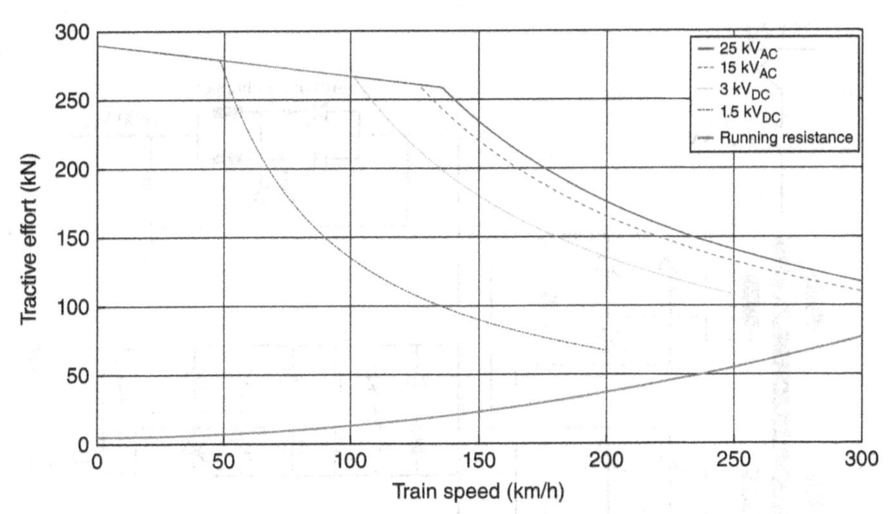

Figure 11.17 Tractive effort diagram for a multisystem train.

be used in lower voltage systems at reduced power. A study of these aspects reveals that these are universal pantographs suitable for operation at high voltages and at high currents.

In general, there will therefore be more pantographs installed onboard.

In any case, there is differentiation between the two different AC and DC systems downstream of the pantograph, which must employ different protection devices.

Through a system of switches a voltage sensor configures the correct circuit and enables the automatic closing of either the 25 kV circuit breaker or the 3 kV high-speed circuit breaker based on the power supply system present, thus avoiding the possibility of incorrect maneuvers. The drive is automatically configured for operation with 3 or 25 kV system as further explained. After the lifting of one of the two pantographs, the system recognizes the catenary voltage by means of a suitable sensor, configures the input stage for DC or AC operation, and closes the correspondent circuit breaker.

At the closing of the line circuit breaker, regardless of the configuration assumed by the locomotive, the converters of the first stage automatically start, and later, those of auxiliary services and the charger. The configuration change is automatically carried out by a suitable combiner.

The input stage conversion system may function as a chopper with the DC power supply or as a 4Q converter with an AC power supply.

The scheme adopted allows, in many cases, the semiconductor of the input stage to be of identical constructions and interchangeable with those of the traction inverters.

The input stage of the conversion system will generate and maintain a constant DC-link voltage.

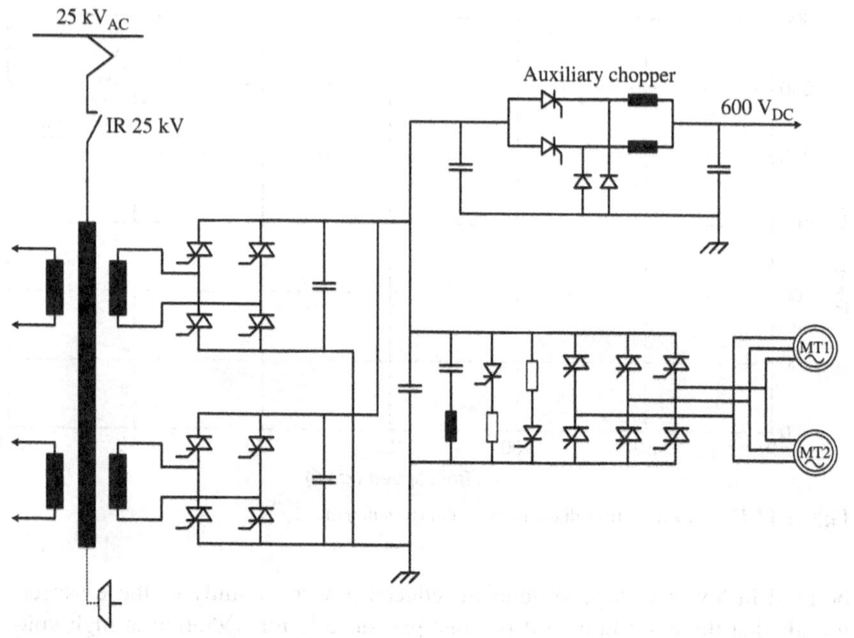

Figure 11.18 Drive with the configuration at 25 kV AC.

Operation in AC The configuration of the AC drive circuit is shown in Figure 11.18.

Each locomotive is provided with two banks with two 4Q converters, which power the DC-link in parallel.

The 4Q converters are driven in such a way that the power drawn from the contact line is virtually only active power, while maintaining the power factor very close to the unit, for a wide field of use. The 4Q converter requires that the intermediate DC circuit voltage is always greater than the instantaneous value of the secondary voltage of the transformer; for this reason the 4Q can be seen as a step-up, reversible inverter, whose modulation varies in time with a sinusoidal law to maintain the balance between power absorbed and power demand in order to maintain the DC voltage DC-link constant.

Each 4Q converter uses the leakage inductance of the winding, from which it is powered as a switching inductance. The interleaving of the converters and PWM logic prevent the installation of line filters.

The previously discussed tuned LC filter is present in the DC-link, generally shared between two 4Q converters. If the vehicle is adapted for operation with different frequency systems, such a filter will have to be retuned to double the frequency value of that of the power supply, suitably varying the parameters L and C.

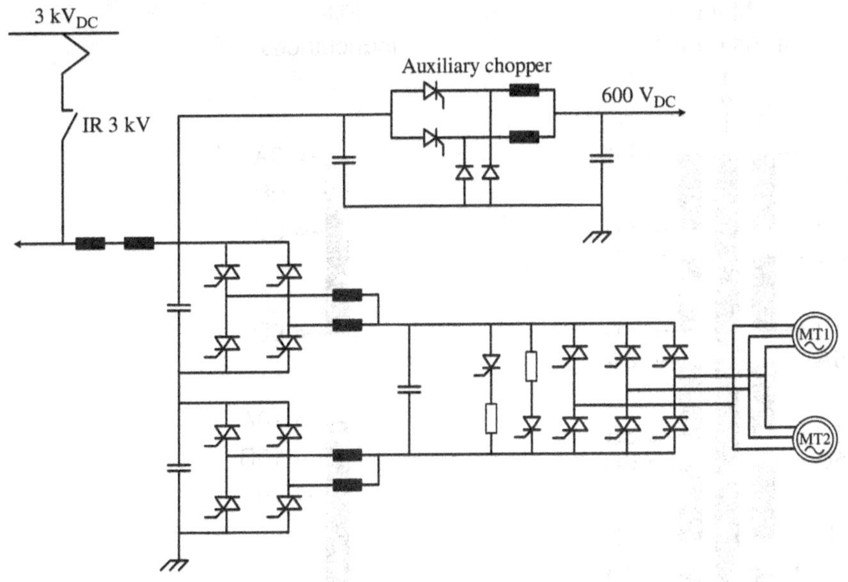

Figure 11.19 Operation of the drive at 3 kV DC.

Operation in DC The configuration of the DC drive circuit is shown in Figure 11.19.

In this configuration, the power semiconductor modules that were connected in the 4Q converter as H-bridges are now used as bidirectional DC–DC converters connected in parallel to each other on both the positive and negative of the DC link.

Short-circuiting the primary side of the transformer, the four power supply windings of the 4Q converters are now used as independent switching inductances of the step-down chopper.

The previously used inductance in the LC filter is now connected upstream as a line filter.

In this manner, all the electromechanical and electronic power components are used in both the AC and DC configuration without needing to double the components.

11.3.2 Example of a Transformer in Multisystem Vehicles

A multisystem locomotive, as mentioned, is equipped with four 4Q converters, each powered by its own winding. The transition from one voltage and frequency system (V'_1, f'_1) to a voltage and frequency (V''_1, f''_1) requires changing the number of secondary coils to maintain constant voltage V_2 at the input to the 4Q

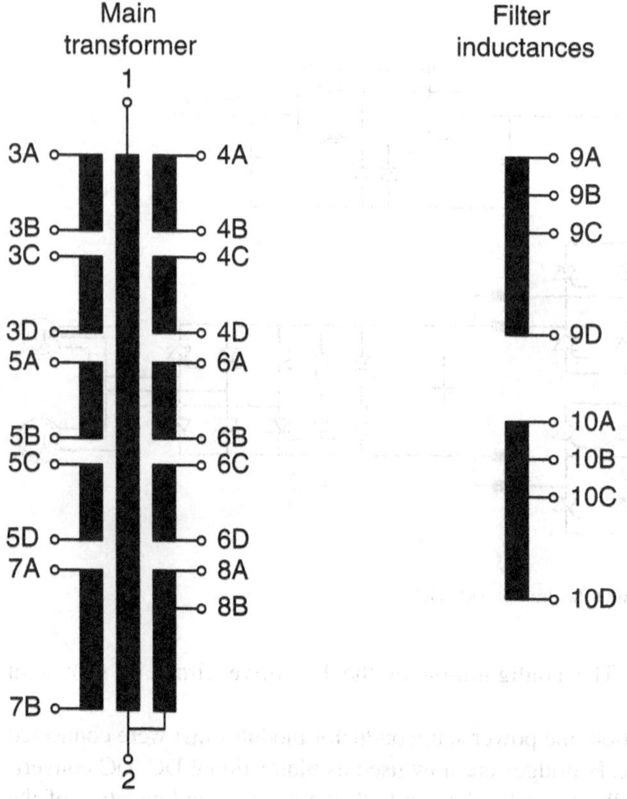

Figure 11.20 Typical configuration of the main transformer and inductors onboard a multisystem vehicle.

converters. It is important to respect this condition because the voltage V_2 is strictly related to that of the DC link for the correct operation of the PWM logic.

If the locomotive is adapted to tow passenger trains, an additional winding is also present for the power supply of the HV feeder along the train. In the car-body that contains the transformer, there are also two inductors: one for each semibench, divided into three sections that serve for the implementation of the LC filters in AC operation or of the line filter in DC operation.

These elements are shown in Figure 11.20.

Table 11.1 shows an example of the reconfiguration of the transformer and the inductors for three different power supply systems.

Table 11.1 Possible Configurations of the Main Transformer and the Inductors for Three Different Power Supply Systems

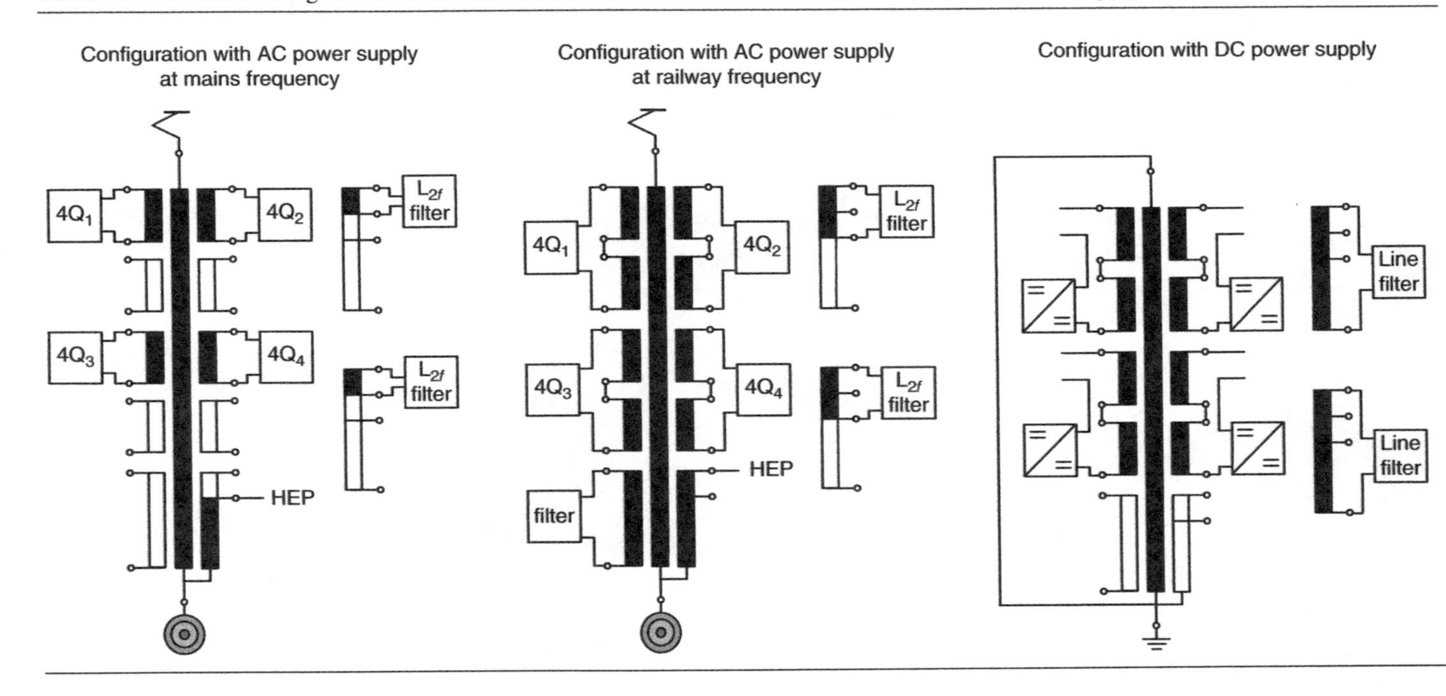

Chapter 12

Self-Propelled Vehicles

Even today most railway lines are still not electrified, especially those characterized by low levels of traffic. This raises the need to use self-propelled rolling stock, not powered through a contact line.

With current technologies, these vehicles use diesel engines, although the environmental aspects to which various administrations are becoming increasingly sensitive are such as to stimulate research toward other solutions based on bimodal vehicles or fuel cells.

12.1 DIESEL–ELECTRIC TRACTION

For vehicles powered by diesel engines, there is problem of achieving a tractive effort diagram similar to that that is obtainable with the traction systems examined so far – that is, with a constant starting tractive effort and constant power at higher speeds.

The diesel engine by itself does not accommodate these needs. The diagram $T(\Omega)$ is in fact rigid, as the torque varies little as rotation speed varies. Furthermore, the motor cannot be started under load, bears minimal overloads, and cannot reverse rotation direction.

A transmission is therefore necessary between the motor and wheels that is able to start the engine under load, reverse gears, and utilize diesel power with variable speed and tractive effort within wide limits.

Possible transmission systems are as follows:

- mechanical transmission, with clutch and gearbox
- hydromechanical transmission, with manual transmission and hydraulic couplings
- hydraulic transmission, with hydraulic couplings and torque converters
- electric transmission

Electrical Railway Transportation Systems, First Edition. Morris Brenna, Federica Foiadelli, and Dario Zaninelli.

Mechanical and hydromechanical transmission have been used in few, low-power applications, with preference being given to hydraulic and electric transmission for higher power. Nowadays, however, electric transmission is increasingly preferred due to both its greater efficiency and the possibility of constructing bimodal vehicles.

The torque converter is a hydraulic device that can increase or decrease the torque generated at the output shaft, connected to the wheels, with respect to the input shaft connected to the prime motor. It consists of a toroidal chamber, composed of a pump (impeller), a turbine, and a stator. The centrifugal pump connected to the crank shaft confers a certain pressure to the flow of fluid contained within it, while the turbine, joined to the part being driven, collects the fluid flow and produces a rotation, which it transmits to the wheels.

The stator or reactor, mounted on a freewheel that prevents it from rotating in the direction opposite to that of the pump, changes the direction of the flow of fluid leaving the turbine before returning to the impeller. This third element allows the converter, at peak, to have a torque output to the turbine that is greater than the input torque to the pump. This principle is represented in Figure 12.1,

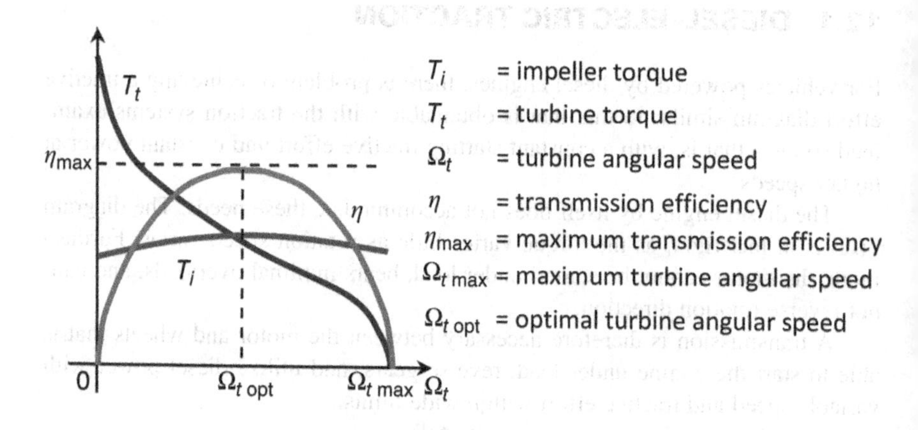

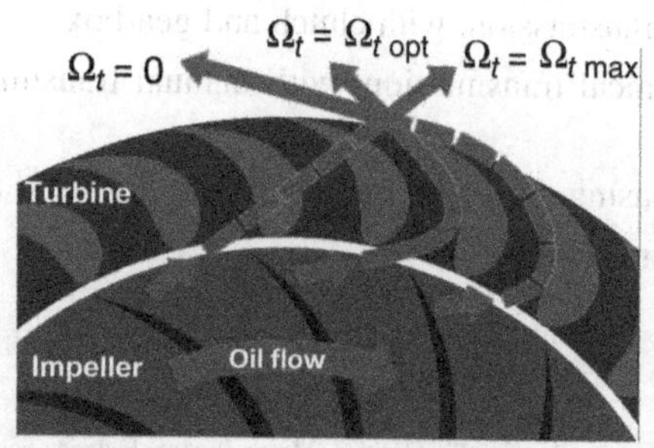

Figure 12.1 Operating characteristics of a torque converter.

wherein you can see how this device is able to notably increase peak torque with respect to prime motor torque.

However, the torque converter presents several flaws. First and foremost is the impossibility of gear reversal, which therefore requires the interposition of another mobile gear in the transmission. The second issue is the impossibility of towing the vehicle without first deactivating the torque converter. In fact, during tow, the turbine acts as a pump, increasing fluid pressure and temperature, which may damage the converter itself. Another aspect to consider is the braking of the vehicle, which must always be carefully evaluated in order to prevent the kinetic energy of the train from being converted into heat within the converter and therefore leading to its breakage. Finally, one of the aspects that is leading to its progressive abandonment on modern vehicles is its poor efficiency, especially with regard to rotation speeds much different from those of the motor.

12.1.1 Characteristics of the Diesel Engine

The characteristic curves in Figure 12.2 represent power P, torque T, and the specific fuel consumption C_S of the diesel engine according to rotation speed Ω, where Ω_0 is the minimum operating speed.

The torque T depends on the average effective pressure, and therefore on the amount of fuel injected into the cylinders. Torque can be adjusted using the injection pump control lever, which controls the fuel supply; in a certain lever

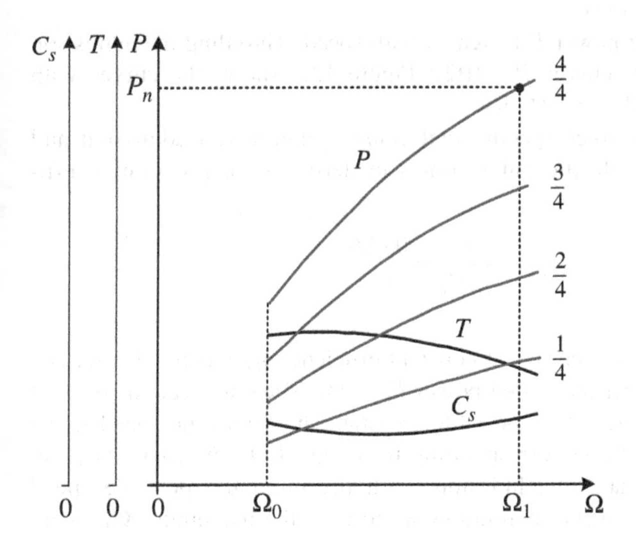

Figure 12.2 Diesel engine operating features. P: power, with different degrees of admission (4/4, 3/4, 2/4, 1/4); T: torque; C_S = specific fuel consumption.

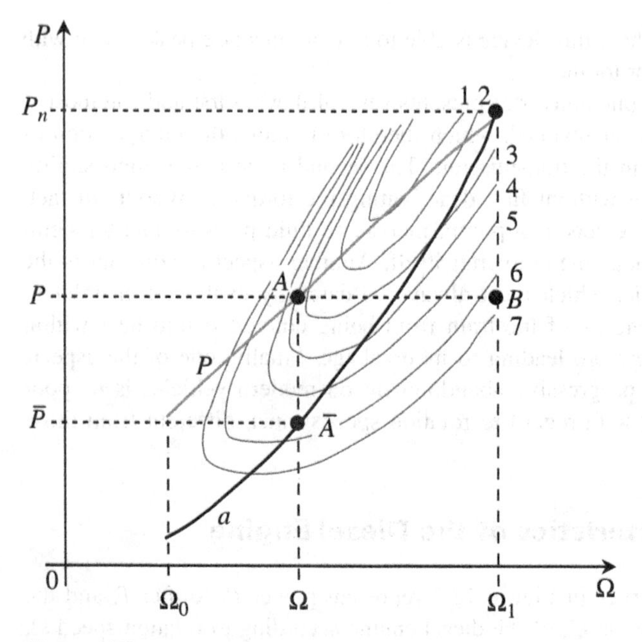

Figure 12.3 Curves at constant efficiency 1, 2, 3, . . . at efficiency $\eta_d = \cos t$ $(\eta_1 > \eta_2 > \eta_3, \ldots)$. *a*: throttle curve (power regulation).

position, for example, corresponding with maximum admission, torque slightly varies with rotation speed, in relation to the variation in the efficiency and quantity of fuel injected per cycle.

The corresponding power P increases with speed. Throttling the admission causes different power curves $P = f(\Omega)$: Figure 12.2 shows the curves with degree of admission 1/4, 2/4, 3/4, 4/4.

If C_s (kg/kWh) denotes specific fuel consumption at full admission and q (kWh/kg) the fuel calorific value, one can derive the expression of efficiency η_d:

$$\eta_d = \frac{1}{q \cdot C_s} = \frac{0.086}{C_s}$$

Figure 12.3 represents the curves at constant efficiency η_d = const. Diesel can develop a power P lower than rated power P_n, with different speed, degrees of admission, and efficiency. It is possible to establish, within the speed range AB, the operating condition corresponding to the greatest efficiency. At point A, the engine works at full admission, with the minimum possible speed $\Omega < \Omega_1$; at point B, conversely, rotation speed is at the maximum (Ω_1), with minimal degree of admission.

Railway diesel engines have a nominal power between hundreds of kilowatts and 4000 kW, with a number of cylinders up to 12–16–20. They are normally supercharged by exhaust gas turbochargers, which allow an increase in the supply air pressure and density, and therefore the quantity of fuel injected into the cylinders. For greater power, two-stage supercharging is adopted, with intermediate air cooling.

For modern supercharged high-power engines, we have $C_s \approx 0.20$–0.21 kg/kWh, $\eta_d = 0.41 - 0.43$.

One of the most important technological advances for diesel motors has been the common rail (CR) electronic injection system.

Unlike traditional systems, common rail involves pressure generation being decoupled from injection; this means that the pressure is generated independent of the rotation speed and quantity of fuel and can be selected within a predetermined range (currently from 15 to 180 MPa). The component enabling decoupling is the high-pressure accumulator; injector opening is controlled electronically by solenoid. The system is managed entirely by electronic controller and allows for the generation of multiple injections (pilot, post-injection management).

Today, diesel engines compete in terms of performance with spark-ignition engines, while ensuring significantly lower specific consumption. The introduction of the CR system allows the following:

- High injection pressures, up to 160 MPa, in already industrialized systems, and up to 180 MPa in new-generation systems.
- Control over injection pressure and parameters (free mapping) independent of motor speed and as a function of a large number of motor and operating parameters.

For the purposes of increasing power per unit of mass and volume, particularly advantageous for motors that must be installed on board railway vehicles, rotation speed is of notable importance. In this regard, it must be kept in mind that with increasing average piston speed v, mechanical stress (proportional to v^2) and moving part wear increase. The current limits are around 10–11 m/s, to which rotation speeds of 1000–1500 rpm correspond; attempt is made to reach the speed of 1500 rpm up to higher powers, also to unburden the electric generator.

Nowadays, much attention is paid to diesel engine emissions, especially with regard to particulates or ultrafine particles. One possible solution involves DPF (diesel particulate filter), a dust-catching system. However, the first types of filter often became clogged with particulates, necessitating a regeneration process for the filter. New PF solutions that do not use additives are based on the raising of exhaust gas temperature up to 600–650 °C. The temperature is raised through a series of post-injections with consequent post-combustions, which partly take place in the exhaust collectors and oxidizing catalysts; these temperature values are more than sufficient to completely burn up particulate matter that has accumulated in the filter.

To facilitate the particulate combustion process, noble metals that work as catalysts are inserted on the filter walls. The system without additives has the advantage of not requiring additive refilling, which, in addition to being dangerous to human health, increases maintenance costs. The filter without additives works at higher regeneration-triggering temperatures.

Permissible emission values are constantly monitored and lowered in relation to the available technology by various government agencies such as the Environmental Protection Agency (EPA) in the United States.

12.1.2 Diesel Engine and Transmission Regulation

Diesel engine operating conditions are determined by the degree of fuel admission and rotation speed.

In diesel–electric traction, speed regulation is normally adopted; the engine has a centrifugal regulator, in early applications, replaced by an electronic one in more modern applications.

A diesel engine allows for overloads that are highly reduced and for a brief time; to avoid them, one must operate in a way that the electric transmission never requires a torque greater than that corresponding to maximum admission. To achieve this, it is necessary to improve the diesel's speed regulation with control over the power absorbed by the transmission.

12.1.3 Electric Transmission

The electric transmission consists of the interposition of a generator and one or more electric traction motors between the prime motor and the wheels.

The system offers great advantages in terms of speed regulation, transmission efficiency, and the layout of devices within the vehicle. In fact, by not requiring a mechanical connection, they can be positioned as a function of the available space or for better weight distribution.

The traditional layout of the electric transmission requires a DC generator directly coupled to the traction motors, as well as these to a DC commutator motor (Figure 12.4a). This solution has the advantage of being able to use the generator as the diesel's starter motor, powering it with onboard batteries. Another variant involved the replacement of the direct current generator with a three-phase synchronous motor and diode bridge, according to the layout of Figure 12.4b.

A more modern variant, the employment of which has been possible with the advent of power electronics, involves the use of three-phase traction motors controlled through inverters according to the basic scheme seen in Figure 12.5.

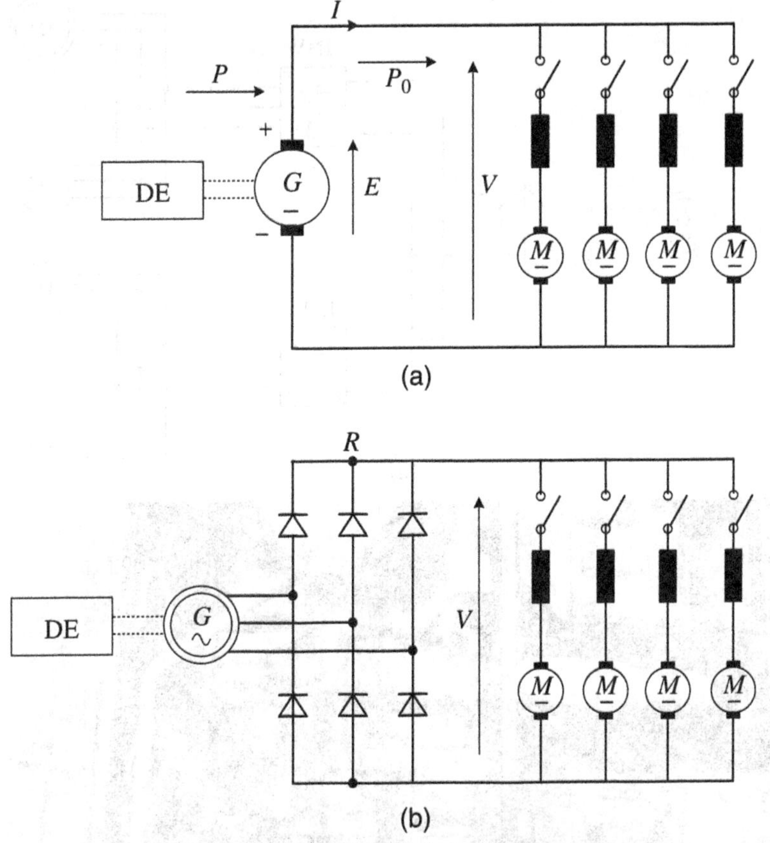

Figure 12.4 Basic scheme of an electric transmission with DC traction motors. (a) Transmission with DC generator. (b) Transmission with alternator and bridge rectifier.

Assuming that the diesel engine works at nominal speed Ω_1 = const and P_n is the power supplied by it, at full admission. From this, the power P_a absorbed by auxiliary services must be deducted, around 10% of P_n.

The power available for traction (net input), usable to power traction motors and referred to the nominal angular speed Ω_1, is given by

$$P_1 = P_n - P_a = T_1 \cdot \Omega_1 \tag{12.1}$$

The vehicle cannot use the power P_1 for any gear speed, but only in a certain speed range, outside of which the transmission absorbs a power $P < P_1$.

Given η_g is the efficiency of the generator, the DC side electric power is given by

$$P_g = \eta_g \cdot P$$

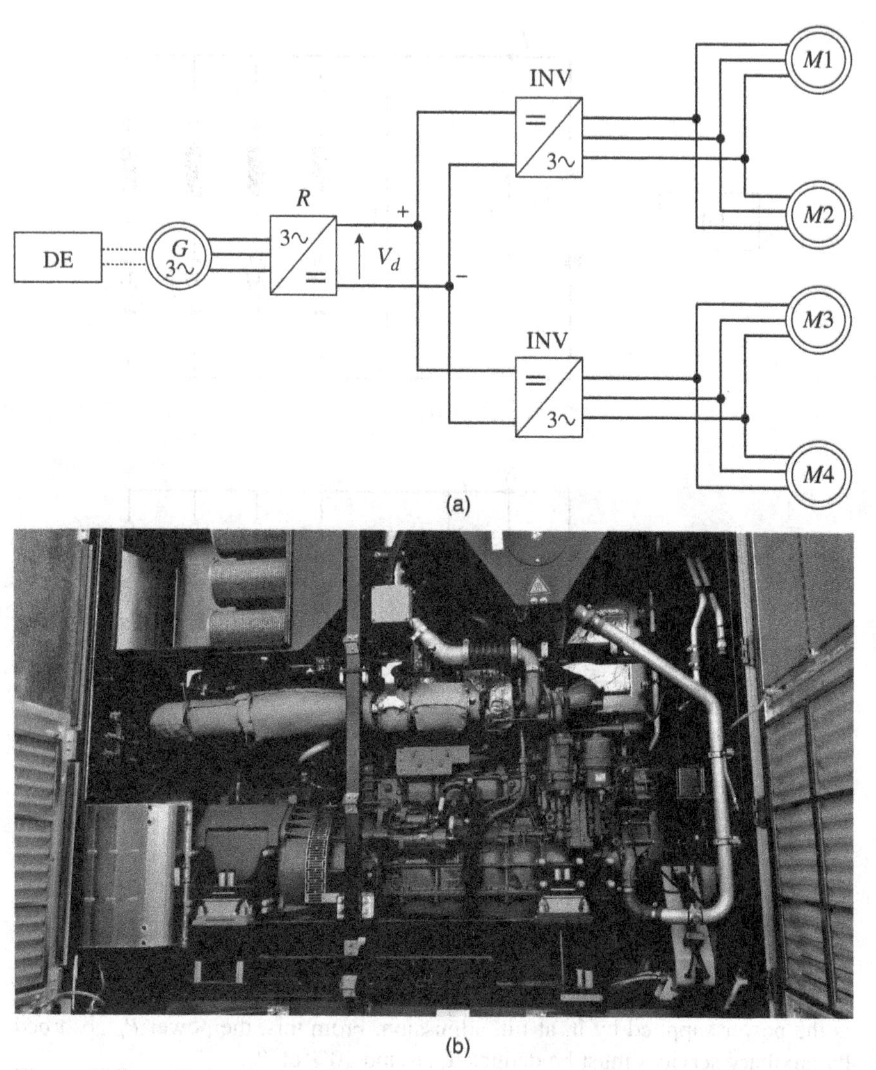

Figure 12.5 (a) Basic scheme of an asynchronous, three-phase electric voltage–source transmission. DE: diesel engine; *G*: three-phase synchronous generator, directly coupled with the diesel engine; *R*: rectifier; INV: voltage–source inverters; *M*: asynchronous traction motors. (b) Diesel generator group.

From this, the traction power is derived:

$$P_t = \eta_m \cdot \eta_i \cdot P_g = \eta \cdot P \tag{12.2}$$

- η_m = efficiency of the traction motors
- η_i = efficiency of the reduction gearboxes
- $\eta = \eta_g \cdot \eta_m \cdot \eta_i$ = total efficiency of the electric transmission

12.1.3.1 Transmissions with direct current traction motors

For this type of transmission, it is useful to refer to the motor traction power supply characteristic $V = f(I)$, referring to the terminals of the generator G, if direct current, or downstream of the rectifier, if the generator is alternating current (Figure 12.4). To use the available power P_1, this direct current side external characteristic, at rotation speed $\Omega = \text{const}$, should be an equation hyperbola:

$$P_{g1} = \eta_g \cdot P_i = \hat{V} \cdot I = \text{const} \tag{12.3}$$

DC or AC generators do not present characteristics with this trend by their nature; it is therefore necessary that their excitation system be appropriately regulated so that the adjusted characteristic $V = f(I)$ approaches the ideal hyperbola in one more or less wide area (12.3). To this end, the transmission regulation system connected to the diesel engine regulator suitably varies the excitation of the generator.

Generators used in traction systems have a nonlinear characteristic (Figure 12.6 – curve b), such as to spontaneously limit both the maximum current I_M and the maximum voltage V_M. One characteristic of this kind intersects the equation hyperbola (12.3) in points B (abscissa I_B) and C (abscissa I_2).

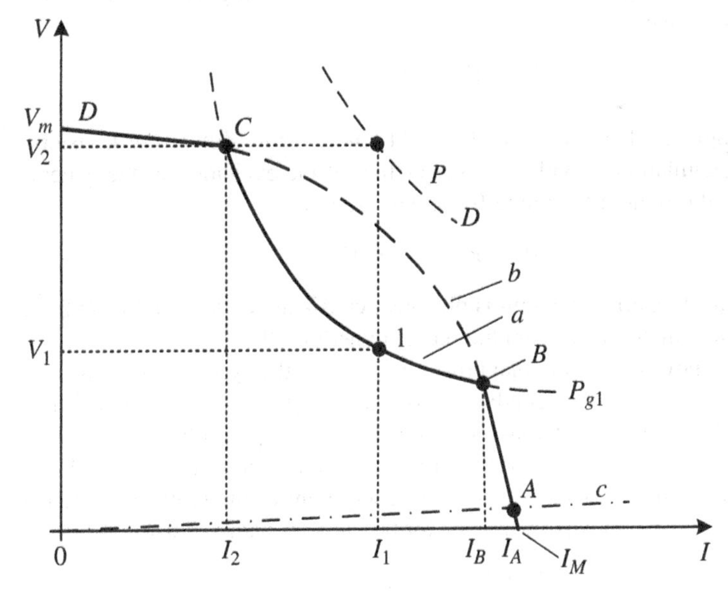

Figure 12.6 Regulated direct current external diagram, for $\Omega_1 = \text{const}$. a: equation hyperbola – $P_{g1} = \overline{U} \cdot I = \text{const}$; b: natural external diagram of the generator; c: traction motor internal voltage drop.

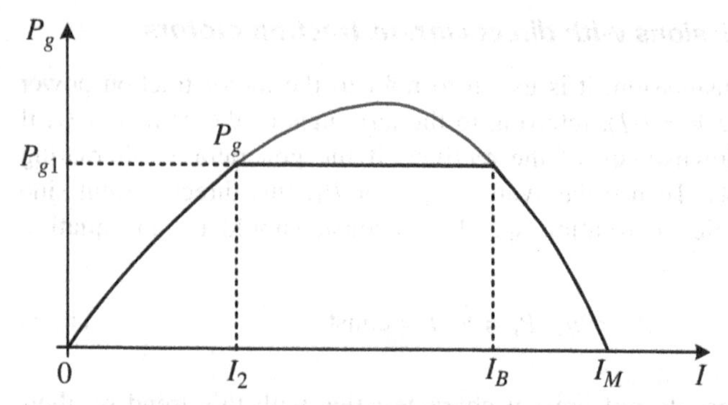

Figure 12.7 Power used by electric transmission.

Given $P_g = V \cdot I$ the electric power corresponding to the natural characteristic $V = f(I)$, for $I < I_2$ and $I_B < I < I_M$, we have (Figure 12.7)

$$P_g < P_{g1}$$

that is, the diesel engine requires a power $P = P_g/\eta_g$ lower than that available for traction (12.1). Because the motor tends to accelerate, its regulator reduces admission, thus the torque, such that the result is $T \cdot \Omega_1 = P$.

For $I_2 < I < I_B$, the generator would require a power higher than that available for traction, that is,

$$P_g > P_{g1}; \quad V > \hat{V}$$

The diesel engine, when overloaded, would tend to slow down; the transmission's power regulation provides for a reduction of the excitation of the generator, such that, throughout the range I_2–I_B, the result is

$$P_g = P_{g1}; \quad V = \hat{V}$$

Thus, a regulated external diagram is obtained, consisting of A–B and C–D from the natural diagram and B–C from the power hyperbola P_{g1}.

When less power than nominal power is required, the system makes the diesel engine work in one of the conditions corresponding to the throttle curve a of Figure 12.3; the regulated external diagram is lowered in proportion.

To achieve good use of adhesion, the m traction motors in electric diesel engines are normally connected in parallel. Each point of the regulated external characteristic $V = f(I)$ thus corresponds, for the motors, with the sizes:

$$I_m = {}^{I}/_{m}, \quad V_m = V$$

In particular, upon starting, traction motors absorb the current I_A, near I_M, corresponding to the intersection of $V = f(I)$ with the straight line of equation

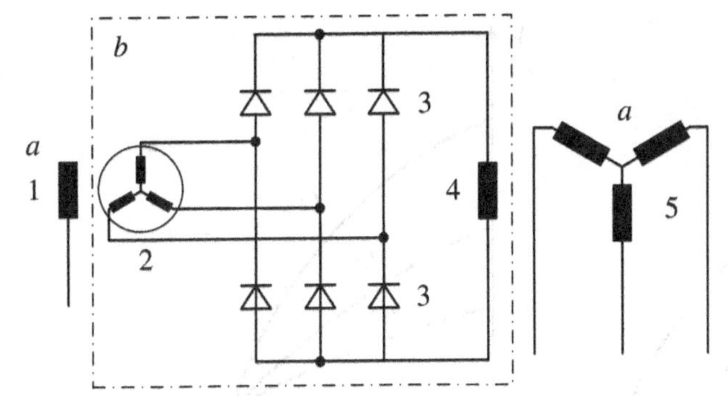

Figure 12.8 Basic diagram of an alternator with incorporated three-phase exciter and rotating rectifier. *a*: stator windings; *b*: rotor windings and equipment mounted on the rotor; 1: excitement windings of the exciter; 2: exciter armature; 3: rectifier bridge; 4: excitation winding of the alternator; 5: alternator armature.

$V = r_t \cdot I$ (point *A* of Figure 12.6), in which r_t is the total equivalent resistance of the *m* motors.

12.1.3.2 Three-phase generators

Salient-pole machines are used, in which the excitation windings are powered, through the slip ring, by an AC exciter with a rectifier. The excitation power is considerably higher than that required by a generator, at equal nominal power.

The so-called brushless machines, that is, not having slip rings and brushes, may also be used by incorporating a three-phase synchronous exciter and the relative semiconductor rectifier bridge in the rotor of the generator. The elimination of the slip ring is made possible by the exciter's armature as part of the rotor (Figure 12.8) and field winding mounted on the stator. The diode bridge, which is also mounted on the rotor, therefore directly feeds the field winding of the main alternator.

In the alternator's three-phase stator winding, normally connected by star scheme, EMF frequency $f = \frac{p}{2}\Omega$ and rms phase voltage are induced, $E = K \cdot N \cdot f \cdot \Phi$, where

- p = number of poles,
- N = number of active-phase conductors,
- Φ = flux per pole, and
- K= Kapp factor.

Line-to-line voltages are $V = \sqrt{3} \cdot E$ applied to a three-phase diode bridge (Figure 12.4b), giving rise to a rectified voltage of average value $V_d = 1.35 \cdot V$.

By effect of the leakage reactance, the alternator works with a lower per unit power factor and demagnetizing armature reaction; the result is an external

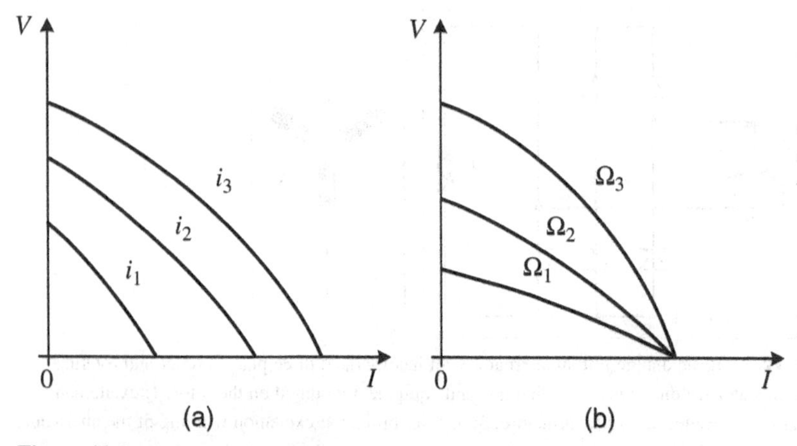

Figure 12.9 External characteristics of a three-phase alternator for traction: downstream of the rectifier. (a) Characteristics for: Ω_1 = const., with excitation $i_1 < i_2 < i_3$. (b) Constant excitation characteristics, with rotation speed: $\Omega_1 < \Omega_2 < \Omega_3$.

characteristic $V = f(I)$, downstream of the rectifier bridge, sagging, similar to that of a direct current generator. In Figure 12.9, the constant speed and constant excitation characteristics are represented.

Alternating current generators, in addition to allowing the transmission of the maximum powers achievable with modern diesel engines, have advantages with respect to direct current generators in their maintenance, not having a commutator, and for their lesser mass at equal power.

Because alternator sizing is slightly influenced, in cases of interest, by the preselected rated voltage value, it is possible in this case to set the optimal values for the sizing of traction motors.

It may also be convenient to divide the armature of high-power machines into separate windings; in this case, each winding feeds, through its own rectifier bridge, one or two traction motors.

In low-power diesel–electric vehicles, an induction generator with a VSC rectifier may be preferred due to its more compact size.

12.1.3.3 Transmissions with AC traction motors

The three-phase transmission is particularly advantageous when used in diesel–electric vehicles; the converter, in fact, works under conditions more favorable than in the case of contact line power. In fact, a voltage value more appropriate to the converter can be chosen, maximum power is not generally subject to overloads as it is supplied by the diesel engine, and there are generally no overvoltages of external origin.

While in traditional transmissions the generator must be oversized with respect to rated power P_{g1}, with the three-phase transmission the task of

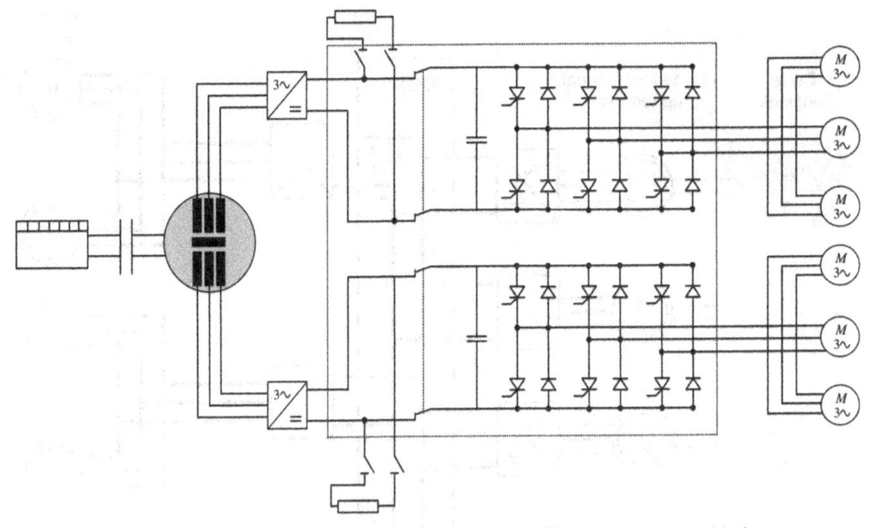

Figure 12.10 Basic diagram of a high-power C_0C_0 diesel–electric locomotive with the generator's armature subdivided into two separate windings.

adapting the voltage and current to the functional requirements of the motors is carried out by the inverter; the generator may therefore be sized as a function of only active rated power P_{g1}.

As seen, AC drives are naturally reversible and thus allow the actuation of the vehicle's electric braking. Because braking power may not be injected upstream into a network nor absorbed by the diesel engine, it must be dissipated through appropriate rheostats connected in the DC link.

Depending on the actual absorbed powers, the inverter/motor complex may be divided into "modules." For example, in the principle diagram shown in Figure 12.5, relative to a B_0B_0 locomotive, the equipment comprises two inverters, each one powering the two traction motors in parallel of a motor bogie. A more complete scheme for high-power locomotives is that represented in Figure 12.10, in which the generator's armature is subdivided into two separate windings, each one of which powers a three-phase drive with three motors in parallel and to which the respective braking rheostat is connected.

12.1.4 Multiengine Systems

In high-power vehicles, it may be useful to divide the generation power into multiple modules in order to start only the necessary diesel engines that may, in this way, work with higher utilization factors and thus higher efficiency.

One possible scheme for the implementation of this solution is that of Figure 12.11, in which two or more generator sets are connected to a common DC link that feeds the traction converters and auxiliary services. The DC link

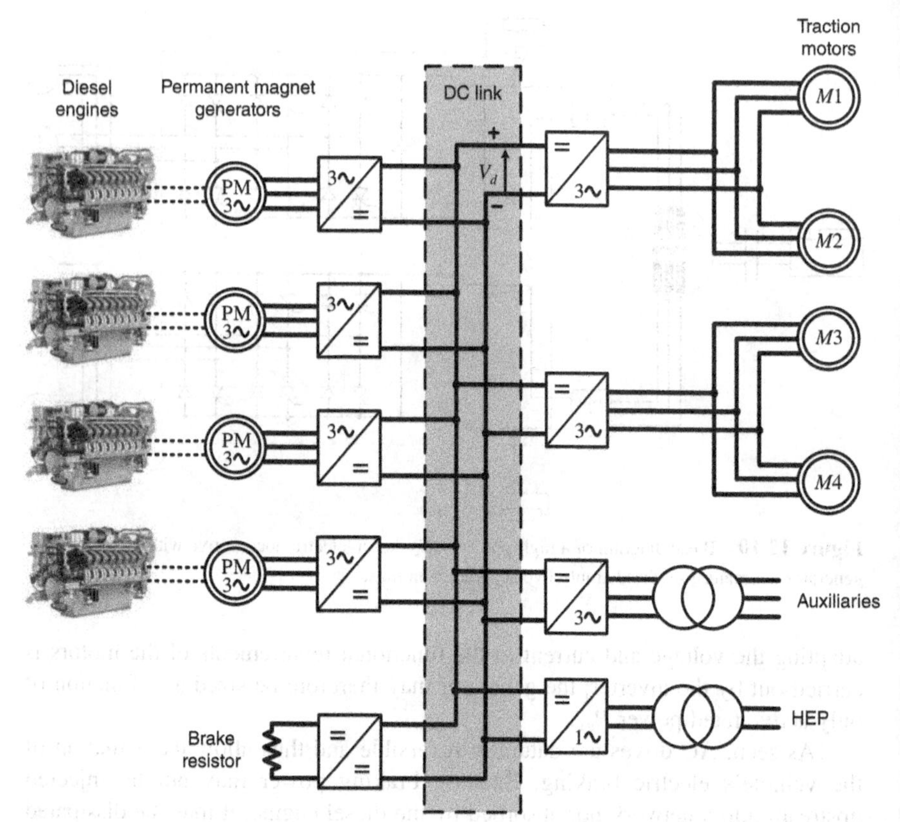

Figure 12.11 Basic diagram of a multiengine diesel locomotive.

may also be connected to the braking rheostat and power supply of the train line (head end power (HEP)), through special converters.

In this case, faster machines with permanent magnet synchronous generators may be used, which are very compact and reliable.

One modern innovation is the inclusion of storage systems in the DC link that can recover part of the braking power, especially in urban services characterized by many restarts.

12.1.5 Dual-Power Vehicles

In services that include electrified and nonelectrified lines, it may be useful to have trains capable of running purely electrically, taking energy from the contact line, or like the railcar, powered by a diesel engine.

This solution allows for greater efficiency and reduced environmental impact, especially in urban contexts, thanks to its purely electric operation.

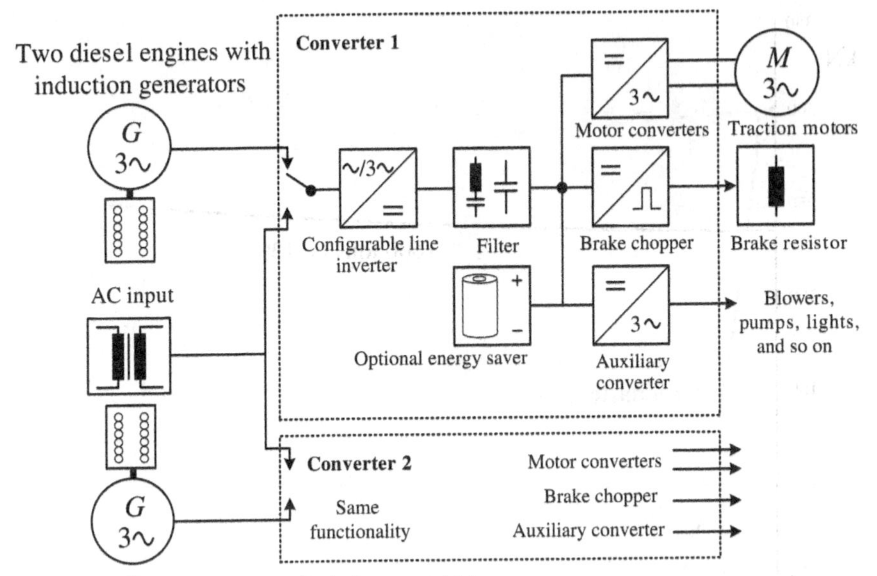

Figure 12.12 Basic diagram of a dual-power vehicle.

The diagram is shown in Figure 12.12, in which the DC link is fed through VSC converters connected to a diesel generator set or reconfigured as 4Q converters in electric mode.

The power available in diesel mode is about half of that in electric mode.

Locomotives of this kind have already been created for France, the United States, and Canada.

A special case of dual-power locomotive is the so-called last-mile diesel module vehicle. These vehicles can be considered electric vehicles for all cases in which a small diesel generator is installed, with power around 100–150 kW, capable of powering the low-speed vehicle for short periods such as in depots, nonelectrified freight terminals, or wherever there are contact line interruptions.

The last mile module may be supplemented with batteries capable of increasing the available power.

Figure 12.13 represents the mechanical characteristic of the last mile module compared with that of electrical operation. As can be observed, maximum peak tractive effort is maintained for the movement of even heavy freight trains. However, given the reduced power available, the maximum speed that can be reached on level ground is around 25 km/h.

12.2 FUEL CELL TRAINS

Nowadays, increasing fuel prices and stricter exhaust and noise emission regulations pose a challenge for railway operators that rely on diesel propulsion, so

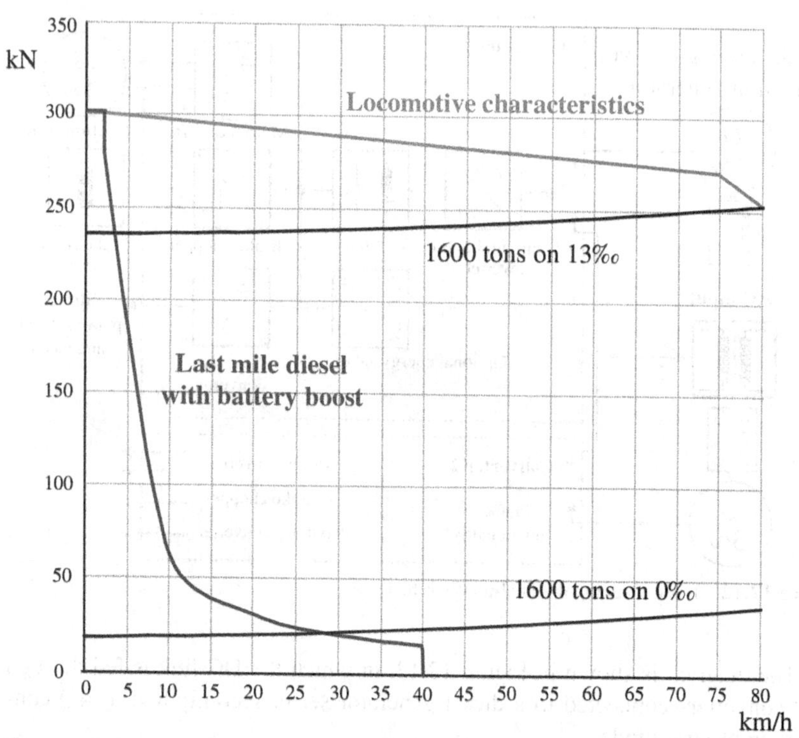

Figure 12.13 Mechanical characteristic of the last mile module compared with that of electrical operation.

they are directing their attention toward alternative propulsion systems for the future.

A solution that caught the attention of train manufacturers is represented by the replacement of the diesel engine with a fuel cell generator.

The fuel cell is a device used to directly generate electrical energy via chemical reaction without any combustion, and as a result has environment-friendly and efficient properties that has led to its intensive development in the energy field.

A proton-exchange membrane fuel cell (PEMFC) has lower operating temperature (50–100 °C) and faster start-up, high-power density, low corrosion that makes it attractive in many fields, such as distributed stationary and mobile applications, electric vehicles, and railway. On the contrary, a PEMFC is sensitive to fuel impurities, so it requires high-quality hydrogen as fuel.

Fuel cell locomotives with PEMFC as a power source are accompanied by a number of devices, such as the compressor, reflux pump, coolant pump, radiator, and electric control equipment. Thermal management is essential to its optimal operational condition. As matter of fact, the temperature affects the performance and reliability of fuel cells. When a chemical reaction between hydrogen and

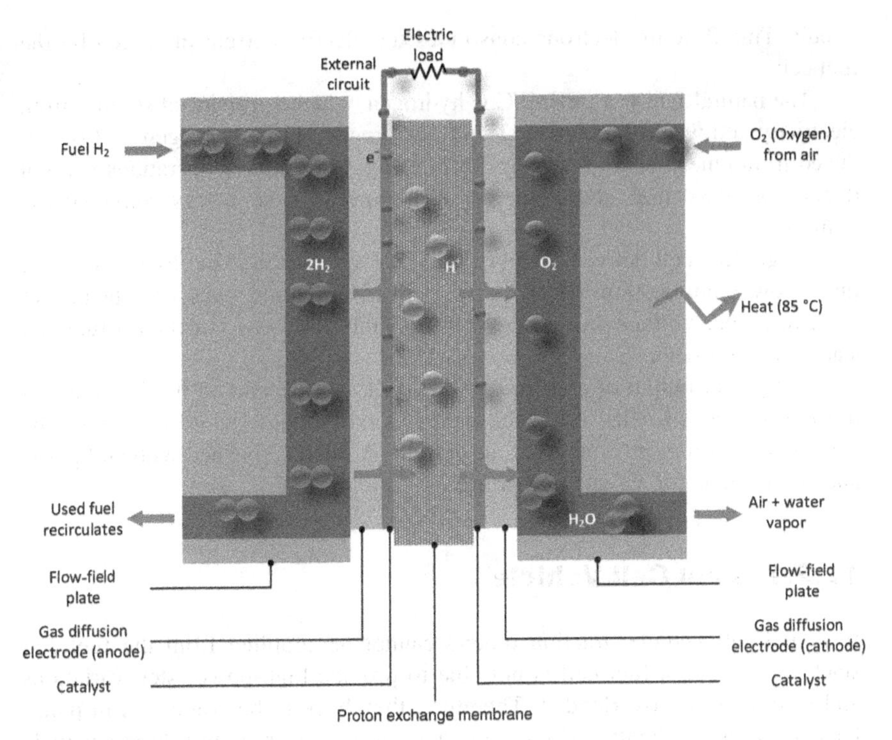

Figure 12.14 Structure of proton-exchange membrane fuel cell.

oxygen occurs, which is just an exothermic reaction, the temperature in cells rises. The high temperature accelerates the chemical reaction and activity of the water molecules in the cells, and then augments membrane moisture and improves the conductivity of the membrane. Thus, it helps to perfect the performance of the cell. Nevertheless, elevated temperature could make the membrane easily suffer mechanical damage and degrade the catalyst layer due to elevated platinum transport, resulting in a loss in fuel cells lifespan.

The basic operation of a PEMFC is described in Figure 12.14.

Starting from hydrogen and oxygen, a PEMFC is able to produce an electric current that flows in an external circuit, producing only water vapor as exhaust and heat. The chemical reactions that govern the operation of PEMFC are the following:

Anode: $H_2 \rightarrow 2H^+ + 2e^-$

Cathode: $\frac{1}{2}O_2 + 2H^+ + 2e^- \rightarrow H_2O$

Overall reaction: $H_2 + \frac{1}{2}O_2 \rightarrow H_2O$

The membrane between the anode and cathode gives access only to the hydrogen ions H^+ and stops the electrons that are forced to flow in an external circuit. This flow of electrons constitutes the electric current produced by the fuel cell.

The natural fuel for a PEMFC is hydrogen, which is produced starting from electricity and water through an electrolytic process. If the hydrogen fuel is produced from renewable or nuclear primary energy, locomotive operations will not depend on fossil fuel, and so the carbon footprint of the energy conversion is reduced.

Single fuel cell delivers about 0.5–1 V voltage, which is too low voltage for most power applications. Just like batteries, individual cells are stacked to achieve higher voltage and power. This assembly of cells is called a fuel cell stack, or just a stack.

The power output of a given fuel cell stack will depend on its size. Increasing the number of cells in a stack increases the voltage, while increasing the surface area of the cells increases the current. A stack is finished with end plates and connections for ease of further use.

12.2.1 Fuel Cell Vehicle

In a fuel cell vehicle, traction motors cannot be supplied from the fuel cell stack only, since a fuel cell is not able to provide high-power step variations and cannot be too overloaded. Therefore, they have to be connected in parallel with a battery bank that provides the power peak required during vehicle acceleration. Thus, fuel cell stacks are sized for the base load, while the batteries are sized to provide the power during acceleration or to absorb it during the braking phase. Instead, the capacity of the hydrogen tanks defines the driving distance of the vehicle. As a consequence, fuel cell vehicles are considered hybrid vehicles.

The hydrogen is stored in suitable tanks at a pressure of 70 MPa (700 bar).

Since railway vehicles are much heavier than road ones, they require more energy to travel, even if the specific energy in terms of passenger·km or tons·km is lower. Therefore, it is not possible to install only one battery sized for the total energy, otherwise it becomes too big and too heavy to be installed onboard. This is the reason why the combination between fuel cells, sized for the base load, hydrogen tanks, sized for the energy, and chemical batteries, sized for the accelerating power, is an optimal solution for a modern self-propelled railway vehicle.

The basic structure for a fuel cell railway vehicle is presented in Figure 12.15.

This structure is similar to those described in Figure 12.11 for multiengine locomotives, in which the diesel engines have been replaced with fuel cell stacks. Every element is connected to a common DC link through its own DC/DC converter that behaves as an energy hub.

Compared to a diesel–electric vehicle, the presence of a bulk storage systems provided by the batteries allows for the exploitation of regenerative braking

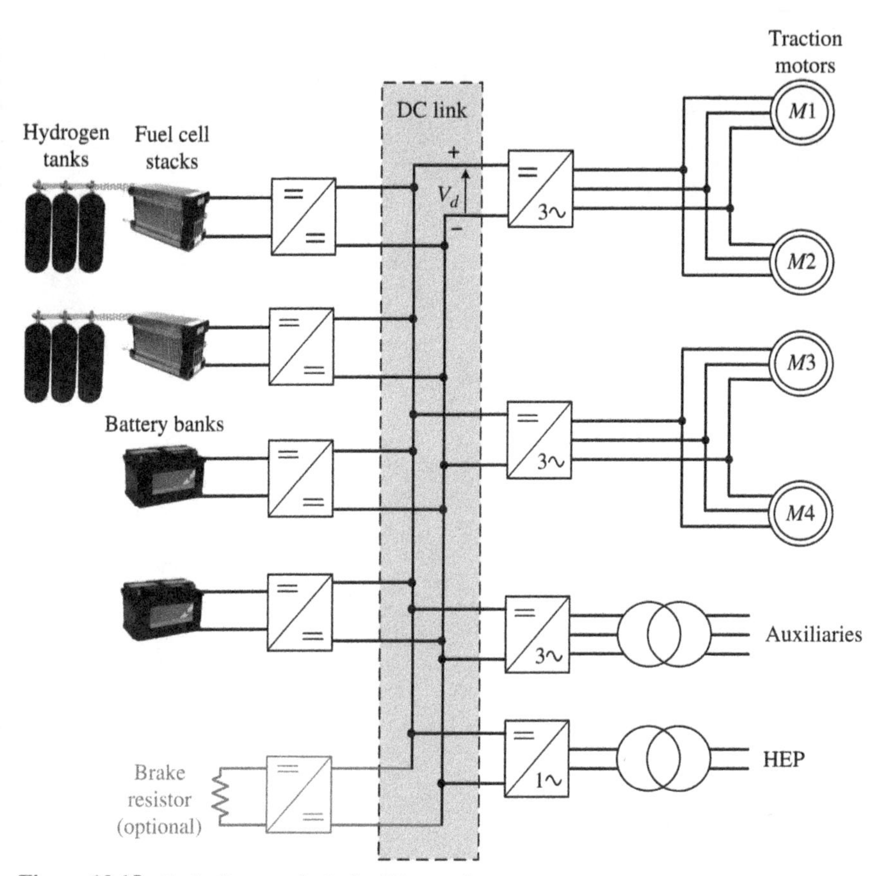

Figure 12.15 Basic diagram of a fuel cell locomotive.

until they are fully charged. For this reason, keeping the brake resistor that allows the electric braking even when the batteries are fully charged can be useful.

Some fuel cell shunting locomotive prototypes are equipped with 240 kW PEMFC and can reach a transient power of 1 MW, thanks to the batteries.

Other types of railway vehicles, such as EMU, are now ready for commercial service using hydrogen fuel cells as primary generators. These vehicles are able to travel up to 800 km with a maximum speed of 140 km/h; therefore, they are suitable for low-traffic regional lines when electrification is not convenient.

Index

Electrical Railway Transportation Systems, First Edition. Morris Brenna, Federica Foiadelli, and Dario Zaninelli.

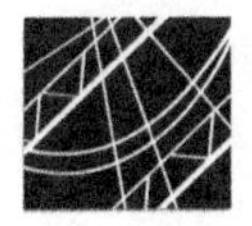

IEEE Press Series on Power Engineering

Series Editor: M. E. El-Hawary, Dalhousie University, Halifax, Nova Scotia, Canada

The mission of IEEE Press Series on Power Engineering is to publish leading-edge books that cover the broad spectrum of current and forward-looking technologies in this fast-moving area. The series attracts highly acclaimed authors from industry/academia to provide accessible coverage of current and emerging topics in power engineering and allied fields. Our target audience includes the power engineering professional who is interested in enhancing their knowledge and perspective in their areas of interest.

1. *Principles of Electric Machines with Power Electronic Applications, Second Edition*
M. E. El-Hawary

2. *Pulse Width Modulation for Power Converters: Principles and Practice*
D. Grahame Holmes and Thomas Lipo

3. *Analysis of Electric Machinery and Drive Systems, Second Edition*
Paul C. Krause, Oleg Wasynczuk, and Scott D. Sudhoff

4. *Risk Assessment for Power Systems: Models, Methods, and Applications*
Wenyuan Li

5. *Optimization Principles: Practical Applications to the Operations of Markets of the Electric Power Industry*
Narayan S. Rau

6. *Electric Economics: Regulation and Deregulation*
Geoffrey Rothwell and Tomas Gomez

7. *Electric Power Systems: Analysis and Control*
Fabio Saccomanno

8. *Electrical Insulation for Rotating Machines: Design, Evaluation, Aging, Testing, and Repair, Second Edition*
Greg Stone, Edward A. Boulter, Ian Culbert, and Hussein Dhirani

9. *Signal Processing of Power Quality Disturbances*
Math H. J. Bollen and Irene Y. H. Gu

10. *Instantaneous Power Theory and Applications to Power Conditioning*
Hirofumi Akagi, Edson H. Watanabe, and Mauricio Aredes

11. *Maintaining Mission Critical Systems in a 24/7 Environment*
Peter M. Curtis

12. *Elements of Tidal-Electric Engineering*
Robert H. Clark

13. *Handbook of Large Turbo-Generator Operation and Maintenance, Second Edition*
Geoff Klempner and Isidor Kerszenbaum

14. *Introduction to Electrical Power Systems*
Mohamed E. El-Hawary

15. *Modeling and Control of Fuel Cells: Distributed Generation Applications*
M. Hashem Nehrir and Caisheng Wang

16. *Power Distribution System Reliability: Practical Methods and Applications*
Ali A. Chowdhury and Don O. Koval

17. *Introduction to FACTS Controllers: Theory, Modeling, and Applications*
Kalyan K. Sen and Mey Ling Sen

18. *Economic Market Design and Planning for Electric Power Systems*
James Momoh and Lamine Mili

19. *Operation and Control of Electric Energy Processing Systems*
James Momoh and Lamine Mili

20. *Restructured Electric Power Systems: Analysis of Electricity Markets with Equilibrium Models*
Xiao-Ping Zhang

21. *An Introduction to Wavelet Modulated Inverters*
S.A. Saleh and M.A. Rahman

22. *Control of Electric Machine Drive Systems*
Seung-Ki Sul

23. *Probabilistic Transmission System Planning*
Wenyuan Li

24. *Electricity Power Generation: The Changing Dimensions*
Digambar M. Tigare

25. *Electric Distribution Systems*
Abdelhay A. Sallam and Om P. Malik

26. *Practical Lighting Design with LEDs*
Ron Lenk and Carol Lenk

27. *High Voltage and Electrical Insulation Engineering*
Ravindra Arora and Wolfgang Mosch

28. *Maintaining Mission Critical Systems in a 24/7 Environment, Second Edition*
Peter Curtis

29. *Power Conversion and Control of Wind Energy Systems*
Bin Wu, Yongqiang Lang, Navid Zargari, and Samir Kouro

30. *Integration of Distributed Generation in the Power System*
Math H. Bollen and Fainan Hassan

31. *Doubly Fed Induction Machine: Modeling and Control for Wind Energy Generation Applications*
Gonzalo Abad, Jesus Lopez, Miguel Rodrigues, Luis Marroyo, and Grzegorz Iwanski

32. *High Voltage Protection for Telecommunications*
Steven W. Blume

33. *Smart Grid: Fundamentals of Design and Analysis*
James Momoh

34. *Electromechanical Motion Devices, Second Edition*
Paul C. Krause, Oleg Wasynczuk, and Steven D. Pekarek

35. *Electrical Energy Conversion and Transport: An Interactive Computer-Based Approach, Second Edition*
George G. Karady and Keith E. Holbert

36. *ARC Flash Hazard and Analysis and Mitigation*
J. C. Das

37. *Handbook of Electrical Power System Dynamics: Modeling, Stability, and Control*
Mircea Eremia and Mohammad Shahidehpour

38. *Analysis of Electric Machinery and Drive Systems, Third Edition*
Paul Krause, Oleg Wasynczuk, S. D. Sudhoff, and Steven D. Pekarek

39. *Extruded Cables for High-Voltage Direct-Current Transmission: Advances in Research and Development*
Giovanni Mazzanti and Massimo Marzinotto

40. *Power Magnetic Devices: A Multi-Objective Design Approach*
S. D. Sudhoff

41. *Risk Assessment of Power Systems: Models, Methods, and Applications, Second Edition*
Wenyuan Li

42. *Practical Power System Operation*
Ebrahim Vaahedi

43. *The Selection Process of Biomass Materials for the Production of Bio-Fuels and Co-Firing*
Najib Altawell

44. *Electrical Insulation for Rotating Machines: Design, Evaluation, Aging, Testing, and Repair, Second Edition*
Greg C. Stone, Ian Culbert, Edward A. Boulter, and Hussein Dhirani

45. *Principles of Electrical Safety*
Peter E. Sutherland

46. *Advanced Power Electronics Converters: PWM Converters Processing AC Voltages*
Euzeli Cipriano dos Santos Jr. and Edison Roberto Cabral da Silva

47. *Optimization of Power System Operation, Second Edition*
Jizhong Zhu

48. *Power System Harmonics and Passive Filter Designs*
J. C. Das

49. *Digital Control of High-Frequency Switched-Mode Power Converters*
Luca Corradini, Dragan Maksimovic, Paolo Mattavelli, and Regan Zane

50. *Industrial Power Distribution, Second Edition*
Ralph E. Fehr, III

51. *HVDC Grids: For Offshore and Supergrid of the Future*
Dirk Van Hertem, Oriol Gomis-Bellmunt, and Jun Liang

52. *Advanced Solutions in Power Systems: HVDC, FACTS, and Artificial Intelligence*
Mircea Eremia, Chen-Ching Liu, and Abdel-Aty Edris

53. *Power Grid Operation in a Market Environment: Economic Efficiency and Risk Mitigation*
Hong Chen

54. *Current Signature Analysis for Condition Monitoring of Cage Induction Motors: Industrial Application and Case Histories*
William T. Thomson and Ian Culburt

55. *Model Predictive Control of Wind Energy Conversion Systems*
Venkata Yaramasu and Bin Wu

56. *Electric Power System Basics for the Nonelectrical Professional, Second Edition*
Steven W. Blume

57. *Energy Production Systems Engineering*
Thomas H. Blair

58. *Understanding Symmetrical Components for Power System Modeling*
J. C. Das

59. *High-Power Converters and AC Drives, Second Edition*
Bin Wu and Mehdi Narimani

60. *Practical Lighting Design with LEDs, Second Edition*
Ron Lenk and Carol Lenk

61. *Introduction to Electric Power and Drive Systems*
Paul Krause, Oleg Wasynczuk, Timothy O'Connell, and Maher Hasan

62. *Instantaneous Power Theory and Applications to Power Conditioning, Second Edition*
Hirofumi Akagi, Edson Hirokazu Watanabe, and Mauricio Aredes

63. *Introduction to AC Machine Design*
Thomas A. Lipo

64. *Advances in Electric Power and Energy Systems: Load and Price Forecasting*
Mohamed E. El-Hawary

65. *Electricity Markets: Theories and Applications*
Jeremy Lin and Fernando H. Magnago

66. *Multiphysics Simulation by Design for Electrical Machines, Power Electronics, and Drives*
Marius Rosu, Ping Zhou, Dingsheng Lin, Dan Ionel, Mircea Popescu, Frede Blaabjerg, Vandana Rallabandi, and David Staton

67. *Electrical Railway Transportation Systems*
Morris Brenna, Federica Foiadelli, and Dario Zaninelli